COMPUTER AIDED DESIGN

A Conceptual Approach

COMPUTER AIDED DESIGN

A Conceptual Approach

Jayanta Sarkar

CRC Press
Taylor & Francis Group
Boca Raton London New York

CRC Press is an imprint of the
Taylor & Francis Group, an **informa** business

ANSYS, ANSYS Workbench, Ansoft, AUTODYN, CFX, EKM, Engineering Knowledge Manager, FLUENT, HFSS, and any and all ANSYS, Inc. brand, product, service and feature names, logos and slogans are trademarks or registered trademarks of ANSYS, Inc. or its subsidiaries located in the United States or other countries. ICEM CFD is a trademark used by ANSYS, Inc. under license. CFX is a trademark of Sony Corporation in Japan. All other brand, product, service, and feature names or trademarks are the property of their respective.

CRC Press
Taylor & Francis Group
6000 Broken Sound Parkway NW, Suite 300
Boca Raton, FL 33487-2742

First issued in paperback 2017

ISBN-13: 978-1-4822-0879-5 (hbk)
ISBN-13: 978-1-138-88544-8 (pbk)

Visit the Taylor & Francis Web site at
http://www.taylorandfrancis.com

and the CRC Press Web site at
http://www.crcpress.com

Knowledge can be in any form and worthwhile.

Jayanta Sarkar

For my parents and my sister, who support me unconditionally.

Contents

Preface

CAD is now becoming a necessity for the design of any equipment or any system. Designers think and create new ideas for products. CAD helps to create ideas, make visualizations, and produce drawings. Often, CAD simulates engineering problems to see the extent of its effect, which may not be possible in real life or without such low cost and convenience.

As industrial competition increases day by day, more optimized designs within lesser time are becoming a continual goal. The design process needs skill, accuracy, engineering expertise, and the ability of mathematical calculations. Obviously, creativity is the generation of the human mind; however, CAD helps to produce faster designs with great accuracy. Nowadays, CAD programs contain various automation functionalities to make customized command sequences or user-defined programs. These user-defined programs are useful to produce 3-D models or drawings with the selection of predefined objects.

This book illustrates effective ways to use CAD in the design process. The various outputs of CAD like drawing generation, automation, and dimensional analysis can be used in effective ways to achieve the designer's goal. Readers are advised to use any advanced software to practice; however, readers may get a clear idea of design based on CAD.

The readers of this book will have a clear idea about the use of CAD in effective ways to lessen design time with the aid of automation and achieving optimized design. Additional material is available from the CRC Web site: http://www.crcpress.com/product/isbn/9781482208795.

Author

Jayanta Sarkar works as an assistant manager in the Heat Exchanger Division of Paharpur Cooling Towers Limited, Kolkata, West Bengal, India, managing day-to-day engineering design activity of various industrial products related to power plants using various advanced CAD software and FEA packages. He has expertise in piping and pressure vessel design and design and development of mechanical systems. He served more than 12 years in various industries, including foundry, material handling, and power plant equipment manufacturing.

Jayanta Sarkar received a degree from the A.M.I.E. from the Institute of Engineers (India); a diploma in mechanical engineering from Murshidabad Institute of Technology, West Bengal, India; and a post-diploma in advance tool design from Central Tool Room and Training Center, Kolkata, India. He presented *Operation of Air Cooled Condenser* at Trichy (March 29–31, 2011) during the seminar *Residential (In-Plant)Training Program (RTP) on Sugar Mill Cogeneration* in Tamil Nadu sponsored by Ministry of Renewable Energy, Government of India, New Delhi. He has participated in many workshops, including "Rejection Control in Castings" by The Institute of Indian Foundrymen and ISO 9001:2008 Awareness Workshop, a workshop on the Theory of Constraints (TOC).

1

Introduction

Computer-aided design (CAD) is a computerized system to assist designers in design, development, and revision work. Designers are needed to design new products and modify existing products as per requirements. They often need to make optimized designs to cut down on product costs. In addition, reduction of design time is also a vital requirement to meet the project schedule. CAD provides excellent facilities to designers to fulfill their objective. Its many facilities include excellent visualization of the designed product for better understanding, analysis at different load conditions, and faster drawing generation. Due to its various facilities, CAD became an essential tool to designers for creation of optimized design within the scheduled time. CAD is utilized for component as well as system designs. Process and piping designers are also using CAD from the beginning of projects to the end. CAD facilitates working on different engineering disciplines in the same project and ensures integration of all the data for sharing, verification, and drawing generation. CAD adds some advantages, including automation for reduction of repeated work, mechanism analysis and tracing of curves, and animation for presentation.

Its capabilities are making CAD an unconditional significant requirement to designers.

1.1 CAD System Tools

It is obvious that not all product designs demand the same assistance from CAD. In other words, different product designers demand different sets of tools of CAD for assisting in the creation or modification of their design. A simple product with fewer components or a single component may need only faster 2-D drawing and design tools, whereas a robust design with a large number of components demands a 3-D drawing with advanced simplified presentation tools. Various CAD software is also available with different features and tools to suit the designer's needs of different design requirements. Software focusing on creation of mainly 2-D drawing without the help of 3-D provides plenty of 2-D editing tools. A designer conceptualizes the product and plots orthogonal or isometric views in a 2-D plane. Detail dimensioning and annotations are placed after finalizing the design, and the final drawing is prepared. In the case of comparatively larger or robust designs, several designers or engineers from different disciplines may be required to work on the same product or project. Partially completed 3-D designed models from different designers are integrated into a system and verified or analyzed for collision/ease of access. Designers also need other CAD facilities like analysis and mechanism at various levels of design, that is, analysis of each component to optimize the cost or analysis of large assembled structures exposed to different load conditions. In such cases, requirement of tools also varies at different levels of design or designers. Normally, high-end CAD software provides most of the tools required for every level of design. Obviously, all the available tools are not required for a single designer or design of a single product or system, and it is also an inconvenience to designers if all the tools are provided on a single screen. Therefore, sets of tools are categorized into modules for specific applications or to suit a designer's specific needs. This way, CAD software provides a choice of selection of modules to customers to suit their specific requirements, thus reducing the cost. Companies making small products use 3-D and 2-D packages with standard available tools, whereas large companies having products in different disciplines use various modules according to their application. Normally, standard packages of 3-D CAD software contain operational design tools for solid modeling. Sheet metal industries use a sheet metal module as an add on. Similarly, surfacing, mechanism, and analysis are also available as add-on modules. The right selection of CAD software for the required work is also very significant.

1.2 CAD Libraries

If the usability of an existing drawing or design increases, a great amount of time and effort is obviously saved. It is always suggested to keep usable drawings or designs in an organized manner to form a library and reuse the items whenever applicable. Modern software is now provided with many libraries like a material library and symbol library. Classification and naming of libraries may also be different in different software. General classifications of libraries in commonly used 3-D software are described in this book. The following libraries are important for the subject's point of view:

1. Material library
2. 2-D symbol libraries
3. 3-D component libraries
4. Customized library

TABLE 1.1

Organizations for Standardization

Abbreviation	Description
ANFOR	Association Française de Normalisation
ANSI	American National Standard Institute
AS	Australian Standard
BSI	British Standard Institution
CNS	National Bureau of Standards for the Republic of China (Taiwan)
CSN	Cechoslovakian Office for Standards and Measurements
DIN	Deutsches Institute for Normung
EN	European Standards
GB	Guojio Biaozhun (China)
GOST	State Standards of Soviet
IS	Indian Standard
ISO	International Organization for Standardization
JIS	Japan Industrial Standards
KS	Korean Standards
PN	Polish Standard
SFS	Finnish Standards
SS	Swedish Standard
STN	Slovakian Standard
UNI	Italian Organization for Standardization

Material, symbol, and component libraries are also available in different units and different countries' standards and codes of engineering practices. Table 1.1 shows some of the most used different countries' codes.

1.2.1 Material Library

Material assignment is an essential requirement especially to 3-D modeled components. When assigning materials to various components, it is well recommended to create and store all the properties of material into a database, thus creates material library. When a material is assigned to a component, all the properties assigned to the material are used for the component. The properties of a material can be subcategorized as follows:

1. Structural properties
2. Thermal properties
3. Miscellaneous properties
4. Graphical/aesthetic properties
5. User-defined properties

TABLE 1.2
Material Properties

Sl. No.	Mechanical	Thermal	Miscellaneous	Graphical	User Defined
1	Poisson's ratio	Thermal conductivity	Hardness	Color	Unit cost
2	Modulus of elasticity	Specific heat capacity	Bend Y factor		Supplier
3	Coefficient of thermal expansion				
4	Ultimate tensile stress				
5	Yield stress				

Subcategorization of material properties may vary and can be different in different software. There can also be many subcategories of properties like chemical, optical, and so forth. Table 1.2 shows some typical properties of materials that are used in most of the 3-D software.

Structural and thermal properties are used for structural and thermal finite element analyses, respectively. Miscellaneous properties like Y factor are used in sheet metal for calculation of developed length. Graphical properties are used for identification and photo-rendering purposes. User-defined properties are very useful for automation. Uses of properties are discussed with practical examples in the latter part of this book.

A material library usually accompanies the software itself. Different sets of data for different materials are stored in the database and available to the software itself. Comprehensive separate material database software is also available in the market, for example, Key to Steel.

1.2.2 2-D Symbol Library

A symbol library consists of 2-D drawings, sketches, or pictures. Most of the 2-D drawings contain a large number of standard 2-D objects. Annotation symbols for welding, standard views for piping components, or structural elements are the most used symbols in 2-D drawings. Organized storing and categorization of these symbols form useful symbol libraries. Further, symbol libraries can be classified as follows:

1. Structural symbol library
2. Fastener symbol library
3. Piping symbol library
4. Annotative symbol library

There can also be various symbol libraries depending upon the application requirement. The mentioned symbol libraries are important from the subject's point of view and discussed in this book. Classification of all the products in a standard library is beyond the scope of this book. However, uses of commonly required library items for mechanical engineers are demonstrated throughout the examples of this book.

- *Structural symbol library*: A structural symbol library contains various 2-D views of structural elements. Some of the most used structural components are shown in Figure 1.1.
- *Fastener symbol library*: A fastener symbol library contains various 2-D orthogonal views of fasteners. Generally, fasteners are axisymmetric; therefore, top, bottom, and front views are sufficient for their presentation in drawings. True views of threads and minor details are not necessary in assembly drawings; therefore, simplified views are shown in Figure 1.2.

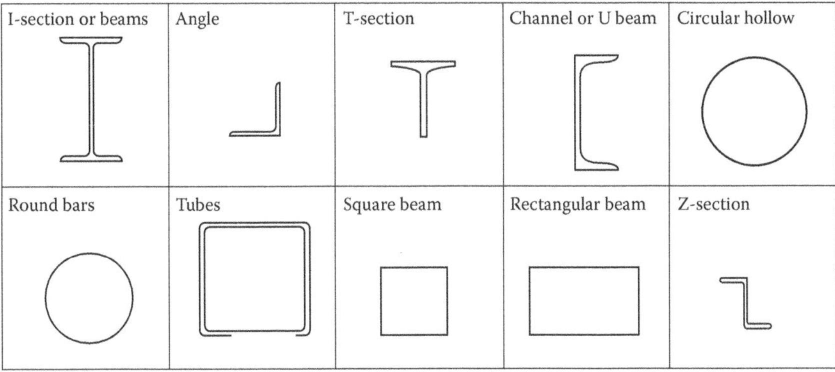

FIGURE 1.1
Commonly used structural sections.

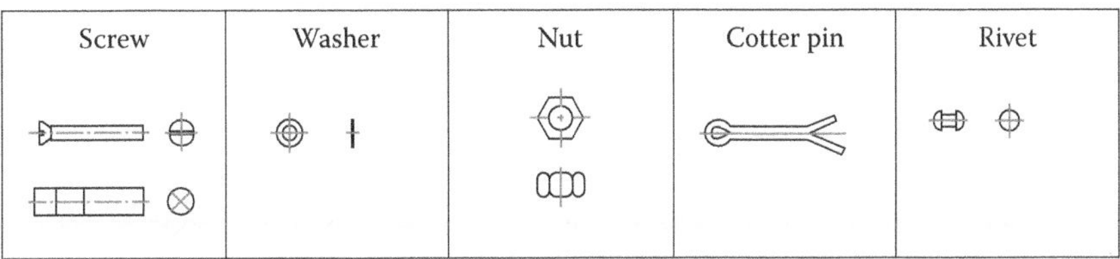

FIGURE 1.2
Simplified representations of fasteners.

- *Piping symbol library*: Piping symbols are very useful for pipe layout and isometric drawing preparation. A scaled drawing of the piping layout is a mandatory requirement for piping design; therefore, the true size of fittings in orthogonal views is needed to check the clearance or interference in piping brunches. Figure 1.3 shows some of the piping fitting and equipment symbols.
- *Annotative symbol library*: Welding symbols and machining symbols are a most used annotation in fabrication and machining production drawings. Figures 1.4 and 1.5 show commonly used welding and machining symbols, respectively. Details of annotations will be covered in Chapter 8.

1.2.3 3-D Component Library

3-D component libraries contain 3-D models of various components. These libraries are often subdivided into several categories, similar to 2-D symbol libraries. Figure 1.6 shows some of the commonly available parts in a 3-D library.

1.2.4 Customized Library

Customized libraries are user-defined libraries and specific to user needs. These libraries are not supplied with the software itself. Users need to build up their own customized library to suit their application or requirement. Original equipment manufacturer (OEM) companies often make libraries

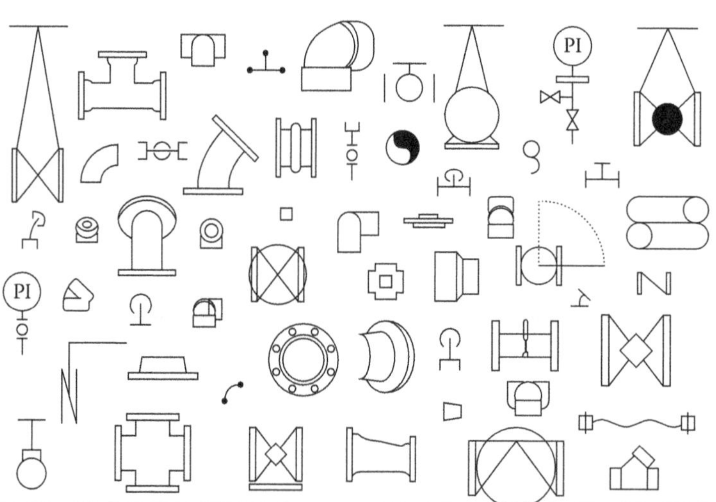

FIGURE 1.3
Piping symbols.

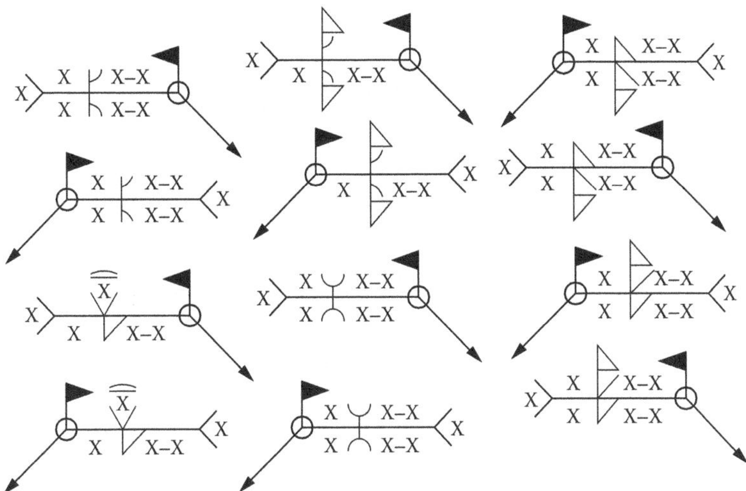

FIGURE 1.4
Welding symbols.

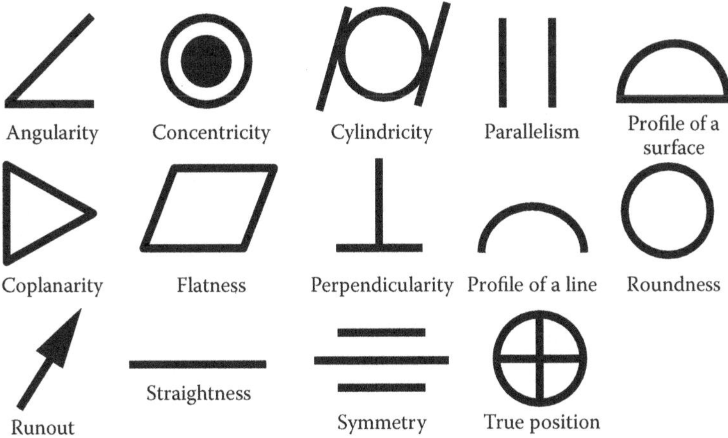

FIGURE 1.5
Machining symbols.

for their own products. Many of them share their 3-D product libraries on the Internet. It helps the designers to use original 3-D products and thus increases accuracy and reduces time of creation. Customized program libraries are very useful for automation, where repeated execution of the same programs is needed. Customized library creation will be covered in detail in the automation part of this book.

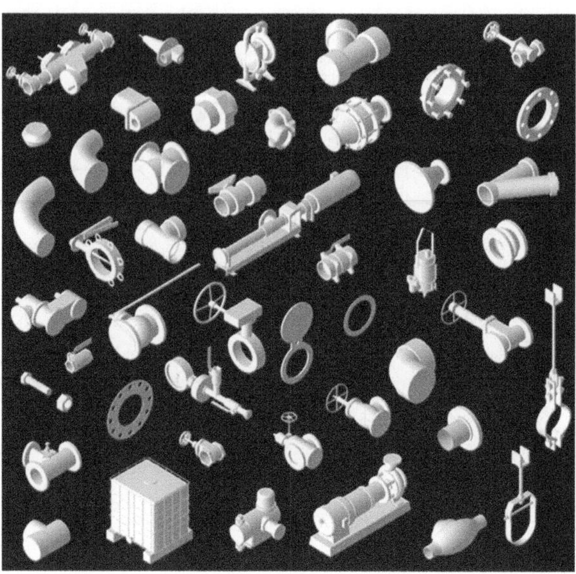

FIGURE 1.6
3-D component library.

1.3 Application-Oriented CAD Systems

Frequency of use of any tool of CAD is dependent on the engineering application or product design. Different applications demand different sets of tool for ease of use or faster work. Structural engineering application demands more structural operational tools, whereas piping engineers need tools to create accurate and quick pipe routing. Software that provides sets of tools for all engineering application fields in a single window may not be user friendly and economical; thus, software companies are developing CAD software targeted to specific engineering application fields. Application-based modules are also available for different engineering fields in many kinds of high-end software like Creo, CATIA, Unigraphics, and so forth. Some of the commonly available application-based software or modules are listed as follows.

1.3.1 Structural Modeling and Drafting

Structural modeling and drafting follow the sequence depicted in Figure 1.7. Upon finalizing the requirement, structural designers conceptualize and create the wire-frame model. Mostly, a prior analysis is needed to verify and optimize the model. Once the wire-frame model and elements are finalized, structural elements or sections must be assigned to the wire-frame model. Structural software provides various sets of tools to configure the joints.

FIGURE 1.7
Process of structural drawing creation.

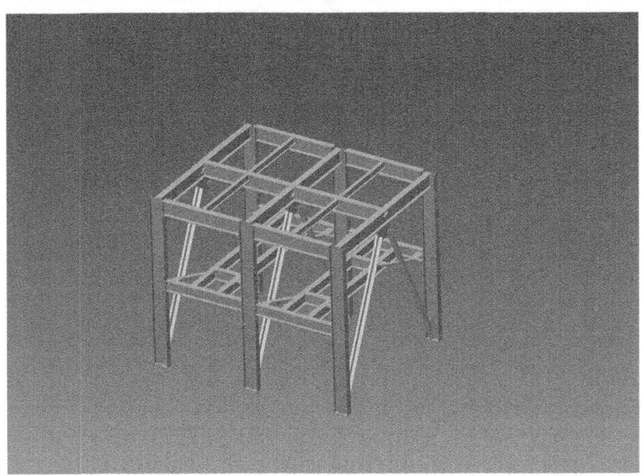

FIGURE 1.8
A structure in a 3-D CAD system.

The structural sections are automatically trimmed or extended to the required length when joint configurations are applied. Once the 3-D model is finalized, a 2-D drawing is generated either automatically or manually. Mostly, structural software provides tools for automatic individual part drawing and bill-of-materials creation. Figure 1.8 shows a structure created in CAD.

1.3.2 Piping Design and Drafting

Piping in CAD can be broadly divided into two categories: non-spec driven and spec driven. Non-specification-driven piping does not require a definite specification to follow and limits its uses for small industrial applications. Spec-driven piping needs specifications, which are defined by the piping designer and are often made specific to a project. The sequential procedure of piping drawing creation is shown in Figure 1.9.

FIGURE 1.9
Process of piping drawing creation.

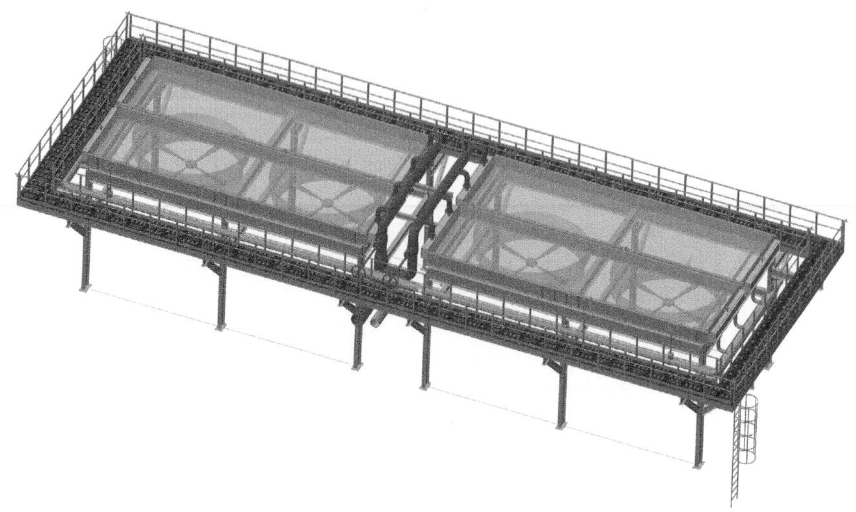

FIGURE 1.10
A plant model in 3-S CAD systems.

A piping and instrumentation diagram (sometimes called a process and instrumentation diagram) contains all the major piping parameters, including size, schedule, and thickness. A piping 3-D model is prepared by piping engineers considering piping rules and many criteria like ease of accessibility, upstream and downstream equipment restrictions, and so forth. A piping layout is generated after finalizing pipe routing. Most of the piping software or modules enable automatic piping isometric drawing creation. Figure 1.10 shows a 3-D model of a plant with the necessary piping made in a CAD system.

1.3.3 Sheet Metal

Sheet metal design is one of the most usable applications in CAD. Apart from drawing generation, automatic developed length calculation is the most advantageous feature of sheet metal design in CAD. A sheet metal module in CAD provides tools for assistance to create punch and die models derived from the final component. Sheet metal operations will be discussed in detail in a latter part of this book. A sheet metal model created in CAD is shown in Figure 1.11.

1.3.4 Plastic Mold

High-end software provides a very useful tool for plastic mold design with mold base and final bill of materials (BOM) creation. The core and the cavity are automatically extracted from the final component. Once the mold bases are created, software provides tools for creation of runners. Figure 1.12 shows a 3-D view of core, cavity, and component extraction.

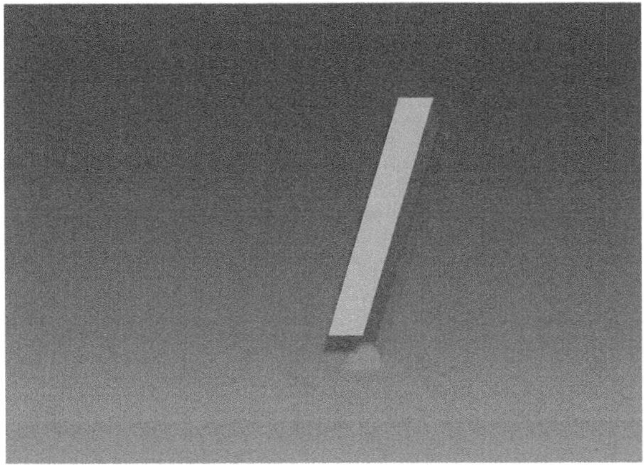

FIGURE 1.11
A sheet metal 3-D model.

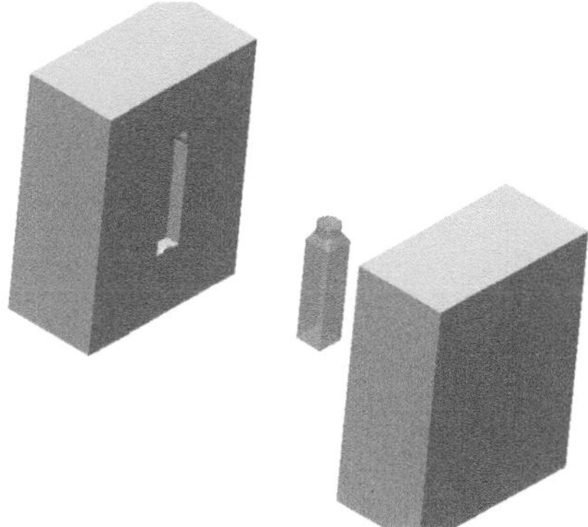

FIGURE 1.12
Core, cavity, and component extraction in 3-D CAD systems.

QUESTIONS

1. Why did CAD become a necessity to designers?
2. Describe the need for CAD libraries.
3. Describe the advantages of CAD libraries.
4. Why is an application-oriented CAD system necessary?
5. How can we reduce time by effectively using CAD?

2

Advantages of CAD

2.1 Introduction

Computer-aided design (CAD) has many advantages in engineering design.

Utilization of CAD is increasing continuously to reduce time, to simulate robust design, for feasibility analysis, and to optimize design. To use CAD in an effective way, we should have a clear idea on the basic concept of using CAD.

Drawing is an engineering language. So the best way to express engineering concepts and exchange engineering information is by using drawings. For example, refer to Figure 2.1; if any component has to be specified, we need the following information:

1. All the dimensions of the component to specify the component
2. Material description
3. Other parameters respective to the application like weight, tensile strength, and so forth

So fundamentally, it requires dimensions that have to be either in the metric or the imperial system, depending on the use and preference. All this information can be expressed in the best way to an engineer if it is depicted in a drawing.

As far as 2-D and 3-D drawings are concerned, they can be drawn manually by board and drafter. The time consumed by manually drafting is much more compared to use of CAD. This is simply because of handling tangible and consumable resources like sheets, pencils, and so forth. However, in the case of CAD, resources like sheets or drawing space are more than sufficient. Also, in manual drafting, mass properties and other important parameters are not calculated automatically.

It is obvious that a design is not a once-through system. Several periods of development and modification and rectification are necessary to carry out a fruitful design. Therefore, it is always fruitful not to use any tangible resources since these are all going to be wasted in the evolution of the design process.

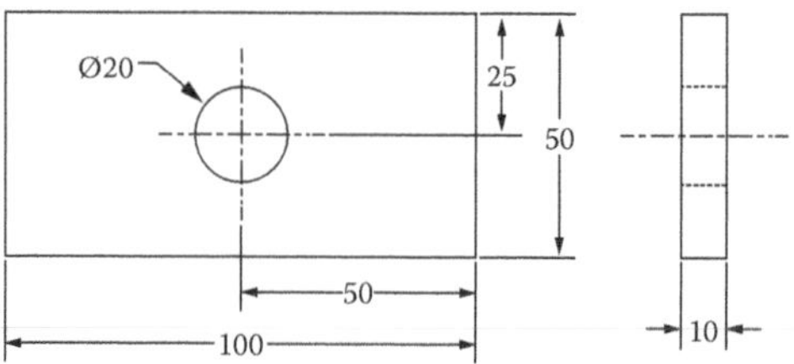

QTY: 2 pcs.
Material: IS 2062 GR. B

FIGURE 2.1
Drawing of a simple component. Note: All dimensions are in mm, unless otherwise specified.

CAD users are easily connected by the net in an office or over the Internet. This allows many designers to work on the same project component simultaneously. This particular method is required in the case of large component designs where engineers of several disciplines work on a project, and their components are related to each other. For example, in the case of plant design, the supports of the main components are also utilized as support for small auxiliary components like electrical item mounting and so forth. Thus, different engineers are involved in the same project.

CAD is the best solution since a designer can think and visualize in various ways to improve a design to the furthest possible extent.

2.2 Advantages

The application and area of CAD is wide. The advantages of CAD can be utilized in many areas of work. Figure 2.2 shows the distinct advantages of CAD.

2.2.1 Viewing

The most effective advantage of CAD is viewing or, in other words, view generation. When there was no CAD system, people used to draw the component views manually. A manual drawing of a view of a component requires much time. The drawing quality also depends on the person and the quality of the tools that are used for the drawing. Prior to the CAD era, the draftsman created mainly orthogonal views. 3-D views also could be created, but

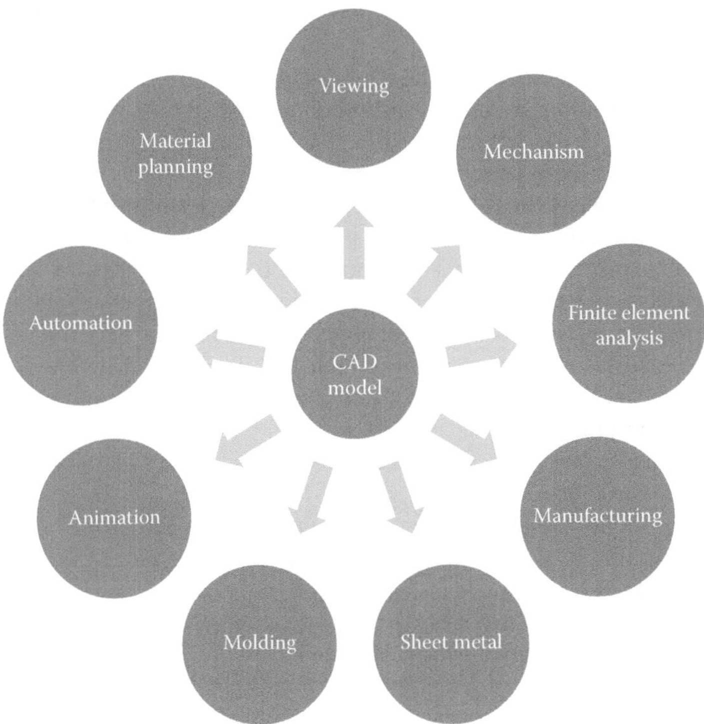

FIGURE 2.2
Advantages of a CAD model.

it required a lot of time and effort. People had to decide first what views were required to completely specify the requirement.

Now, in the era of CAD, we can create a 3-D model of almost any object within much less time. If the component has very fewer features, then we can create it within a few minutes.

If a conceptual drawing is available, then we can easily model with dimensions. When the model is created, we can rotate and generate different views from different angles. The views from very odd angles are also possible through CAD.

Sectional views are a distinct advantage of CAD. It is really very easy to create as much of a section as one wants to view and to understand the component from a different angle. When creating sectional views in CAD, every small edge is traced and drawn in the screen. This enables a designer to have an absolute understanding of the component.

Only 2-D drawings in hard copies are currently used in most of the manufacturing shop floors. When 2-D generation is required, CAD provides excellent advantages in generating 2-D orthogonal and 3-D views. It is now made easy to provide 3-D views in hard copies for quick and better understanding of the product on the shop floor.

It is obvious, in 2-D sheets, if the drawing goes wrong; probably, creating a new drawing becomes a viable solution. However, in CAD, the model of the components can be easily removed or erased or can be modified so easily. An interactive design with checking at every stage is possible.

Nowadays, most of the 3-D models can be viewed without the parent software by using many free views. This allows engineers to communicate better about the component with a 3-D model of the product.

3-D models are very useful for presentation. An obvious example is when a real component, especially for larger sizes of components, may not be always available. Also, when a component is such that it is created for the first time, the 3-D views are very useful for making everyone understand the concept of the new component.

3-D files can be stored and transferred very easily and in less time through e-mail, eTransmit, or FTP.

3-D components are always made in true scale. So the model size can be seen in comparison to other models, and the interference or the size difference can easily be viewed.

Multiple copies of 3-D models can be made very easily, which helps designers to view the patterns of a model together.

Certain manipulation of 3-D models can also be possible, which help to generate and share ideas from different aspects, like cost, feasibility, and so forth.

Different colors can be applied on the surface of 3-D models, so the best one can be chosen without any real expense of painting.

3-D models can be scaled and viewed as required by the user for better understanding.

In a 3-D model, the dimensions and distances can be measured from any point to any other point with minimum effort. This enables a designer to make an optimized model and see the variation of one dimension over the other. Angular dimensions between surfaces also can be measured with minimum effort.

Mass properties can be checked and found very easily in 3-D models. Normally, in the case of large assemblies, it is very difficult to calculate mass properties, like moment of inertia, center of gravity, and so forth. However, in CAD, the center of gravity can be exactly located and measured from any other point. Mass properties, like weight and moment of inertia, can be calculated within seconds. Moment of inertia about any axis can be determined easily.

Figure 2.3 shows a 3-D model of a tank. It is obvious that by making only 2-D engineering drawings, it is not possible for anyone to justify aesthetic aspects of the prototype. CAD can generate realistic views of a product in any direction vector. Also, design variations are made possible in CAD. Therefore, a few obvious advantages are as follows:

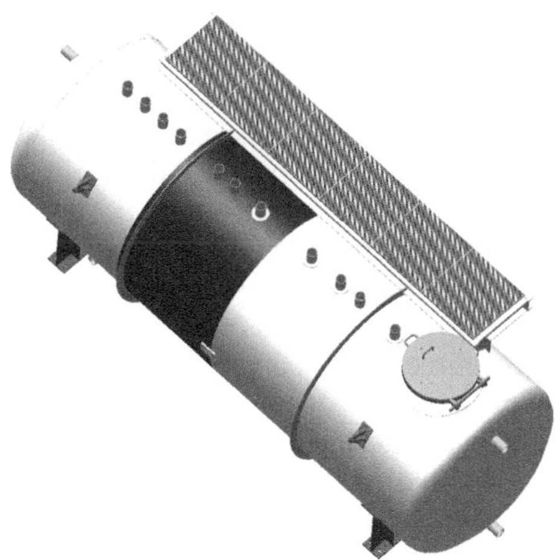

FIGURE 2.3
A tank modeled in a 3-D CAD system.

1. Dimensional checking and checking for clearances
2. Aesthetic design
3. Accurate calculation like mass, volume, surface area, and so forth
4. Section generation
5. Physical properties generation like center of gravity (CG), moment of inertia, and so forth

2.2.2 Mechanism Design

Mechanism design gets direct help from CAD systems. Designers can create a 3-D model of each of the components to create the mechanism. All the mechanisms can be assembled one by one with certain joint configuration so as to simulate the real condition. When all the joints are defined, the mechanism is formed. The mechanism can be manually dragged to verify how it works. Prime movers are assigned to the driver ends of the mechanism to simulate the mechanism in real conditions.

Mechanism design can be divided into two types, kinematic design and dynamic design. In kinematic design, the forces are not considered in the design, and the forces and its effects are considered in the dynamic analysis.

Kinematic design allows a designer to animate the model. This enables excellent visualization and verification of a mechanism in different locations relative to the various elements of the mechanism. In kinematic design analysis, it is often required to trace the curve of any point relative to the position of a point of other elements. CAD allows excellent tools to trace the curves, which is done in positional analysis where the position of each element in the mechanism is recorded with respect to time frame.

The prime mover may be assigned in the form for a motor, where the velocity is input parameter. The velocity can be linear or may be a function of time. Various predefined functions are available in advanced CAD software; it even allows user-driven functions. The velocities of any element or any point can be determined in the CAD system. This velocities analysis requires tremendous calculations if it is to be done in hand. However, CAD provides excellence in determining the absolute and relative velocities of any link or point with respect to other links or points. This saves a tremendous amount of effort and time. However, for mechanism analysis of large assemblies, high-end computer resources may be required, which are often justified with respect to the output.

Dynamic analysis is even more critical when done in hand. If the forces and its effects are considered in the analysis, the calculation becomes too large, where a single error may lead to a large difference in result. CAD allows the application of forces or torque at the driver end, which simulates the synchronization of the speed or velocity profile that will be generated in actual conditions. The friction at various joints can be defined in the process of defining the joints. Therefore, when the forces are applied and links of any mechanism turn or move, the frictional forces act. The other parameters like inertia of all the links in the mechanism are automatically taken into account in the process of analysis. Determination of mass properties like mass, inertia, and so forth, itself takes much more time for large assemblies where the sections are varied. Therefore, a huge task is reduced due to automatic calculation and consideration of these properties in analysis.

The results of dynamic analysis like the forces at various points are also a most important output. Dynamic analysis enables fatigue analysis, which may be used to determine the fatigue life cycle of a component and endurance limit. The effect of dynamic analysis is that forces at various points, surfaces, or members can be stored and transferred to finite element analysis, which simulates the stresses and deflection of the member under forces from the mechanism.

Thus, CAD helps a mechanism designer to create, simulate, and analyze the effect of forces on all the members in a mechanism. This helps designers to optimize and improve the design so that the quality can be improved and maintained with reduced cost.

2.2.3 Finite Element Analysis

In the extensive use of CAD, finite element analysis is one of the popularly used tools to simulate a member under various loading conditions, which is considered as a part of computer-aided engineering (CAE). CAE is comprised of using computer programs in various engineering areas. It involves assisting in designing, manufacturing, planning, and so forth.

When a force is applied on a material, some stresses are generated. If the stresses are under the elastic limit, the material will not deform permanently;

however, if the forces are beyond the elastic limit, the material is deformed permanently. The properties of the materials also can be nonlinear which invokes the use of nonlinear analysis. The stresses and deflection can also be found by the use of conventional equations of engineering. However, the formulas are derived mostly for regular shapes. If irregular shapes are loaded, the use of conventional engineering formulas become difficult and may require extensive engineering skills.

Finite element analysis is an excellent tool provided in CAD software, which enables the user to analyze the model under different loading conditions with excellent visualization aids.

In CAD, before finite element analysis, the finite element model (FEM) needs to be made. A CAD model is converted. The process sequence of finite element analysis is shown in Figure 2.5. First, the CAD model is imported into the finite element analysis environment. Often, CAD models are made without assigning the material and its associated material properties. An engineering component may consist of different parts made of different materials. Individual parts also can be assigned with different materials; this allows the model to be analyzed in the loading of a real condition. The model is then divided into a finite number of small parts called elements. These elements can be of the same or different types. The elements are basically assigned approximating functions defined in terms of field variables of the specified points, which are called nodes or nodal points. First, the elements are to be defined, and then the selected parts or region is to be meshed with those elements. The selection of elements depends on the shape of the model and the purpose. For example, a single dimensional point can be solved with 1-D analysis, whereas a 3-D problem may require 3-D elements like a brick element. In some cases, line elements also can be used in the simplified model of a large structure, which is called idealization. A common example of idealization is analysis of trusses. It is obvious that trusses are the assemblies of a large number of beam elements and components. So the line elements are used to model the truss with preassigned properties as in the beam elements. This simplifies the model and saves a lot of effort and time. A boundary condition is where the action comes into sequence after meshing. Boundary conditions are defined as the applying of constraints on the boundary geometries like external surfaces, edges, or vertices. Constraints are defined in terms of locking the degree of freedom. It is often seen that the boundary condition may not be provided as per real conditions. In those cases, the boundary conditions are imagined and provided in conceptualized position. The last step before analysis is assigning the loads. The loads are assigned to the model's geometries like surfaces, edges, and so forth. The types of loads are varied, depending upon the type of analysis. For example, the loads for structural analysis are mainly forces, pressure, gravity, temperature rise, and so forth, whereas in the case of thermal analysis, the loads may be the temperatures at different surfaces. Thus, a CAD model becomes an elementary model with its own property matrix, boundary condition, and loads, which is called an FEM. Analysis is the process of solving the matrix of equations and determination of unknown variables like displacement and so forth at various nodes. It often consumes much time for large and complex geometries, which is also dependent on computer resources. Postprocessing is the last and final step, which involves the review of analysis results. Results can be seen in the form of graph, animation, and so forth. In a window, the results are scaled into color maps, and various colors are displayed in the model according to the intensity of the result parameters in the color scale. It really helps to quickly locate the maximum stress area. Designers can see the developed stresses and deflection in the various areas of the component, which helps them to optimize the design to the maximum extent. Finite element analysis provides sufficiently accurate results. The sequence of finite element analysis from a model preparation is shown in Figure 2.4.

The FEM of an impeller is shown in Figure 2.5. Finite element analysis of such critical parts has numerous differential equations to solve, which might require large calculation steps. Nowadays, finite element analyses are becoming popular due to their capability of analysis in complex large parts with accuracy.

The forces and moments at different node points can be measured, which helps designers to improvise the design to a maximum extent.

2.2.4 Manufacturing

CAD provides a distinctive advantage in manufacturing. The extent of use of CAD in the production or manufacturing is computer-aided manufacturing (CAM). The most used areas of CAM in production are as follows:

1. Manufacturing through CNC machines
2. Computerized measuring or inspection
3. Computerized robotics application

FIGURE 2.4
A mechanism to transmit power through spur gear.

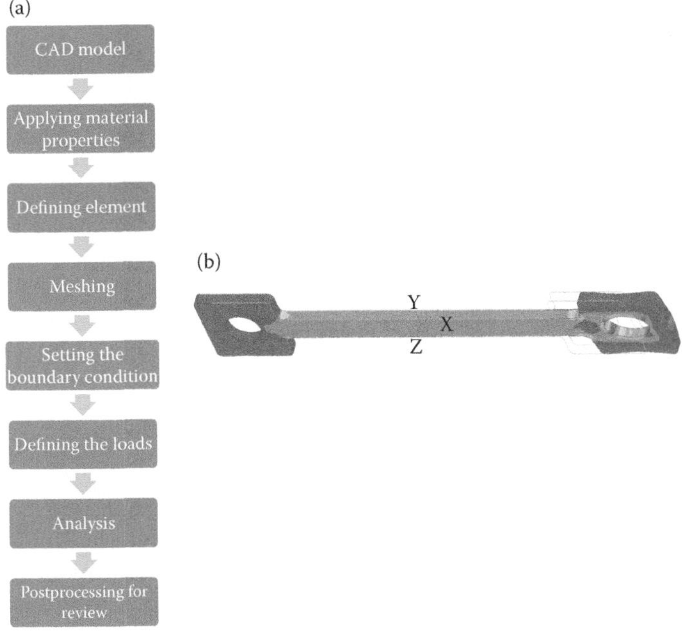

FIGURE 2.5
(a) Sequence of finite element analysis. (b) Finite element analyzed 3-D model.

Nowadays, computerized machine controls are popular in industries. CNC machines are used for manufacturing precision parts and complex profile operation with great accuracy and considerably low tolerances. Accurately dimensioned 3-D complex geometries are made possible by cutting operations through CNC machines. In the CNC machines, the tool is the most used varied profile creator. These profiles are called tool paths. Tool paths must be created first, followed by the code generation procedure. The code file for the CNC is a series of programmed codes, which are actually instructions to the machine. The machining is finally carried out by the use of the codes.

CAD software often provides an integrated module to generate a tool path and associated codes for the respective machines. The codes may be different for different machine controls, including Fanuc and Sinumeric. The CAD model is imported first, followed by the defining of the workpiece from which the model will be cut. Certain parameters of the machine have to be entered in setting the machine configuration, which specifies the limits of the machines. The workpiece shall be located with respect to the machine axis by a coordinate system. Most applicable tools are to be preserved for the machines. Various tool parameters need to be specified. In some cases, when more than one tool is required for an operation, the tools must be carefully selected as in real operation. Tool paths can be generated for a few selected

FIGURE 2.6
Machining tool path.

surfaces or for the whole part automatically by the CAD. Often, complex geometries require careful generation of tool paths. Simulations of tool paths are also available in the program, which simulates the motion of tools in actual operation. Various important derived parameters are generated from tool path simulation, like machining time, power requirements, and so forth. Some software allows inspection of the real model, generated from the machining for inspection, which can be compared with the real CAD model, so that a designer can understand whether the requirement of the product's surface finish is satisfied. Therefore, after finishing the tool path generation, the codes are generated respective to the machine control system, which can be transmitted to the machine for real operation.

Figure 2.6 shows tool cutting profiles for complex 3-D surface geometry. Such complex profiles are very time consuming to program in a CNC input panel.

2.2.5 Sheet Metal

Sheet metal is an advanced module in CAD, which is customized for sheet metal and press work. Sheet metal parts can be imagined as surfaces with very small thickness compared to its other dimensions. In CAD, sheet metal modules have various sets of handy tools that are used to create most used sheet metal features. Thus, they increase user-friendliness and reduce the time to create the sheet metal model. One of the best advantages of sheet metal is automatic generation of developed blanks. In other words, a sheet metal may contain many types of operations, including punching, bending, and notching. When the bending operation is carried out, the internal and external fibers are stressed, and the neutral plane shifts in accordance with the bending procedure. This shifting of the neutral plane can be specified in the system so that the development length, which is the length of the neutral

FIGURE 2.7
3-D model of a simple sheet metal component.

plane, can be calculated easily. When there are many bending features on a sheet metal part with different bending angles, calculation of developed length becomes complex. CAD provides an excellent solution by providing an automatically developed model.

CNC programming for sheet metal works is also available in CAD. This allows progressive or time-dependent sequential operation of machines. This eliminates the user interface and reduces the chance of human error.

Some CAD software enables automatic insertion of press tool components; however, automatic press tools creation can be programmed in CAD, so that after creating the die or punch, the whole sheet metal assembly automatically fits. It reduces the time of creation of a press tool design.

New sheet metal products can be modeled and imagined with enhanced tools for sheet metal creation in advanced CAD software. Designers are allowed to visualize the parts and estimate the cost of making before declaring the design. Often, the feasibility of placing various operations on sheet metal parts can be seen; thus, it reduces the cost of making a prototype.

It is always a comparatively large exercise to calculate the mass properties of sheet metal parts with a large number of features. Accurate volumes, masses, and weights of sheet metal parts can be calculated easily for optimization.

Figure 2.7 shows a sheet metal model. Development of such sheet metal models is a complex job. In CAD software, sheet metal modeling is much easier, and a developed model can be easily traced.

2.2.6 Molding

Molding is greatly enhanced in CAD with built-in handy tools. Plastic or metal flow can be simulated in extensive use of CAD or CAE. In the case of finite element simulation, casting models must be prepared first.

Metal casting: This is created by adding gates, runners, risers, and so forth to the product, which is actually molded inside boxes. Adding all necessary components where the fluid flows makes a complete casting model. The

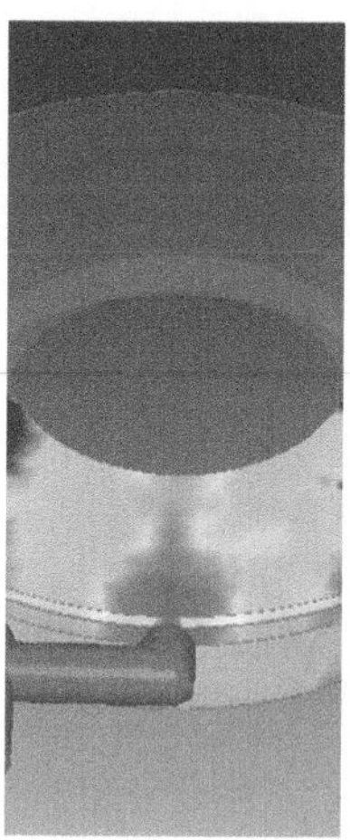

FIGURE 2.8
Casting analysis model.

model is simulated by the finite element method. The main advantage of casting simulation is that the blowholes inside the volumes can be found, which are difficult to determine without a destructive test. This extends the usability of CAD in the casting design field. A casting analysis model is shown in Figure 2.8.

Plastic molding: In the case of plastic molding, the final models are created first. The part is placed inside the workpiece from which the core and cavities are extracted. The parting plane must be defined before the extraction of the core and the cavity that actually separates the mold parts. The entry point must be defined in the component, which allows simulation of plastic flow to prepare the model. Various important parameters, like temperature and shrinking, can be seen from the results. Once the insertion location of plastic is defined, the sprue, runner, and gates are cut into the mold. The standard mold base can be assigned to prepare the complete mold. This plastic molding allows the designer to find flaws in the design and to prepare the production drawing of the complete plastic mold within much less time compared to hand drawing. A plastic analysis model is shown in Figure 2.9.

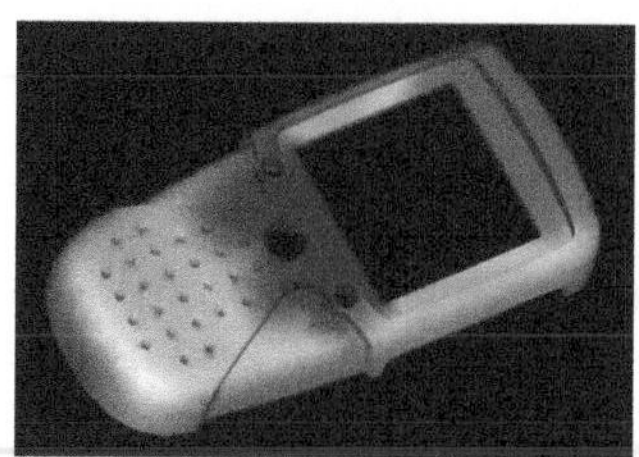

FIGURE 2.9
Plastic analysis model.

2.2.7 Animation

Animation is the most effective module for presentation and understanding of component functions or processes. Animation is a series of frames played continuously in order to depict the function.

In the case of mechanisms, the animation of mechanisms is very helpful to understand the function and effectiveness of the component. In most of the mechanisms, more than one member moves relative to each other. Often, the path of the members in a mechanism is difficult to understand by simple drawing. Even in completely specified drawings, the relative motion of different components and the functions are not well understood. The animation simulates the mechanism and shows the behavior of all the components in a mechanism.

In the case of comparatively large assemblies with many components, the sequence of assembling the component is often a critical criterion, which needs to be followed; otherwise, the component cannot be assembled. In such cases, animation of a component from exploded views to assembled views is very helpful in order to understand the sequence of assembly. In the case of a large component, like pant engineering, the erection of the components takes much time. The proper sequence of erection can be found from exploding the CAD model and its animation, which, in turn, helps to reduce the time and cost.

Postprocessing in finite element analysis itself provides animation of deflection or stresses on loading on the member. It increases the understandability of the effect of forces or moments.

In the manufacturing area, the tools path and component material removing operations are well animated. This allows analysis and optimization of the tool path to reduce cost and time. Process sequences for sheet metal progressive dies can be animated, which reduces the chance of error.

In other words, animation helps in better understanding and improves presentation in all fields.

Figure 2.10 shows an animation of a piston cylinder mechanism. Different types of animation like the assembling method of any product can be made by animating exploded views.

FIGURE 2.10
(See color insert.) Piston cylinder mechanism.

FIGURE 2.11
A 3-D model of a gear created in an automation program.

2.2.8 Automation

It is obvious that CAD can reproduce the similar designs and drawings. If sufficient programming can be made on CAD, automatic component generation is very much possible. In most engineering products, there are many rules that are standardized; a designer has to carry out these exercises of calculation for every new component. These rules can be imposed in the CAD model itself, such that the real variables are kept open to variation and all dimensions are related to those variables. In this way, the model can be regenerated with different values of open variables, and thus, a new product shape is formed. Once the automation is done on the 3-D models, the 2-D drawing shapes are automatically changed with the drawing. Thus, it also invokes automation of 2-D drawings.

Comparatively small and simple components can be atomized to maximum extent, and that may be 100% where no manual intervention is required in the case of repetitive design, whereas full automation on large components may not be possible.

Any amount of automation certainly increases productivity. It reduces manual effort and time. If the rules are precisely specified, the automation certainly produces highly accurate output and reduces human error. Figure 2.11 shows a gear created in an automation program. An example of automation of a gear is shown in Chapter 13.

2.2.9 Material Utilization

When several components can be cut from a single sheet, auto nesting enables us to ensure maximum utilization of materials or a workpiece for the required number of materials of the same shape or different shapes. Components of various shapes can be placed and adjusted in a single sheet, which reduces the waste materials. Placing of these components is a time-consuming job. It is obvious that if the components are in tensile loading, the gain direction also plays an important role when placing and adjusting the

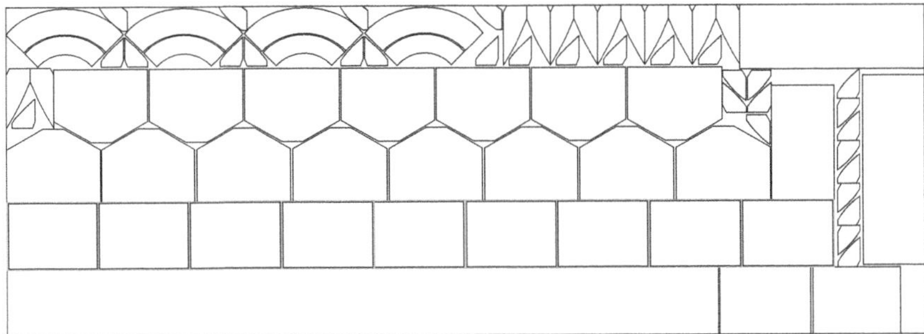

FIGURE 2.12
Example of material utilization.

components in a sheet. CAD provides automatic evolution of and adjustment in a different pattern so that the material utilization can be increased, thus reducing the cost. Figure 2.12 shows a material cutting diagram for optimum use of materials and reducing wastage.

QUESTIONS

1. Broadly describe the advantages of CAD.
2. Why is finite element analysis essential?
3. Describe the advantages of using automation.
4. Describe the applications of sheet metal and modeling.
5. Site an example where mechanisms in CAD would be helpful.
6. Describe the advantages of simulation of a tool path in CAD.
7. How can the CAD system be effective in reducing wastage in material planning?

3

Structure of Computer-Aided Design Modeling

3.1 Introduction

A computer-aided design (CAD) model of any component consists of various entities and features like datums, solids, surfaces, and so forth. Various operations are needed to prepare these features, entities, and final CAD model. These operations are carried out by the executable commands available in the CAD software. CAD software can be imagined as a set of predefined executable programs in an arranged way. These programs or commands are made available to users in the form of menus or icons. It is obvious that CAD software helps designers to create final production drawings from conceptual designs. Preparation of a conceptual design and final product design and production drawing need various types of operations. Therefore, a large number of menus or icons are provided to a user based on the designer's requirement. These icons or menus are also logically structured in a way to provide the user with maximum accessibility.

Location of icons and menus may vary depending on the software; however, principally, most of the CAD software allows the commands in the same pattern.

3.2 Window Identification

In general, graphically, parametric CAD software has few basic areas: a graphic section, a command and menu bar section, and a history tree. It is obvious that the graphic section is the larger section for displaying parts in 3-D or 2-D. The location of the history tree and the command section may change with different software depending on the user interface. A typical view of a Creo user interface is displayed in Figure 3.1. Most CAD software allows customizing of these individual sections. In the area provided for graphic sections or menu bars, ribbon areas can also be adjusted. Sometimes, this customization of the user interface is also useful for improving visualization or increasing accessibility. For example,

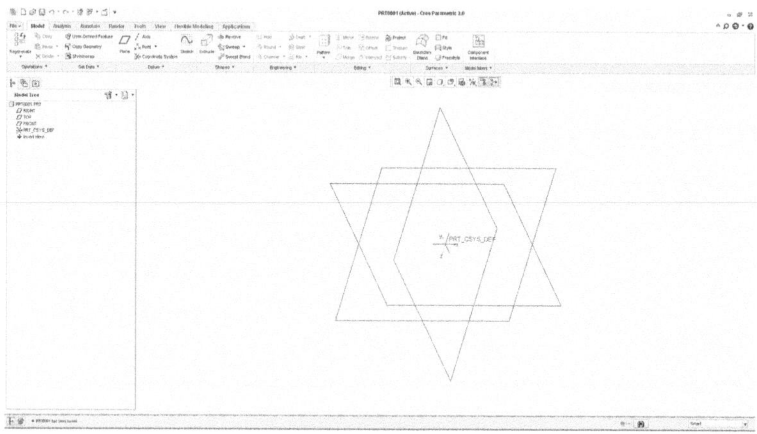

FIGURE 3.1
Graphical user interface screen of the software Creo.

if someone works with a product of aluminum using the real color of aluminum, he or she will find a dark background useful. An added set of icons for map keys or macros may need to be suitably placed when working with automation.

It is recommended for users to know about customization of the user interface, which may increase the ease of work depending on the application.

Any action for the feature of parts creation or any assembly operations are controlled from the command sections or menu bar. The history tree provides the sequence and properties of the features or parts used for the model. Some CAD software provides some common actions with a right mouse button (RMB) click on the history tree or graphic area. Since most CAD software includes a large number of commands, they are made available only on requirement. For example, sheet metal options like wall-creation and bend–unbend operations are not available in part-creation mode. This increases simplicity and increases user accessibility.

3.3 Unit Settings

Most CAD software contains predefined unit sets. Designers or CAD users should set the unit before starting any work. Also, most of the software provides user-defined unit sets to use according to user needs.

A unit system mainly consists of primary units, which are length, mass, and time. Sometimes, other primary units like temperature* that are needed in analysis may be present. The main unit systems are as follows:

* Temperature is not required for preparation of CAD model unless used for analysis or information purpose.

1. SI (Système International d'Unités)
2. MKS (meter kilogram second)
3. IPS (inch pound second)
4. FPS (foot pound second)
5. CGS (centimeter gram second)
6. Custom units with example

Often, it is necessary to put the dimension in different units; therefore, a detailed correlation chart of different units is given.

For relations between primary units, a comparison chart is given in Table 3.1.

For relations between secondary units, a comparison chart is given in Table 3.2.

Most software has the built-in functionality to change the unit systems from one to another. If the functionality is not available in the CAD software, one can easily scale the model with the respective scale factor to obtain the desired size of the model. However, it is to be kept in mind that all

TABLE 3.1

Primary Units

Dimension	SI/MKS	IPS	FPS	CGS
Length	1 m	39.37 in.	3.28 ft.	100 cm
Mass	1 kg	lbf s^2/in.	0.0685 slug	1000 g
Time	1 s	Second	Second	Second
Temperature	1K	[1.8(K – 272.15) + 32]°F	[1.8(K– 272.15) + 32]°F	–272.15°C

TABLE 3.2

Secondary Units

Dimension	SI/MKS	IPS	FPS	CGS
$\text{Force} = \dfrac{\text{Mass} \times \text{Length}}{\text{Time}^2}$	1 newton	0.2248 lbf	0.2248 lbf	105 dyne
$\text{Acceleration} = \dfrac{\text{Length}}{\text{Time}^2}$	1 m/s^2	39.37 in./s^2	3.28 ft./s^2	100 cm/s^2
$\text{Pressure} = \dfrac{\text{Force}}{\text{Area}^2}$	1 Pa, N/m^2	0.000145 lbf/in.^{2}	0.020885 lbf/ft.^{2}	10 dynes/cm^2
Energy = (Force × Length)	1 joule (J)	8.8507 lbf in.	0.7375 ft. lbf	10^7 J erg (g cm^2/s)
$\text{Density} = \dfrac{\text{Mass}}{\text{Length}^3}$	1 kg/m^3	lbf s^2/in.^{4}	slug/ft.^{3}	g/cm^3
Angle	1°	0.0174 rad	0.0174 rad	1
$\text{Power} = \dfrac{\text{Energy}}{\text{Time}}$	1 W	8.8507 lbf in./s	0.7375 ft. lbf/s	10^7 erg/s
$\text{Velocity} = \dfrac{\text{Length}}{\text{Time}}$	1 m/s	39.37 in./s	3.28 ft./s	100 cm/s
Volume = Length^3	1 m^3	61023.7 in.^{3}	35.31 ft.^{3}	10^7 cm^3

corresponding dimensions may need to change to the corresponding unit if the model is used for analysis or any other application.

It is to be kept in mind that before starting a design, the unit system shall be fixed, and in an assembly, all components shall be in the same unit system to avoid any mismatch and future problems.

3.4 Sketch Entities, Objects, and Classification

A 3-D CAD model may consist of many entities or objects like sketches, solids, annotations, and so forth. Sketches are required to generate solid or surface models, whereas annotations are provided to display required process information or dimensions. A CAD model can be defined as a combination of various objects. All the objects in a CAD model facilitate many applications like production drawing generation, analysis, manufacturing sequences, and so forth. The objects in a CAD model are as follows:

- Datums
- Sketch entities
- Solids
- Surfaces
- Annotations
- Parameters

3.5 Datums

Datum entities are those that do not have any physical existence or mass properties; however, they have an important function in creation of sketches or solids or in other applications like analysis, automation, and so forth. Most datum entities are very useful to maintain relationships with parameters and with proper naming provide sensible information for automation and programming.

A parametric sketch can be best drawn on a flat plan; however, when starting the creation of a model, there would not be any flat surface on which sketches can be made. Planes are necessary to fulfill this need.

3.5.1 Plane

A plane is an imaginary or real surface such that when end points of a joining straight line lie on the plane, every point that lies on the line also lies on the plane. Planes are mostly used as sketch planes. However, planes are also

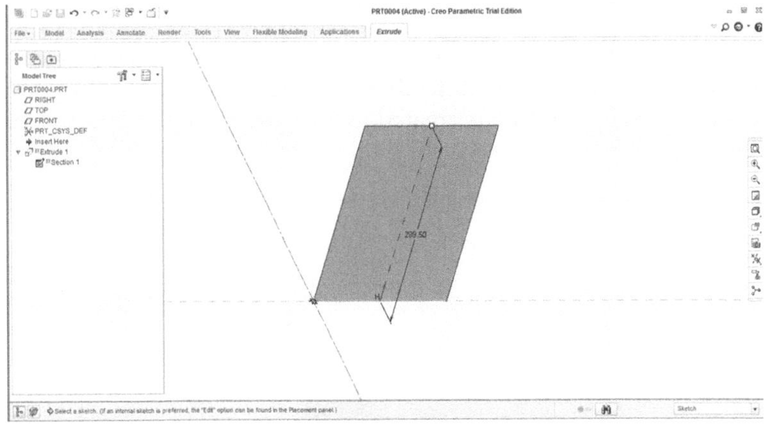

FIGURE 3.2
Example of using datum plane as sketch plane.

being used as a reference element for measuring distance or for orientation and view creation. Normally, software provides three planes and a coordinate system by default to start with. Figure 3.2 shows an example where the plane is used as a sketch plane.

3.5.2 Axis

An Axis is basically an imaginary infinite straight line in space. Axes are also used as reference geometry for sketching or locating or being an axis of rotation. Figure 3.3 shows an axis, and the plane tool is used to create another plane at an angle to the first plane.

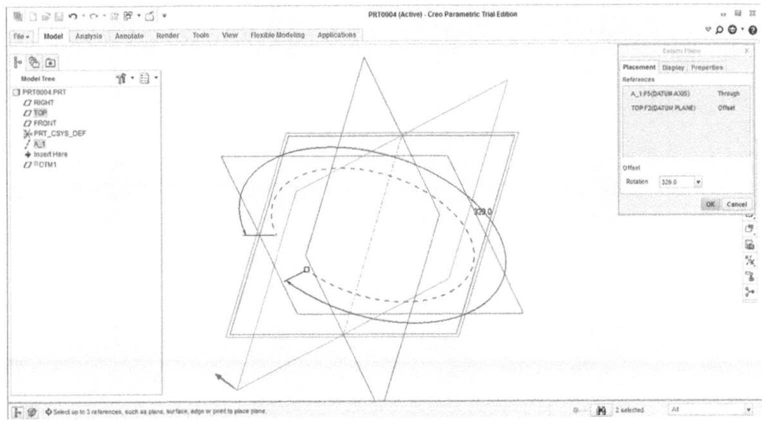

FIGURE 3.3
Example of creation of plane through axis.

3.5.3 Point

Points can be created in a sketch or directly in 3-D space. Points are used to create various geometries or to facilitate measurement and dimensioning. Figure 3.4 shows a point that is created to measure from the middle point of an edge.

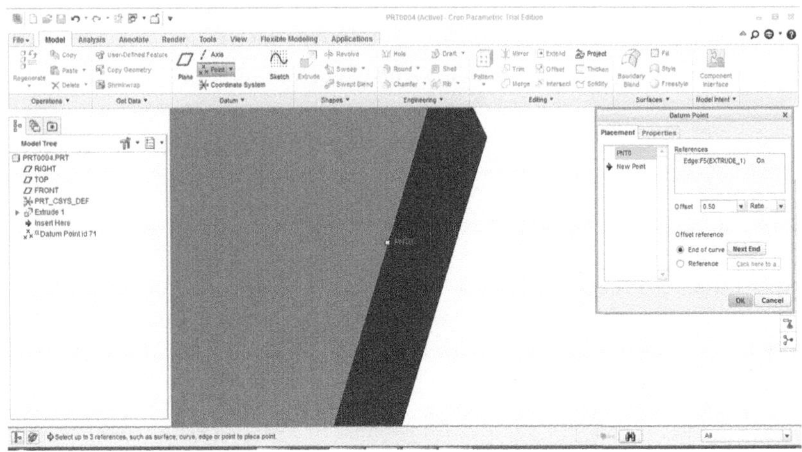

FIGURE 3.4
Example of creation of point.

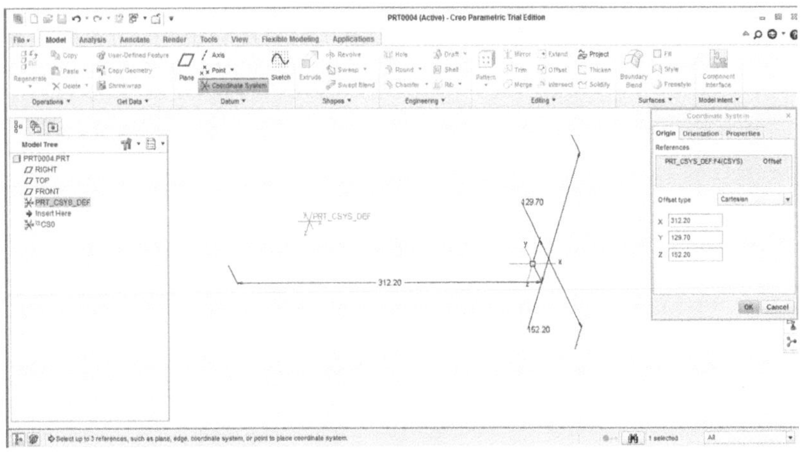

FIGURE 3.5
Example of creation of coordinate system.

3.5.4 Coordinate System

A coordinate system serves the ultimate origin references in the CAD model. A coordinate system has three perpendicular axial directions: x, y, and z. The need for any other references often can be fulfilled by selecting a coordinate system as a reference. Figure 3.5 shows a typical selecting coordinate system as a reference to create a coordinate system.

3.6 Sketch Entities

In engineering, large assemblies are mostly combinations of simple or complex individual models or solids. These solid are also created from geometries or curves. When simple geometries are constrained in relation to each other, they form parametric sketches. Most CAD software provides common geometry tools for sketching. Few kinds of advanced CAD software also provide a free hand geometry creation tool. The detail equations of each of the sketch entities are discussed in the latter part of this book. In engineering, the common geometries used for sketching are as follows.

3.6.1 Point

A point is the simplest and lightest geometry. A point is useful for construction of complex geometry and to provide relations and snapping functions. A point can be drawn in sketcher mode, which is in 2-D sketches, or in 3-D space directly defining the distances from reference objects. The functions of point are as follows:

- To create a dimensional reference
- To provide constraints with different objects and entities
- To be used as reference geometry for creation of other objects like axes, planes, coordinate systems, and so forth

Figure 3.6 shows an example of application of points.

3.6.2 Line

The most used sketch entity is a line or straight line. Lines in a sketch plane are needed to create sectional geometries that create one or more straight edge. Lines are also used for constructional geometries and dimensional references.

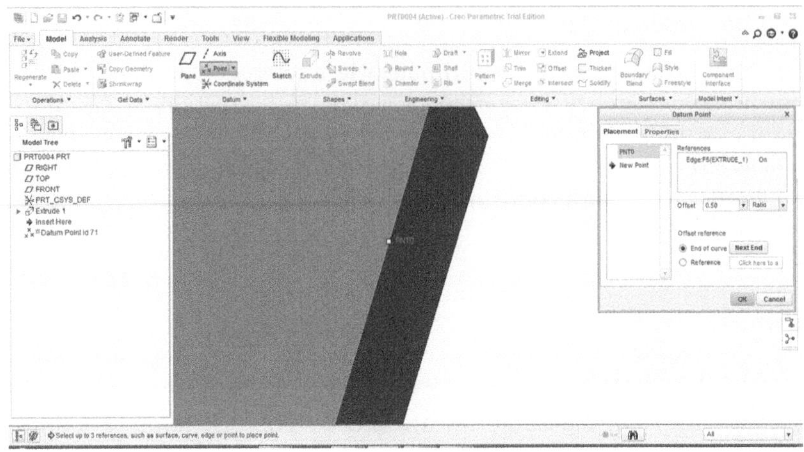

FIGURE 3.6
Example of application of points.

The functions of lines are as follows:

- To create sectional geometries to create a straight edge
- To create constructional geometries
- Dimensional references

3.6.3 Arc or Circle

Arc or circular segments are used in drawing of sectional geometries of cylindrical objects. Circles are widely used to extrude round-shaped sections, whereas arcs are often used to create fillets at the corners.

3.6.4 Rectangle

A rectangle is a polygon having four edges including four right angles. A rectangle will always have at least two sets of equal edges. If all the edges of a rectangle are equal, the rectangle will become a square. Drawing of rectangles is widely used in CAD for creating the base sketches of boxes, plates, and so forth. The following figure shows an example of a rectangle.

3.6.5 Parallelogram

A parallelogram can be defined as a polygon of two sets of equal, opposite, and parallel edges. If the parallelogram contains a right angle between the adjacent edges, it becomes rectangle. The following figure shows an example of a parallelogram.

3.6.6 Polygon

A polygon is a plane figure closely bounded by straight planes. Polygons are used to create specific shapes. For example, a hexagonal bolt is widely used in industry and has a hexagonal-shaped polygon. The following figure shows an example of an arc and a polygon.

3.6.7 Ellipse

An ellipse is a curve of two fixed focal points such that summations of distance of any point on the curve to the focal points are always same. An ellipse is used to create a special profile section for lofted or blended geometries. The following figure shows an example of an ellipse.

3.6.8 Spline

A spline is a curve mathematically defined by smooth polynomial function, drawn through various sets of selected points. In CAD, a spline is required to draw irregular shapes and for manual development of sheet metal parts. The following figure shows an example of a spline.

3.7 3-D Curves

If all the points on the curve do not lie in a single plane, then the curve is called a three-dimensional curve or 3-D curve. 3-D curves are essentially varied in three coordinates and cannot be drawn in a single plane. 3-D curves can be created by one of the following methods:

- By equation
- By special available tools like 3-D curve, helical sweep, and so forth
- By intersection of two intersecting surfaces
- By tracing of a point in a mechanism

The explicit applications of 3-D curves are as follows:

- To create an object
- To create blended surfaces
- To create the path of a point, which may be created from a mechanism
- To visualize the intersections

Figure 3.7 shows an example of using a 3-D curve for the creation of a 3-D spring.

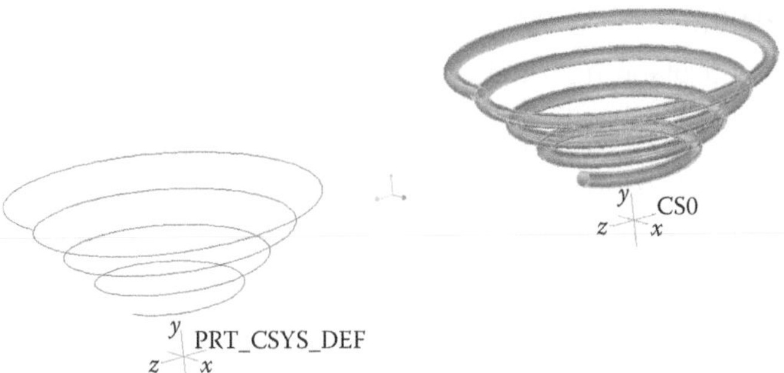

FIGURE 3.7
Example of use of 3-D curve in CAD.

3.8 Surfaces

A surface can be imagined as the skin of a solid. Surfaces are widely used in designing of shape, such as a vehicle's body, wings of airplanes, and so forth, where shapes of exterior surfaces are extremely important. When a 3-D solid is modeled, automatically, the boundary surfaces are created. Therefore, the sections to create the solid are derived from the boundary surfaces. CAD provides an excellent tool to create the surfaces first and later fill the space between surface boundaries. In this way, the exterior surface can be designed and shaped directly when such need arrives. Figure 3.8 shows a surface created from a boundary blend operation by selection boundary curves. It is obvious that by changing the boundary curves, the surface will change; therefore, the boundary curves

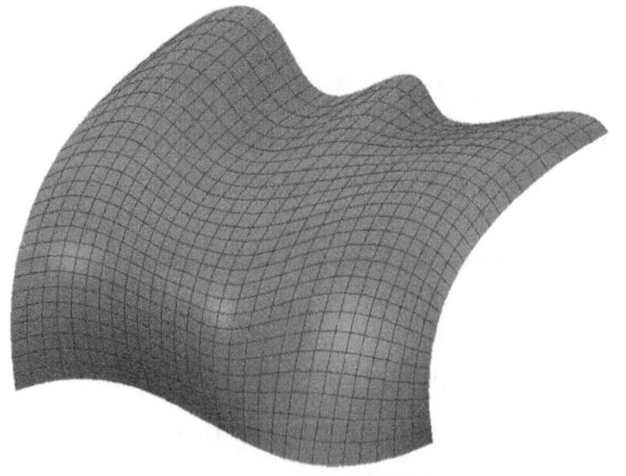

FIGURE 3.8
Example of a surface created by boundary blend operation.

directly control the shape of the surface, which enhances the facility of the designer to create a surface as per design need.

3.9 Solids

The most widely created object in CAD modeling is 3-D solids. Solids are created by operating primitives or protruding the sections in a path. Solids are mainly used for the following purposes:

- To visualize component shape and size
- To analyze a model
- To generate a production drawing

3.10 Regional Operations

Regional operations are mainly manipulation of regions to create a desired shape. A desired shape of a product may not be generated easily by single creation of any entity or object. In such cases, modification of entities and objects is needed to create the desired shape. Boolean operations are used for constructive solid geometry creation. It is mainly used in primitive approaches. Where the solid of typical shapes like cones, spheres, pyramids, and so forth is available, Boolean operation in solid modeling can be broadly divided into three categories as follows:

1. Union/addition
2. Difference/subtraction
3. Intersection

These three types of Boolean operations are shown in simple form in Figure 3.9. Let the volume of the cube and spheres be $V1$ and $V2$.

Let the sphere intersection of the volumes be dV.

Hence,

- The final volume after union operation will be $V1 + V2 - dV$.
- The final volume after subtraction will be $V1 - dV$.
- The final volume after intersection will be dV.

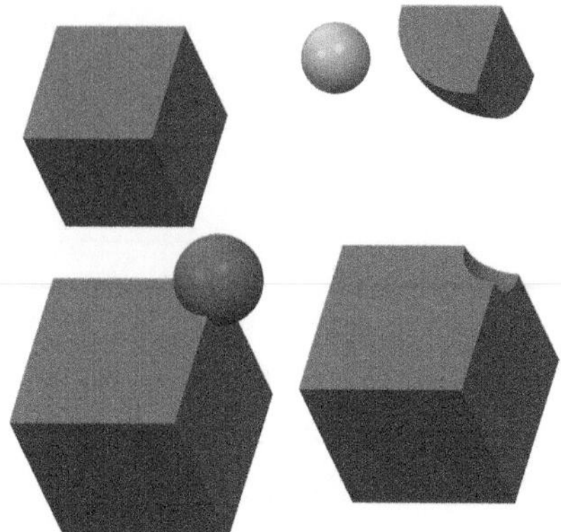

FIGURE 3.9
Example of Boolean operations in 3-D model.

3.11 Sketching Operations

Sketching operations are those that must take place in the 2-D sketching modes. It is required to create the desired sections, which are combinations of various sketch entities. The sketching operations are as follows:

- Trimming
- Extending
- Deleting
- Copying
- Patterning
- Transformation

These sketching operations are described in the latter portion of this book.

3.12 Use of Model Tree and Sequence of Operation

Most engineering components are combinations of various parts; these individual parts may be a created using a single feature or a combination of various features. In parametric software, when creating parts of a sequence of operations or creating features, the features are interrelated, with references taken from each other in a chronological manner. It is obvious that if any modifications on any operation are made, the subsequent changes are also made in succeeding operations. In parametric CAD software, the sequences are made accessible through a model tree. The model tree displays the operations in chronological order. The interconnecting relationship between operations is termed as a *parent–child relationship*. If any parent operation is modified, the children's operations may be affected; however, if the children's operation is modified, the parent operation will not change.

In parametric software, the objects can be linked to their predecessors. Parametric software provides facilities to modify these objects even after creation. It is natural; the modification of any object also affects its dependent objects. A model tree stores and displays the objects and operations in a sequential manner. It also facilitates the accessibility of modifying the objects and shows the relationship between various objects. Figures 3.10 and 3.11 show an example of modification through model tree operations. Two

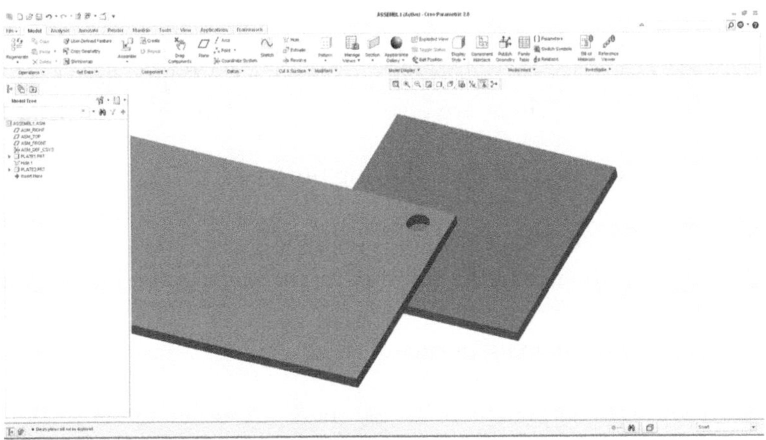

FIGURE 3.10
Example of model tree operation—hole before a plate.

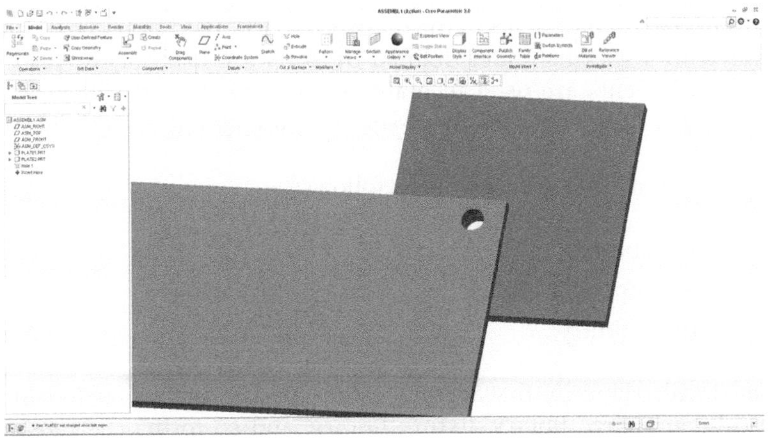

FIGURE 3.11
Example of model tree operation—hole after a plate.

plates overlap each other. When a hole is made in assembly before the addition of the second plate, the hole does not cut through the second plate. When the hole is dragged or repositioned after the addition of second plate, it cuts through it.

3.13 Accessing of CAD Libraries

CAD libraries serve a database of readily available standard components to the user. It is always suggested to the user to increase the use of CAD library components as much as possible to reduce time and effort. Users are also advised to create their own library of repeatedly used components. Most standard components are made available in CAD libraries to the user by the software itself. There are also many companies that provide CAD libraries.

CAD libraries are used for the following purposes:

- To reduce the modeling of standard components
- To maintain the use of the same parts throughout the industry for standardization
- To maintain the name sequence

Apart from 3-D component libraries, there are other kinds of libraries that are used in CAD, as discussed earlier. The use of other libraries will be described in the latter part of this book.

Most component libraries are created in family table options along with standard interfaces. Many kinds of CAD software provide automatic insertion of standard components in the required space. The library components are stored in a specific folder defined by software itself or the user. The library components are used in many assemblies; however, they are not stored in every folder where the assemblies are stored, unless overruled by the user.

The most used CAD libraries are as follows:

1. Fastener library
2. Piping component library
3. Structural library

There are also libraries that are application specific, such as the tool library, electrical component library, fixture library, and so forth. The most used libraries for mechanical engineers are described in this book.

3.13.1 Fastener Library

Fastener library components are used for fastening of materials. Many variations of these components are available and used; however, only broad classification is covered in this book. The most used fastening components are shown in Figure 3.12.

The fasteners are commonly designated by the parameters shown in Table 3.3.

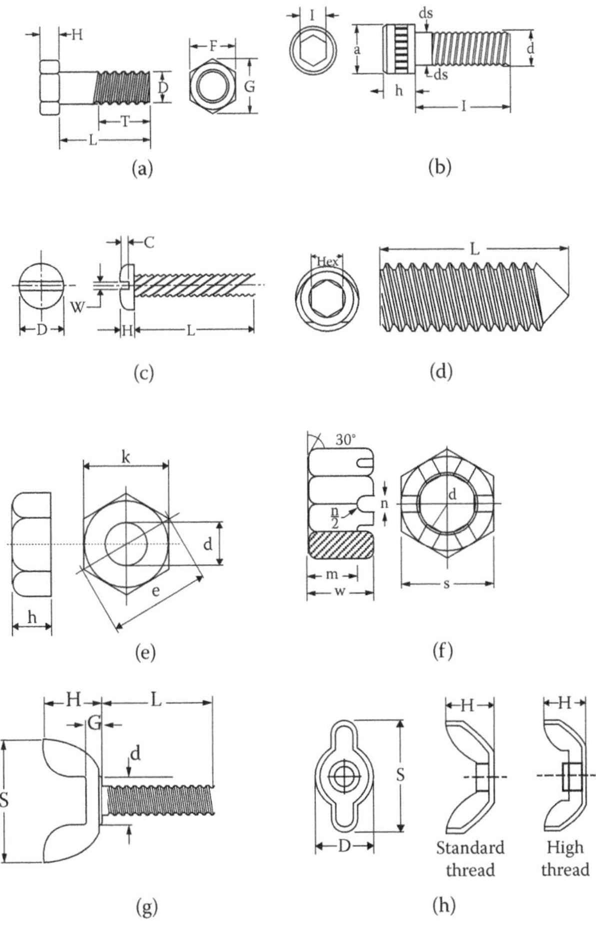

FIGURE 3.12
Example of some common fasteners: (a) Standard hex bolt and hex nut. (b) Hexagon socket head cap screw. (c) Slotted head screw. (d) Grub screws. (e) Standard hexagonal nut. (f) Castle nut. (g) Wing screw. (h) Wing nut.

TABLE 3.3
Parameter Designation

Designated Parameters:	Note:
1. Major diameter of thread 2. Length 3. Thread length 4. Pitch 5. Type of head The nut is designated by the major diameter of thread and pitch.	1. The screw and nuts are used to make a temporary joint. 2. Head selection is dependent on certain parameters like accessibility, standardization, availability, and so forth.

3.13.2 Piping Component Library

Piping components are widely used in spec-driven and non–spec-driven piping. In spec-driven piping, all the piping components are manufactured based on applicable and available standards. Specifications are prepared and drive the use of respective components in the piping. The most used piping components are as follows.

Valve: The valves are used for isolation, controlling, and safety purposes in the pipeline or pressure vessel. Common designated parameters of valves are as follows:

1. Type
2. Size
3. Material
4. Pressure rating/class

A sketch of a valve is shown in Figure 3.13.

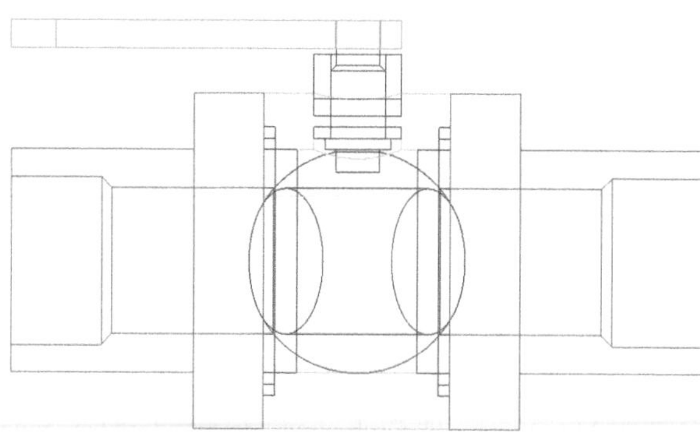

FIGURE 3.13
Sketch of valve.

Branch: The tees or branches are used to divide or unite the flow. Common designated parameters of branches are as follows:

1. Size
2. Material
3. Pressure rating/schedule
4. Type (equal/reduced)

A sketch of a tee is shown in Figure 3.14.

Elbow and return: The elbows are used to change the direction of the flow. Common designated parameters of elbows are as follows:

1. Size
2. Material
3. Pressure rating/schedule
4. Type (long radius/short radius)
5. Angle

Sketches of an elbow and a return are shown in Figure 3.15.

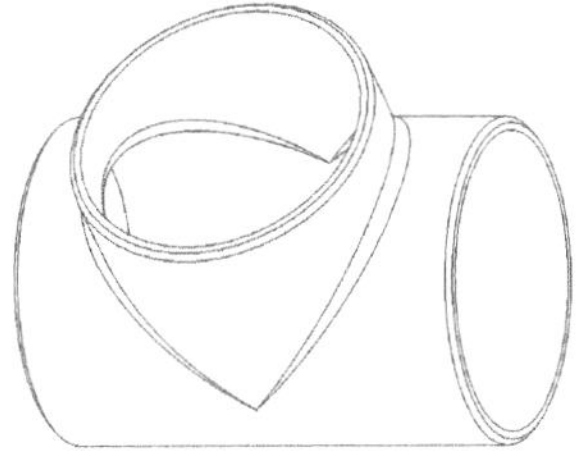

FIGURE 3.14
Sketch of branch or tee.

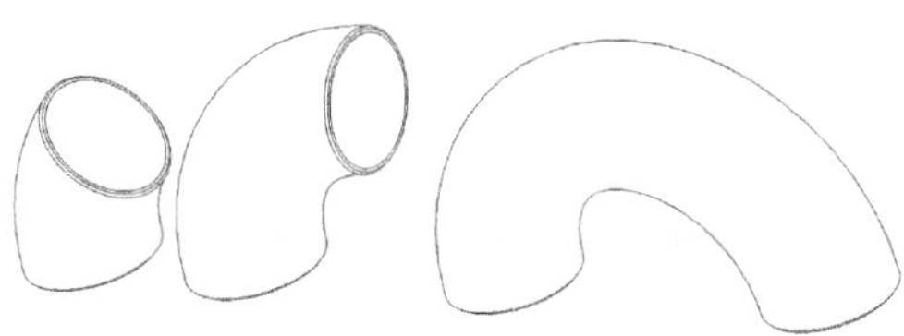

FIGURE 3.15
Sketch of elbow and return.

Reducer: Common designated parameters of reducers are as follows:

1. Size
2. Material
3. Pressure rating/schedule
4. Type (eccentric/concentric)

The reducers are used to change the pipe diameter. Sketches of reducers are shown in Figure 3.16.

Cap: The caps are used to isolate the pipeline. A sketch of a cap is shown in Figure 3.17.

Common designated parameters of a cap are as follows:

1. Size
2. Material
3. Pressure rating/schedule

Flange: Flanges are widely used for making temporary joints in the piping and piping components. Sketches of a flange are shown in Figure 3.18.

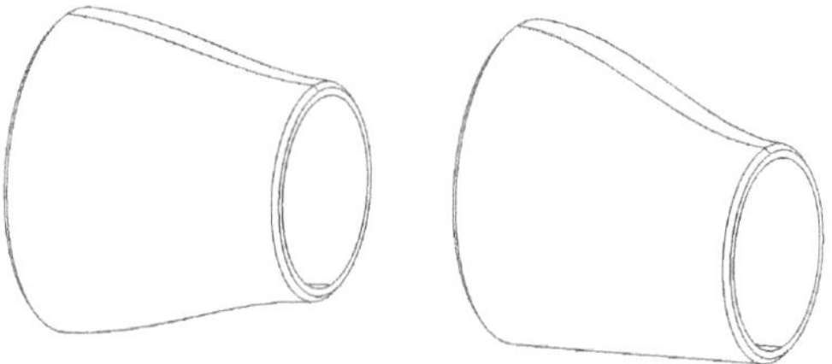

FIGURE 3.16
Sketch of reducers.

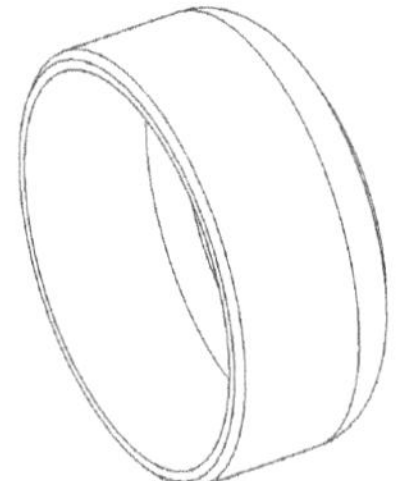

FIGURE 3.17
Sketch of cap.

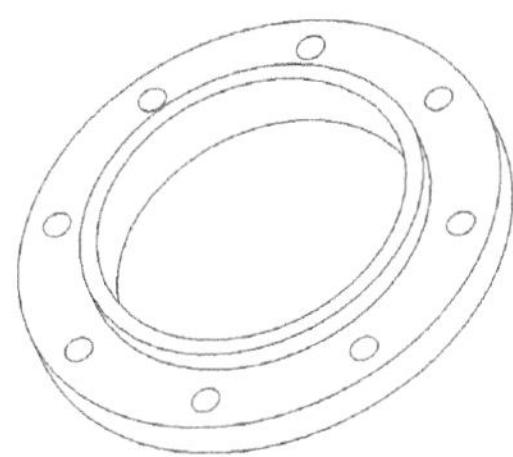

FIGURE 3.18
Sketch of flanges.

Common designated parameters of flange are as follows:

1. Size
2. Material
3. Pressure rating/schedule
4. Type (socket/slip-on, etc.)
5. Face finish (raised face/full face)

Coupling and half coupling: Couplings are used to join pipelines and, similarly, to branch out a small-bore pipeline from bigger-size pipe. Sketches of couplings are shown in Figure 3.19.

Common designated parameters of coupling and half coupling are as follows:

1. Size
2. Material
3. Pressure rating
4. Type (half/full)
5. Joining (thread/socket weld)

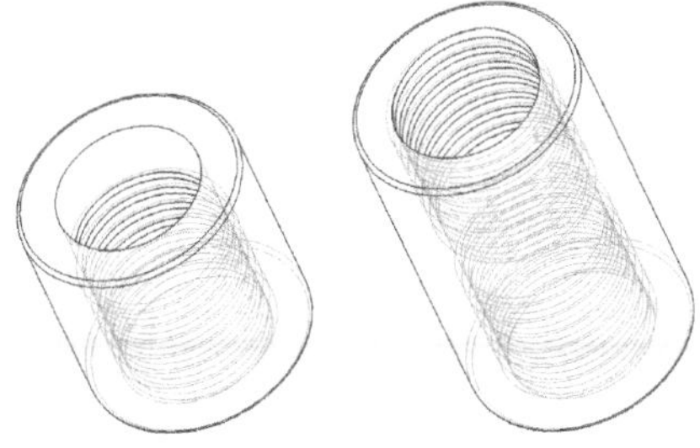

FIGURE 3.19
Sketch of couplings.

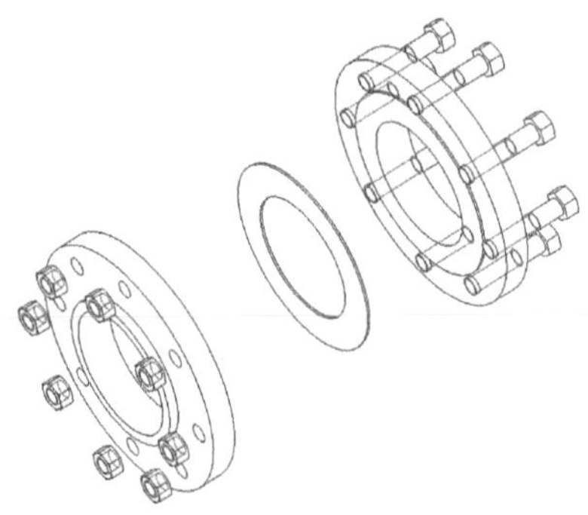

FIGURE 3.20
Sketch of gasket and hardware.

Gasket and hardware: Gaskets are used as flow material that is to be passed between two flanges in order to prevent any leakage.

Hardware for flanges commonly includes a stud/machine bolt with nuts and a washer. A sketch for a gasket and hardware is shown in Figure 3.20.

Common designated parameters of a gasket and hardware are as follows:

1. Size
2. Material
3. Pressure rating
4. Type
5. Thickness and so forth

3.13.3 Structural Library

Structural library items commonly include I sections; C sections and so forth are the most used in industry.

Joist/beam
I section (Figure 3.21) commonly designated parameters:

a. Height
b. Width
c. Thickness flange
d. Thickness of web

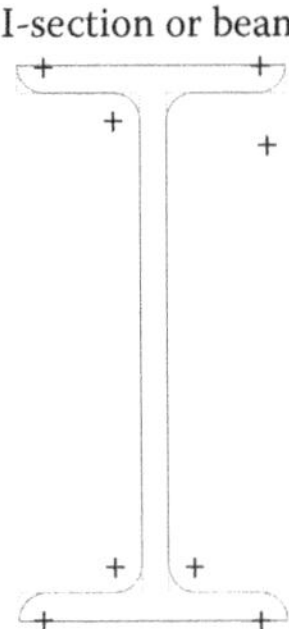

FIGURE 3.21
Sketch of I sections/joist/beam.

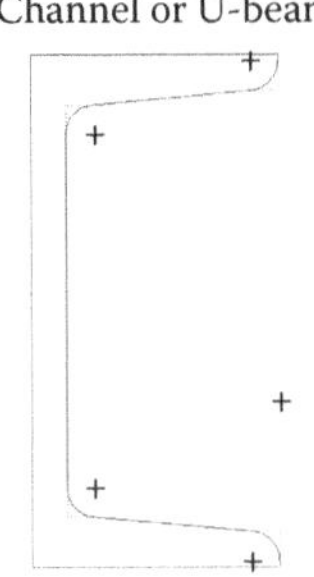

FIGURE 3.22
Sketch of channel or U beam.

Channel/U beam (Figure 3.22) commonly designated parameters:

a. Height
b. Width
c. Thickness flange
d. Thickness of web

Angle (Figure 3.23) commonly designated parameters:

a. Length of each leg
b. Thickness

Circular hollow and round bar (Figure 3.24) commonly designated parameters:

a. Outer diameter
b. Thickness

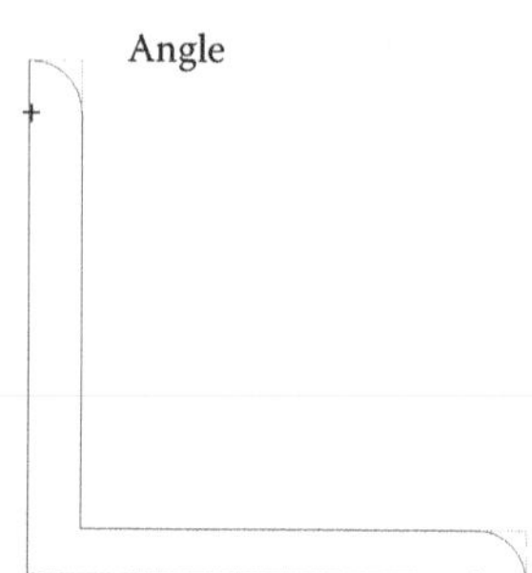

FIGURE 3.23
Sketch of angle.

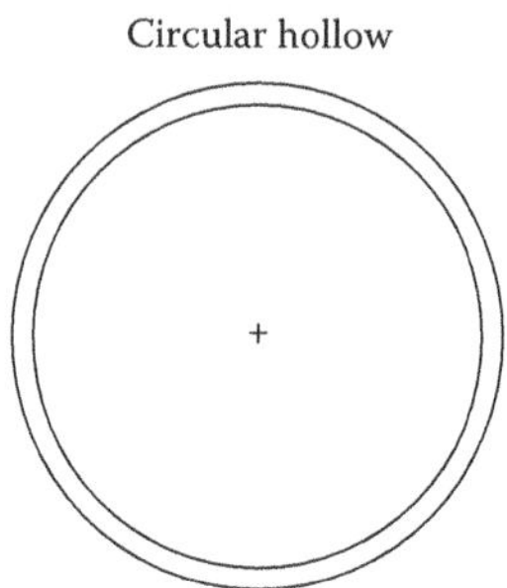

FIGURE 3.24
Sketch of circular hollow.

Square tube (Figure 3.25) commonly designated parameters:

a. Outer dimensions
b. Thickness

Tee section (Figure 3.26) commonly designated parameters:

a. Height
b. Width
c. Thickness flange
d. Thickness of web

Rectangular beam (Figure 3.27) commonly designated parameters:

a. Length
b. Width

FIGURE 3.25
Sketch of square hollow tube.

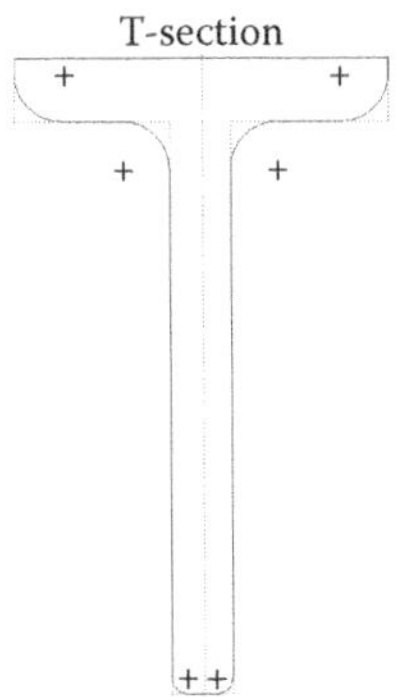

FIGURE 3.26
Sketch of T section.

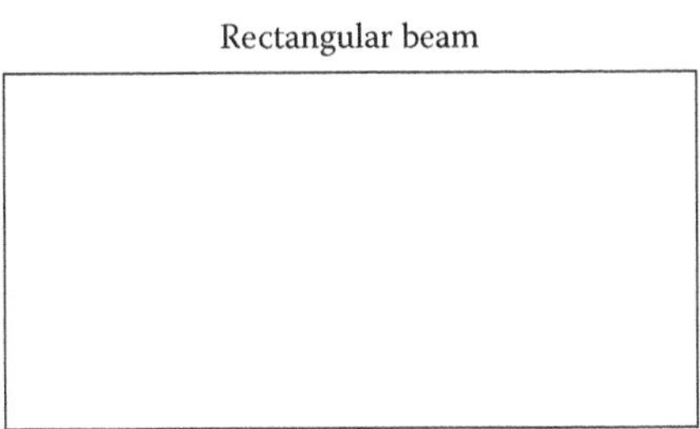

FIGURE 3.27
Sketch of rectangular beam.

Engineering components are designated with major and important parameters. For example, a commonly used hex bolt is designated by major diameter of thread, length, thread length, and pitch. All the libraries' items are also designated with the major and important dimensions or parameters. The required library parts are being filtered with the specific parameter value and sent to the CAD model. The widely used accessible parameters for each of the libraries are adjacent to Figures 3.13 through 3.27.

QUESTIONS

1. What are the various unit systems?
2. What are the advantages of using libraries?
3. Cite an example of Boolean operation.
4. Why are the sketch operations required?
5. Describe the use of a model tree.
6. Describe the use of surface modeling.
7. Why are libraries essential?
8. What would be the impact of modeling if the surface creation option is not available?
9. Write down the designation parameters of joist and angle.
10. Write down the difference in designation parameter between half coupling and full coupling.

4

Curve Entities

4.1 Introduction

Curves are geometric entities of analytical and synthetic nature. Analytical curves include lines, circles, ellipses, parabolas, and so forth. Synthetic curves include splines, B-spline, Bezier curves, and so forth. The basic geometric properties like shape, size, and so forth of any computer-aided design (CAD) model must be driven by some curves. Therefore, curves are the most essential elements in the creation of any type of CAD model. A simple CAD model of orthogonal shapes is basically a function of analytical curves, whereas a complex surface may be generated from synthetic curves. To learn CAD modeling effectively, it is essential to have a basic idea of curves.

4.2 Curves

Curves are mainly divided into two categories: analytical curves and synthetic curves. Analytical curves have their parametric equation based on analytical functions of some real variables. It is preferred to have a concept of deriving the parametric equation of each of the analytical curves to understand CAD better. In the design of any system, it is very common to have many revisions, which require numerous interpolations of the basic sketch, made by a combination of curves. Also, it is often required to have programmed curves for automation modeling. Readers are advised to generate parametric equations of each of the curves in 2-D and 3-D coordinates with respect to real variables as outlined in this book.

Coordinate system: The method of defining the position of geometry in space is called a coordinate system. There are three types of coordinate systems most often used in CAD:

- Cartesian coordinate system
- Cylindrical coordinate system
- Spherical coordinate system

TABLE 4.1
Defining Method in Coordinate Systems

	Cartesian	Cylindrical	Spherical
General form	X, Y, Z	R, θ, Z	R, φ, θ
Parametric form	$X = C$	$R = C$	$R = C$
	$Y = C$	$\theta = C$	$\varphi = C$
	$Z = C$	$Z = C$	$\theta = C$

Table 4.1 shows the defining method in three types of coordinate systems. Every geometry can be defined by these coordinate systems. Modern CAD software allows curve creation using an equation formulated with the respective coordinate system.

Point: A point is the simplest form of CAD geometry. A point has only its coordinates with respect to its origin.

Example 4.1

In 2-D space or an XY plane, a point has its coordinates x, y. The distance of the point from its origin is

$$\sqrt{x^2+y^2}. \tag{4.1}$$

In 3-D space, the distance of the point from its origin is

$$\sqrt{x^2+y^2+z^2}. \tag{4.2}$$

Example 4.2

There are two points in the space. The coordinates of the points are $(x1, y1)$ and $(x2, y2)$, respectively.

In 2-D space or the XY plane, the distance between the points is

$$\sqrt{(x1-x2)^2+(y1-y2)^2}. \tag{4.3}$$

In 3-D space, the distance between the points is

$$\sqrt{(x1-x2)^2+(y1-y2)^2+(z1-z2)^2}. \tag{4.4}$$

With respect to origin, the 2-D space or XY plane is divided into four regions, as shown in Figure 4.1: first quadrant, second quadrant, third quadrant, and fourth quadrant. It is clear from the figure that when the sign of the abscissa or ordinates alters, the point lies in different quadrants. Similarly, in 3-D space, due to alteration of the sign of its coordinates, the point lies in eight different regions.

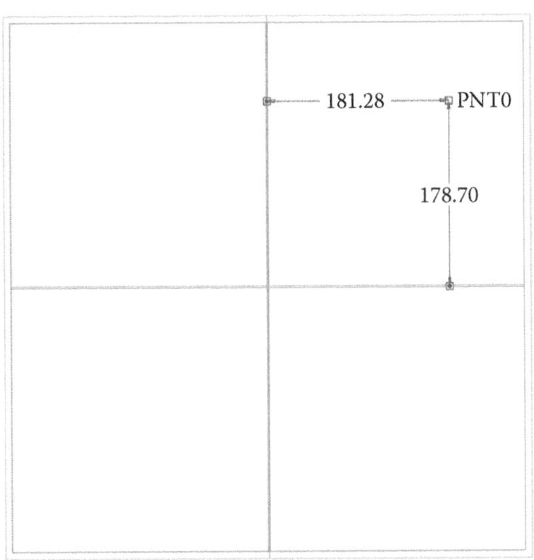

FIGURE 4.1
Point in 2-D space.

4.3 Lines

A straight line is the most useful and simple geometry used in the creation of any straight edge in the solid model. A straight line can be defined as such geometry in which the distance between any two points lying on the line is minimum if measured with the path of that line.

The general form of the equation of the line in two dimensions is $y = mx + C$, where $m = \frac{Y}{X}$ and C is a constant.

Method: Find the equation of a line that connects two points ($x1$, $y1$) and ($x2$, $y2$).

The slope of the line $m = \frac{y2 - y1}{x2 - x1}$. Putting the value of m and the coordinates of any point in the general equation of line, we get the required equation. In CAD, the equations are represented as a parametric equation.

If t is the real variable, then the equation of a straight line will be that which passes through the abscissa:

$$X = x_1 + t(x_2 - x_1),\ 0 \geq t \geq 1. \tag{4.5}$$

A similar equation can be derived for the line passing though the ordinate:

$$Y = y_1 + t(y_2 - y_1),\ 0 \geq t \geq 1. \tag{4.6}$$

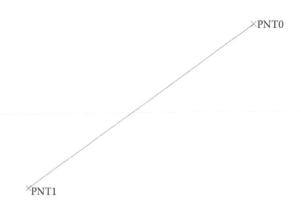

FIGURE 4.2
A line in space.

Hence, referring to Figure 4.2, the line can be written as

$$X = a + f(t)$$

$$Y = b + g(t)$$

where a and b are constant and t is a real variable, $0 \geq t \geq 1$.

In 3-D, the equation would be

$$X = a + f(t) \tag{4.7}$$

$$Y = b + g(t) \tag{4.8}$$

$$Z = d + h(t). \tag{4.9}$$

If any of the functions of t is not linear, then the straight line will become a curve.

Hence, the equation of the line P can be written as

$$P(t) = [x(t), y(t), z(t)]^T. \tag{4.10}$$

This is the parametric equation of curve.

Example 4.3

Find the equation of a line that connects two points (5,2) and (9,25).

The slope of the line $m = \frac{y2 - y1}{x2 - x1} = \frac{25 - 2}{9 - 5} = \frac{23}{4}$. Putting the value of m in the general equation of the line m, we have, $y = \frac{23}{4}x + c$. Now, putting in the coordinates of one of the points, we have $2 = \frac{115}{4} + C$, or $C = -\frac{107}{4}$. Hence, the equation of the line is $y = \frac{23}{4}x - \frac{107}{4}$, or $23x - 4y - 107 = 0$.

The parametric equation of the line for this example would be

$$\begin{aligned} X &= x_1 + t(x_2 - x_1), \quad 0 \geq t \geq 1 \\ &= 5 + t(9 - 5) \\ &= 5 + 4t. \end{aligned}$$

Similarly,

$$Y = 2 + 23t.$$

Now, the above-mentioned equations, $X = 5 + 4t$ and $Y = 2 + 23t$ can be imagined in the form of

$$X = a + f(t)$$

$$Y = b + g(t).$$

Here, a and b are simply representing the coordinate of the starting point of the line. The functions $f(t)$ and $g(t)$ represent the slope of the line. When an equation of a line is specified in the form of $Ax + By + C = 0$, the equation can be written in its parametric form:

$$X = a - B(t)$$

$$Y = b + A(t).$$

Now, the coordinate of starting points a and b cannot be chosen abruptly to satisfy the value C.

$$Aa + Bb = C$$

$$b = \frac{Aa - C}{B}$$

Similarly,

$$a = \frac{Bb - C}{A}.$$

Therefore, for given equation of a line, only one coordinate needs to be specified to draw the line by parametric equation.

Hence, the parametric equation for a given equation of a line $Ax + By + C = 0$ becomes

$$X = a - B(t)$$

$$Y = \frac{Aa - C}{B} + A(t) \tag{4.11}$$

where a is the input value in the X direction and $0 \geq t \geq 1$.

Similarly, for a 3-D coordinate system, if a line represented by the equation $Ax + By + Cz = 0$ and $A_1x + B_1y + C_1z = 0$, the parametric form of equation would be

$$X = a - \left(B - \frac{CB_1}{C_1}\right) * t \tag{4.12}$$

$$X = b + \left(A - \frac{CA_1}{C_1}\right) * t \tag{4.13}$$

$$X = c + \left(B - \frac{AB_1}{A_1}\right) * t \tag{4.14}$$

where a, b, and c are the coordinates of the starting point of the line. And the relations between coordinates are as follows:

$$b = -\left\{\left(D - \frac{CD_1}{C_1}\right) + \left(A - \frac{CA_1}{C_1}\right)\right\} / \left(B - \frac{CB_1}{C_1}\right)$$

$$c = -\left\{\left(D - \frac{AD_1}{A_1}\right) + \left(B - \frac{AB_1}{A_1}\right)\right\} / \left(C - \frac{CA_1}{A_1}\right)$$

In a cylindrical coordinate system, the parametric equation is represented as

$$R = f(t) \tag{4.15}$$

$$\theta = C$$

$$Z = F(t). \tag{4.16}$$

The functions in Equations 4.15 and 4.16 must be linear.

In a spherical coordinate system, the parametric equation is represented as

$$R = f(t) \quad (4.17)$$

$$\varphi = C \quad (4.18)$$

$$\theta = C. \quad (4.19)$$

4.4 Circles

A circle is a locus of a point that maintains a constant distance from a fixed point when moving. A circle or segments of a circle are the most used entity to construct most of the geometry or section after the use of a line.

The general equation of a circle in the Cartesian system is $x^2 + y^2 + 2gx + 2fy + c = 0$, where the center is $(-g, -f)$ and the radius is $\sqrt{g^2 + f^2 - c}$. In parametric form, the equations can be written as follows.

In the Cartesian coordinate system,

$$X = r\cos\theta \quad (4.20)$$

$$Y = r\sin\theta \quad (4.21)$$

$$\theta = 360{*}t. \quad (4.22)$$

The equation of a circle in the cylindrical coordinate system will be

$$r = C \quad (4.23)$$

$$\theta = 360{*}t \quad (4.24)$$

$$Z = C. \quad (4.25)$$

The equation of a circle in the spherical coordinate system will be

$$r = C \quad (4.26)$$

$$\theta = C \quad (4.27)$$

$$\varphi = 360{*}t. \quad (4.28)$$

Hands-On Exercise, Example 4.1: Creating a Circle by Equation (Software Used: Creo Parametric)

Steps:

- Create a part file and rename it as Circle. (Other names may be given, but it is advisable to provide a relevant name.)
- Select Solid on the right hand side of the box. This will activate the solid mode and environment.
- Uncheck the box of the default template and click OK to create a new solid file.
- The template selection box will appear.
- Select mmns_part_Solid. This will allow the use of the solid template file of the following units:
  - mm—millimeter for length unit
  - n—newton for force unit
  - s—second for time unit
- Once the solid part is created, go to File > Prepare > Model properties to check the units. In the upper sections, the units of the model will be displayed.
  - Note: The units usually are displayed in the summary section of any CAD software. The setting of the unit usually is in the tools or settings.
- Click on the datum drop-down menu from the Model tab of the ribbon interface.
- Click on Curve > Curve from equation.
- Cartesian shall be selected by default.
- Click on the Reference tab and select the default coordinate system PRT_SYS_DEF.
- Click on Equation; the equation window shall appear.
- Click on Local parameter, add a parameter, and rename it as "r."
- Select Real number as the type.
- Enter the value 100 as the radius.
- In the equation window, write down the following equations:
  - $\theta = 360*t$
  - $x = r*\cos(\theta)$
  - $y = r*\sin(\theta)$
- Click on Verify on the right upper side of the relation window.
- A notification box with the message "Relations have been successfully verified" should appear.
- Click OK to close the relation window.
- Click OK to close the equation window.
- Click OK to create the curve feature.

4.5 Ellipses

An ellipse is a locus of a point that keeps a constant of less than one of the distance from a fixed point to a fixed straight line when moving. A circle or segments of a circle are the most used entity to construct most of the geometry or section after the use of a line. However, ellipses or elliptical curves are often required to form geometry in CAD.

Mathematically, when the center point of the ellipse lies on the origin, the equation of the ellipse is as follows:

$$\frac{x^2}{a^2}+\frac{y^2}{b^2}=1 \tag{4.29}$$

where x and y are the coordinates of any point on the ellipse and a and b are the radii on the x and y axes, respectively.

If the coordinates of the center point of the ellipse are h and k, the equation of ellipse will be

$$\frac{(x-h)^2}{a^2}+\frac{(y-k)^2}{b^2}=1. \tag{4.30}$$

The equation of an ellipse in parametric form can be defined as

$$x = a\cos(t) \tag{4.31}$$

$$y = b\sin(t) \tag{4.32}$$

where a and b are the radii of the major and minor axis and t varies from 0 to 2π.

Hands-On Exercise, Example 4.2: Creating an Ellipse by Equation (Software Used: Creo Parametric)

Steps:

- Create a part file and rename it as Ellipse. (Other names may be given but it is advisable to provide a relevant name.)
- Select Solid on the right-hand side of the box. This will activate the solid mode and environment.
- Uncheck the box of the default template and click OK to create a new solid file.
- The template selection box will appear.
- Select mmns_part_Solid. This would allow the use of the solid template file of the following units:
    - mm—millimeter for length unit
    - n—newton for force unit
    - s—second for time unit

- Once the solid part is created, go to File > Prepare > Model properties to check the units. In the upper sections, the units of the model will be displayed.
  - Note: The units usually are displayed in the summary section of any CAD software. The setting of the unit usually is in the tools or settings.
- Click on the datum drop-down menu from the Model tab of the ribbon interface.
- Click on Curve > Curve from equation.
- Cartesian shall be selected by default.
- Click on the Reference tab and select the default coordinate system PRT_SYS_DEF.
- Click on Equation; the equation window shall appear.
- Click on Local parameter, add a parameter, and rename it as "a."
- Select Real number as the type.
- Enter the value 100 as the radius on the major axis.
- Click on Local parameter, add a parameter, and rename it as "b."
- Select Real number as the type.
- Enter the value 50 as the radius on the minor axis.

- In the equation window, write down the following equations:
  - $\theta = 360^*t$
  - $x = a^*\cos(\theta)$
  - $y = b^*\sin(\theta)$
- Click on Verify on the right upper side of the relation window.
- A notification box with a message "Relations have been successfully verified" should appear.
- Click OK to close the relation window.
- Click OK to close the equation window.
- Click OK to create the curve feature.

4.6 Parabolas

A parabola is a locus of a point that maintains equal distance from a fixed point to a fixed straight line when moving. A parabola has many uses in the real world. Parabolic curved panels are used in headlights of cars, torches, and so forth. The light is placed in the focus of a parabolic mirror so that the reflected lines are parallel. Therefore, these lights are very strong.

The equation of a parabola in its general form is

$$Ax^2 + Bxy + Cy^2 + Dx + Ey + F = 0 \tag{4.33}$$

where x and y are the coordinates of the point lying on the parabola.

The simplest equation of a parabola is

$$y = x^2. \tag{4.34}$$

In parametric form, it can be shown as

$$x = t, y = t^2. \tag{4.35}$$

Hands-On Exercise, Example 4.3: Creating a Parabola by Equation (Software Used: Creo Parametric)

Steps:

- Create a part file and rename it as Parabola. (Other names may be given, but it is advisable to provide a relevant name.)
- Select Solid on the right hand side of the box. This will activate the solid mode and environment.
- Uncheck the box of the default template and click OK to create a new solid file.
- The template selection box will appear.
- Select mmns_part_Solid. This will allow the use of the solid template file of the following units:
  - mm—millimeter for length unit
  - n—newton for force unit
  - s—second for time unit
- Once the solid part is created, go to File > Prepare > Model properties to check the units. In the upper sections, the units of the model will be displayed.
  - Note: The units usually are displayed in the summary section of any CAD software. The setting of the unit usually is in the tools or settings.
- Click on the datum drop-down menu from the Model tab of the ribbon interface.
- Click on Curve > Curve from equation.
- Cartesian shall be selected by default.
- Click on the Reference tab and select the default coordinate system PRT_SYS_DEF.
- Click on Equation; the equation window shall appear.
- In the equation window, write down the following equations:
  - $x = 4*t$
  - $y = 7*t^2$
- Click on Verify on the right upper side of the relation window.
- A notification box with a message "Relations have been successfully verified" should appear.
- Click OK to close the relation window.
- Click OK to close the equation window.
- Click OK to reate the curve feature.

4.7 Hyperbolas

A hyperbola is a locus of a point that maintains a constant of greater than one of the distance from a fixed point to a fixed straight line when moving.

The equation of a hyperbola can be written as

$$\frac{x^2}{a^2} - \frac{y^2}{b^2} = 1. \tag{4.36}$$

Hyperbolas are used in natural draft cooling towers' structure, which can resist high wind forces with minimum material. Hyperboloids used for skew helical gears are also an application of the hyperbola curve in three dimensions, used in power transmission.

In parametric form, the equation of hyperbola can be shown as follows:

$$x = \frac{a}{\cos(t)}, \quad y = b\tan(t). \tag{4.37}$$

4.8 Conics

Conics are conical curves that are generated from cutting a right-angle cone by a plane at a different angle. Lines, ellipses, parabolas, and hyperbolas are all forms of conics. A circle is generated when a plane perpendicular to the axis of the right-angle cone cuts the boundary surface, whereas ellipses, parabolas, and hyperbolas are generated by cutting by an angular plane.

4.9 Cubic Splines

Mathematically, a spline is a piecewise defined smooth polynomial function. A cubic spline is the most used, that is, of order 3.

Let us have a table of points $[X_i, Y_i]$ for $i = 0,1,\ldots,n$ for the function $y = f(x)$. For the function, it makes n intervals, hence, $n + 1$ points. The cubic function is a piecewise continuous curve passing through each of the values of the user defined table.

The equation of the spline becomes

$$S_i(x) = C_3(x - x_i)^3 + C_2(x - x_i)^2 + C_1(x - x_i) + C. \tag{4.38}$$

In parametric form, the equation of the spline can be defined as

$$S(t)=\sum_{i=0}^{3} C_i t^i,\quad 0 \le t \le 1 \tag{4.39}$$

where t is the parameter and C_i are the polynomial coefficients.
The equation in its simplest form can be written as

$$x(t) = C_{3x}t^3 + C_{2x}t^2 + C_{1x}t + C_{0x} \tag{4.40}$$

$$y(t) = C_{3y}t^3 + C_{2y}t^2 + C_{1y}t + C_{0y} \tag{4.41}$$

$$z(t) = C_{3z}t^3 + C_{2z}t^2 + C_{1z}t + C_{0z}. \tag{4.42}$$

4.10 Bezier Curves

A Bezier curve is obtained by defining a polygon by defining a set of data points. Bezier curves were used for automobile surfaces in 1962 by French engineer Pierre Bezier in his software system called UNISURF. The designers used it to create outer panels of automobiles. The advantages of the Bezier curve are as follows:

1. The shape of the Bezier curve is controlled by its defining points. Unlike the cubic spline, tangent vectors are not used, thus facilitating better control for designers through input points and output curves.
2. The degree of the Bezier curve is variable and related to the number of points used for defining it. An nth-degree curve is defined by $n + 1$ defined points, unlike the cubic spline, where the order is always cubic for the spline.

The equation of the Bezier curve is given by

$$S(t)=\sum_{i=0}^{3} P_i B_i,\ n^t \quad 0 \le t \le 1 \tag{4.43}$$

where $S(t)$ is a point on the curve, S_i are the control points, and $B_{i,n}$ are the Bernstein control points.

The main disadvantage of the Bezier curve is a lack of local control. Increasing control points adds little local control on the curve.

4.11 B-Spline Curves

A B-spline curve is similar to a Bezier curve, involving more information with the influence of each vertex over a defined range.

Mathematically, the b-spline curve is as follows:

$$S(t) = \sum_{i=0}^{3} P_i N_i, p^t \quad 0 \le t \le 1. \tag{4.44}$$

4.12 Project Work

Hands-On Exercise, Example 4.4: Creating Conical Curves through Cross Section of Cylinder (Software Used: Creo Parametric; Figure 4.3)

Steps:

- Create a part file and rename it as Cylinder. (Other names may be given, but it is advisable to provide a relevant name.)
- Select Solid on the right-hand side of the box. This will activate the solid mode and environment.
- Uncheck the box of the default template and click OK to create a new solid file.
- The template selection box will appear.

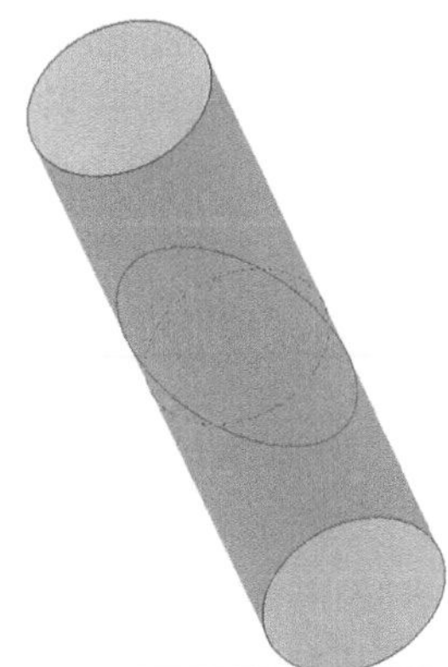

FIGURE 4.3
Example of conical curves on cylinder.

- Select mmns_part_Solid. This will allow the use of the solid template file of the following units:
  - mm—millimeter for length unit
  - n—newton for force unit
  - s—second for time unit
- Once the solid part is created, go to File > Prepare > Model properties to check the units. In the upper sections, the units of the model will be displayed.
  - Note: The units usually are displayed in the summary section of any CAD software. The setting of the unit usually is in the tools or settings.
- Click on Extrude from the Model tab in the ribbon interface.
- Click on the Placement tab and select Define to define the sketch.
- Select the front plane as the sketch plane.
- The right plane shall be automatically taken as reference and orientation as right.
- Click on Sketch.
- The sketcher model will be activated.
- Draw a circle of diameter 100 placing the center point at the intersection of horizontal and vertical references.
- Click OK to create and exit the sketch.
- Select Mid plane extrusion.
- Enter the value 400 mm as the extrusion depth.
- Click on OK to create the extrusion.

- Click on Datum axis from the Model tab of the ribbon interface.
- Select the Top and Right planes as references by pressing the control key.
- Click OK to create the datum axis.
- Click OK datum plane from the Model tab of the ribbon interface.
- Select the axis and Top plane as reference.
- Set the angle 45 with Top plane.
- Click OK to create the datum plane named DTM1.

- Click on Section from the View tab of the ribbon interface.
- Select the plane Top.
- Click OK to create the section named XSEC0001.

- Similarly, click on Section again from the View tab of the ribbon interface.
- Select the plane DTM1.
- Click OK to create the section named XSEC0002.

- Click on the datum drop-down menu from the Model tab of the ribbon interface.
- Select Curve > Curve from cross section.
- Select cross section XSEC0001.
- Click OK to create the curve from cross section.

- Similarly, again select Curve > Curve from cross section.
- Select cross section XSEC0002.
- Click OK to create the curve from cross section.

- Click on Appearance gallery from the View tab of the ribbon interface.
- Right-click on the model area and select new.
- Rename the color as 1.
- Set transparency to 90.
- Click OK to create new color.
- Select the color 1 from the appearance gallery and select the part Cone.
- Press the middle button to assign the transparency on the model.
- Right-click on curve 1 and select Properties.
- Click on color and select the color green.
- Similarly, select the color red for curve 2.
- Notice the curves on different section planes.
- Save and exit the model.

Hands-On Exercise, Example 4.5: Creating Conical Curves through Cross Section (Software Used: Creo Parametric; Figure 4.4)

Steps:

- Create a part file and rename it as Cone. (Other names may be given, but it is advisable to provide a relevant name.)
- Select Solid on the right-hand side of the box. This will activate the solid mode and environment.
- Uncheck the box of the default template and click OK to create a new solid file.
- The template selection box will appear.

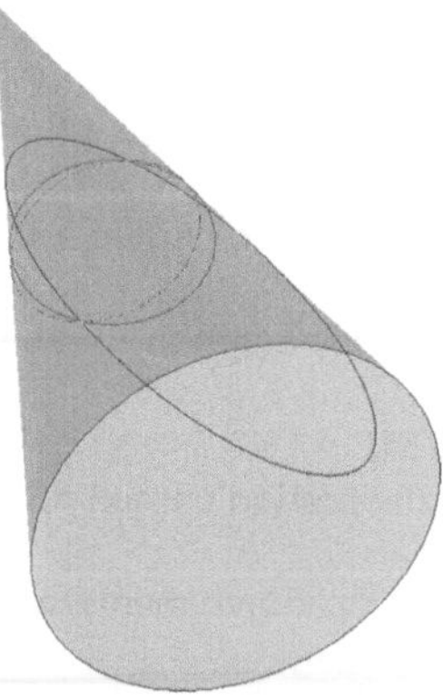

FIGURE 4.4
Example of conical curves created through the cross section of an ellipse.

- Select mmns_part_Solid. This would allow the use of the solid template file of the following units:
  - mm—millimeter for length unit
  - n—newton for force unit
  - s—second for time unit
- Once the solid part is created, go to File > Prepare > Model properties to check the units. In the upper sections, the units of the model will be displayed.
  - Note: The units usually are displayed in the summary section of any CAD software. The setting of the unit usually is in the tools or settings.
- Click on Revolve from the Model tab in the ribbon interface.
- Click on Placement tab and select Define to define the sketch.
- Select the front plane as the sketch plane.
- The right plane shall be automatically taken as reference and orientation as right.
- Click on Sketch.
- The sketcher model will be activated.
- Draw the sketch as displayed in Figure 4.5.
- Click OK to create and exit the sketch.
- Ensure that the revolve angle shall be 360°.
- Click on OK to create the revolve feature.

- Click on Datum axis from the Model tab of the ribbon interface.
- Select the Top and Right planes as references by pressing the control key.
- Click OK to create the datum axis.

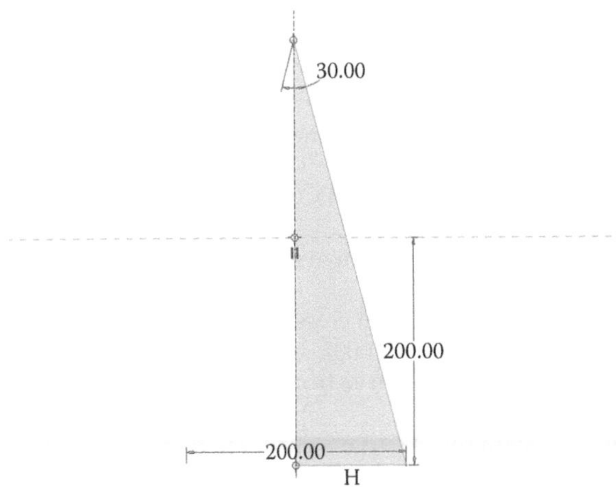

FIGURE 4.5
Sketch for a revolution for creating a cone.

- Click OK datum plane from the Model tab of the ribbon interface.
- Select the axis and Top plane as reference.
- Set the angle 15 with Top plane.
- Click OK to create the datum plane named DTM1.

- Click OK datum plane from the Model tab of the ribbon interface.
- Select the axis and Top plane as reference.
- Set the angle 45 with Top plane.
- Click OK to create the datum plane named DTM2.

- Click OK datum plane from the Model tab of the ribbon interface.
- Select the axis and Top plane as reference.
- Set the angle 60 with Top plane.
- Click OK to create the datum plane named DTM3.

- Click on Section from the View tab of the ribbon interface.
- Select the plane Top.
- Click OK to create the section named XSEC0001.

- Similarly, click on Section again from the View tab of the ribbon interface.
- Select the plane DTM1.
- Click OK to create the section named XSEC0002.

- Similarly, click on Section again from the View tab of the ribbon interface.
- Select the plane DTM2.
- Click OK to create the section named XSEC0003.

- Similarly, click on Section again from the View tab of the ribbon interface.
- Select the plane DTM3.
- Click OK to create the section named XSEC0004.

- Right-click on the section XSEC0004 and select Deactivate.

- Click on the datum drop-down menu from the Model tab of the ribbon interface.
- Select Curve > Curve from cross section.
- Select cross section XSEC0001.
- Click OK to create the curve from cross section.

- Similarly, again select curve > curve from cross section.
- Select cross section XSEC0002.
- Click OK to create the curve from cross section.

- Similarly, again select Curve > Curve from cross section.
- Select cross section XSEC0003.
- Click OK to create the curve from cross section.

- Similarly, again select Curve > Curve from cross section.
- Select cross section XSEC0004.
- Click OK to create the curve from cross section.

- Click on Appearance gallery from the View tab of the ribbon interface.
- Right-click on the model area and select New.
- Rename the color as 1.
- Set transparency to 90.
- Click OK to create new color.
- Select color 1 from the appearance gallery and select the part Cone.
- Press the middle button to assign the transparency on the model.
- Right-click on curve 1 and select Properties.
- Click on Color and select the color green.
- Similarly, select the colors red, yellow, and blue for curves 2, 3, and 4, respectively.
- Notice the curves on different section planes.
- Save and exit the model.

QUESTIONS

1. Describe the various types of coordinate systems.
2. Compare the various type of coordinate system.
3. What curve entities are used in CAD?
4. Describe the uses of the various curve entities.
5. Why are lights of automobiles so strong?
6. What are the inputs required for creating any point in CAD?
7. Why do the natural draft cooling towers have a hyperbola shape?
8. Write down the equation of an ellipse that is generated by cutting a cylinder of diameter 200 mm by a plane at an angle 30° to the axis of revolution.

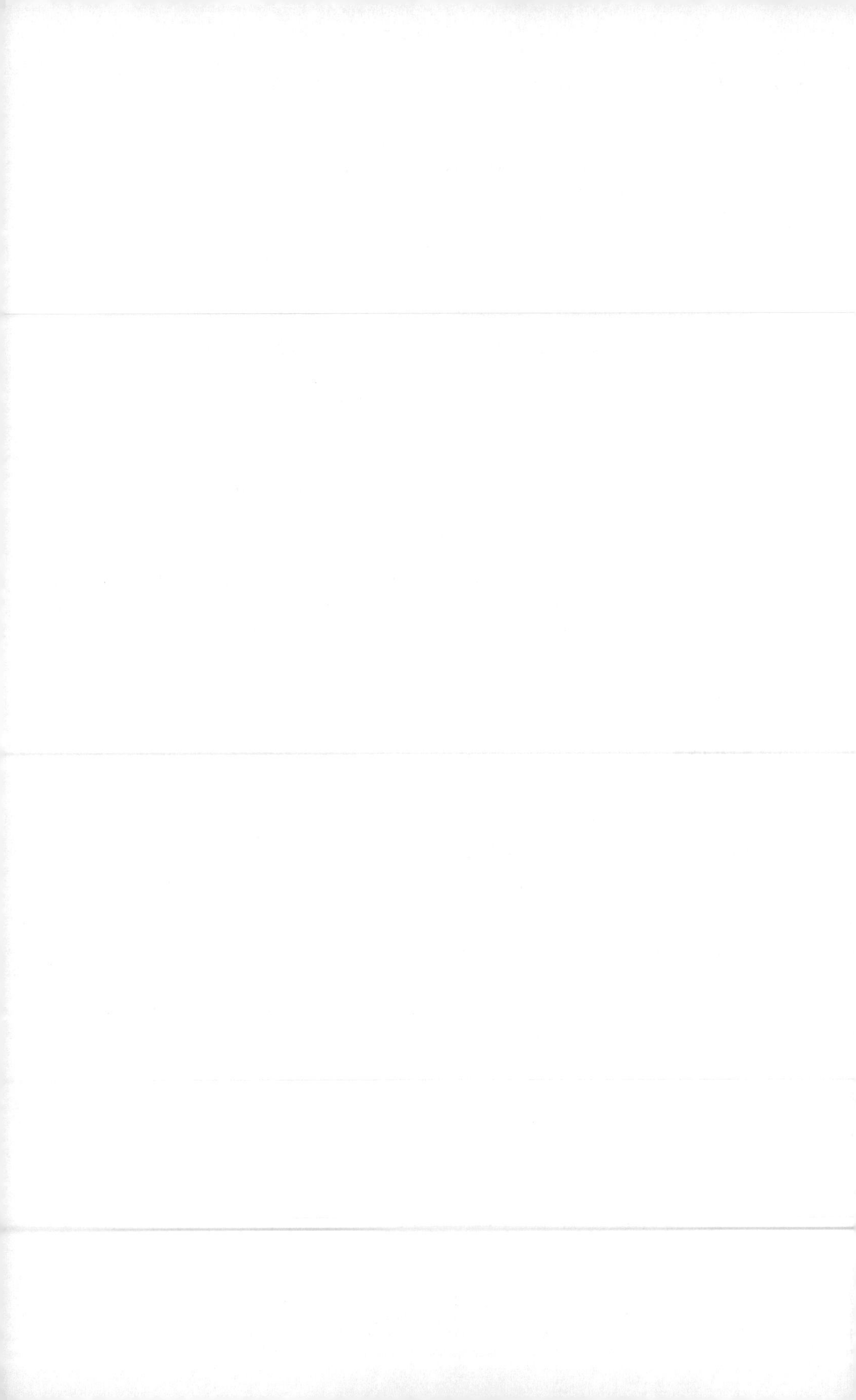

5

Sketching

5.1 Introduction

Sketching is the most basic function of computer-aided design (CAD) modeling. A sketch is required to create any parametric model that governs the shape of the component. A sketch consists of curves, which are discussed in Chapter 4. Most of the modern CAD software provides various sets of tools in sketcher mode to create curves or sketches to suit our requirements. The set of tools may be displayed on demand depending on the user interface of the software. Apart from 3-D solid creations, sketches are also used for various purposes like finding dimensions, creation of an analytical model, and so forth. Creation of a sketch follows a sequence of selecting a sketch plane or flat surface and references. The references are required to define the origin. The origin is the point from which all the curve entities will be defined. References can be taken from any predefined entity like a point, edge, line planes, axis, and so forth.

A sketch also can be formed or made by many methods; for example, a half circle can be drawn with the arc command or trimming it from a circle. It is advisable to use the minimum sketch entities and commands to create any geometry to save time and reduce human error by keeping unnecessary entities.

Sketch entities can be of two types: constructional and solid. Constructional entities are those that are not used for creating any edges directly. They are required to construct the solid geometry for creation of required sections. They also become unavailable outside the sketches. Constructional geometries are mainly used to create any geometry using many variables.

5.2 Sketching Tools

CAD software provides a wide set of tools for creation of any sketch entities. Few kinds of software also provide a command prompt interface, ribbon interface, and so forth. Depending on the interface, the input method may

TABLE 5.1

Defining Method in Coordinate Systems

	Cartesian	Cylindrical	Spherical
General form	X, Y, Z	R, θ, Z	R, φ, θ
Parametric form	$X = C$	$R = C$	$R = C$
	$Y = C$	$\theta = C$	$\varphi = C$
	$Z = C$	$Z = C$	$\theta = C$

get changed. Whatever the interface, the input and output always remain the same. Availability of these sketching tools also depends on the software. A minimum required set of tools is always provided for the creation of sketches in any CAD software. Table 5.1 shows general and parametric form of inputs in different coordinate systems.

5.2.1 Point

A point is the simplest geometry. It can be entered by putting coordinates with respect to an origin or by clicking a mouse button. A vertex can also act as a point; however, it may depend on its parent geometry, like lines and so forth, whereas a point can be made independent and also made available outside the sketch or in 3-D space without any proceeding geometry. Points are also very useful in parametric modeling where any handling of actual geometry is not possible.

Uses: Points are used as a reference element, especially in 3-D. Point-to-point piping is a very useful method to create a nonspec pipe more quickly.

Example: Create points of all vertices of a 3-D cube of dimension 1000 × 1000 × 1000.

Hands-On Exercise, Example 5.1: Point Array (Software Used: Creo Parametric; Figure 5.1)

POINT ARRAY

Steps:

- Create a part file and rename it as Point_array. (Other names may be given, but it is advisable to provide a relevant name.)
- Select Solid on the right-hand side of the box. This will activate the solid mode and environment.
- Uncheck the box of the default template and click OK to create a new solid file.
- The template selection box will appear.

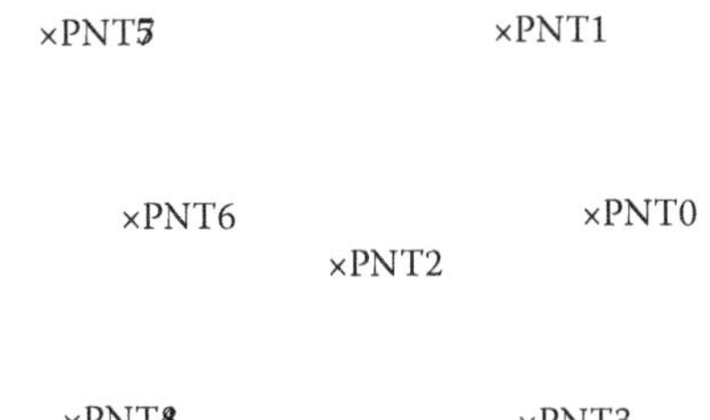

FIGURE 5.1
Array of point.

- Select mmns_part_Solid. This will allow us to use the solid template file of the following units:
  - mm—millimeter for length unit
  - n—newton for force unit
  - s—second for time unit
- Once the solid part is created, go to File > Prepare > Model properties to check the units. In the upper sections, the units of the model will be displayed.
  - Note: The units usually are displayed in the summary section of any CAD software. The setting of units usually is in the tools or settings.
- Click on the drop-down arrow beside Point from the Model tab of the ribbon interface.
- Click on Offset coordinate system.
- Select the default coordinate system PRT_CSYS_DEF.
- Click on the table area to add the point.
- Enter the *X Y Z* offset value as displayed in Figure 5.2.
- Click OK to create all the vertex points for a cube of dimension 1000 × 1000 × 1000.
- Save and exit the file.

	Name	X Axis	Y Axis	Z Axis
1	PNT0	1000.00	0.00	0.00
2	PNT1	1000.00	1000.00	0.00
3	PNT2	1000.00	1000.00	1000.00
4	PNT3	1000.00	0.00	1000.00
5	PNT4	0.00	0.00	1000.00
6	PNT5	0.00	1000.00	0.00
7	PNT6	0.00	0.00	0.00
8	PNT7	0.00	1000.00	0.00
9	PNT8	0.00	0.00	1000.00

FIGURE 5.2
Display of offset values in three axes.

5.2.2 Line

Lines are used for creating any straight edge or for construction geometry. The simplest method of creating any line is by clicking the mouse and selecting two points.

Uses: Straight lines are used to form geometry for features like extrude, revolve, sweep, and so forth. Some profiles may consist of only single-line entities, and some are of multiple-line elements.

Example: Create a rectangular surface of dimension 100 × 200.

Hands-On Exercise, Example 5.2: Creation of Flat Surface (Software Used: Creo Parametric; Figure 5.3)

LINE SURFACE

Steps:

- Create a part file and rename it as Line_Surface. (Other names may be given, but it is advisable to provide a relevant name.)
- Select Solid on the right-hand side of the box. This will activate the solid mode and environment.
- Uncheck the box of the default template and click OK to create a new solid file.
- The template selection box will appear.
- Select mmns_part_Solid. This will allow us to use the solid template file of the following units:
  - mm—millimeter for length unit
  - n—newton for force unit
  - s—second for time unit
- Once the solid part is created, go to File > Prepare > Model properties to check the units. In the upper sections, the units of the model will be displayed.
  - Note: The units usually are displayed in the summary section of any CAD software. The setting of units usually is in the tools or settings.
- Click on Extrude from the Model tab in the ribbon interface.

FIGURE 5.3
Example of a flat surface.

- Click on the Placement tab and select Define to define the sketch.
- Select the front plane as the sketch plane.
- The right plane shall be automatically taken as reference and orientation as right.
- Click on Sketch.
- Activate the surface creation option.
- The sketcher model will be activated.
- Draw a line of dimension 100.
- Click OK to create and exit the sketch.
- Enter the value 200 mm for extrusion depth.
- Click on OK to create the extrusion.
- Save and exit the file.

5.2.3 Circle and Arc

A circle and an arc can be defined in many ways. The concept is to supply sufficient references to define or satisfy a circular circle. A circle can be drawn by many methods. The following methods are most often used.

1. *Center point and radius or diameter*: This method is mostly used when drawing any circle and can be used by translating or copying the object.
   Figure 5.4 shows the creation of a circle by placing the center point and allocating the radius dimension.
2. *Three points*: A circle can be drawn through three points, as shown in Figure 5.5. This method is used in the case of sketching complex geometries.
3. *Two points assuming the center at the mid*: This method is also used for creating circles in complex geometries, as shown in Figure 5.6.
4. *Tangent with two lines and radius or diameter*: This method is very useful since placing a circle tangent with one line requires lot of effort. This method is shown in Figure 5.7.

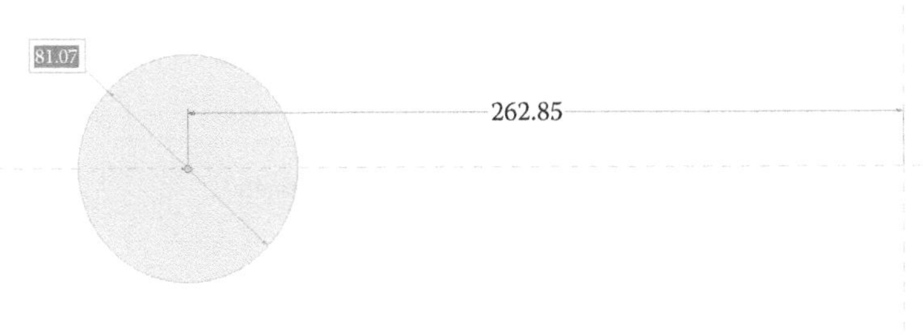

FIGURE 5.4
Sketch of a circle, defining by center point and radius.

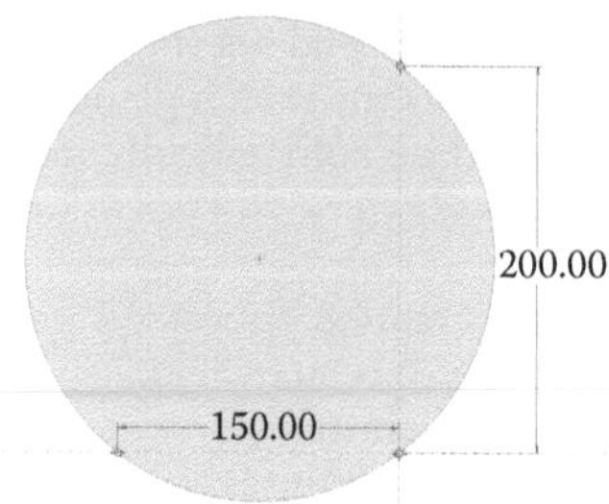

FIGURE 5.5
Sketch of a circle, defining by three points.

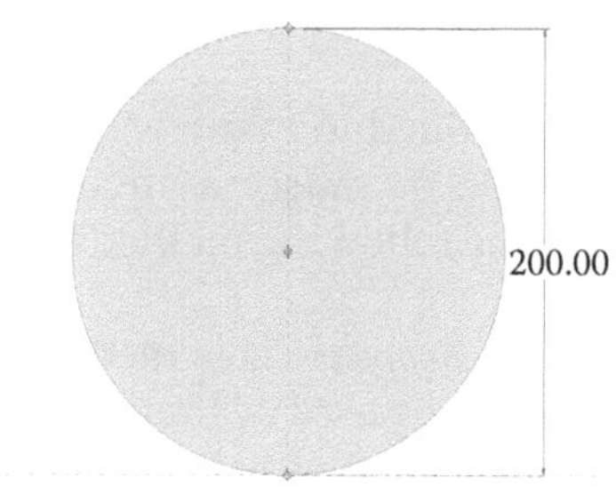

FIGURE 5.6
Sketch of a circle, defining by two points.

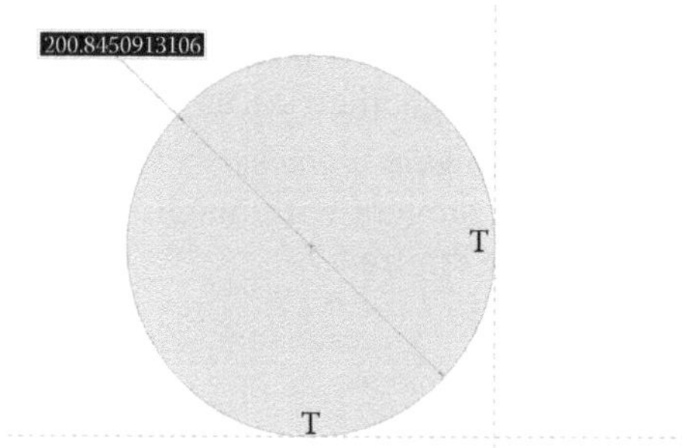

FIGURE 5.7
Sketch of a circle, defining by tangency with two lines and diameter.

5.2.4 Rectangle

The rectangle is only a combination of two sets of orthogonal lines with an end constant. Many times, it is easy to draw a rectangle and then delete one of its elements to obtain geometry. A rectangle bears a relation of perpendicularity in between all of its adjacent hands. Thus, when creating a rectangle, a user need not create or provide those relations in the sketch. When geometry is such that all the relations of a rectangle are needed when losing one hand, it becomes very useful in 3-D modeling.

Hands-On Exercise, Example 5.3: Open-Face Rectangle (Software Used: Creo Parametric; Figure 5.8)

OPEN-FACE RECTANGLE

Steps:

- Create a part file and rename it as Open_face_Rectangle. (Other names may be given, but it is advisable to provide a relevant name.)
- Select Solid on the right-hand side of the box. This will activate the solid mode and environment.
- Uncheck the box of the default template and click OK to create a new solid file.
- The template selection box will appear.
- Select mmns_part_Solid. This will allow us to use the solid template file of the following units:
  - mm—millimeter for length unit
  - n—newton for force unit
  - s—second for time unit
- Once the solid part is created, go to File > Prepare > Model properties to check the units. In the upper sections, the units of the model will be displayed.
  - Note: The units usually are displayed in the summary section of any CAD software. The setting of units usually is in the tools or settings.

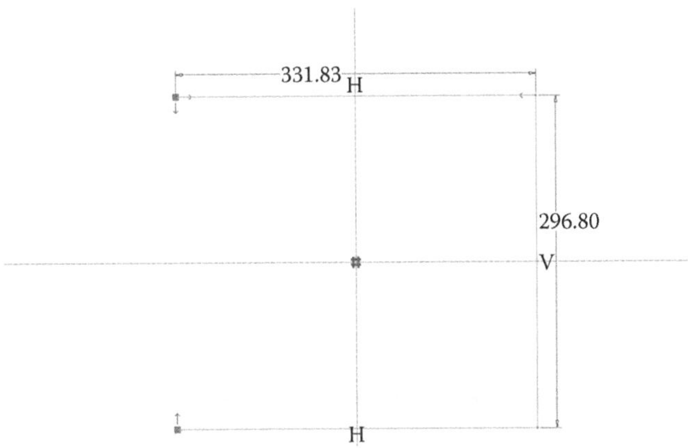

FIGURE 5.8
Sketch of an open-face rectangle.

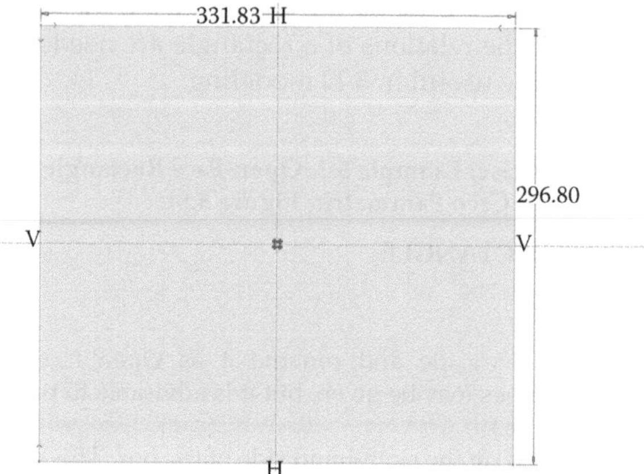

FIGURE 5.9
Sketch of a rectangle.

- Click on Sketch from the Model tab of the ribbon interface.
- Click on the Placement tab and select Define to define the sketch.
- Select the front plane as the sketch plane.
- The right plane shall be automatically taken as reference and orientation as right.
- Click on Sketch.
- Take the references of the top and front right planes as horizontal and vertical references.
- Click on Rectangle from the Sketch tab of the ribbon interface.
- Click on the top left corner to bottom right corner so that it considers automatic symmetry relations, as shown in Figure 5.9.
- Delete the left hand and notice that the relations are still retained.
- Click OK to create and exit the sketch.
- Save and exit the file.

5.2.5 Ellipse

An ellipse is required to draw any elliptical profile or sketch.

Hands-On Exercise, Example 5.4: Elliptical Fillet (Software Used: Creo Parametric; Figure 5.10)

ELLIPSE

Steps:

- Create a part file and rename it as Ellipse. (Other names may be given, but it is advisable to provide a relevant name.)
- Select Solid on the right-hand side of the box. This will activate the solid mode and environment.

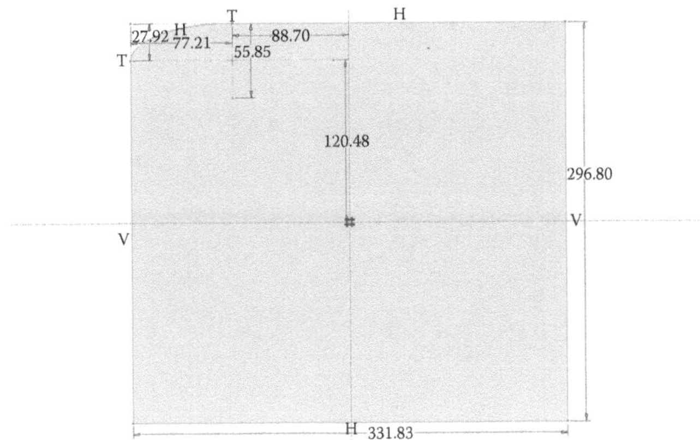

FIGURE 5.10
Example of uses of elliptical fillet in a sketch.

- Uncheck the box of the default template and click OK to create a new solid file.
- The template selection box will appear.
- Select mmns_part_Solid. This will allow us to use the solid template file of the following units:
  - mm— millimeter for length unit
  - n—newton for force unit
  - s—second for time unit
- Once the solid part is created, go to File > Prepare > Model properties to check the units. In the upper sections, the units of the model will be displayed.
  - Note: The units usually are displayed in the summary section of any CAD software. The setting of units usually is in the tools or settings.
- Click on Sketch from the Model tab of the ribbon interface.
- Click on the Placement tab and select Define to define the sketch.
- Select the front plane as the sketch plane.
- The right plane shall be automatically taken as reference and orientation as right.
- Click on Sketch.
- Take the references of the top and front right planes as horizontal and vertical references.
- Click on the drop-down arrow of the fillet and select elliptical Trim.
- Click on the left and top hands of the rectangle in Rx and Ry position, as shown in Figure 5.11.
- Click OK to create and exit the sketch.
- Save and exit the file.

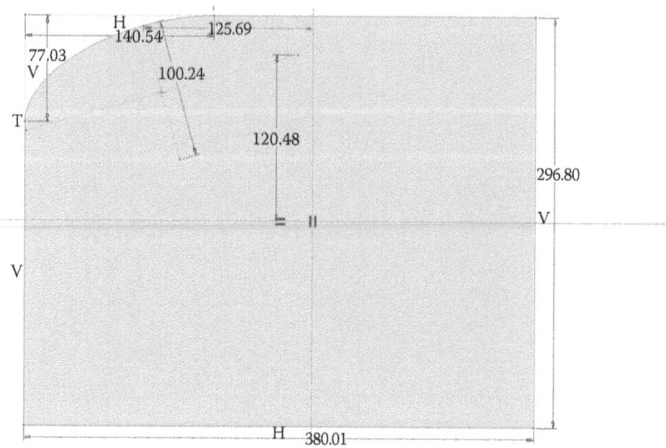

FIGURE 5.11
Example of uses of elliptical fillet in a sketch.

5.2.6 Spline

A spline is often used to create a curve with points, as in the case of sheet metal development and so forth. Equations of splines are described in Chapter 4. Splines are drawn through created points by selection.

5.2.7 Poly Line

A poly line is a combined line with various interconnected segments. The segments are may be straight line or curve. In AutoCAD, closed poly lines are essential to form region.

5.3 Modification Tools

5.3.1 Trim

Trim is a very useful tool. The required profile geometry is often not drawn at once. Many times, standard sketch entities are used in the profile and then trimmed to the desired sketch. The trim command requires a boundary to trim the object and the object to trim. Side trimming is to be selected and will be deleted after trimming.

The following is an example of the use of trim to create profile or section sketches.

Hands-On Exercise, Example 5.5: Use of Trim to Obtain Desired Sketch (Software Used: Creo Parametric; Figure 5.12)

Steps:

- Create a part file and rename it as Example_1. (Other names may be given, but it is advisable to provide a relevant name.)
- Select Solid on the right-hand side of the box. This will activate the solid mode and environment.
- Uncheck the box of the default template and click OK to create a new solid file.
- The template selection box will appear.
- Select mmns_part_Solid. This will allow us to use the solid template file of the following units:
  - mm—millimeter for length unit
  - n—newton for force unit
  - s—second for time unit
- Once the solid part is created, go to File > Prepare > Model properties to check the units. In the upper sections, the units of the model will be displayed.
  - Note: The units usually are displayed in the summary section of any CAD software. The setting of units usually is in the tools or settings.
- Click on Sketch from the Model tab of the ribbon interface.
- Click on the Placement tab and select Define to define the sketch.
- Select the front plane as sketch plane.
- The right plane shall be automatically taken as reference and orientation as right.
- Click on Sketch.

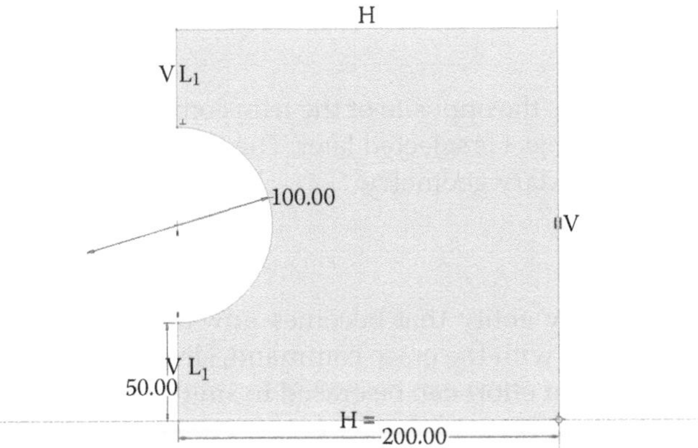

FIGURE 5.12
A sketch created by trim operation.

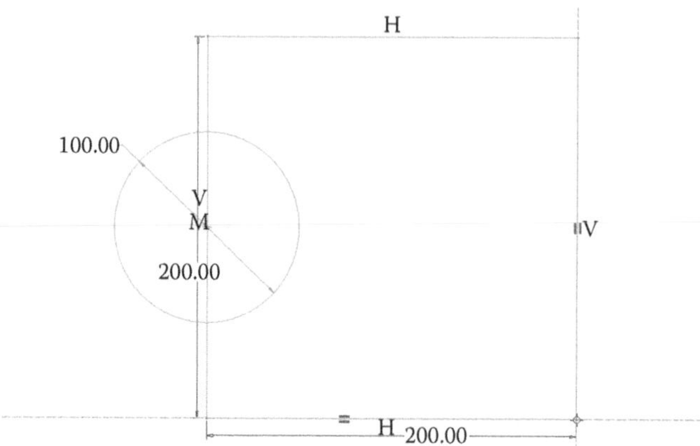

FIGURE 5.13
A sketch with a rectangle and a circle.

- Take the references of the top and front right planes as horizontal and vertical references.
- Method 1:
  - Draw the lines as displayed in Figure 5.13, followed by placing the arc.
  - Provide the dimension, as shown in Figure 5.13.
- Method 2:
  - Draw a rectangle of dimension 200 × 200 and a circle of diameter 100, placing the center at the middle of the left hand of the rectangle, as shown in Figure 5.13.
- Click Delete segment in the Sketch tab of the ribbon interface.
- Select Unwanted geometries.
- Click OK to create and exit the sketch.
- Save and exit the file.

5.3.2 Extend

The extend command is the opposite of the trim command. The boundary is selected first, and the object is selected later. The extension of the object must intersect with the boundary geometry.

5.3.3 Erase

Erase is deletion of any entity that becomes unwanted in the sketch. One must always be careful with the erase command, since the erased objects do not come back and lot of effort can be erased in single click.

5.3.4 Mirror

Mirror is the mirroring of geometry about an axis or line. In 3-D software line Creo, a centerline is always required to execute the mirror command.

5.3.5 Pattern or Array

Pattern is the creation of several components at a regular incremental distance or angle. Linear array requires the direction of the array, whereas polar array requires a point assuming the axis perpendicular to the sketch plane at that point.

5.3.6 Divide

Divide command divides a curve into equal segments. The usefulness of divide command can be greatly utilized in case of the curves which are difficult to measure or divide into equal segment *y* measuring.

5.4 Annotation Tools

5.4.1 Orthogonal Dimension

Horizontal and vertical dimensioning of any object is termed *orthogonal dimensioning*. Orthogonal dimensioning is mostly used in drawings since in a workshop, it is easy to create markings on the orthogonal dimension from horizontal and vertical references. An example of orthogonal dimensioning is shown in Figure 5.14.

Here the dimension of an inclined line can also be directly given as length; however, the terminal points are used in order keep all the dimensions orthogonal.

5.4.2 Aligned Dimension

In aligned dimensioning, the dimensions are aligned with the object. It is useful to define the length of the inclined object or specify the normal distance

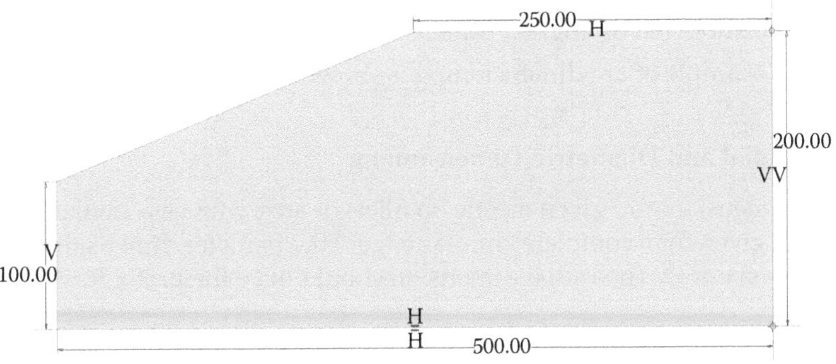

FIGURE 5.14
Example of orthogonal dimensioning.

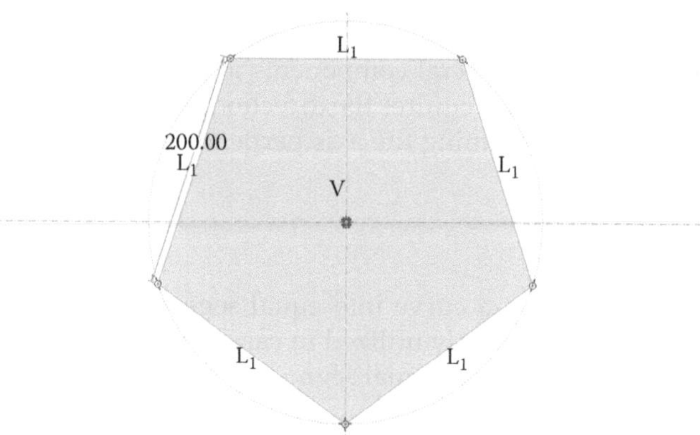

FIGURE 5.15
Example of aligned dimensioning.

from an inclined line. An example of aligned dimensioning is shown in Figure 5.15.

5.4.3 Arc Dimensioning

An arc basically needs to be defined with following dimensions:

1. Radius of the arc
2. Dimension of center point with respect to horizontal and vertical references
3. Dimension of two end points

   Whenever the radius dimension is given, R = specified dimension would be the common method for understanding radial dimension. The two end points of an arc can be dimensioned with the orthogonal or aligned dimensioning method.

   An example of arc dimensioning is shown in Figure 5.16.

5.4.4 Radial and Diametric Dimensioning

Radial dimensions are given mostly on fillets or arcs, whereas circular dimensions are given for a complete circular edge. The diameter dimension should have a prefix of Ø. The radial dimension should have the prefix R.

5.4.5 Incremental and Ordinate Dimension

Incremental dimensioning is used where dimension from one feature to another is more important, whereas ordinate dimensioning is used where

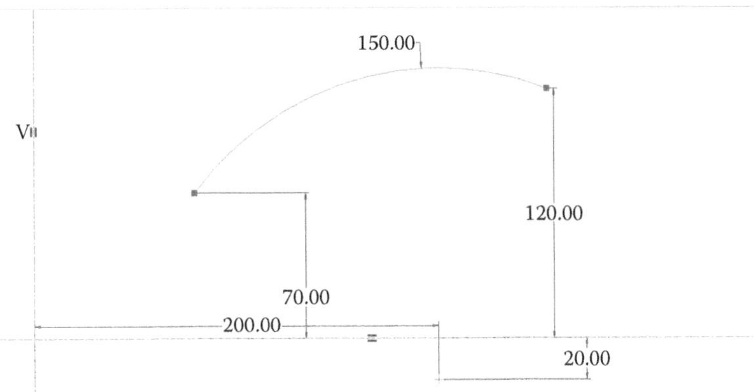

FIGURE 5.16
Example of arc dimensioning.

dimension from a reference line is more significant. Mostly, levels in industrial plants are dimensioned with ordinate dimensioning. Figure 5.17 shows an example of incremental dimensioning.

5.4.6 Text Creation

Texts are an important annotation in a drawing. A lot of information can be given by texts, which may or may not be possible to express by drawing entities. Texts can be of many styles, and underlined text lines are useful for highlighting important texts.

Texts can be horizontal, vertical, or inclined and can even be aligned with an arc.

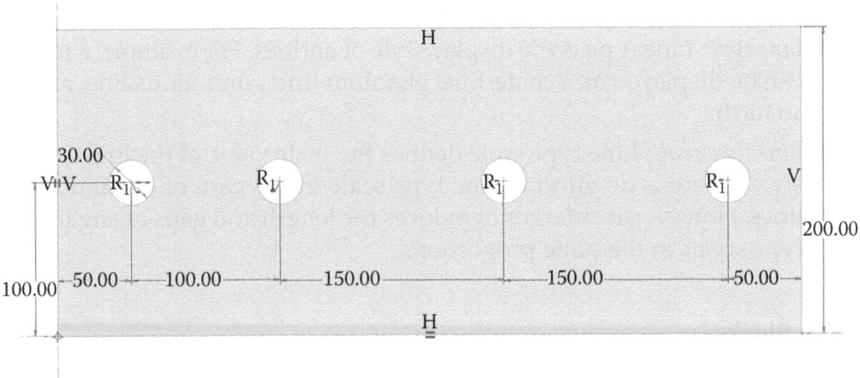

FIGURE 5.17
Example of incremental dimensioning.

5.5 Advanced Tools

5.5.1 Layers

Layers are a very useful tool, especially in 2-D. Layers are basically creating a group having a special attribute to certain entities. Operations like blank, freeze, lock, and so forth can be done on the entities under specific layers. This provides aid to work with complex and multiple overlapping geometries without deleting any entities. The commonly used layer functions are as follows:

1. *Lock*: This function locks the layer. The entities within the locked layer will be displayed but cannot be selected for any operation. This function is especially useful while creating a 2-D drawing while taking references from an existing drawing.
2. *Layer off*: The layer off function turns off the display of entities of a layer. This function is used while taking print or generating a final drawing while turning off the reference entities.
3. *Freeze*: Layer freeze is similar to layer off, except that freeze function can turn off the same entities in many view ports while displaying other view ports.

5.5.2 Properties

Entity properties are as follows:

1. *Color*: Color is an important property of an entity, which is very useful to distinguish the different entities in a drawing.
2. *Layer*: As discussed earlier, layers have some definite functions. Therefore, entities need to be specified under different layers for facilitating the advantages of layers.
3. *Line type*: Line type is the display style of entities. For example, a line can be displayed as a centerline, phantom line, continuous line, and so forth.
4. *Line type scale*: Line type scale defines the scale value of the line type style. There is no effect of line type scale in the case of continuous lines. However, it enlarges or reduces the length and gaps of any line type styles in the same proportion.

5.5.3 Blocks

Blocks are groupings of entities with some special advantages like counting, auto change, and so forth. Blocks can be easily copied into any drawing from other drawings.

5.6 Parametric Sketching

Parametric sketching is sketching while keeping the relation with the entities of the sketch such that when the dimension of an entity changes, the corresponding entities are also propagated and changed. An example of a parametric sketch is shown in Figure 5.18a and b:

The constraints or relation used in a parametric sketch are as follows:

1. *Horizontal*: Make the line horizontal with respect to references in the sketch plane. These constants always keep the selected line horizontal with respect to selected references. Therefore, it has to be remembered that if the references change, the orientation of the generated geometry may shift from its original position.

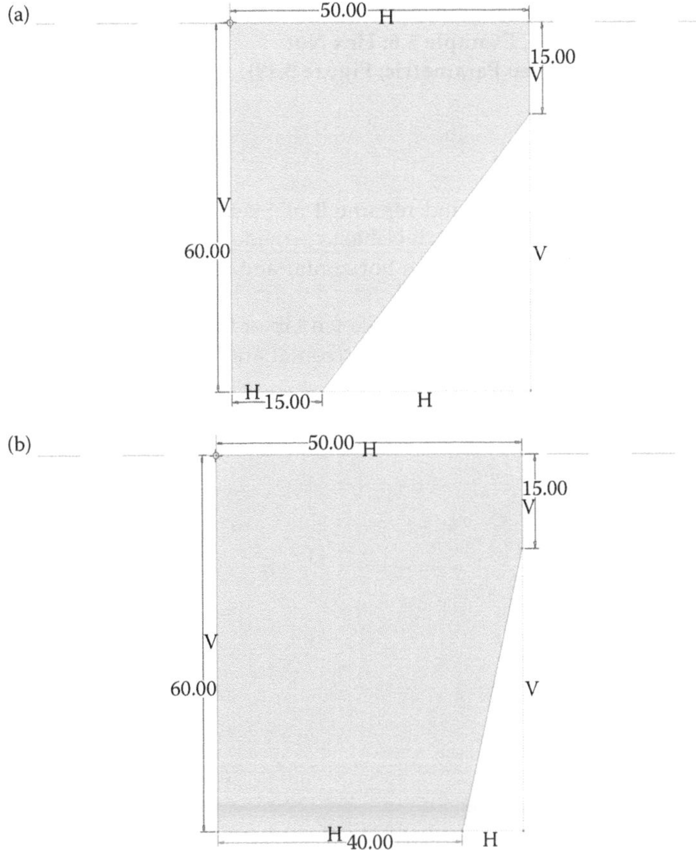

FIGURE 5.18
(a) Example of parametric sketch. (b) Example of modification of parametric sketch.

2. *Vertical*: Make the line horizontal with respect to references in the sketch plane.
3. *Coincident*: Coincide two points or lines.
4. *Parallel*: Make the line parallel to another line.
5. *Perpendicular*: Make the line perpendicular to another line or reference line.
6. *Tangent*: Make any line tangent to any circle or circular curve.
7. *Concentric*: Make the circle concentric to another circle or reference circle or circular curve.
8. *Equal*: Make one line equal to another line.
9. *Symmetric*: Create a symmetric constant between two points about any centerline.
10. *Midpoint*: Make the point at the middle of any line.

Hands-On Exercise, Example 5.6: Hex Nut (Software Used: Creo Parametric; Figure 5.19)

SKETCH 1

Steps:

- Create a sketch file and rename it as Sketch_1. (Other names may be given, but it is advisable to provide a relevant name.)
- Place two centerlines, one horizontal and one vertical line, as displayed.
- Click on the circle from the Sketch tab of the ribbon interface. Click on the intersection of horizontal and vertical references

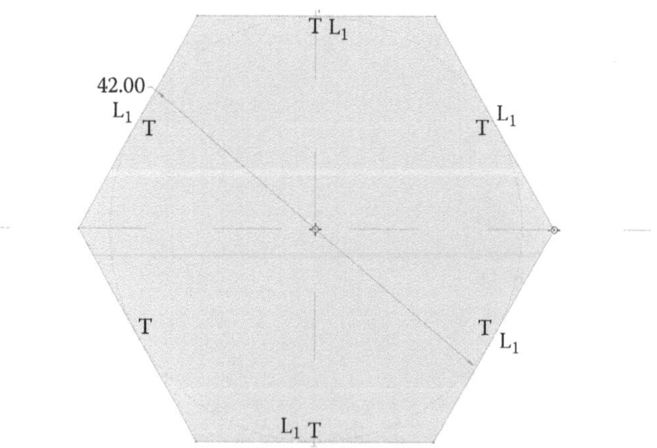

FIGURE 5.19
Sketch of a hexagon.

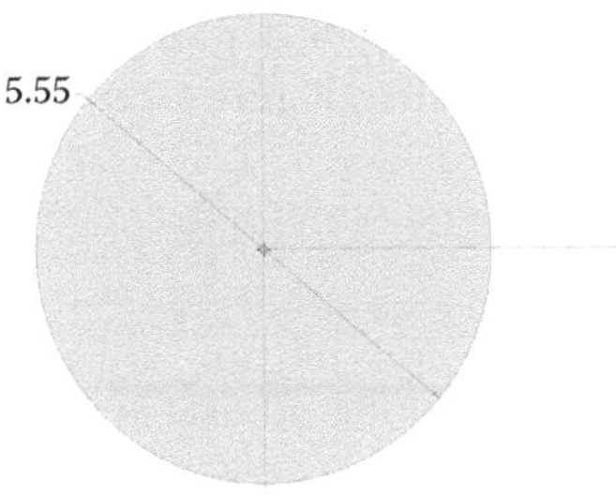

FIGURE 5.20
Sketch of a circle.

to place the center point at the intersection point as displayed in Figure 5.20.

- Double click on the diameter and change it to 42.
- Click on the circle and right click on it; select Construction to set the circle entity to construction geometry.
- Select Straight line from the Sketch tab of the ribbon interface.
- Create the lines, as shown in Figure 5.21.
- Click on Tangent constraint from the Sketch tab of the ribbon interface.
- Click on any line and the circle.
- Similarly, make all the lines tangent to the circle.
- Click on the constraint equal from the Sketch tab of the ribbon interface.
- Select the top horizontal line and the line attaché to the top horizontal line at the left corner.
- Save and exit the file.

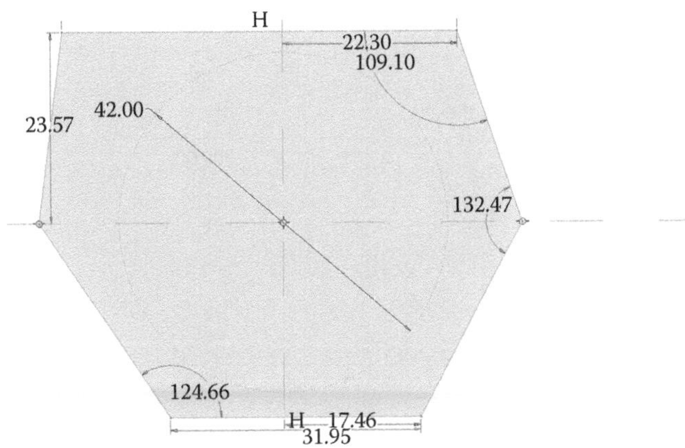

FIGURE 5.21
Sketch of an unconstrained hexagon.

Hands-On Exercise, Example 5.7: Dish End Sketch (Software Used: Creo Parametric; Figure 5.22)

DISH END

Steps:

- Create a sketch file and rename it as Sketch_4. (Other names may be given, but it is advisable to provide a relevant name.)
- Place two centerlines, one horizontal and one vertical line, as displayed.
- Click on the circle from the Sketch tab of the ribbon interface. Click on the horizontal line to place the center point of the circle to the left side of the vertical centerline.
- Click on the dimension from the Sketch tab of the ribbon interface.
- Click once on the circle and press the middle button to place the radial dimension.
- Double click on the radius dimension and enter 900 as the radius of the circle.
- Click on the line from the Sketch tab of the ribbon interface.
- Click on the vertical centerline and connect it to the circle, as shown in Figure 5.23.
- Click on the delete segment from the Sketch tab of ribbon interface.
- Delete segments of the circle, as shown in Figure 5.24.
- Select Fillet from the Sketch tab of the ribbon interface.

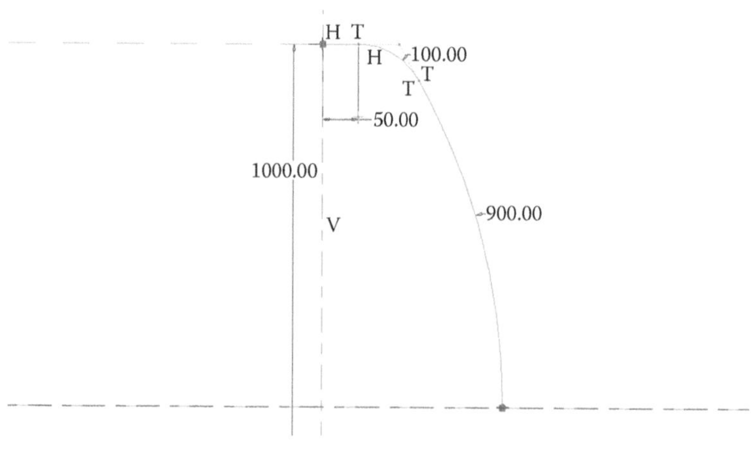

FIGURE 5.22
Sketch for an arc of a dish end.

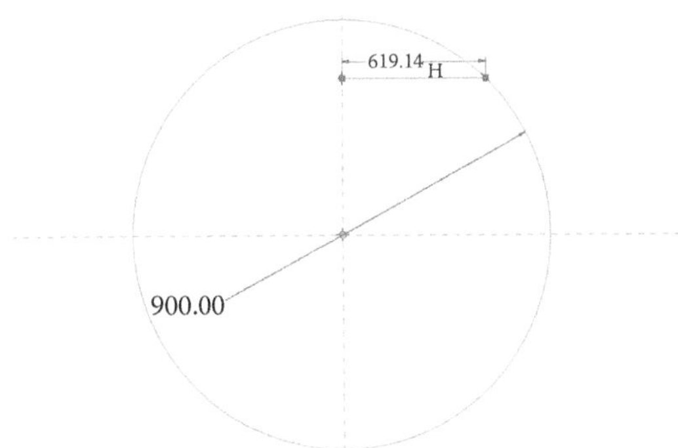

FIGURE 5.23
Sketch of constructional geometries for dish end.

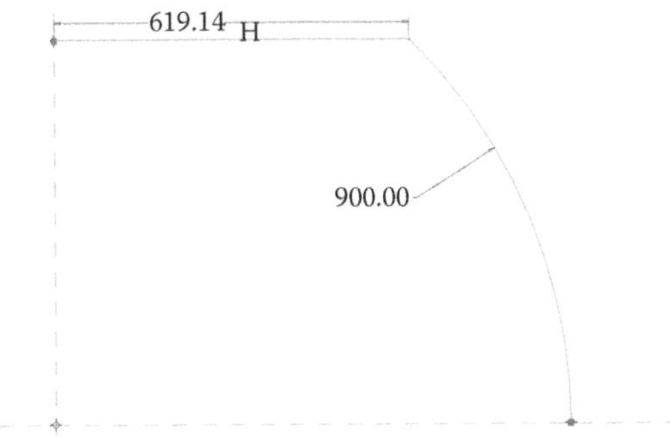

FIGURE 5.24
Sketch of trimmed entities.

- Click on the straight line and the circular segment to create a fillet on that.
- Click on the dimension from the Sketch tab of the ribbon interface.
- Select Straight line segment and press the middle button.
- Double click on the dimension and change it to 50.
- Similarly, click on the fillet radius and change it to 100.
- Select Dimension again from the Sketch tab of the ribbon interface, if not selected.
- Click on the straight line, horizontal centerline, and straight line again to create symmetric dimension.
- Double click on the dimension and change it to 1000.
- Save and exit the file.

5.7 Project Work

Hands-On Exercise, Example 5.8: Sketch_1 (Software Used: Creo Parametric; Figure 5.25)

SKETCH 1

Steps:

- Create a sketch file and rename it as Sketch_1. (Other names may be given, but it is advisable to provide a relevant name.)
- Place two centerlines, one horizontal and one vertical line, as displayed.
- Click on the circle from the Sketch tab of the ribbon interface. Click on the intersection of horizontal and vertical references to place the center point at the intersection point, as displayed in Figure 5.26.
- Double click on the diameter and change it to 80.
- Similarly, click on the circle again and place the center point at the horizontal reference and at the right-side vertical reference line at a certain distance, as shown in Figure 5.27.
- Click on Line from the Sketch tab of the ribbon interface.
- Click on the edge of one circle and then click on the edge of the other circle, as displayed in Figure 5.28.
- Click on the tangent constraint from the Sketch tab of the ribbon interface.
- Click on the first circle and the line to make the line tangent to the first circle.
- Similarly, make the line tangent to the other circle too.
- Similarly, create another tangent line with both the circles at the bottom side of the horizontal reference line.

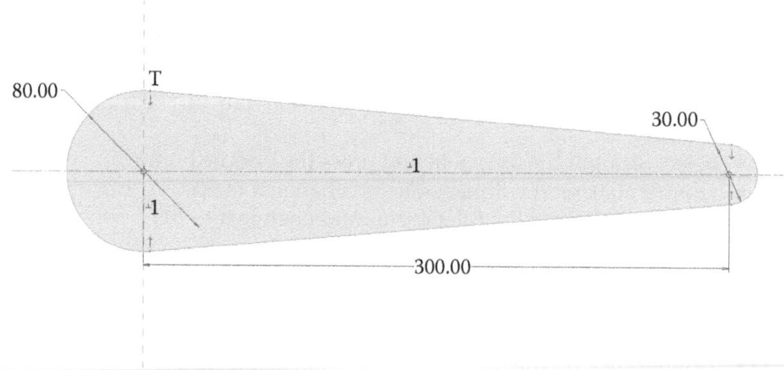

FIGURE 5.25
Constructional sketch of handle of pulley.

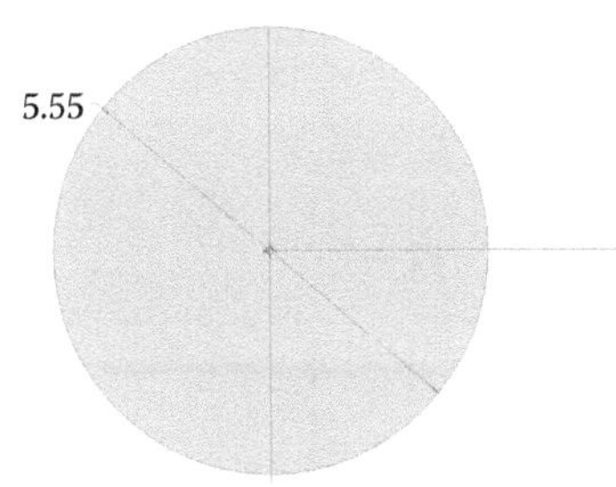

FIGURE 5.26
Sketch of a circle.

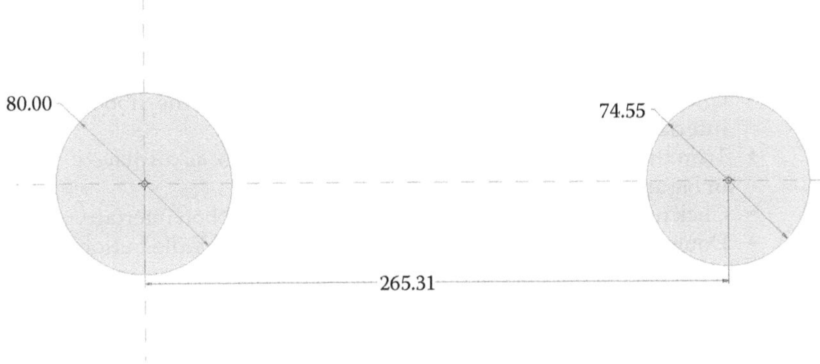

FIGURE 5.27
Sketch of two circles.

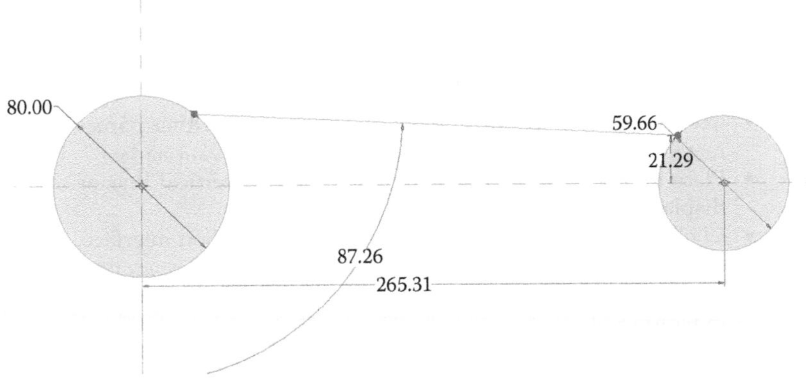

FIGURE 5.28
Constructional sketch of handle of pulley.

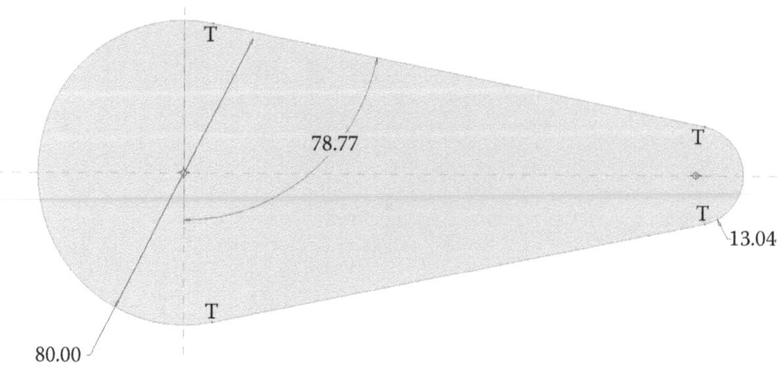

FIGURE 5.29
Sketch of handle of pulley.

- Click on Delete segment from the Sketch tab of the ribbon interface.
- Trim the inside circle segment of both the circles, as displayed in Figure 5.29.
- Click on Dimension from the Sketch tab of the ribbon interface.
- Double click on the second circle and press the middle button to create the diametric dimension.
- Enter 80 as the diameter of the first circle by double clicking on it.
- Similarly, enter 30 as the diameter of the second circle by double clicking on it.
- Double click on the center-to-center distance and enter 300.
- Save and exit the file.

Hands-On Exercise, Example 5.9: Sketch_2 (Software Used: Creo Parametric; Figure 5.30)

SKETCH 2

Steps:

- Create a sketch file and rename it as Sketch_2. (Other names may be given, but it is advisable to provide a relevant name.)
- Place two centerlines, one horizontal and one vertical line, as displayed.
- Click on the circle from the Sketch tab of the ribbon interface. Click on the intersection of horizontal and vertical references to place the center point at the intersection point, as displayed in Figure 5.31.
- Double click on the diameter and change it to 70.
- Draw a vertical line from the center to beyond the edge of the circle at the top side, as shown in Figure 5.32.
- Click on Line from the Sketch tab of the ribbon interface.

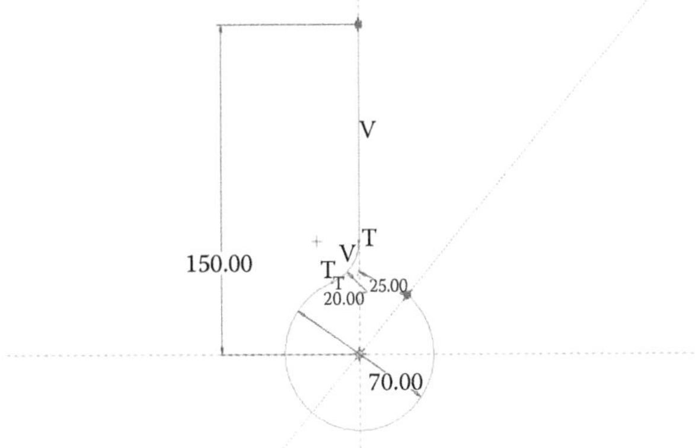

FIGURE 5.30
Sketch of industrial hook.

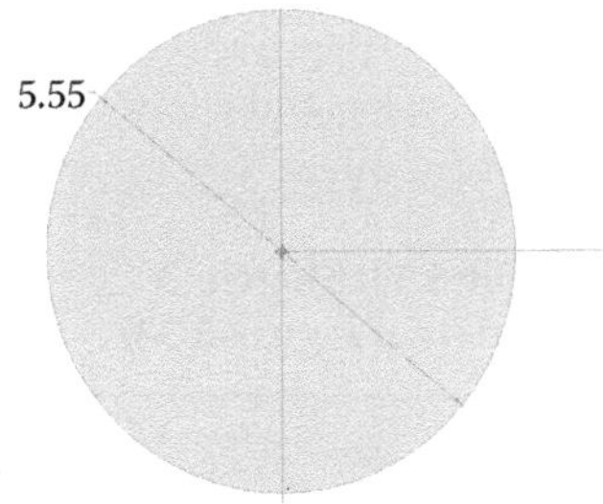

FIGURE 5.31
Sketch of a circle.

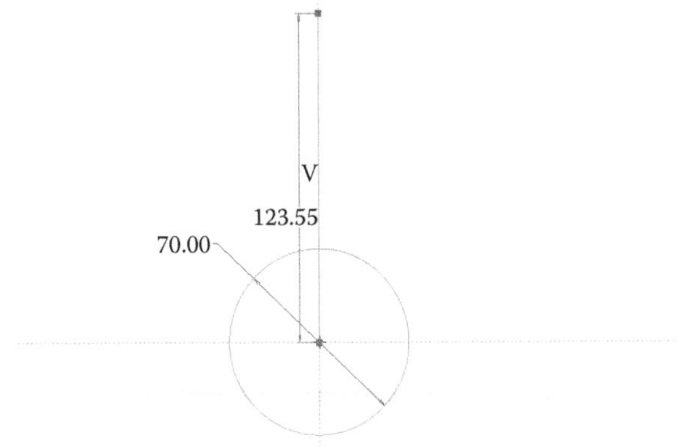

FIGURE 5.32
Constructional sketch of an industrial hook.

- Click on Delete segment from the Sketch tab of the ribbon interface.
- Delete the segment of the straight line inside the circle.
- Click on Fillet from the Sketch tab of the ribbon interface.
- Click on the line and the left portion of the circle.
- Click on Center line from the Sketch tab of the ribbon interface.
- Click on the intersection of horizontal and vertical references and place it at an angle of approximately 30° with horizontal reference line.
- Click on Delete segment from the Sketch tab of the ribbon interface.
- Delete the segment of the circle between the fillet and the centerline.
- Delete the line segment between the fillet and the circle.
- Click Dimension from the Sketch tab of the ribbon interface.
- Click on the centerline and extended fillet intersection.
- Double click and change the dimension to 25.
- Click Dimension from the Sketch tab of the ribbon interface.
- Click on the horizontal centerline and the furthest end point of the line segment.
- Double click on the dimension and change it to 150.
- Save and exit the file.

Hands-On Exercise, Example 5.10: Sketch_3 (Software Used: Creo Parametric; Figure 5.33)

Steps:

- Create a sketch file and rename it as Sketch_3. (Other names may be given, but it is advisable to provide a relevant name.)
- Place two centerlines, one horizontal and one vertical line, as displayed.

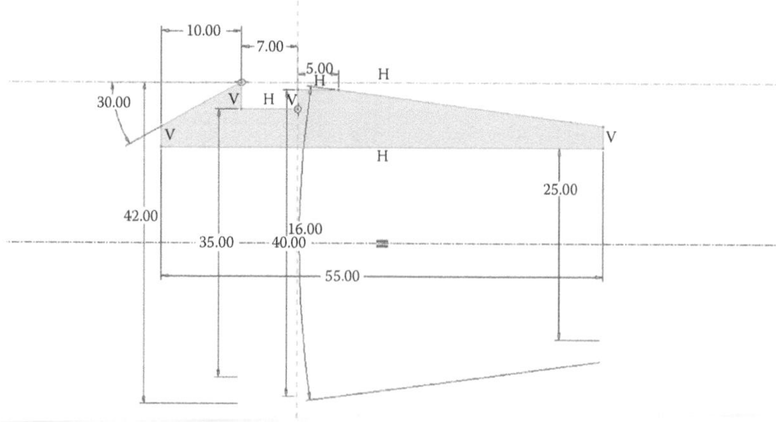

FIGURE 5.33
Sketch of collet.

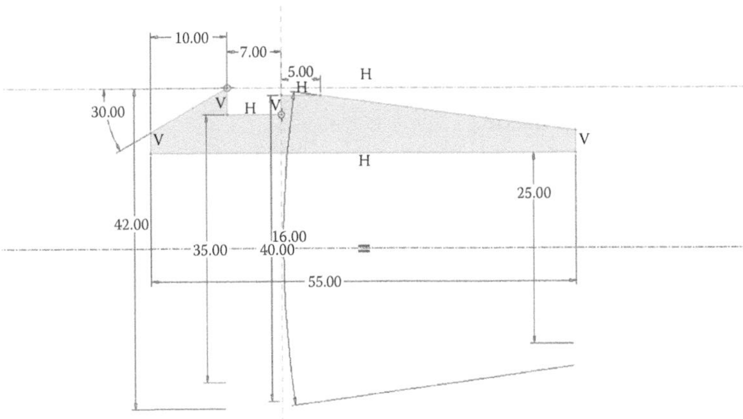

FIGURE 5.34
Constructional sketch of collet.

- Click on Line from the Sketch tab of the ribbon interface.
- Draw a line parallel to the horizontal centerline. The line should be extended more at the right side of the vertical centerline.
- Double click on the length dimension and change it to 55.
- Select Dimension from the Sketch tab of the ribbon interface.
- Click on Line, click on the centerline, again click on the line, and press the middle button to create symmetric dimension for diameter reference.
- Double click on the symmetric dimension and enter 25 as the diameter.
- Select Line from the Sketch tab, click on the left end to start the line, and click Vertical little upward to create a line on the left section.
- Similarly, create another line on the right end.
- Select the line again from the Sketch tab of the ribbon interface.
- Draw as shown in Figure 5.34.
- Enter all the dimensions, as shown in Figure 5.33.
- Save and exit the file.

QUESTIONS

1. Describe the use of sketching.
2. Describe the use of splines in a sketch.
3. Describe the poly lines and their uses.
4. Differentiate the commands trim and extend.
5. Describe the use of mirror and array.
6. Describe the use of divide.
7. Why are aligned dimensions used?

8. Describe the differences between incremental and ordinate dimensioning.
9. State the uses of different layers.
10. Why is the block used?
11. Write down the function of parametric sketching.

6

Part Modeling

6.1 Introduction

Part modeling is one of the most time-consuming jobs in computer-aided design (CAD). Part modeling or solid modeling happens sequentially after sketching. A component that can be manufactured from a single work piece without any assembly operation can be termed *part*. Any intensive properties of any part will always be the same across the geometry. No engineering component can be made without a part. Therefore, part modeling is one of the essential steps before continuing to the creation of any assembly.

It is obvious that any part creation consists of all the operations that provide the shape and size of the part.

Parts can be created or modeled with two approaches: primitives and features. In the primitive approach, the basic solid shapes like blocks, cylinders, cones, spheres, and so forth are available via predesigned definable templates. The final model is created by Boolean operation like addition or subtraction of primitives to prepare the final solid model. In the feature approach, the solid is directly created by generating the sections. Feature-based modeling is the most used nowadays in most of the modern CAD software like PRO/E, CATIA, Solid Works, Unigraphics, and so forth. Since feature-based modeling is the most used in most of the CAD industry in mechanical component design, it is only discussed in the latter portion of this book.

Obviously a 3-D solid model can be created by one or many sets of combinations of features. When creating any operation, CAD does not understand the practical limit of creating the operation, unless any warning limit system is otherwise introduced. For example, a too-narrow slot with excessive depth can be practically difficult; however, it is easy in CAD. Therefore, one may end up creating a CAD model that may not be possible or that may be too expensive to manufacture. It is always advisable to create the component in conjunction with the actual manufacturing sequence whenever possible. For example, a plate with a hole is displayed in Figure 6.1. The displayed model can be generated by sketching a rectangle with a hole followed by an extrusion command or by creating a solid block followed by a hole operation. When

FIGURE 6.1
Example of a plate with a hole.

manufacturing methods are considered, in the former method, the component may be molded to create the model, and in the other method, a simple hole operation is required in some metal piece that may be cut from a strip. It is obvious that the former method is more expensive.

So it is recommended to use the features sequentially with the manufacturing processes that may be involved in the production of the part in real life.

The following are the advantages to creating a model as per practical manufacturing operations:

1. Designers will never have the problem of practical component creation.
2. The creation method of the component or part can be shown step by step for understanding the manufacturing procedure.
3. The model in various stages can be used for different purposes, if the final part or component is modeled by the practical creation method sequence. For example, if a part is created from casting and followed by machining, the part without the machining operation can be used for pattern creation drawing.
4. It helps for optimization among different operation sequences of creation of the part.

However, there are some disadvantages, which designers sometimes ignore, to creating parts as per manufacturing sequences. They are as follows:

1. When creating a mechanism or complex assembly design, designers need to see the usefulness of various shaped parts for which following practical operation sequences is not viable before finalizing the component or part.
2. It requires more time; therefore, for simple parts, designers often ignore practical sequences.
3. Sometimes, creating a part with the actual operation sequence may be difficult and is not advised.

Nowadays, most parts are created using feature-based approaches. Therefore, it is essential to know about the various available features in CAD.

6.2 Feature

In CAD, a feature can be defined as an object or entity or a created region capable of possessing many properties but may or may not have an effect on mass properties of the model. Every feature can also include its own properties like name, parameters, and so forth. The following features are essential in the creation of parts.

6.2.1 Datum Features

Datum features are nonsolid features with null mass properties. They often are required to create the solid model and/or act as reference geometry.

- *Plane*: As described earlier, planes are imaginary flat surfaces, used as a sketch plane or reference geometry. Planes can be created by many methods, the objective being to select a sufficient entity to satisfy the equation of a plane.
  - *Three-point plane*: The equation of a plane through a given point can be given by $ax + by + cz + d = 0$. If the places pass through a point $(x1,y1,z1)$, we have $ax1 + by1 + cz1 + d = 0$.

    Subtracting these equations from each other, we get

    $$a(x - x1) + b(y - y1) + c(z - z1) + d = 0. \tag{6.1}$$

    Let there be three points whose coordinates are as follows: M(4,3,5), N(8,2,9), and O(10,–4,8).

    Hence, the equation of the plane becomes

    $$a(x - 4) + b(y - 3) + c(z - 5) = 0. \tag{6.2}$$

    If the plane passes through the other points, then the equation becomes $a(8 - 4) + b(2 - 3) + c(9 - 5) = 0$ and $a(10 - 4) + b(-4 + 3) + c(8 - 5) = 0$, that is, $4a - b + 4c = 0$ and $6a - b + 3c = 0$.

    Solving these equations, we get $a = \frac{b}{12} = \frac{c}{6}$.

    Substituting, these values to Equation 6.2, we get

    $$(x - 4) + 12(y - 3) + 6(z - 5) = 0$$

    or $x + 12y + 6z - 70 = 0$, which is the required equation of the plane.

- *Parallel planes*: If there are two planes and the equations of the planes are

$$ax + by + cz + d = 0$$

and

$$a1x + b1y + c1z + d1 = 0,$$

if the planes are parallel, their normals are also parallel; hence, the condition would be

$$\frac{a}{a1} = \frac{b}{b1} = \frac{c}{c1}.$$

Hence, the equation of any plane parallel to the plane

$$ax + by + cz + d = 0$$

will be

$$ax + by + cz + k = 0$$

where k is the constant.

For example, if the equation of a plane is

$$5x + 4y + 9z + 10 = 0,$$

the equation of parallel plane to this plane passing through point (–9,–10,–5) would be as follows:

$$5x + 4y + 9z + k = 0$$

$$5(-9) + 4(-10) + 9(-4) + k = 0$$

$$-4 - 40 - 36 + k = 0$$

$$k = 80$$

Hence, the required equation would be

$$5x + 4y + 9z + 80 = 0.$$

Perpendicular to a straight line and on a point: It is often required in CAD to draw a plane that is perpendicular to a line and passing through a point. If the line is joining two points (5,8,10) and (2,7,9) and the passing point of the plane is (1,–5,–9), then we know the equation of the plane be

$$a(x-1)+b(y+5)+c(z+9)=0.$$

The direction ratios of the joining line would be 5–2,8–7,10–9 or 3,1,1.

Hence, the required equation of the plane would be

$$3(x-1)+1(y+5)+1(z+9)=0$$

$$3x+y+z+11=0$$

- *Lies on an axis and a point*: A plane can be defined such that it is lying on an axis and a point and the point must not lie on the axis. When the plane is lying on the axis, all the points lying on the axis are lying on the plane. Hence, if we consider any two points lying on the axis, the equation can be derived easily by considering three-point-plane equations.
- *Rotated through an axis with regard to a plane or surface*:

  Let the equation of a plane be $4x + 5y + 9z + 2 = 0$.

- *Axis*: An axis is an imaginary line that can be used as a reference entity for construction of a sketch or other entity. Similar to a plane, to create an axis, the selection shall be sufficient to satisfy the equation of a straight line.
  - *Selecting two points*:

    The equation of a line in its symmetrical form is

$$\frac{x-x1}{l}=\frac{y-y1}{m}=\frac{z-z1}{n}. \tag{6.3}$$

If the line passes two points ($x1,y1,z1$) and ($x2,y2,z2$), the direction ratios become ($x2 - x1$), ($y2 - y1$), ($z2 - z1$), and the line is passing through the point $x1,y1,z1$.

Hence, the equation of the line becomes

$$\frac{x-x1}{x2-x1}=\frac{y-y1}{y2-y1}=\frac{z-z1}{z2-z1}. \tag{6.4}$$

    - *Selecting two intersecting planes*: The easiest way to create an axis in CAD is selecting two planes so that the axis is created in the intersection of two planes, say,

$$ax + by + cz + d = 0 \quad \text{and} \quad a1x + b1y + c1z + d1 = 0,$$

    which together are called the equation of a line in general form.

    Perpendicular to a plane and passing through a point: Let us say the equation of a plane is $4x + 5y + z + 10 = 0$, and the line is piercing the plane at a point (2,3,–33). Hence, the direction ratio normal to the plane is (4,5,1). Hence, the equation of the line is

$$\frac{x+2}{4} = \frac{y+3}{5} = \frac{z-33}{1}.$$

    - Parallel to a line and passing through a point.
- *Point*: Points can be created by selecting a sufficient entity to satisfy single coordinates. The common methods of creation of points are as follows:
    - *Intersection of three planes*: If two planes are intersecting, a line is formed, and if a third plane is intersecting that line, the point is found.

    Example: Find the coordinate of a point where three planes are intersecting whose equations are as follows:

$$5x + 1y + 2z + 10 = 0$$

$$7x - 4y + 9z + 1 = 0$$

$$2x + 3y + z - 11 = 0$$

    Solving the equations, we get the coordinates of the point as (–246/43, 410/43, 287/43).
    - Intersection of two axes.
    - Dimensional offset from an existing point.
    - Sketched points.
- *Coordinate system*: A coordinate system can be created by selecting similar references with points, except additional selection may require defining the direction of the axes. A coordinate system can be defined by one of the following methods:

- Offsetting from a coordinate system
- Selecting three planes and directional references
- Selecting a point and directional references

6.2.2 Extrude

Extrusion is a method of creation of a protruded region in a particular direction. Some CAD software provides extrusion to a particular definable direction vector, whereas much CAD software provides an option only for creation in a normal direction of the sketch plane. Extrusion can be used for addition or subtraction of materials. The following options are available while extruding:

1. *Blind extrusion*: It is the extrusion process with definite value to the normal direction. This is the most used in modeling. A user has to calculate the values of extrusion each time when creating any extrusion with blind values.
2. *Extrude up to surface or selected feature*: When the need is such that the depth of extrusion is derived by another surface that is available for selection, this feature can be used. This system is particularly useful in creation of a parametric model where the depth of several features of common depth can be linked to a common datum; hence, the selection up to the surface or plane is made so that when the datum moves, it changes the depth of all features. However, it is to be kept in mind that the selection entity for this feature must be available; otherwise, the entire feature related to the selection will miss the references, and feature failure can occur.
3. *Mid plane extrusion*: Mid plane extrusions create the depth of the part symmetrical to both sides of the sketch plane.
   a. *Advantages*:
      i. This creates symmetrical depth of the feature across the sketch plane.
      ii. The sketch plane is readily available as the mid plane if reference is required for assembly or measurement and so forth.
   b. *Disadvantages*:
      i. It may be difficult to understand the sides of the feature if the part is symmetrical about the sketch plane. Therefore, if the part is placed several times in an assembly with different orientations, any addition of the feature can appear different or opposite in the assembly for different instances of the same component.
4. *Extrude with draft*: This feature is available in some CAD software and is useful to model taper parts especially for casting models where a draft is necessary.

Hands-On Exercise, Example 6.1: Create a Block of Dimension 500 × 500 × 200 with Blind Depth and Mid Plane Option (Software Used: Creo Parametric)

Steps:

Assume a part file is created with three default planes, RIGHT, TOP, and FRONT.

- Select the Extrude feature from the ribbon at the top. Clicking on the feature opens the selection box for all other options.
- You may change the name of the feature to be created; it helps to trace back the feature in the future so that one easily can understand which feature is used where.
- Click on the Placement tab, and select Define sketch. A predefined sketch also can be selected in this box. However, as no predefined sketch is available now, click on Define.
- As Define is clicked, a small toolbox appears for placement of the sketch. Select the top plane for the sketch. The references for orientation (deciding orientation of sketch plane) may need to be selected if perpendicular planes of the sketch plane are not available. Once the references are available, click on the Sketch button to define the sketch.
- The references to locate the sketch should have been automatically created; however, if the references are not taken, click on the Reference icon at the top left corner and select the references. Perpendicular planes automatically may be taken as references. However, axes, points, or other geometries also can be taken as references. For our case, select the right and front planes for references not being automatically taken.
- Select Axis from the sketch ribbon to keep the original sketch symmetrical to the axis. Draw one vertical and one horizontal axis along the references of the right and front planes.
- Once the axes are drawn, select Rectangle from the sketch ribbon, and click on the top left quadrant to locate the left corner. Then click again at the right bottom quadrant to an approximate distance such that it can form a symmetrical rectangle with respect to the axes. Automatically, the symmetrical constraint is taken by the system. This practice saves time for creating constraints.
- Two weak dimensions may be displayed if the weak dimension display is activated. Enter the dimension 500 to both fields by double-clicking the weak dimensions. Select OK.
- When creating any sketch, make sure there is no weak dimension. This ensures that sections or sketches are not overlooked.
- Once finished with the sketch, enter the value of 200 in the blind field. The flip option for changing the direction of extrusion is beside the blind depth value box.
- The mid plane option also can be selected via the drop-down menu beside Blind.

- The top plane will act as a symmetrical mid plane if the mid plane option is selected.
- Click OK to finally create the feature.
- Save the file.

Hands-On Exercise, Example 6.2: Use of the Up to the Selection Option (Software Used: Creo Parametric)

Steps:

Assume a part file is created with three default planes, RIGHT, TOP, and FRONT.

- Select the plane creation option in the Model tab of the ribbon interface.
- Select the top plane as reference and create a plane by offset value of 500 upward from the top plane. The plane will be automatically named DTM1.
- Select Axis creation of option from the Model tab of the ribbon interface.
- Create an axis by selecting the right plane and DTM1 plane. The axis will be named A_1.
- Select the plane creation option again. Create a plane by selecting the axis A_1 and plane DTM1. The plane shall be at an angle of 45° to the plane DTM1.
- Select the Extrude feature from the ribbon at the top. Clicking on the feature opens the selection box for all other options.
- You may change the name of the feature to be created; it helps to trace back the feature in the future so that one easily can understand which feature is used where.
- Click on the Placement tab, and select Define sketch. A predefined sketch also can be selected in this box. However, as no predefined sketch is available now, click on Define.
- As Define is clicked, a small toolbox appears for placement of the sketch. Select the top plane for the sketch. The references for orientation (deciding orientation of sketch plane) may need to be selected if perpendicular planes of the sketch plane are not available. Once the references are available, click on the Sketch button to define the sketch.
- The references to locate the sketch should have been automatically created; however, if the references are not taken, click on the Reference icon at the top left corner and select the references. Perpendicular planes automatically may be taken as references. However, axes, points, or other geometries also can be taken as references. For our case, select the right and front planes for references not being automatically taken.
- Select Axis from the sketch ribbon to keep the original sketch symmetrical to the axis. Draw one vertical and one horizontal axis along the references of the right and front planes.

- Once the axes are drawn, select Rectangle from the sketch ribbon, and click on the top left quadrant to locate the left corner. Then click again at the right bottom quadrant to an approximate distance such that it can form a symmetrical rectangle about the axes. Automatically, the symmetrical constraint is taken by the system. This practice saves time for creating constraints.
- Two weak dimensions may be displayed if the weak dimension display is activated. Enter the dimension 500 to both fields by double-clicking the weak dimensions. Select OK.
- When creating any sketch, make sure there is no weak dimension. This ensures that sections or sketches are not overlooked.
- Once finished with the sketch, select the option To-selected from the drop-down menu beside the blind depth field.
- This will allow the selecting of any reference geometry to calculate the depth of the feature. Select DTM1; the extrusion depth of 500 mm will be displayed.
- Select the plane DTM2 instead of DTM1 in the To-selected field. The feature will be tapered at the extruded end. This will allow creation of a feature up to a surface with the end shape as allowed by the surface.
- The To-selected option is also used to create many features having depth to common reference.
- A most effective example of the To-selected option is described. Creation of this example needs knowledge of top–down assembly creation; however, it is shown here for practical application of the To-selected option.
- Create an assembly from the Find new menu and the template mmns_asm_design.
- The assembly will contain three planes, ASM_RIGHT, ASM_TOP, and ASM_FRONT.
- Rename the top plane as FFL_0m. This signifies the plane at a finished floor level at 0 m.
- Create another plane by offsetting the plane FFL_0m to 4500 mm. This signifies a level elevation of 4.5 m. Rename the plane as concrete_top_4500.
- Select Create a component in assembly mode, and create a part named column1.
- Select the option Locate default datums and click OK. Select the assembly planes ASM_Right, FFL_0m, and ASM_FRONT. The part column1 will be automatically activated; select Extrude and create a square of 500 × 500. Click OK. When prompted for depth, select the Up to select option and select the plane concrete_top_4500. Click OK to create the feature.
- Activate the main assembly by right-clicking on it and select Activate.
- Create another column, column2, using the same procedure. Make column2 a distance of 5000 mm from column1 and with dimensions of 1000 × 1000 mm.

- Create another column, column3, using the same procedure. Make column3 a distance of 10,000 mm from column1 and with dimensions of 2000 × 2000 mm.
- All the columns must be created with the To-selected option and by selecting the plane concrete_top_4500 plane.
- Now all the three columns are of the same height. Right-click on the plane concrete_top_4500 and select Edit. Change the value to 10,000 mm and click on Regenerate. All the columns will be regenerated to 10,000 mm length.

6.2.3 Revolve

Revolve is a method of generating a region by rotating a section around an axis. Any kind of axisymmetric section is suitable for a revolve operation. A solid shape generated from a revolve operation may also be generated from an extrude operation; however, an extrude operation may not be logical and may be time consuming when the section can be generated by revolving about a single axis. The choice is to be made logically through a manufacturing sequence. If the section is such that it is generated from such a rotating manufacturing procedure, the model should be made by the revolve feature.

Axis of revolution: The essential entity in the revolve feature is axis of rotation. Axis of rotation should be drawn or selected first when creating rotating sections.

Similar to the extrude feature, the revolve feature also has the same options like blind revolving through a user-entered angle with one or both directions or the To-selected option.

Hands-On Exercise, Example 6.3: Basic Revolve Option (Software Used: Creo Parametric)

Steps:

Assume a part file is created with three default planes, RIGHT, TOP, and FRONT.

- Select the Revolve feature from the ribbon at the top. Clicking on the feature opens the selection box for all other options.
- You may change the name of the feature to be created; it helps to trace back the feature in the future so that one easily can understand which feature is used where.
- Click on the Placement tab, and select Define sketch. A predefined sketch also can be selected in this box. However, as no predefined sketch is available now, click on Define.
- As Define is clicked, a small toolbox appears for placement of the sketch. Select the front plane for the sketch. The references for orientation (deciding orientation of sketch plane) may need to be selected if perpendicular planes of the sketch plane are

not available. Once the references are available, click on the Sketch button to define the sketch.

- The references to locate the sketch should have been automatically created; however, if the references are not taken, click on the Reference icon at the top left corner and select the references. Perpendicular planes automatically may be taken as references. However, axes, points, or other geometries also can be taken as references. For our case, select the right and top planes for references not being automatically taken.
- Select Axis from the sketch ribbon, and draw a vertical axis.
- Once the axes are drawn, select Rectangle from the sketch ribbon, and make the section.
- When creating any sketch, make sure there is no weak dimension. This ensures that sections or sketches are not overlooked.
- Once finished with the sketch, enter the value of 360 in the blind field. The flip option for changing the direction of extrusion is beside the blind depth value box.
- The mid plane option also can be selected via the drop-down menu beside Blind.
- Click OK to finally create the feature.
- Create a plane offset of the right plane by a value of 300 mm.
- Create an axis by selecting the front and newly created plane DTM1.
- Select the revolve feature in the model tree and drag it to the bottom after the newly created axis.
- This will make the axis available for selection when editing the revolve feature.
- Select the revolve feature, right-click, and select Edit definition.
- Click on the Placement tab and select the newly created axis in the axis selection field.
- The section will be now revolved about the newly created axis.
- Set a value for rotational degree, and click OK to create the feature.
- Save the file.

Hands-On Exercise, Example 6.3a: Revolving a Machining Section

Steps:

Assume a part file is created with three default planes, RIGHT, TOP, and FRONT.

- Select the Revolve feature from the ribbon at the top. Clicking on the feature opens the selection box for all other options.
- You may change the name of the feature to be created; it helps to trace back the feature in the future so that one easily can understand which feature is used where.
- Click on the Placement tab, and select Define sketch. A predefined sketch also can be selected in this box. However, as no predefined sketch is available now, click on Define.

- As Define is clicked, a small toolbox appears for placement of the sketch. Select the front plane for the sketch. The references for orientation (deciding orientation of sketch plane) may need to be selected if perpendicular planes of the sketch plane are not available. Once the references are available, click on the Sketch button to define the sketch.
- The references to locate the sketch should have been automatically created; however, if the references are not taken, click on the Reference icon at the top left corner and select the references. Perpendicular planes automatically may be taken as references. However, axes, points, or other geometries also can be taken as references. For our case, select the right and top planes for references not being automatically taken.
- Select Axis from the sketch ribbon, and draw a vertical axis.
- Create the section of a rectangle.
- When creating any sketch, make sure there is no weak dimension. This ensures that sections or sketches are not overlooked.
- Once finished with the sketch, enter the value of 360 in the blind field. The flip option for changing the direction of extrusion is beside the blind depth value box.
- The mid plane option also can be selected via the drop-down menu beside Blind.
- Click OK to finally create the feature.
- Save the file.

Hands-On Exercise, Example 6.4: Creating a Flywheel (Software Used: Creo Parametric; Refer to Figure 6.2)

Steps:

- Create a part file and rename it as Flywheel. (Other names may be given, but it is advisable to provide a relevant name.)
- Select Solid on the right-hand side of the box. This will activate the solid mode and environment.
- Uncheck the box of the default template and click OK to create a new solid file.
- The template selection box will appear.
- Select mmns_part_Solid. This will allow us to use the solid template file of the following units:
    - mm—millimeter for length unit
    - n—newton for force unit
    - s—second for time unit
- Once the solid part is created, go to File > Prepare > Model properties to check the units. In the upper sections, the units of the model will be displayed.
    - Note: The units usually are displayed in the summary section of any CAD software. The setting of units usually is in the tools or settings.
- The flywheel as shown in Figure 6.2 is to be created.

FIGURE 6.2
3-D flywheel created in CAD.

- Click on Revolve from the Model tab in the ribbon interface.
- Click on the Placement tab and select Define to define the sketch.
- Select the front plane as the sketch plane.
- The right plane shall be automatically taken as reference and orientation as right.
- Click on Sketch.
- The sketcher model will be activated.
- Draw the sketch as displayed in Figure 6.3.
- Click OK to create and exit the sketch.
- Ensure the revolve angle shall be 360°.
- Click on OK to create the revolve feature.
- Click on Extrude from the Model tab in the ribbon interface.
- Click on the Placement tab and select Define to define the sketch.

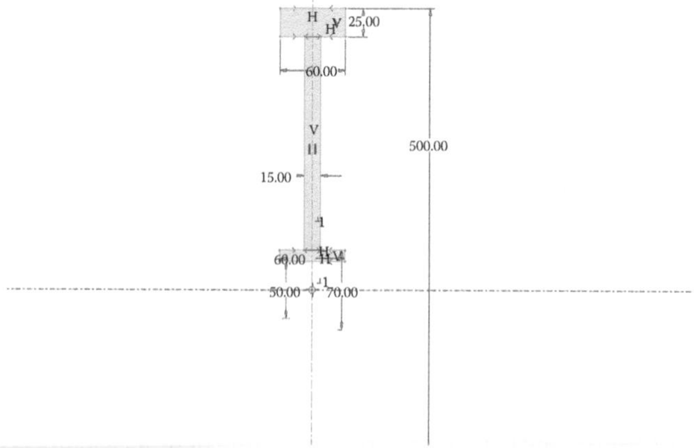

FIGURE 6.3
Constructional sketch of a flywheel.

- Select the middle flat surface as the sketch plane.
- The front plane shall be automatically taken as reference and orientation as right.
- Click on Sketch.
- The sketcher model will be activated.
- Draw the sketch as shown in Figure 6.4.
- Click OK to create and exit the sketch.
- Select the Through all option in the field of depth selection.
- Select the option Material removal feature.
- Flip the orientation toward the body of the revolved feature if required.
- Click on OK to create the extrusion cut.
- Click on Round feature from the Model tab of the ribbon interface.
- Set the value of fillet to 10 mm.
- Select all four internal edges of the newly created feature.
- Click OK to create the feature.
- Select the newly created extrusion cut and round from the model tree by pressing the Control key.
- Right-click and select Group to make the feature group.
- Select the newly created group in the model tree.
- Right-click and select Pattern.
- Select Axis in the Type of pattern selection field.
- Select the central axis.
- Enter the quantity 8 for patterned feature.
- Select the option to make the pattern equally spaced within 360°.
- Click OK to create the patterned feature.
- Click on the auto round feature from the Model tab of the ribbon interface.
- Set the value of fillet to 3 mm.
- Select the shat bore edges that are not required to be round.
- Click OK to create the feature.
- Click on Extrude from the Model tab in the ribbon interface.
- Click on the Placement tab and select Define to define the sketch.
- Select the flat shaft face as the sketch plane.
- The front plane shall be automatically taken as reference and orientation as right.
- Click on Sketch.
- The sketcher model will be activated.
- Draw a square of 5 × 5 mm placing the bottom corners at the top side of the inside circular edge.
- Click OK to create and exit the sketch.
- Select the Through all option in the field of depth selection.
- Select the option Material removal feature.
- Flip the orientation toward the body of the revolved feature if required.
- Click on OK to create the extrusion cut.
- Save and exit the file.

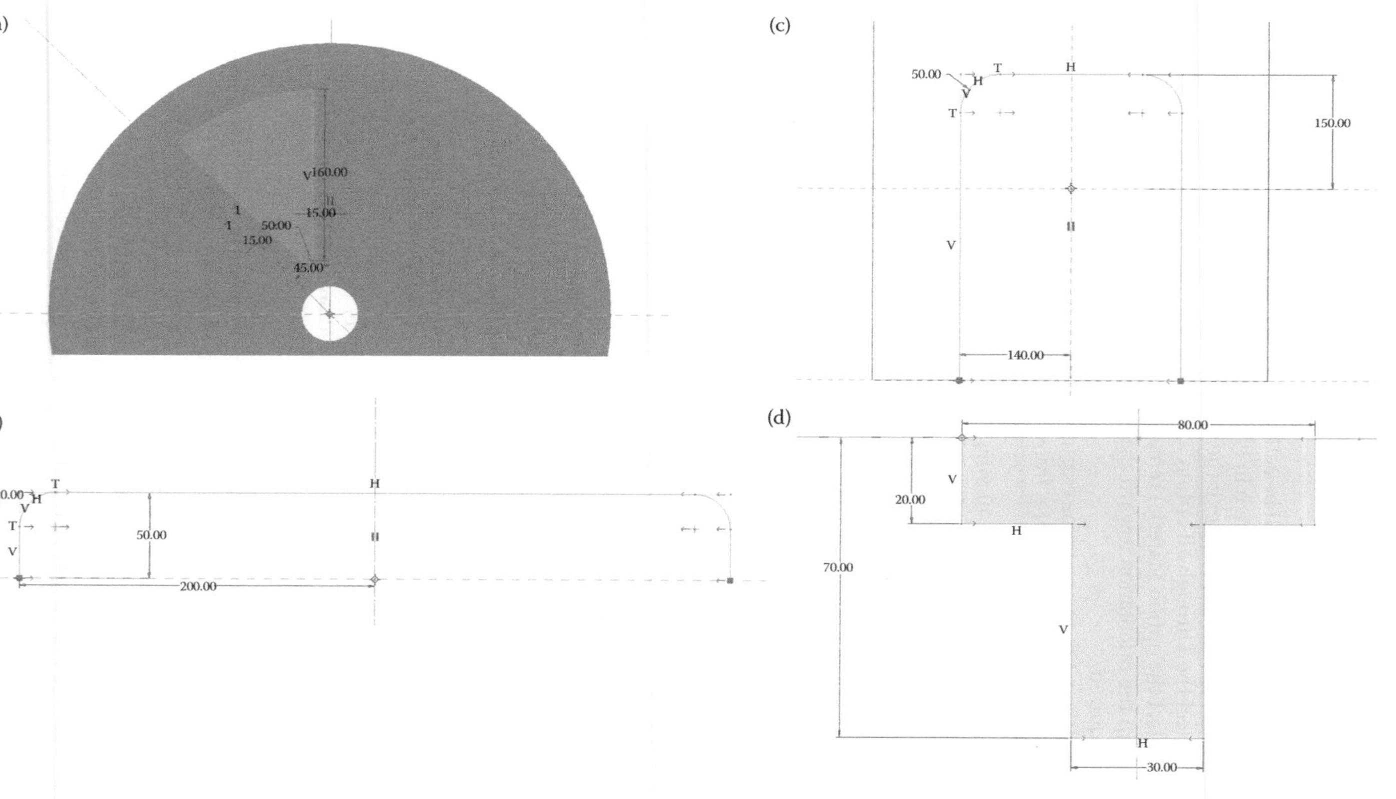

FIGURE 6.4
Constructional sketch of the material removal feature in a flywheel. (a) Constructional sketch of Fly wheel. (b) Constructional sketch of sweep trajectory of handle. (c) Constructional sketch of sweep trajectory of T-slot. (d) Constructional sketch of sweep profile of T-slot.

6.2.4 Sweep

Sweep is an operation of generating a region by moving a section through a curved path. Curve paths are called trajectories. A guide curve can also be straight or curved. A straight guide curve will result in a model shape similar to extrusion. A sweep can be of constant section or variable section.

Hands-On Exercise, Example 6.5: Creating a Handle and T-Slot

Steps:

Assume a part file is created with three default planes, RIGHT, TOP, and FRONT.

- Select the Sketch feature from the ribbon at the top. Clicking on the feature opens the selection box for all other options.
- You may change the name of the feature to be created; it helps to trace back the feature in the future so that one easily can understand which feature is used where.
- Click on the Placement tab, and select Define sketch. A predefined sketch also can be selected in this box. However, as no predefined sketch is available now, click on Define.
- As Define is clicked, a small toolbox appears for placement of the sketch. Select the front plane for the sketch. The references for orientation (deciding orientation of sketch plane) may need to be selected if perpendicular planes of the sketch plane are not available. Once the references are available, click on the Sketch button to define the sketch.
- The references to locate the sketch should have been automatically created; however, if the references are not taken, click on the Reference icon at the top left corner and select the references. Perpendicular planes automatically may be taken as references. However, axes, points, or other geometries also can be taken as references. For our case, select the right and top planes for references not being automatically taken.
- Select Axis from the sketch ribbon to keep the original sketch symmetrical to the axis. Draw one vertical and one horizontal axis along the references of the right and front planes.
- Create the sketch as displayed in Figure 6.4a.
- When creating any sketch, make sure there is no weak dimension. This ensures that sections or sketches are not overlooked.
- Click OK.
- Select the Sweep feature from the ribbon at the top. Clicking on the feature opens the selection box for all other options.
- Select the previously created sketch as a trajectory in the Reference tab.
- Select the Sketch button in the ribbon interface.
- Draw a circle of diameter 10 mm placing the center of the circle at the cross section of the references.
- Click OK to finally create the feature.
- Save the file.

T-SLOT PROFILE

Assume a part file is created with three default planes, RIGHT, TOP, and FRONT.

- Select the Extrude feature from the ribbon at the top. Clicking on the feature opens the selection box for all other options.
- You may change the name of the feature to be created; it helps to trace back the feature in the future so that one easily can understand which feature is used where.
- Click on the Placement tab, and select Define sketch. A predefined sketch also can be selected in this box. However, as no predefined sketch is available now, click on Define.
- As Define is clicked, a small toolbox appears for placement of the sketch. Select the top plane for the sketch. The references for orientation (deciding orientation of sketch plane) may need to be selected if perpendicular planes of the sketch plane are not available. Once the references are available, click on the Sketch button to define the sketch.
- The references to locate the sketch should have been automatically created; however, if the references are not taken, click on the Reference icon at the top left corner and select the references. Perpendicular planes automatically may be taken as references. However, axes, points, or other geometries also can be taken as references. For our case, select the right and front planes for references not being automatically taken.
- Select Axis from the sketch ribbon to keep the original sketch symmetrical to the axis. Draw one vertical and one horizontal axis along the references of the right and front planes.
- Once the axes are drawn, select Rectangle from the sketch ribbon, and click on the top left quadrant to locate the left corner. Then click again at the right bottom quadrant to an approximate distance such that it can form a symmetrical rectangle with respect to the axes. Automatically, the symmetrical constraint is taken by the system. This practice saves time for creating constraints.
- Two weak dimensions may be displayed if the weak dimension display is activated. Enter the dimension 500 to both fields by double-clicking the weak dimensions. Select OK.
- When creating any sketch, make sure there is no weak dimension. This ensures that sections or sketches are not overlooked.
- Once finished with the sketch, enter the value of 200 in the blind field. The flip option for changing the direction of extrusion is beside the blind depth value box.
- The mid plane option also can be selected via the drop-down menu beside Blind.
- The top plane will act as a symmetrical mid plane if the mid plane option is selected.
- Click OK to finally create the feature.
- Select the Sketch feature from the ribbon at the top. Clicking on the feature opens the selection box for all other options.

- You may change the name of the feature to be created; it helps to trace back the feature in the future so that one easily can understand which feature is used where.
- Click on the Placement tab, and select Define sketch. A predefined sketch also can be selected in this box. However, as no predefined sketch is available now, click on Define.
- As Define is clicked, a small toolbox appears for placement of the sketch. Select the top surface of the blind feature for the sketch. The references for orientation (deciding orientation of sketch plane) may need to be selected if perpendicular planes of the sketch plane are not available. Once the references are available, click on the Sketch button to define the sketch.
- The references to locate the sketch should have been automatically created; however, if the references are not taken, click on the Reference icon at the top left corner and select the references. Perpendicular planes automatically may be taken as references. However, axes, points, or other geometries also can be taken as references.
- Select Axis from the sketch ribbon to keep the original sketch symmetrical to the axis. Draw one vertical and one horizontal axis along the references.
- Create the sketch as displayed in Figure 6.4b.
- When creating any sketch, make sure there is no weak dimension. This ensures that sections or sketches are not overlooked.
- Click OK.
- Select the Sweep feature from the ribbon at the top. Clicking on the feature opens the selection box for all other options.
- Select the previously created sketch as a trajectory in the Reference tab.
- Select the Sketch button in the ribbon interface.
- Draw the T profile as per Figure 6.4c.
- Click OK to finally create the feature.
- Save the file.

Hands-On Exercise, Example 6.6: Creating a Helical Sweep and Spring

A helical sweep is sweeping a cross section along a helix or helical curve. The helical curve can be created separately or as part of a helical sweep as desired by the user. Helical sweeps are very useful to create helical profiles like springs, threads, and so forth.

- Select the Sweep feature from the ribbon at the top. Clicking on the feature opens the selection box for all other options.
- You may change the name of the feature to be created; it helps to trace back the feature in the future so that one easily can understand which feature is used where.
- Click on the References tab, and select Define helix profile. A predefined sketch also can be selected in this box. However, as no predefined sketch is available now, click on Define.

- As Define is clicked, a small toolbox appears for placement of the sketch. Select the front plane for the sketch. The references for orientation (deciding orientation of sketch plane) may need to be selected if perpendicular planes of the sketch plane are not available. Once the references are available, click on the Sketch button to define the sketch.
- Select Axis from the Sketch toolbar and create a vertical axis in the vertical reference line, which will act as the axis of the spring.
- Select Line and create a vertical line of length 500 and 100 offset from the vertical centerline starting from horizontal centerline upward.
- When creating any sketch, make sure there is no weak dimension. This ensures that sections or sketches are not overlooked.
- Click OK.
- Select the sketch from the helical sweep interface ribbon.
- As the sketch interface is reopened, draw a circle of diameter 5 mm, placing the center at the bottom of the sketched line.
- Click OK to finish the sketch.
- Set the pitch value to 30 mm.
- Click OK to finally create the feature.
- Save the file.

6.2.5 Loft/Blend

The loft or blend feature is used to create a shape by multiple cross sections with or without guide curves. It uses include creating transitions between dissimilar profiles.

Hands-On Exercise, Example 6.7: Creation of Rectangle-To-Round Transition (Software Used: Creo Parametric)

Steps:

Assume a part file is created with three default planes, RIGHT, TOP, and FRONT.

- Select the Sketch feature from the ribbon at the top. Clicking on the feature opens the selection box for all other options.
- You may change the name of the feature to be created; it helps to trace back the feature in the future so that one easily can understand which feature is used where.
- Click on the Placement tab, and select Define sketch. A predefined sketch also can be selected in this box. However, as no predefined sketch is available now, click on Define.
- As Define is clicked, a small toolbox appears for placement of the sketch. Select the top plane for the sketch. The references for orientation (deciding orientation of sketch plane) may need to

be selected if perpendicular planes of the sketch plane are not available. Once the references are available, click on the Sketch button to define the sketch.

- The references to locate the sketch should have been automatically created; however, if the references are not taken, click on the Reference icon at the top left corner and select the references. Perpendicular planes automatically may be taken as references. However, axes, points, or other geometries also can be taken as references. For our case, select the right and front planes for references not being automatically taken.
- Select Axis from the sketch ribbon to keep the original sketch symmetrical to the axis. Draw one vertical and one horizontal axis along the references of the right and front planes.
- Once the axes are drawn, select Rectangle from the sketch ribbon, and click on the top left quadrant to locate the left corner. Then click again at the right bottom quadrant to an approximate distance such that it can form a symmetrical rectangle with respect to the axes. Automatically, the symmetrical constraint is taken by the system. This practice saves time for creating constraints.
- Two weak dimensions may be displayed if the weak dimension display is activated. Enter the dimension 500 to both fields by double-clicking the weak dimensions. Select OK.
- Select Plane from the ribbon interface and create a plane at an offset of 400 mm from the top plane.
- Select Sketch from the ribbon and set the newly created plane DTM1 as the sketch plane.
- Select References and select the four corners of the rectangle to get the vertex points as references.
- Select Axis and create two axes diagonally passing through the points.
- Select Circle and create a circle of diameter 300 mm placing the center at the cross section of two axes or centerlines.
- Select the Divide option from the sketch editing ribbon, and place the break point at all the points where the centerlines cross the circle. This divides the circle into four parts.
- Click OK to create the sketch.
- Select the feature Blend from the ribbon.
- Select the Sections tab and select the Selected sketch option.
- Select the rectangle and the circle with a click insert in between.
- Click OK to create the feature.
- Select Shell command from the ribbon and set the thickness to 8 mm.
- Click on the Reference tab and select the circular and rectangular surface to remove.
- Click OK to create the feature.
- Select the Blend feature from the model tree and select Edit by right-clicking on it.

- Change the dimension of any of the rectangle's hands from 500 to 900.
- Click on Regenerate.

Hands-On Exercise, Example 6.8: Creation of An Eccentric Reducer (Software Used: Creo Parametric)

Eccentric reducers are often required to be created in engineering applications.

Steps:

Assume a part file is created with three default planes, RIGHT, TOP, and FRONT.

- Select the Blend feature from the ribbon at the top. Clicking on the feature opens the selection box for all other options.
- You may change the name of the feature to be created; it helps to trace back the feature in the future so that one easily can understand which feature is used where.
- Click on the References tab and select Sketched feature.
- Click on Define, select the sketch plane, and click On sketch.
- The horizontal and vertical references are to be taken automatically by the system.
- Select Circle and place the circle center somewhere above the origin point and in the vertical axis.
- Click on the constraint tangent and make the circle tangent to the horizontal reference line.
- Make the diameter of the circle 1000 mm.
- Click OK to finish the sketch.
- Click Insert to sketch another section and provide 950 as the offset from the secion1 option (offset dimension option).
- Click on Sketch to sketch another circle.
- The horizontal and vertical references are to be taken automatically by the system.
- Select Circle and place the circle center somewhere above the origin point and in the vertical axis.
- Click on the constraint tangent and make the circle tangent to the horizontal reference line.
- Make the diameter of the circle 750 mm.
- Click OK to finish the sketch.
- Click on the thin feature option and set the value of 10 mm as the thickness.
- Click OK to create the feature and save the file.

6.2.6 Draft Creation

A draft is mostly required in preparation of CAD models for casting applications. The vertical surfaces need to be tapered to facilitate removal of the component from the mold. The final component is made first without any draft on the vertical surfaces. The draft is then manually added on every surface that needs to be tapered. Draft creation principally requires the following inputs or selections:

- Vertical surfaces that need to be tapered
- A hinge surface from which the angular draft creation will be started
- Angular value for the draft

6.2.7 Cosmetic Features

Cosmetic features are used when a symbolic representation of any feature with complex geometry needs to shown. For example, true representation or creation of thread geometry may not be required in the case of a large part containing several holes. However, it is required to indicate the thread of screw holes on the part. A simple dotted line offsetting the hole edges would be sufficient to show in the drawing. If the full thread profile is created on every hole, unnecessarily, the mode creation time will increase. This would also need sufficient memory, which can be a constant.

Cosmetic features are very useful to represent complex and time-consuming geometry by symbolic representation. A thread is commonly used to show a symbolic representation by cosmetic feature.

Hands-On Exercise, Example 6.9: Cosmetic Feature—Internal Thread (Software Used: Creo Parametric; Refer to Figure 6.5)

Steps:

- Create a part file and rename it as Base.
- Select Solid on the right-hand side of the box. This will activate the solid mode and environment.
- Uncheck the box of the default template and click OK to create a new solid file.
- The template selection box will appear.

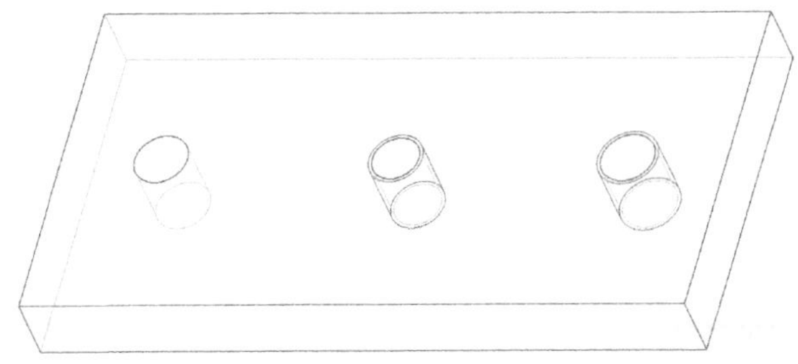

FIGURE 6.5
Example of a cosmetic thread in a hole surface.

- Select mmns_part_Solid. This will allow us to use the solid template file of the following units:
  - mm—millimeter for length unit
  - n—newton for force unit
  - s—second for time unit
- Once the solid part is created, go to File > Prepare > Model properties to check the units. In the upper sections, the units of the model will be displayed.
  - Note: The units usually are displayed in the summary section of any CAD software. The setting of units usually is in the tools or settings.
- The model as displayed in Figure 6.5 will be created.
- Click on Extrude from the Model tab in the ribbon interface.
- Click on the Placement tab and select Define to define the sketch.
- Select the top plane as the sketch plane.
- The right plane shall be automatically taken as reference and orientation as right.
- Click on Sketch.
- The sketcher model will be activated.
- Click on drop-down menu beside the rectangle and select Center rectangle.
- Place the rectangle center point in the intersection of the horizontal and vertical references.
- Draw a rectangle of dimensions 100 × 50 mm.
- Click OK to create and exit the sketch.
- Select mid plane extrusion.
- Enter the value 20 mm for extrusion depth.
- Click on OK to create the extrusion.
- Click on the feature hole from the Model tab of the ribbon interface.
- Change the diameter of the hole to 16 mm.
- Click on the drop-down menu for depth selection and select Through all to create the hole throughout the available material.
- Click on the Placement tab.
- Select the top flat face of the extrusion as the placement surface.
- Set the reference type to linear.
- Click on Offset reference area.
- Select the front plane and left perpendicular surface of the hole placement surface by pressing the Control key.
- Set the offset value with the front plane to 0 mm or make it align.
- Enter the value of 30 mm in the offset field with the selected surface.
- Click OK to create the hole.
- Click on the feature hole from the Model tab of the ribbon interface.
- Change the hole type to standard hole.
- Select ISO—M16X2 hole size.
- Click on the drop-down menu for depth selection and select Through all to create the hole throughout the available material.

- Click on the Shape tab.
- Set the thread to Through all.
- Click on the Placement tab.
- Select the top flat face of the extrusion as the placement surface.
- Set the reference type to linear.
- Click on Offset reference area.
- Select the front plane and right plane by pressing the Control key.
- Set the offset value with the front plane to 0 mm or make it align.
- Set the offset value with the top plane to 0 mm or make it align.
- Click OK to create the hole.
- Click on the feature hole from the Model tab of the ribbon interface.
- Change the diameter of the hole to 16 mm.
- Click on the drop-down menu for depth selection and select Through all to create the hole throughout the available material.
- Click on the Placement tab.
- Select the top flat face of the extrusion as the placement surface.
- Set the reference type to linear.
- Click on Offset reference area.
- Select the front plane and right perpendicular surface of the hole placement surface by pressing the Control key.
- Set the offset value with the front plane to 0 mm or make it align.
- Enter the value of 30 mm in the offset field with the selected surface.
- Click OK to create the hole.
- Now click on the engineering drop-down menu from the Model tab of the ribbon interface.
- Click on Standard thread to ensure the standard available thread in the hole surface.
- Click on the Placement tab and select Inner surface of lastly created hole.
- Click on the Depth tab and select the top surface as the start thread surface.
- Click on the drop-down menu of the Extrusion depth option.
- Click to select an option and select the bottom surface of the extrude feature.
- Click OK to create the cosmetic thread.
- Click on Front view from the icon at the top of the graphical area.
- Set the presentation to Hidden line from the top of the graphical area.
- It can be seen that the simple hole does not include any thread surface.
- The second hole has a purple dotted line around the hole indicating the thread representation.
- In the same manner, the third hole, which is a simple hole, is displayed with the indicated thread representation due to the cosmetic thread feature as displayed in Figure 6.5.
- Save and exit the file.

Hands-On Exercise, Example 6.10: Creating a Plug (Software Used: Creo Parametric; Refer to Figure 6.6)

Steps:

- Create a part file and rename it as Plug. (Other names may be given, but it is advisable to provide a relevant name.)
- Select Solid on the right-hand side of the box. This will activate the solid mode and environment.
- Uncheck the box of the default template and click OK to create a new solid file.
- The template selection box will appear.
- Select mmns_part_Solid. This will allow us to use the solid template file of the following units:
    - mm—millimeter for length unit
    - n—newton for force unit
    - s—second for time unit
- Once the solid part is created, go to File > Prepare > Model properties to check the units. In the upper sections, the units of the model will be displayed.
    - Note: The units usually are displayed in the summary section of any CAD software. The setting of units usually is in the tools or settings.
- The plug as shown in Figure 6.6 is to be created.
- Click on Extrude from the Model tab in the ribbon interface.
- Click on the Placement tab and select Define to define the sketch.
- Select the front plane as the sketch plane.
- The right plane shall be automatically taken as reference and orientation as right.
- Click on Sketch.
- The sketcher model will be activated.
- Draw a hexagon circumscribed on a circle of diameter 42 mm.
- Right-click on the circle and select Construction to make the circle a construction entity.
- Click OK to create and exit the sketch.
- Flip the material orientation upward.
- Enter the value 15 mm for extrusion depth.

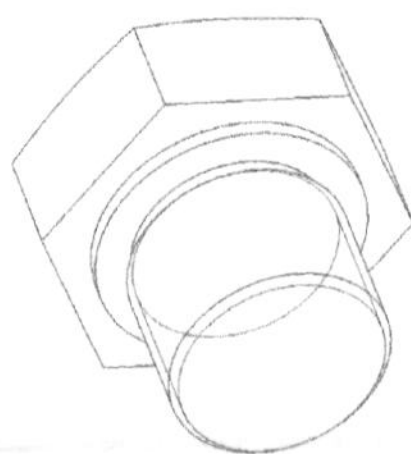

FIGURE 6.6
Example of a plug with a cosmetic thread.

- Click on OK to create the extrusion.
- Click on Revolve from the Model tab in the ribbon interface.
- Click on the Placement tab and select Define to define the sketch.
- Select the front plane as the sketch plane.
- The right plane shall be automatically taken as reference and orientation as right.
- Click on Sketch.
- The sketcher model will be activated.
- Draw the sketch as displayed in Figure 6.7.
- Click OK to create and exit the sketch.
- Ensure the revolve angle shall be 360°.
- Click on OK to create the revolve feature.
- Click on Revolve again from the Model tab in the ribbon interface.
- Click on the Placement tab and select Define to define the sketch.
- Select the right plane as the sketch plane.
- The top plane shall be automatically taken as reference and orientation as left.
- Click on Sketch.
- The sketcher model will be activated.
- Draw the sketch as displayed in Figure 6.8.
- Click OK to create and exit the sketch.
- Ensure the revolve angle shall be 360°.
- Click to activate the material removal option.
- Click on OK to create the revolve feature.
- Now click on Cosmetic thread from the engineering drop-down menu from the Model tab of the ribbon interface.
- Click on Standard thread and Unified National Thread (UNF) 1 1/8.
- Click on the Placement tab and select Major diameter outer surface of revolved feature.

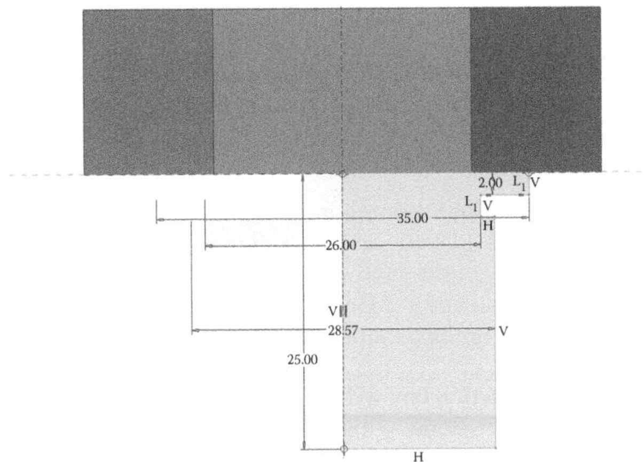

FIGURE 6.7
Constructional sketch of a plug.

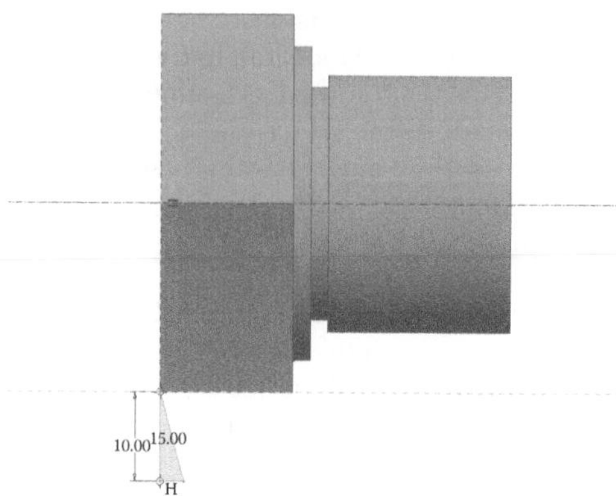

FIGURE 6.8
Constructional sketch of the head of a plug.

- Click on the Depth tab and select the top surface as the start thread surface.
- Select the flat surface of the groove as the end surface.
- Click OK to create the cosmetic thread.
- Save and exit the file.

6.2.8 Pattern

Pattern is an operation to create several instances of any feature with incremental values. The incremental values can be the same or different. If the incremental values are different, table patterns need to be used; these may be available only in advanced CAD software like Creo.

Hands-On Exercise, Example 6.11: Creating a Fan Screen (Software Used: Creo Parametric)

Steps:

- Create a part file and rename it as Fan screen. (Other names may be given, but it is advisable to provide a relevant name.)
- Select Solid on the right-hand side of the box. This will activate the solid mode and environment.
- Uncheck the box of the default template and click OK to create a new solid file.
- The template selection box will appear.
- Select mmns_part_Solid. This will allow us to use the solid template file of the following units:
    - mm—millimeter for length unit
    - n—newton for force unit
    - s—second for time unit

- Once the solid part is created, go to File > Prepare > Model properties to check the units. In the upper sections, the units of the model will be displayed.
  - Note: The units usually are displayed in the summary section of any CAD software. The setting of units usually is in the tools or settings.
- The fan screen as shown in Figure 6.9 is to be created.
- Click on Revolve from the Model tab in the ribbon interface.
- Click on the Placement tab and select Define to define the sketch.
- Select the front plane as the sketch plane.
- The right plane shall be automatically taken as reference and orientation as right.
- Click on Sketch.
- The sketcher model will be activated.
- Draw the sketch as displayed in Figure 6.10.
- Click OK to create and exit the sketch.
- Ensure the revolve angle shall be 360°.

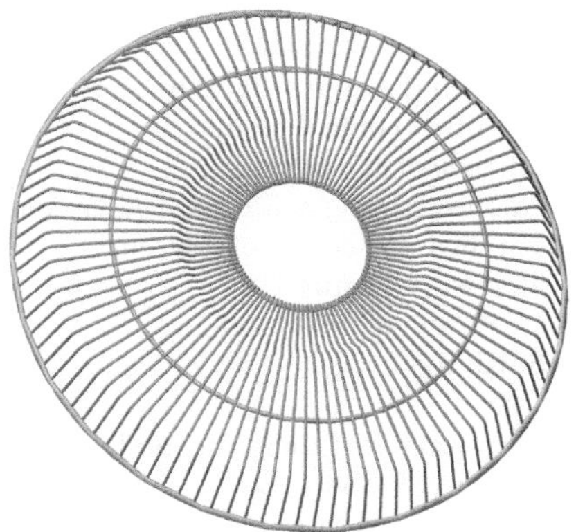

FIGURE 6.9
(See color insert.) 3-D model of a fan screen.

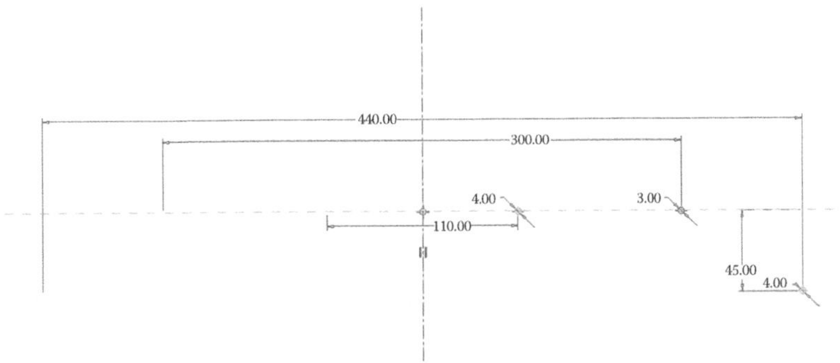

FIGURE 6.10
Constructional sketch of a fan screen model.

- Click on OK to create the revolve feature.
- Select Sketch from the Model tab of the ribbon interface.
- A sketch window for selection of sketch plane will appear.
- Select the front plane as the sketch plane.
- The right plane shall automatically be selected as reference and the orientation is to be right.
- Click on Sketch.
- The sketcher window will open up and the reference selection window appears for selection of references.
- The two references (i.e., top plane as horizontal reference and right plane as vertical reference) shall be automatically taken; however, if not, please select the same.
- Create a sketch as displayed in Figure 6.11.
- Click OK to create the sketch.
- Click on Sweep from the Model tab of the ribbon interface.
- Select the lastly created sketch as a trajectory.
- Click on Sweep from the Model tab of the ribbon interface.
- Graphically select the sketch as a trajectory.
- The system assumes the profile sketch point and shows it by displaying the directional arrow.
- Graphically click on the arrows of feature creation to draw the sweep profile in other control points.
- Click on Sketch to sketch the profile.
- The sketcher mode activates with the sketcher environment.
- Draw a circle at the intersection of references.
- Change the diameter to 2 mm.
- Click OK to create and exit the sketch.
- Click OK to finish the sweep.

- Select the newly sweep feature in the model tree.
- Right-click and select Pattern.
- Select Axis in the Type of pattern selection field.
- Select the central axis.
- Enter the quantity 100 for patterned feature.

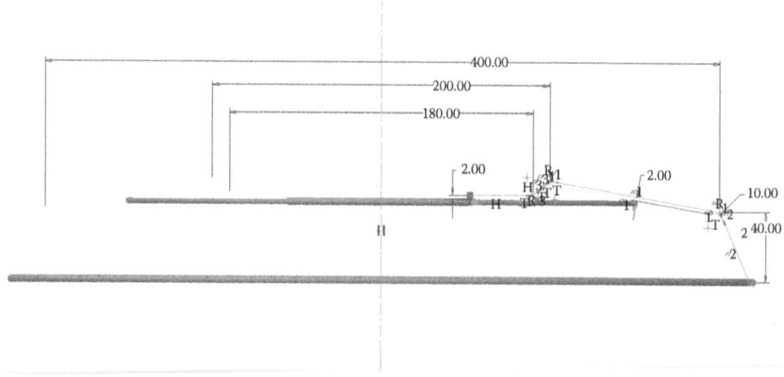

FIGURE 6.11
Constructional sketch of a fan screen model.

- Select the option to make the pattern equally spaced within 360°.
- Click OK to create the patterned feature.
- Click save and exit the file.

Hands-On Exercise, Example 6.12: Creating a Screen (Software Used: Creo Parametric; Figure 6.12)

Steps:

- Create a part file and rename it as Screen. (Other names may be given, but it is advisable to provide a relevant name.)
- Select Solid on the right-hand side of the box. This will activate the solid mode and environment.
- Uncheck the box of the default template and click OK to create a new solid file.
- The template selection box will appear.
- Select mmns_part_Solid. This will allow us to use the solid template file of the following units:
    - mm—millimeter for length unit
    - n—newton for force unit
    - s—second for time unit
- Once the solid part is created, go to File > Prepare > Model properties to check the units. In the upper sections, the units of the model will be displayed.
    - Note: The units usually are displayed in the summary section of any CAD software. The setting of units usually is in the tools or settings.
- The screen as shown in Figure 6.12 is to be created.
- Click on Revolve from the Model tab in the ribbon interface.
- Set the thicken option.
- Enter the thickness 1 mm.
- Click on the Placement tab and select Define to define the sketch.
- Select the front plane as the sketch plane.
- The right plane shall be automatically taken as reference and orientation as right.
- Click on Sketch.
- The sketcher model will be activated.

FIGURE 6.12
3-D model of a screen.

- Draw the sketch as displayed in Figure 6.13.
- Click OK to create and exit the sketch.
- Ensure the revolve angle shall be 360°.
- Click on OK to create the revolve feature.

- Select Sketch from the Model tab of the ribbon interface.
- A sketch window for selection of sketch plane will appear.
- Select internal bottom surface as sketch plane.
- The right plane shall automatically be selected as reference and the orientation is to be right.
- Click on Sketch.
- The sketcher window will open up and the reference selection window appears for selection of references.
- The two references (i.e., top plane as horizontal reference and right plane as vertical reference) shall be automatically taken; however, if not, please select the same.
- Project the internal circular edge.
- Click OK to create and exit the sketch.
- Select Hole from the Model tab of the ribbon interface.
- Enter 5 as the diameter of the hole.
- Click the Through all option to make the hole throughout the available material.
- Click on the Placement tab and select the central axis and bottom inside surface area by pressing the Control key.
- Click OK to create the hole feature.

- Select the newly created hole feature in the model tree.
- Right-click and select Pattern.
- Select Fill in the Type of pattern selection field.
- Select the inside sketch.
- Select the hexagon pattern.
- Enter 9 mm in the field of spacing of patterned object.
- Click OK to create the pattern.
- Save and exit the file.

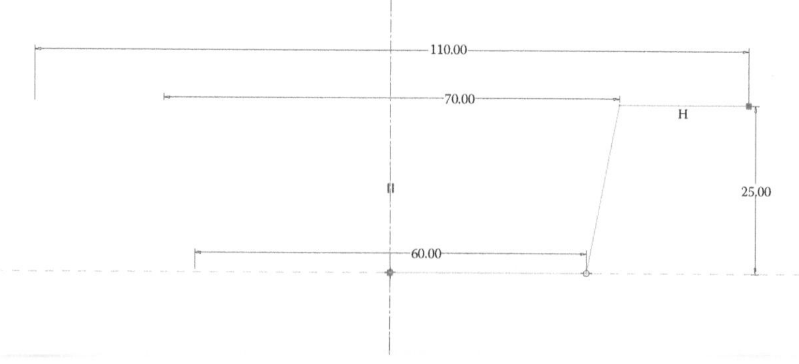

FIGURE 6.13
Constructional sketch of a screen.

6.3 Surface

Typically, a surface can be defined as the skin of any solid having null thickness. A surface cannot have any mass and mass-related properties. It can only have surface area. There is no existence of an independent surface without a solid in the real world. However, very thin sheets often can be modeled as surfaces to reduce the complexity of creating thin solids. When a solid model is created, surfaces are created automatically associated with a solid. The shapes of the surfaces of the solids are controlled by the features used in building the solid. It is obvious that if there is such a requirement of making any complex surface, there will be a requirement of controlling the surface directly. Thus, surface modeling is necessary where the boundary surfaces are modeled first and the associated solids are then created by filling the boundaries of surfaces.

Typically, the surfaces are represented in parametric or nonparametric form. The equation of a surface is given by

$$P = [xyz]^T.$$

The parametric representation includes vector function $P(u, v)$ of two parameters u and v.

$$P(u, v) = [xyz]^T = [x(u, v)\ y(u, v)\ z(u, v)]^T$$

$$u_{min} < u < u_{max},\ v_{min} < v < v_{max}$$

1. *Plane surface*: Plane surfaces are the simplest in surface geometry creation. When a line is extruded in a constant vector, a plane surface is generated. If a plane surface is parallel to the XY plane, the Z is constant. A plane surface is shown in Figure 6.14.
2. *Ruled surface*: This kind of surface can be created by selecting a boundary profile lying on a single plane. The boundaries of the closed sketch can be any curve, and the surface interpolates linearly between the curves. Ruled surfaces are linear in one direction and connected to boundary curves. Rules surfaces are developable surfaces. Cones, cylinders, and symmetrical transitions are examples of ruled surfaces. A ruled surface is shown in Figure 6.15.
3. *Surface of revolution*: These surfaces are generated by revolution of any curve about an axis. A surface generated by revolution of a curve is shown in Figure 6.16.

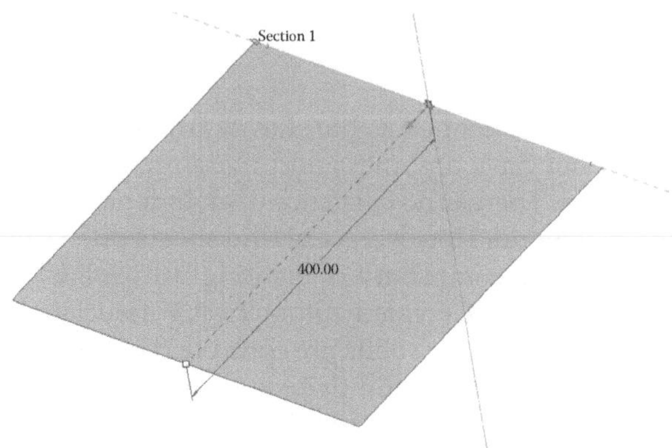

FIGURE 6.14
Example of a plane surface.

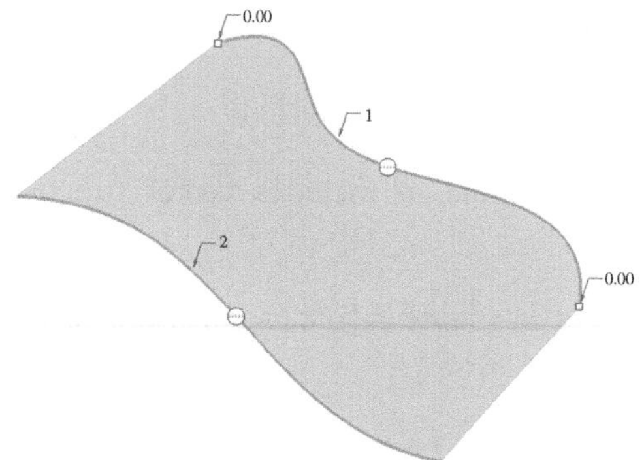

FIGURE 6.15
Example of ruled surface.

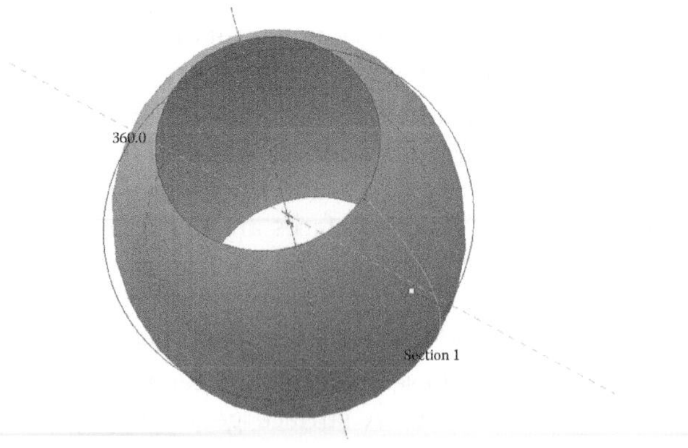

FIGURE 6.16
Example of a surface created by revolution of a curve.

4. *Synthetic surfaces*: The synthetic surfaces are created by approximations or interpolations given input data. The synthetic surfaces can allow only global control, like a Bezier surface, or can also permit local control, like B-spline surfaces. These surfaces are mostly used where very accurate controls on surfaces are required, for example, blade surfaces in critical fan, where the designs of these surfaces are determined by thermal, aerodynamic, and fluid flow simulations.

6.4 Properties and Their Classification

Every feature and model itself possesses some properties. Basically, these properties are the information associated with a respective feature, object, or entity. Some of the information drives the object, whereas the rest are driven out from the object. Parametric software provides facilities and options to change its various properties, which can actually control the object associated with it. For example, a dimension can be a property of a hole feature, and by changing the dimension, the size of the hole can be varied, and thus, mass properties of the model also vary. Therefore, dimension is the property that derives the feature, whereas mass property is driven out from the model.

Thus, properties can be broadly classified as two categories driving properties and derived properties.

1. *Driving properties*: Driving properties can be defined as the properties of an object that preserve the various other properties of the object and by changing which the model is affected. Most geometric dimensions are main driving parameters. Varying a dimension definitely results in varying the shape of the associated feature. Therefore, it is always required predict the desired effect in varying the driving properties.
2. *Derived properties*: Derived properties are those that are preserved by the object and a direct change of which is not possible. Geometrical properties like perimeter, area, and so forth are derived properties. Direct change of this perimeter is normally not possible in CAD.

   Example: Draw a circle of diameter 100 mm as shown in Figure 6.17. If the diameter gets changed to any value, the perimeter will automatically get changed. The diameter of the circle is a driving property here, whereas the perimeter is the derived geometrical property.
3. *Mass properties*: By definition, mass is the amount of matter of any object, and the properties of the matter are called mass properties. The mass properties are calculated from the shape of the model

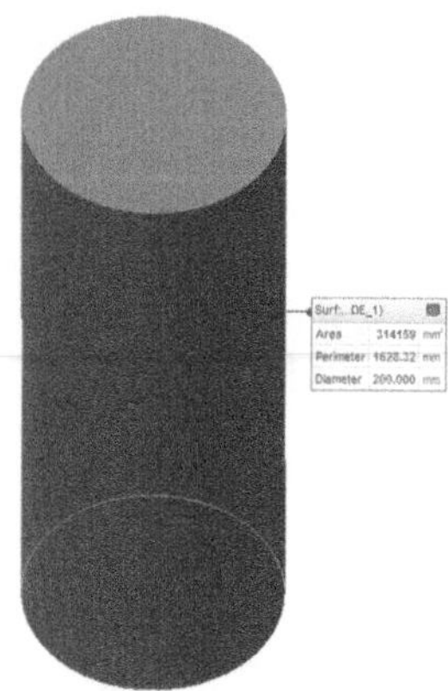

FIGURE 6.17
Example of derived properties.

and its associated material properties. The commonly derived mass properties are as follows:

a. *Volume*: Volume is a property that is calculated automatically by the CAD system when a solid is created. The dimension of volume is $(\text{length})^3$. Hence, the unit of volume depends on the set unit of the length. The length unit can be set in millimeters, meters, inches, and so forth. Length unit conversion factors are shown in Table 6.1.

   Volume is a derived parameter. It cannot be controlled directly in CAD.

b. *Mass*: Mass is the amount of matter contained in a body or object. Mass is an important parameter in CAD that is often required to be measured when designing any component. Mathematically, mass = volume × density. Density in CAD is to be taken from the material properties in CAD. When working with big assemblies,

TABLE 6.1
Length Unit Conversion Factor

	1 km	1 m	1 cm	1 mm	1 μm	1 in.	1 ft.	1 yd.
1 km	1	1×10^3	1×10^5	1×10^6	1×10^9	39,370.08	3280.84	1093.61
1 m	0.001	1	100	1×10^3	1×10^6	39.37	3.28	1.094
1 cm	1×10^{-5}	0.01	1	10	1×10^4	0.3937	0.0328	0.010936
1 mm	1×10^{-6}	1×10^{-3}	0.1	1	1×10^3	0.03937	0.00328	0.0010936
1 μm	1×10^{-9}	1×10^{-6}	1×10^{-4}	1×10^{-3}	1			
1 in.						1	0.08333	0.02777
1 ft.						12	1	0.333
1 yd.						36	3	1

it may not always be possible to apply the same material in every part. However, if the total assembly is made by materials of the same density, we can easily get the mass by getting the volume from CAD and multiplying it with density manually. Mass is a derived parameter, which cannot be controlled directly in CAD. Dimension or density alteration is required to change the mass.

The most used units of mass in most of the CAD software are as follows:

i. *Kilogram* (*kg*): It is the System International (SI) or meter-kilogram-second (MKS) unit of mass.
ii. *Gram* (*g*): It is the centimeter-gram-second (CGS) unit of mass.
iii. *Pound* (*lb.*): It is the foot-pound-second (FPS) unit of mass.
iv. *Slug*: It is the unit of mass expressed commonly in imperial system.
v. *Tonne*: It is the unit in the metric system for a bigger quantity of mass.

c. *Weight*: It is the amount of force that is experienced by the body due to the gravitational force of the Earth. Mathematically, weight = mass × gravitational force. The weight is mostly required in cases of static or dynamic analysis. If the density and gravitational force are set properly in CAD, accurate weight can be obtained very easily. Weight is also a derived parameter; it depends on mass and gravitational force.
d. *Density*: Density is a secondary unit, which can be derived from the set unit of mass and length, that is, if the set unit of mass in CAD is kilograms and length is meters, then the unit of density would be kg/m^3. The density in a material is a driving parameter, which in turn decides the mass of the material of particular volume.
e. *Surface area*: Surface area is the area of the part exposed to the air. Surface area is also dependent on the unit of length.
f. *Cross-sectional area*: Cross-sectional area is the area of a part that is cut by a particular plane.
g. *Center of gravity*: Center of gravity can be imagined as the point through which the weight of the whole body acts. For a part of one material, the center of gravity depends on the shape and size of the component and its features. The center of gravity is measured in a linear dimensional unit from a reference point in three axes (x, y, and z). The center of gravity is often required to be determined, since it is the key deciding factor to locate the lifting points if required for the part.
h. *Moment of inertia*: The unit of moment of inertia is dependent on the unit of length.

i. *Section modulus.*
j. *Radius of gyration*: Mathematically the radius of gyration is

$$K = \sqrt{\frac{I}{M}}$$

where

I = moment of inertia

M = mass of the body

4. *Aesthetic properties*: Aesthetic or graphical properties are the properties that control the display of the CAD model. Aesthetic properties are required due to the following reasons:
   a. To make a proper presentation of the model
   b. To differentiate the parts and surfaces
   c. To make the decal of the various surfaces
5. *Material properties*: Most mass properties like density, mass, weight, and so forth are dependent on material properties.
6. *User-driven properties*: Most advanced parametric CAD software enables facilities to link the user-driven properties with driving properties. In other words, the user-driven properties control the driving properties; thus, the model varies with user-driven properties.

6.5 Family Tables

Geometrically similarly shaped products with varying dimension values or properties are created in a family table. A family table defines the creation of a range of products from the same family with varying dimension values or properties entered in tabular form. The product properties that can be varied are as follows:

1. *Dimensions*: Family tables are mostly used for the range of components with dimension variations.

Hands-On Exercise, Example 6.13: Creation of a Blind Flange (Software Used: Creo Parametric)

Steps:

Assume a part file is created with three default planes, RIGHT, TOP, and FRONT.

- Select the Revolve feature from the ribbon at the top. Clicking on the feature opens the selection box for all other options.
- You may change the name of the feature to be created; it helps to trace back the feature in the future so that one easily can understand which feature is used where.
- Click on the References tab and select Sketched feature.
- Click on Define, select the sketch plane, and click on Sketch.
- The horizontal and vertical references are to be taken automatically by the system.
- Create a vertical axis at the center point.
- Select Line and draw the geometry as shown in Figure 6.18.
- Click OK to finish the sketch.
- Make sure that the entered angle for feature creation is 360°.
- Click on Hole to create a hole on the flange.
- Select the full face surface of the flange (the surface without the raised face) as the primary placement reference.
- Click on Type and select Diameter for diametric referenced hole.
- Click on Offset reference and select the axis and a plane.
- Enter the pitch circle diameter (PCD) as 190.5 mm.
- Make the hole trough and the diameter to be 16 mm.
- Click on Finish to create the feature.
- Select the hole in the model tree and click on Pattern in the ribbon to create the pattern.
- Select 360 as the total angle and enter the number 8 for creation of 8 holes.
- Go to the Tools tab and click on Family table to make the other components of this family by varying the dimensions.
- Click on the vertical arrow to select the dimensions.
- Select the dimensions (230, 190.5, 16; number of holes; thickness of –22.3; and raised face diameter of 171 mm).

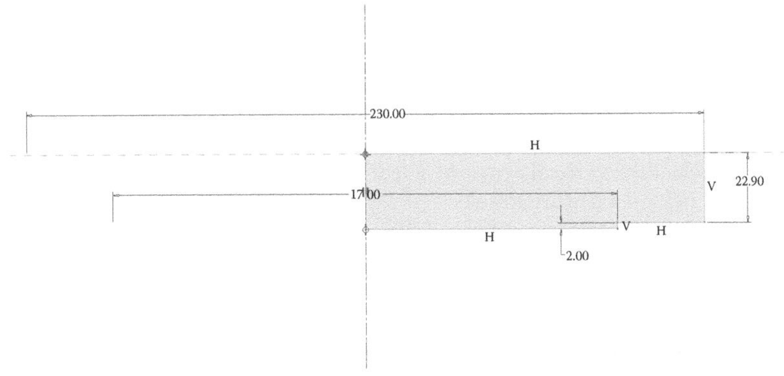

FIGURE 6.18
Constructional sketch of a blind flange.

Type	Instance Name	Common N...	d1	d3	d2	d12	p19	d21	F648 [REVOLVE_2]	F697 [REVOLVE_3]
	FLANGE	flange.prt	230.00	171.00	22.90	190.50	8	16.00	Y	Y
	SORF-100NB	flange.prt_INST	230.00	171.00	22.90	190.50	8	16.00	Y	Y
	SORF-150NB	flange.prt_INST	280.00	219.00	23.90	241.30	8	20.00	Y	Y
	SORF-200NB	flange.prt_INST	345.00	273.00	27.00	298.50	8	20.00	Y	Y
	BLIND-100NB	flange.prt_INST	230.00	171.00	22.90	190.50	8	16.00	N	N
	BLIND-150NB	flange.prt_INST	280.00	219.00	23.90	241.30	8	20.00	N	N
	BLIND-200NB	flange.prt_INST	345.00	273.00	27.00	298.50	8	20.00	N	N

FIGURE 6.19
Dimensional variation of a family table for a blind flange.

- Click on OK and click on the horizontal arrow thrice to create three components.
- Enter all the dimensions as shown in Figure 6.19.
- Click on Verify. The status shall be successful for verification.
- Click on OK to finish the feature.
- Save the file.

2. *Parameters*: Many times, dimensions are in relation and controlled by meaningful parameters; in such cases, parameters are varied in a family table. Apart from that, there are many parameters that may not affect the model geometry; these parameters also can be controlled by a family table. Density can be such a type of parameter that changes due to change of materials.
3. *Features*: A family of a product can also contain some product that has some particular features that are not present; in such cases, it is necessary to control the feature presence by a family table.

Hands-On Exercise, Example 6.14: Creation of a Slip On Raised Face (SORF) Flange (Software Used: Creo Parametric)

- Follow the steps in the example on the use of the Up to the selection option to create the blind flange.
- Click on Revolve in the feature ribbon.
- Make the revolved feature of hub extrusion.
- Select the Revolve feature again and create the inside hole or cut for the pipe.
- Select Family table from the Tools tab.
- Select the vertical arrow to select variations again.
- Click on a feature to select.
- Select the last two revolve features that were created for SORF flange.
- Click OK.
- Enter N for all blind flanges and Y for all SORF flanges in the feature field.
- Click on Verify to see the status; it should be successful for all the instances.
- Save the file.

Primarily, the application of family tables facilitates the use of the same component and reduces the effort of making similar products. Secondarily, the application of family tables enhances automation.

Examples:

1. Bolt
2. Nut
3. Flanges
4. Material alteration
5. Painting calculation

6.6 Use of Macro/Map Key

A macro or map key in Pro-Engineer or Creo is a recording of sequential commands that can be executed to perform similar tasks to save time. Typically, it records the sequential actions of execution of commands. Macros or map keys can also be edited. Macros are saved in a file that is played back by using keyboard commands or a graphical icon. CAD software permits the assignment specific keyboard shortcuts or choosing of a graphical icon of the user's choice. Though a macro or map key can be used to record and playback any sequential commands, the usefulness is limited to certain applications where designers need to perform repetitive work. These are as follows:

1. Creation of geometry
2. Geometrical measurements
3. Changing of parameters

The macro creation method is very simple compared to any other automation process. The use of macros or map keys is as follows (software used: Creo parametric):

1. *Create the macro*: Type Map key in the command search box and click on it. A map key creation box will open. Click on New to create a new map key. Enter the key in the key sequence field. This key will activate the map key after recording. Enter the name and full description of the map key.
2. *Start recording*: Start recording and carry out all the work that can be included in the map key.

3. *Perform the activities sequentially*: Whenever recording for a map key, it is essential to perform sequentially all the operations that will be included in the map key.
4. *Stop recording*: Stop recording if the map key creation is finished. If there is some operation that need not be included in the map key, like editing of geometric dimensions, the recording can be paused and restarted again.
5. *Run the macro*: Once the map key is created, before using it for repetitive purposes, run it to ensure that the performance is up to the requirement.
6. *Edit the macro.*
7. *Save it to file.*
8. *Assign a key sequence or graphical icon.*

Hands-On Exercise, Example 6.15: Map Key (Software Used: Creo Parametric)

Steps:

- Create a part file and rename it as Map key Example. (Other names may be given, but it is advisable to provide a relevant name.)
- Select Solid on the right-hand side of the box. This will activate the solid mode and environment.
- Uncheck the box of the default template and click OK to create a new solid file.
- The template selection box will appear.
- Select mmns_part_Solid. This will allow us to use the solid template file of the following units:
    - mm—millimeter for length unit
    - n—newton for force unit
    - s—second for time unit
- Once the solid part is created, go to File > Prepare > Model properties to check the units. In the upper sections, the units of the model will be displayed.
    - Note: The units usually are displayed in the summary section of any CAD software. The setting of units usually is in the tools or settings.
- Click on the command search field on the upper right hand corner.
- Write Map key to search for the Map key command. This command is not command list.
- Click on the Map key command in the searched result window.
- A map key window appears.
- Click on New in the map key window.
- A Record map key window appears.
- Write "b" in the key sequence field. The map key will be executed by pressing the key "b."

- Write the name of the map key as Block. Any name can be given, but it is always advisable to give a meaningful name so that one can easily remember and it can be understood by anyone other than the creator.
- Write down the purpose of this map key in the description field. In this case, write To make a block in the description field.
- Click on Record to start the recording of command sequences.
- Click on Extrude from the Model tab in the ribbon interface.
- Click on the Placement tab and select Define to define the sketch.
- Select the top plane as the sketch plane.
- The right plane shall be automatically taken as reference and orientation as right.
- Click on Sketch.
- The sketcher model will be activated.
- Click on the drop-down menu beside the rectangle and select Center rectangle.
- Place the rectangle center point in the intersection of the horizontal and vertical references.
- Draw a rectangle of dimensions 500 × 500 mm.
- Click OK to create and exit the sketch.
- Select mid plane extrusion.
- Enter the value 200 mm for extrusion depth.
- Click on OK to create the extrusion.
- Click on the Stop map key button in the record map key window.
- Click OK to create the map key.
- Click Run to run the map key.
- On clicking Run map key, the user is immediately asked to select the sketch plane.
- Select the top face of the model as the sketch plane.
- Draw a center rectangle, again placing the center point in the intersection of the horizontal and vertical references.
- The system will create the extrusion of height 200 mm.
- Right-click on the lastly created extrusion in the model tree.
- Select Edit.
- The dimensions will be displayed.
- Change the dimensions as required.
- Regenerate the model.
- Save and exit the file.

6.7 Parametric Modeling

Parametric modeling is modeling controlled by associated parameters. In other words, the parameters control the geometry of the modeling. If the parameter changes its value, the geometry associated with the parameter will also change; thus, the component shape changes. A model can be nonparametric or parametric. If the requirement is such that the modeled component is very simple with a very small number of features, parametric modeling techniques may not be economical with respect to time. However, for a robust model or in developing product, parametric modeling is extremely useful to save time and achieve accuracy. Parametric modeling has the following advantages:

1. The model is related to meaningful parameters.
2. Changing of any parameter automatically changes the model.
3. The effect of geometries with respect to certain parameters can be seen and plotted, which enables a designer to understand that the behavior of changing parameters can be accomplished.
4. Relating of parameters is easy to understand with meaningful parameter names.
5. The same parameter can be used in various relations.

Apart from its numerous advantages, parametric modeling also has some difficulties because of which users sometimes ignore creation of parametric models. These difficulties are as follows:

1. It requires good knowledge and understanding of the parametric modeling techniques and CAD software.
2. It involves more time to create the parametric model.
3. It involves programming knowledge or knowledge of syntax to create relations among parameters since most of the parametric modeling involves the use of relation among the parameters.
4. Some parameters may lead to impractical values of model dimensions, which in turn lead to feature failure. Most basic CAD users find it difficult to resolve the failed features. Also, feature failure options are specific to different CAD software, so these require good knowledge of the CAD software.
5. It is difficult to edit a robust parametric model if the design changes and, particularly, for other users who were not involved in the creation of the parametric model.
6. Many geometric dimensions are not meaningful or can be related to parameters, which can also lead to feature failure.
7. A set of relations is required to be available before creation of modeling.

Parametric modeling software like Creo is continuously developing and increasing the ease of users' accessibility to reduce the aforementioned difficulties. Instead of these difficulties, parametric modeling significantly contributes to reducing design time and increasing accuracy. It is suggested that users use parametric modeling as much as possible.

In order to create a parametric model, some steps are as follows:

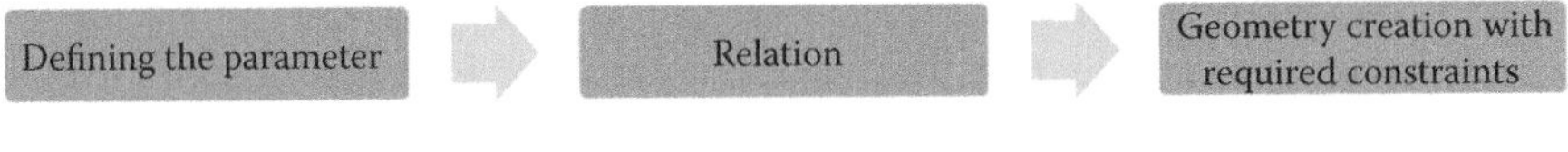

6.8 Parametric Modeling Procedures

The modeling operation defines the creation of 3-D models in the CAD software. It is the most time-consuming work performed in CAD software. Accurate modeling is the most significant work for CAD users since any output from CAD is related to the prepared model. The modeling operation includes the following.

6.8.1 Parameter Creation

Parameters are required, particularly for parametric modeling. It is advisable that users create all the driving or major parameters before creating any object. Highly parametric software like Creo allows defining the attributes of parameters. These include type, dimension, units, accessibility, and so forth. The types of parameters are as follows:

- *Integer*: If a parameter is defined as an integer, the parameter will always store the integer portion of any real number if the value is derived from any relation.

  Example: Parameter N is defined as the quantity of holes arrayed in s strips of x length with spacing of dx. Now the N can be obtained from the formula $N = x/dx$. Now, it is obvious that the quantity of the holes shall be in integers; therefore, the N parameter has to be defined as an integer so that the system will automatically take the integer part of the derived value.

  If

$$x = 500,$$

$$dx = 60,$$

$$N = x/d\text{x} = 500/60 = 8.33333,$$

the integer part of the derived value is 8, so the quantity will be 8.

- *Real number*: A parameter having the attribute of real number shall display the actual calculated value. In case of mass properties, real number attributed parameters are always useful.
- *String*: String is the parameter that is used for alphabetical information, such as product name, project name, description, and so forth.
- *Logical*: A logical parameter stores only either a yes or no value within it. In many cases, to build a large model with relation, logical parameters are essential.

6.8.2 Feature Creation

The first step of creating any object is feature creation. As described earlier, the features can be many, like solid modeling features, surface features, or datum features. In parametric CAD software, every specific feature can store parameters. If the parameters are created before as described earlier, the parameters can be directly entered in the creation of features when prompted for values. In this way, the built relation is kept inside the feature, which may be useful when building large parametric models but difficult to find out. In such cases, keeping all the relations in a part is very useful and easier to understand.

Hands-On Exercise, Example 6.16: Creating a Plug Sheet (Software Used: Creo Parametric)

Steps:

- Create a part file and rename it as Plug sheet. (Other names may be given, but it is advisable to provide a relevant name.)
- Select Solid on the right-hand side of the box. This will activate the solid mode and environment.
- Uncheck the box of the default template and click OK to create a new solid file.
- The template selection box will appear.
- Select mmns_part_Solid. This will allow us to use the solid template file of the following units:
  - mm—millimeter for length unit
  - n—newton for force unit
  - s—second for time unit
- Once the solid part is created, go to File > Prepare > Model properties to check the units. In the upper sections, the units of the model will be displayed.
  - Note: The units usually displayed in the summary section of any CAD software. The setting of units usually is in the tools or settings.
- The plug sheet as shown in Figure 6.20 is to be created.

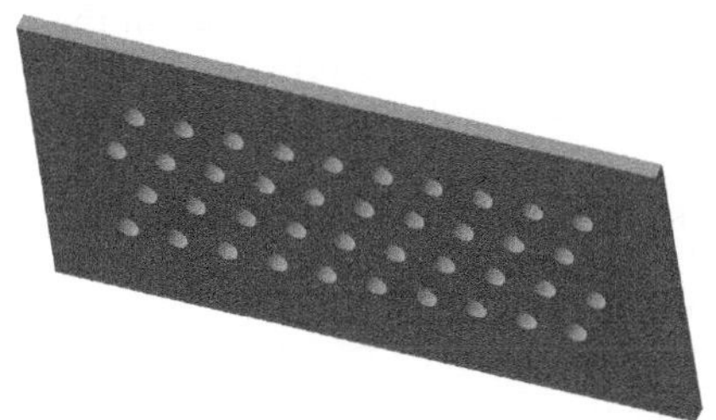

FIGURE 6.20
3-D view of a plug sheet.

- Click Parameters from the Tools tab.
- Create the following parameters.

Name	Type	Value
HOR_PITCH	Real number	60.00
VER_PITCH	Real number	60.00
NO_OF_PLUG_PER_ROW	Integer	10
NO_OF_ROWS	Integer	4
CLEARANCE	Real number	100.00
SHEET_THICKNESS	Real number	20.00
SHEET_LENGTH	Real number	0.00
SHEET_WIDTH	Real number	0.00

- Click OK to create the parameters.
- Click on Relations from the Tools tab of the ribbon interface.
- The relations creation window opens up.
- Write down the following relations.
  - sheet_length = HOR_PITCH*(NO_OF_PLUG_PER_ROW – .5) + CLEARANCE*2
  - sheet_width = Ver_pitch*(NO_OF_ROWS – 1) + CLEARANCE*2
- Click on Verify on the upper right side in the relation window to verify the relations.
- The message "Relations have been successfully verified" should come out.
- Click OK to close the relation window.
- Click on Extrude from the Model tab in the ribbon interface.
- Click on the Placement tab and select Define to define the sketch.
- Select the front plane as the sketch plane.
- The right plane shall be automatically taken as reference and orientation as right.
- Click on Sketch.
- The sketcher model will be activated.

- Draw a rectangle placing the left corner in the intersection of the horizontal and vertical centerlines.
- Double-click the length dimension of the rectangle and enter sheet_length.
- Double-click on the width dimension of the rectangle and enter sheet_width.
- Click OK to create and exit the sketch.
- Select mid plane extrusion.
- Enter the value sheet_thickness in the field of extrusion depth.
- Click on OK to create the extrusion.
- Select Hole from the Model tab of the ribbon interface.
- Select a standard screw hole and include a thread surface.
- Select UNF of size 1 1/8-12.
- Select the Through all option.
- Click on the Placement tab and select the front large face of the sheet.
- Select the left and bottom surfaces as offset references.
- Double-click on the offset dimensions and enter clearance.
- Click OK to create the hole feature.
- Similarly, create another hole of the same size by following the aforementioned procedure on the same face with offset reference from the existing hole.
- Double-click on the offset dimensions and enter ver_pitch for the vertical offset dimension field and hor_pitch/2 for the horizontal pitch dimension field.
- Click OK to create the hole feature.
- Select both the holes in the model tree by keeping the Control key pressed.
- Right-click and select Group.
- Select the newly created group in the model tree.
- Right-click and select Pattern.
- Select Directions in the Type of pattern selection field.
- Select the front plane in the direction 1 field and top plane in the direction 2 field.
- For the incremental dimension field, enter hor_pitch for direction 1 and ver_pitch*2 for direction 2.
- Enter the quantity 2 for patterned feature in both directions.
- Make sure the patterned features are approaching the empty area of the sheet.
- Click OK to create the patterned feature.
- Click on Relations from the Tools tab of the ribbon interface.
- The relations creation window opens up.
- Write down the following relations.
  - p23 (graphically may need to select by clicking the pattern feature, feature number may be different) = NO_OF_PLUG_PER_ROW
  - p24 (graphically may need to select by clicking the pattern feature, feature number may be different) = NO_OF_ROWS/2

- Click on Verify on the upper right side in the relation window to verify the relations.
- The message "Relations have been successfully verified" should come out.
- Click OK to close the relation window.
- Click on Parameters from the Tools tab of the ribbon interface.
- Change any parameters like hor_pitch or NO_OF_PLUG_PER_ROW.
- Click OK to close the parameter window.
- Click on Regenerate and notice the differences.
- Save and exit the file.

6.8.3 Relation Buildup

Relation building is the most significant step in parametric modeling and requires a good knowledge of the product and the CAD software too. As discussed earlier, relationships can be maintained within the feature or at the part level. Both have specific advantages and difficulties.

Advantages of keeping relations within features:

- A structure for relations can be maintained.
- It is useful to keep the relations separate for specific features.
- It consumes less time if the parameters are available during creation of features.
- It is useful for creating parametric models of large assemblies or components.

Disadvantages:

- It requires good knowledge to create and understand the parametric model.
- It is difficult to find out the problem in some relations in the case of feature failure issues.
- Parameters are required to be created before features, which may not be possible all the time.
- It is not very useful for the simple parametric model.

If the relations are kept in the part mode or in the top level in a part, relations are required to be created after creation of the feature. This involves selection of dimension symbols and relating them to parameters. This method has advantages and disadvantages.

Advantages:

- It requires less skill and knowledge to create the parametric model.
- The relations can be easily found in the part mode.

- Since parameters need not be created before features, one can indentify the required parameter after creating the feature and recreate it before creating relations.

Disadvantages:

- Sometimes, features are required to be identified by feature number or internal IDs, which involves the use of specific syntax.

Hands-On Exercise, Example 6.17: Relation in Parametric Modeling (Software Used: Creo Parametric)

Steps:

- Create a part file and rename it as Vertical Shell. (Other names may be given, but it is advisable to provide a relevant name.)
- Select Solid on the right-hand side of the box. This will activate the solid mode and environment.
- Uncheck the box of the default template and click OK to create a new solid file.
- The template selection box will appear.
- Select mmns_part_Solid. This will allow us to use the solid template file of the following units:
  - mm—millimeter for length unit
  - n—newton for force unit
  - s—second for time unit
- Click on the Tools tab in the ribbon interface.
- Click on Parameters.
- The parameters window will appear.
- Add a parameter and rename it as Dia.
- Keep the type as real number.
- Enter the parameter value as 500.
- Add a parameter and rename it as Length.
- Keep the type as real number.
- Enter the length parameter value as 1000.
- Add another parameter and rename it as Volume.
- Keep the type as real number.
- Keep the volume parameter value unchanged at 0.000.
- Click on Extrude from the Model tab in the ribbon interface.
- Click on the Placement tab and select Define to define the sketch.
- Select the top plane as the sketch plane.
- The right plane shall be automatically taken as reference and orientation as right.
- Click on Sketch.
- The sketcher model will be activated.
- Draw a circle keeping the center at the intersection of the horizontal and vertical reference lines.

- Double-click on the diameter dimension and enter Diameter.
- Click OK to create and exit the sketch.
- Select Blind extrusion.
- Enter Length in the field for extrusion depth.
- Click on Relation from the Tools tab of the ribbon interface.
- The relation window opens up.
- Write down the relation "volume = ((PI()/4)*diameter^2)*length."
- Click on Verify on the upper right side of the relation window.
- A notification box with a message "Relations have been successfully verified" should appear.
- Click OK to close the box.
- Click on the Analysis tab of the ribbon interface.
- Click on Mass properties.
- The mass properties analysis window will appear.
- Click on Preview in the mass properties analysis window or middle-click on the mouse button to calculate the mass properties.
- Mass properties will show in the mass properties box. The appeared volume will be the same as calculated volume.
- Now click on Relation from the Tools tab of the ribbon interface.
- The relation window opens up.
- Change the relation to "length = volume/((PI()/4)*diameter^2)."
- Click on Verify on the upper right side of the relation window.
- Notice that the length parameter box is locked now.
- Change the volume to some new value.
- Click on Verify on the upper right side of the relation window.
- Click OK in the bottom of the relation window to close.
- Click on Regenerate in the upper left section.
- The length will be automatically changed.
- Save and exit the file.

Other way: The CAD software itself creates unique symbols for each dimension that is entered into the system. The symbols follow a format as specified in the CAD software, like d0, d1, and so forth, which increases as the user continues building more features and entering values. When creating relations, the parameters have to be linked or relate to each of these dimension symbols; thus, CAD understands the relations between the parameters.

$$\text{Inherent dimension symbols} = f(\text{parameters}).$$

This method is very easy to create; however, sometimes, it is difficult to understand and find symbols for which the relations are made.

The smartest way is to rename the inherent symbols to some meaningful name that allows the user to understand the relationship easily and thus reduces complexity.

6.8.4 Property Assignment

Once the feature creation and relation buildup is completed, the required properties need to be assigned. Sometimes, the properties are also required to be assigned during or before building up the relation. The properties that are required to be assigned are as follows:

- *Material properties*: Material properties automatically get assigned with the assignment of material into the part. The important material properties include the following:
  - *Density*: The ratio of mass to volume
  - *Young's modulus*: Also called modulus of elasticity
  - *Bulk modulus*
  - *Coefficient of thermal expansion*
  - *Poisson's ration*: The ratio of volumetric strain to linear strain
  - *Aesthetical properties*
- *Mass properties*: Mass properties can be derived from the model or can be overwritten as per a user's requirement. The mass properties are as follows:
  - *Mass*: The matter of quantity in any part
  - *Volume*: The space required for the part
  - *Center of gravity*: The point through which the whole mass of the product is acted
  - *Moment of inertia*: The second moment
  - *Section modulus*: The ratio of the moment of inertia to the distance of the fiber from the center of gravity
  - *Radius of gyration*

In many cases, mass properties are required to be used in the relationship to create further parameters. In such cases, mass properties are required to be generated or assigned before building up relationships.

In many cases, the complete drawing for a 3-D model may be unavailable or strategically not given. For example, a company may not give all the drawings to make their product, which are proprietary items. However, they can provide enough data that is actually required for engineering applications like total mass, center of gravity, and so forth. In such cases, overwriting of mass properties is essential, which in turn overwrites the mass properties for calculation as per a user-entered value.

- *Aesthetical properties*: Aesthetical properties are the graphical properties assigned to parts or surfaces. Graphical properties are discussed in detail in the latter part of this book; however, the following are essential in the subject's point of view:

- *Color*: Assignment of color is the simplest way to distinguish different parts and surfaces. In part mode, it is obvious that the use of color is limited to use in surfaces. Some CAD software also allows coloring of individual features. Also, some CAD software automatically provides color on building up different features. This enhances visibility. Color is used to understand different surfaces or make the model more realistic so that users get an idea of how the component will be seen when made in real life.
- *Transparency*: Transparency allows the part beneath to be seen. It helps to model the transparent parts or to see the internal features under the exterior surfaces.
- *Texture*: Textures are assigned as a picture on a specific surface or part. Textures are used in the following cases:
    - When there are painted surfaces with a specified picture, a user can realize the appearance after making the actual component.
    - When the model would be too complicated or can consume too much memory during modeling of full features, like small repetitive features (e.g., gratings, screens). The easiest solution for this situation is to model the part with the simplest detail and use the texture of the actual component or surface, for example, gratings and so forth.
    - Textures are also used for sample models with the company name and details, which might be used for advertising or promotion purposes before manufacturing the actual part.

- *User-defined properties or parameters*: When modeling a component a user may have some specific requirements to measure, which must be calculated or stored in the part; those parameters can be retrieved in a latter stage. For example, sometimes, the actual material cost needs to be calculated. The unit material cost can be placed as a parameter.
    - Let the unit cost of the material be \$$X$/kg. The total calculated mass of the part from the model is M kg. Hence, the total cost of the part shall be \$$MX$. If this relationship can be placed on screen text, the designer can see what the raw material cost involves when building each feature.
    - Let the unit cost of painting be \$$x$. Hence, as the feature increases, it may include or reduce some more of the surface that may need to be painted. If such a relation can be written, the total painting cost = area × \$$x$ = \$$AX$ = total painting cost. Users may build relationships to calculate the total cost including raw materials + manufacturing + material handling + painting and so forth to get the total cost.

6.8.5 Checking/Verifying

Verifying or checking the part is most important from the design point of view before declaring that a part is completed. Verifying can be of two types:

- *Verifying from the CAD point of view*: When the part is parametrically modeled, it is obvious that the assigned parameters may change during finalization of design. Thus, all the features are required to be regenerated as the parameters get changed. It is often seen that features may fail if there is any unsatisfied relation. A part or model with failed features often leads to improper model shape or is the cause of other feature failures or improperly derived parameter generation. Thus, it has to be ensured that the model is free of any failure feature and improper relation.
- *Verifying from the design point of view*: Sometimes, the design may need to be more optimized or can be modified. It is always advisable to stretch on preparing each model to its final design, since further changing the model may affect its dependents, like other related parts in an assembly.

6.8.6 Editing

It may be a requirement to edit the part or relations before declaring them final and releasing them to proceed with further applications. In parametric modeling, when all the features are linked with parameters, relations are to be carefully edited.

6.9 Modeling Techniques

1. *Planning of features*: Creating a simple 3-D model becomes very easy as CAD software increases its user-friendliness. However, creating a complex model always requires some planning before starting the model; otherwise, there will be ample loss of time. Especially in the case of parametric modeling, the planning is important when feature-based relations are to be written. If the model is being prepared from an existing drawing, it is advisable that the feature creation process be followed as per the manufacturing procedure. This helps in visualizing the manufacturing steps involved in the production of the product and always ensures the right way of creation.
2. *Creation of required datum*: Datum items are always required to start with the model. Also datum planes are extremely useful for using reference entities. The uses of datum items are as follows:

a. Datum plane as level or surface
b. Coordinate system as reference entities

The advantages of using reference entities are as follows:

- They do not depend on solid feature regeneration. Sometimes, some surfaces may be excluded due to feature regeneration.
- They can be renamed as required.
- They can be created and used in the template part. Since they do not have any mass properties, they do not have any effect on the mass properties of the final model.
- Their shapes can be adjusted as required.
- They can be patterned or copied.
- They can be put into layers independently so that during presentation, no datum items will be displayed, if required.

3. *Creation of features*: Creation of features is the next step after creation of required datum items. Sometimes, the datum items may need to be created during or after feature creation. It is advisable to rename each feature to some meaningful name when creating a parametric and complex model. The following points are important when creating features in a parametric model.
   a. References shall be taken carefully and be kept in mind such that in any case, the references should not get outdated or deleted during regeneration with changed parameters.
   b. If the relations are entered in features, the proper renaming shall be done for the features with the relation.
   c. Features shall be created in accordance with practical design practice.
   d. If there are some properties to be assigned to the feature, it is advisable to assign the properties just after creation of the feature, if possible.
4. *Checking of final models*: Final model check.

6.10 Application-Based Modeling

It is often seen that the procedure obtained to prepare any CAD model also differs based on application. The applications for which commonly CAD models are used are as follows:

- *Casting or molding*: The modeling required for casting may be derived from the casting applications itself.

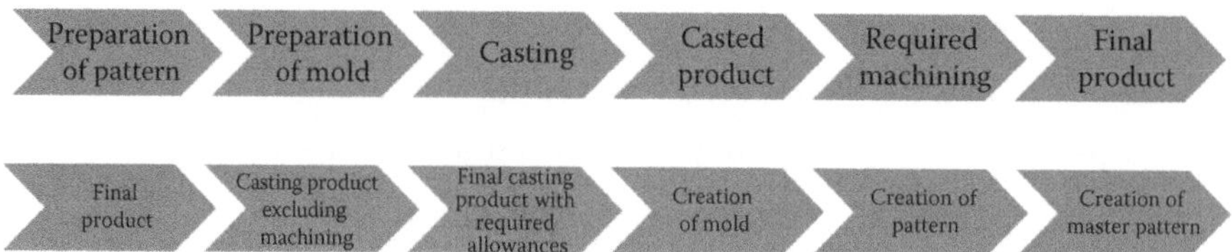

- *Assembly or fabrication*: 3-D models or drawings for assembly or fabrication often do not need fully detailed features. It is often not useful to create a detail model for assembly or fabrication.
- *Mechanism or simulation*: If the purpose of the modeling in CAD is to see or analyze the mechanism or for simulation, sometimes it is beneficial to exclude the irrelevant features from the model, which simplifies the model and also saves time and memory.
- *Analysis*: The type of analysis required for the model or design is also derived from the extent of detailing. If every feature has to be thoroughly checked from the design point of view, it may be called a detailed model.
- *Illustration*: A CAD model is always called for in the creation of a manual. Obviously, the models need not be too detailed for illustration, and sometimes, a similar-looking model serves the purpose.
- *Design automation*: When a design automation is planned for and the frequency of use is much more, then it is advisable to create as detailed a model as possible.
- *Concept generation*: CAD models are very useful in generation of concepts. The outlook of a component after certain operations may require a CAD model. It is obvious that the time available for this type of modeling is not much, and also, a detailed model is not needed.
- *Applicable codes and standards*: In engineering, there is a standard practice in many cases where complicated models are presented with simpler symbols. It is essential to maintain the symbols while excluding detailed modeling of the part.
    1. *Identification of requirements*: When creating any CAD model, it is essential to evaluate the requirement so that the time and effort to prepare the CAD model can be optimized while satisfying the requirement. All the ideas of creating a parametric model are to save time, reduce accurate model with optimized design. Identification of requirements can be determined by the following procedure:
        a. *Count the dependencies on the features*: If the features are too interdependent with specified relations, it is recommended to use a parametric model so that if the design varies at any stage of feature creation, the relative dependent feature also gets changed.

b. *Estimate the time requirement to make the model*: Whether it requires too many man hours to prepare the model and chances of changes in the future will not much affect the model.
c. Count the number of uses to be required for the model. If the model is required to be used many times, it is better to make a highly parametric model. Normally, the requirement of a parametric model may have been called for even only for a model that can be used only for a second time. Therefore, the requirement is for a highly parametric model.
d. The extent of variability of the design is an important criterion and deciding factor for creation of a parametric model. If the product is in the conceptual research and development stage, where the entire set of relations for any feature may get changed, a parametric model may not be economical. However, if the relations for some features have been frozen, a parametric model can be fruitful.
e. To whom the model is to be shown or presented is also an important criterion in modeling. If it is to be presented to a customer, obviously, the model has to be as detailed as possible, unless the customer agrees otherwise.

Once the requirement of a parametric model is stabilized, the next step is to determine the extent of parameterization. It is always desirable to make a complete parametric model, where the user only has to give input and the entire model generates the final finished product. However, the extent of work may not be economical for a complete parametric model. Therefore, this kind of situation can be approached with a semi-parametric model, where the parameterization shall be up to a user's needs.

2. *Creation of relation*: It is essential to understand the design relation of each feature before transferring it to a CAD program. Many times, it is seen that some feature does not have much importance in relation from a design point of view; however, it is required in CAD for a complete parametric model. This feature becomes important since many important features may not get regenerated due to failure of unimportant features during regeneration. All the relations shall be thoroughly checked for accuracy and integrity.
3. *Required parameter creation*: When all the sets of relations created are in a notepad or any text editor, the next job is to implement in CAD. It is recommended to create all the parameters and relations before creating the model. Creation of any relationship

during the feature creation may be overlooked, and the model may run into the problem.

4. *Creation of model*: Once the parameters and relationships are completed, the model shall be created as per planning of features.

Hands-On Exercise, Example 6.18: Creating an Eyebolt (Software Used: Creo Parametric)

Steps:

- Create a part file and rename it as Eyebolt. (Other names may be given, but it is advisable to provide a relevant name.)
- Select Solid on the right-hand side of the box. This will activate the solid mode and environment.
- Uncheck the box of the default template and click OK to create a new solid file.
- The template selection box will appear.
- Select mmns_part_Solid. This will allow us to use the solid template file of the following units:
    - mm—millimeter for length unit
    - n—newton for force unit
    - s—second for time unit
- Once the solid part is created, go to File > Prepare > Model properties to check the units. In the upper sections, the units of the model will be displayed.
    - Note: The units usually are displayed in the summary section of any CAD software. The setting of units usually is in the tools or settings.
- The eyebolt as shown in Figure 6.21 is to be created.
- Click on Sweep from the Model tab of the ribbon interface.
- The sweep trajectory will be created first followed by a sweep operation.

FIGURE 6.21
3-D view of an eyebolt.

- The trajectory can also be created inside the sweep operation; however, the sketch will not be available if the sweep feature is deleted.
- Select Sketch from the Model tab of the ribbon interface.
- A sketch window for selection of sketch plane will appear.
- Select the front plane as the sketch plane.
- The right plane shall automatically be selected as reference and the orientation is to be right.
- Click on Sketch.
- The sketcher window will open up and the reference selection window appears for selection of references.
- The two references (i.e., top plane as horizontal reference and right plane as vertical reference) shall be automatically taken; however, if not, please select the same.
- Create a sketch as displayed in Figure 6.22.
- Click OK to create and exit the sketch.
- Click on Sweep from the Model tab of the ribbon interface.
- Graphically select the sketch as a trajectory.
- The system assumes the profile sketch point and shows it by displaying the directional arrow.
- Graphically click on the arrows of feature creation to draw the sweep profile in other control points.
- Click on Sketch to sketch the profile.
- The sketcher mode activates with the sketcher environment.
- Draw a circle at the intersection of references.
- Change the diameter to 20 mm.
- Click OK to create and exit the sketch.

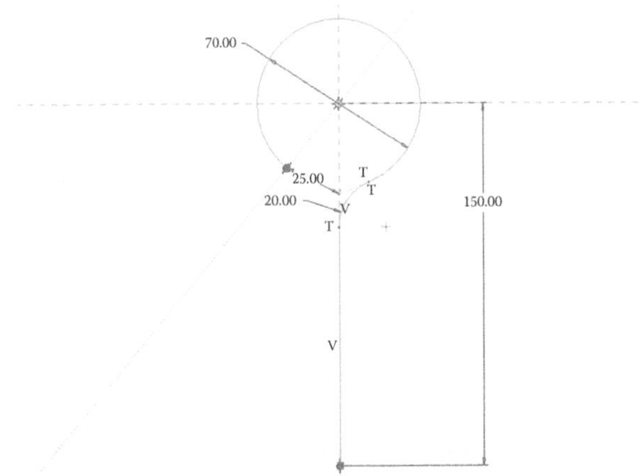

FIGURE 6.22
Constructional sketch of an eyebolt.

- Click OK to finish the sweep.
- Click on Round feature from the Model tab of the ribbon interface.
- Set the value of fillet to 5 mm.
- Select the edges of the circular face of the eye.
- Click OK to create the feature.
- Click on the Helical sweep option from the drop-down menu of Sweep from the Model tab of the ribbon interface.
- Click on the References tab and select Define to define the helix sweep profile.
- Select the front plane as the sketch plane.
- The right plane shall be automatically taken as reference and orientation as right.
- Click on Sketch.
- The sketcher model will be activated.
- Draw a geometry centerline on the vertical reference line in the middle.
- Add the reference of the left outer edge of the straight cylinder part of the hook.
- Draw a vertical line at the reference edge. The bottom point of the line shall coincide with the bottom face with the help of the bottom surface reference.
- Set the length of the line as 70 mm.
- Click OK to create the sketch and exit the sketcher window.
- Click Sketch to create the profile sketch above the Reference tab.
- Create the sketch as shown in Figure 6.23.
- Click OK to create and exit the sketch.
- Enter the pitch value of 3 mm.
- Click on the Right-handed rule icon to create the helical spring clockwise downward.

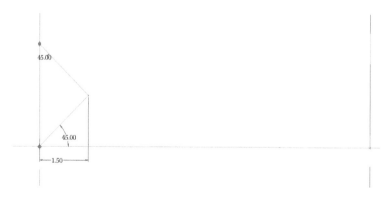

FIGURE 6.23
Constructional sketch of the thread profile section for an eyebolt.

- Click to activate material removal.
- Click OK to create the external thread.
- Click on Extrude from the Model tab in the ribbon interface.
- Click on the Placement tab and select Define to define the sketch.
- Select the front plane as the sketch plane.
- The right plane shall be automatically taken as reference and orientation as right.
- Click on Sketch.
- The sketcher model will be activated.
- Draw a rectangle of 40 mm length and 3 mm width. Place the rectangle such that the bottom edge of the rectangle shall coincide with the straight bottom edge of the hook part and be symmetrical with the vertical reference line.
- Click OK to create and exit the sketch.
- Select mid plane extrusion.
- Click on the Options tab and select Through all in both depth selection fields.
- Click to activate the material removal option.
- Click on OK to create the extrusion.
- Save and exit the file.

6.11 Presentation

Presentation of a single CAD model is different and also very useful. The presentation includes excluding or hiding features, making the model transparent, and so forth. Presentation of any CAD model is very important. The reasons are as follows:

1. It is to be clearly understandable to manufacturing people.
2. The process of design and affectability may need to be explained during verification.
3. The component may need to be advertised for marketing and promotion purposes.

The model can be presented with the following variations:

1. Excluding features
2. Transparency on surfaces
3. Colors

Graphical presentation creation is discussed in detail in Chapter 12.

6.12 Template Creation

Template creation is very useful especially when repetitive parts have some common features or properties like parameters, materials, and so forth. Template creation has the following advantages:

1. Time saving
2. Accuracy
3. Use for automation

6.13 Project Work

Hands-On Exercise, Example 6.19: Creating a Paper Clip Handle (Software Used: Creo Parametric)

Steps:

- Create a part file and rename it as Paper Clip Handle. (Other names may be given, but it is advisable to provide a relevant name.)
- Select Solid on the right-hand side of the box. This will activate the solid mode and environment.
- Uncheck the box of the default template and click OK to create a new solid file.
- The template selection box will appear.
- Select mmns_part_Solid. This will allow us to use the solid template file of the following units:
  - mm—millimeter for length unit
  - n—newton for force unit
  - s—second for time unit
- Once the solid part is created, go to File > Prepare > Model properties to check the units. In the upper sections, the units of the model will be displayed.
  - Note: The units usually are displayed in the summary section of any CAD software. The setting of units usually is in the tools or settings.
- The paper clip handle as shown in Figure 6.24 is to be created.
- Paper clip handle can be created by sweep method.
- Principally, it requires one cross-sectional sweep profile and sweep trajectory.
- Click on Sweep from the Model tab of the ribbon interface.

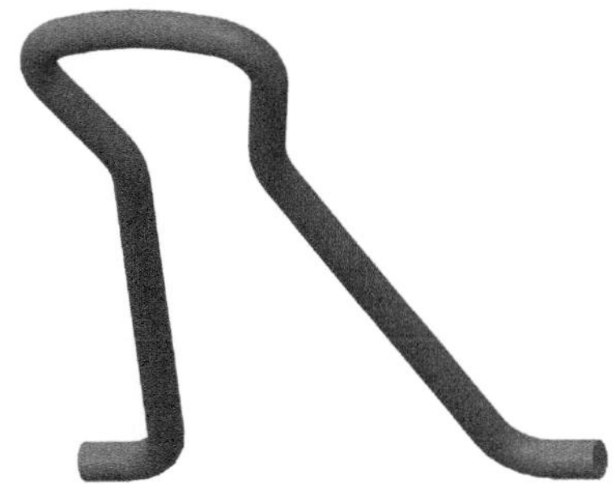

FIGURE 6.24
3-D view of a handle of a paper clip.

- The sweep trajectory will be created first followed by a sweep operation.
- The trajectory can also be created inside the sweep operation; however, the sketch will not be available if the sweep feature is deleted.
- Select Sketch from the Model tab of the ribbon interface.
- A sketch window for selection of sketch plane will appear.
- Select the front plane as the sketch plane.
- The right plane shall automatically be selected as reference and the orientation is to be right.
- Click on Sketch.
- The sketcher window will open up and the reference selection window appears for selection of references.
- The two references (i.e., top plane as horizontal reference and right plane as vertical reference) shall be automatically taken; however, if not, please select the same.
- Create a sketch as displayed in Figure 6.25.
- Click OK to create and exit the sketch.
- Click on Sweep from the Model tab of the ribbon interface.
- Graphically select the sketch as a trajectory.
- The system assumes the profile sketch point and shows it by displaying the directional arrow.
- Graphically click on the arrows of feature creation to draw the sweep profile in other control points.
- Click on Sketch to sketch the profile.
- The sketcher mode activates with the sketcher environment.
- Draw a circle at the intersection of references.
- Change the diameter to 2 mm.
- Click OK to create and exit the sketch.
- Click OK to finish the sweep.
- Save and exit the file.

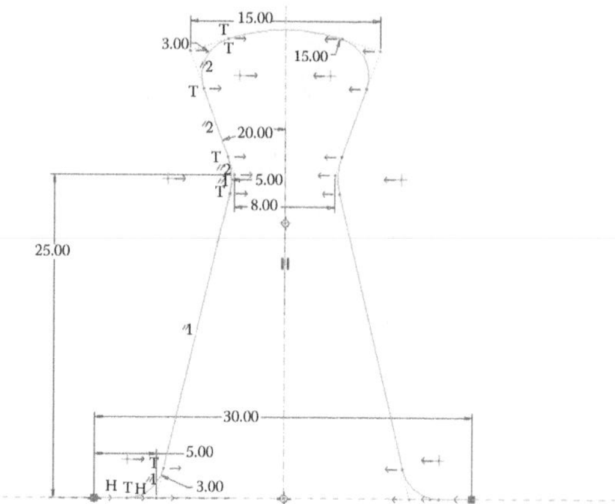

FIGURE 6.25
Constructional sketch for a sweep profile of a paper clip handle.

Hands-On Exercise, Example 6.20: Creating a Board Clip Spring (Software Used: Creo Parametric)

Steps:

- Create a part file and rename it as Board Clip spring. (Other names may be given, but it is advisable to provide a relevant name.)
- Select Solid on the right-hand side of the box. This will activate the solid mode and environment.
- Uncheck the box of the default template and click OK to create a new solid file.
- The template selection box will appear.
- Select mmns_part_Solid. This will allow us to use the solid template file of the following units:
  - mm—millimeter for length unit
  - n—newton for force unit
  - s—second for time unit
- Once the solid part is created, go to File > Prepare > Model properties to check the units. In the upper sections, the units of the model will be displayed.
  - Note: The units usually are displayed in the summary section of any CAD software. The setting of units usually is in the tools or settings.
- The board clip spring as shown in Figure 6.26 is to be created.
- Click on the Helical sweep option from the drop-down menu of Sweep from the Model tab of the ribbon interface.
- Click on the References tab and select Define to define the helix sweep profile.
- Select the front plane as the sketch plane.

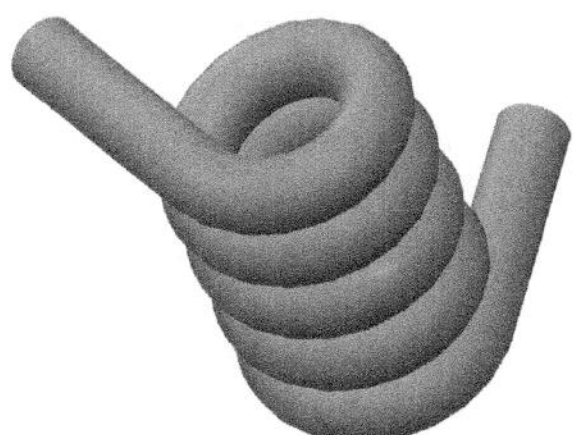

FIGURE 6.26
3-D view of a board clip spring.

- The right plane shall be automatically taken as reference and orientation as right.
- Click on Sketch.
- The sketcher model will be activated.
- Draw a geometry centerline on the vertical reference line in the middle.
- Draw a vertical line at a distance of 2 mm from the vertical centerline.
- Set the length of the line as 8 mm.
- Click OK to create the sketch and exit the sketcher window.
- Click Sketch to create the profile sketch above the Reference tab.
- Create a circle of 1.5 mm diameter placing the center at the bottom end of the vertical line.
- Click OK to create and exit the sketch.
- Enter the pitch value of 1.6 mm.
- Click on the left-handed rule icon to create the helical spring clockwise downward.
- Click OK to create the helical spring.
- Click on Plane from the Model tab of the ribbon interface to create a plane.
- Select the top plane and set the offset value to 10 mm toward the end of the spring.
- Click OK to create the plane.
- Select Extrude from the Model tab of the ribbon interface.
- Click on the Placement tab and select Define to define the sketch.
- Select the newly created plane as the sketch plane.
- The right plane shall be automatically taken as reference and orientation as right.
- Click on Sketch.
- The sketcher model will be activated.
- Create a line from the intersection point of the horizontal and vertical reference lines toward the third quadrant.
- Set the angle of 119° from the horizontal line and length of 4 mm.
- Click OK to create and exit the sketch.
- Click on Blind extrusion and enter the depth of 3.7 mm.

- Flip the orientation if required to make the feature in the opposite direction of material removal.
- Click OK to create the extrude cut feature.
- Select Extrude again from the Model tab of the ribbon interface.
- Click on the Placement tab and select Define to define the sketch.
- Select the bottom cross-sectional face of the spring profile as the sketch plane.
- The right plane shall be automatically taken as reference and orientation as bottom.
- Click on Sketch.
- The sketcher model will be activated.
- Project the circular edge of the spring profile at the bottom end.
- Click OK to create and exit the sketch.
- Click on Blind extrusion and enter the depth of 5 mm.
- Click OK to create the feature.
- Similarly, select Extrude again from the Model tab of the ribbon interface.
- Click on the Placement tab and Select define to define the sketch.
- Select the top cross-sectional face of the spring profile as the sketch plane.
- The surface shall be automatically taken as reference and orientation as right.
- Click on Sketch.
- The sketcher model will be activated.
- Project the circular edge of the spring profile at the top end.
- Click OK to create and exit the sketch.
- Click on Blind extrusion and enter the depth of 5 mm.
- Click OK to create the feature.
- Save the file and exit.

Hands-On Exercise, Example 6.21: Creating a Pulley Handle (Software Used: Creo Parametric)

Steps:

- Create a part file and rename it as Pulley Handle. (Other names may be given, but it is advisable to provide a relevant name.)
- Select Solid on the right-hand side of the box. This will activate the solid mode and environment.
- Uncheck the box of the default template and click OK to create a new solid file.
- The template selection box will appear.
- Select mmns_part_Solid. This will allow us to use the solid template file of the following units:
  - mm—millimeter for length unit
  - n—newton for force unit
  - s—second for time unit

- Once the solid part is created, go to File > Prepare > Model properties to check the units. In the upper sections, the units of the model will be displayed.
  - Note: The units usually are displayed in the summary section of any CAD software. The setting of units usually is in the tools or settings.
- The pulley handle as shown in Figure 6.27 is to be created.
- Click on Extrude from the Model tab in the ribbon interface.
- Click on the Placement tab and select Define to define the sketch.
- Select the front plane as the sketch plane.
- The right plane shall be automatically taken as reference and orientation as right.
- Click on Sketch.
- The sketcher model will be activated.
- Draw the sketch as displayed in Figure 6.28.
- Click OK to create and exit the sketch.
- Select mid plane extrusion.
- Enter the value 30 mm for extrusion depth.
- Click on OK to create the extrusion.
- Select Extrude again from the Model tab of the ribbon interface.
- Click on the Placement tab and select Define to define the sketch.

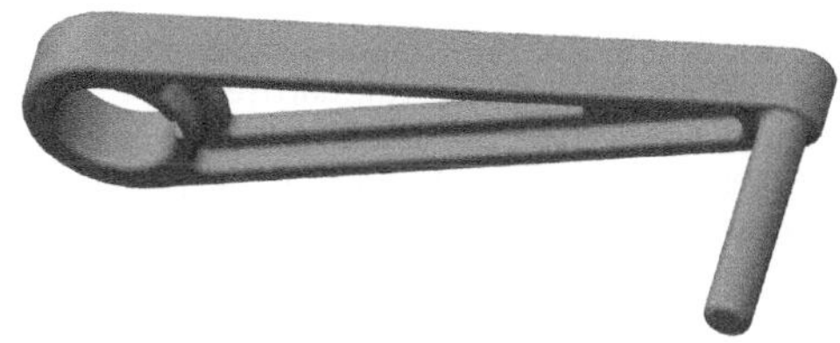

FIGURE 6.27
3-D view of a pulley handle.

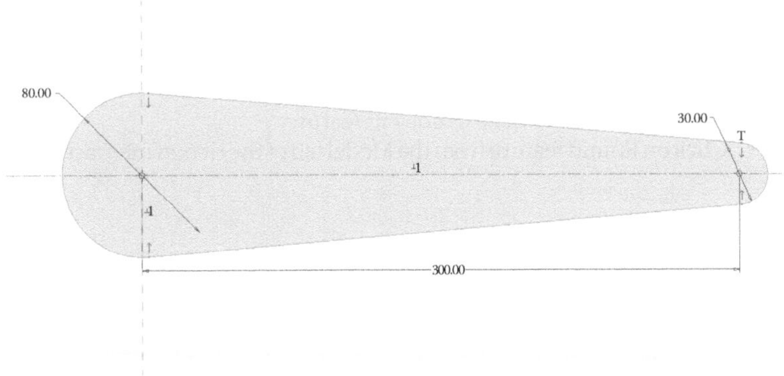

FIGURE 6.28
Constructional sketch of a pulley handle.

- Select the front face as the sketch plane.
- The right plane shall be automatically taken as reference and orientation as right.
- Click on Sketch.
- The sketcher model will be activated.
- Create a circle of diameter 50 mm at the center of the large arc.
- Click OK to create and exit the sketch.
- Click on the Material removal icon to make the extrude cut.
- Flip the orientation of extrusion if required.
- Click on the Through all option to make the cut through all the available depth.
- Click OK to create the extrude cut feature.
- Select Extrude again from the Model tab of the ribbon interface.
- Click on the Placement tab and select Define to define the sketch.
- Select the front face as the sketch plane.
- The right plane shall be automatically taken as reference and orientation as right.
- Click on Sketch.
- The sketcher model will be activated.
- Create a circle of diameter 20 mm at the center of the small arc.
- Click OK to create and exit the sketch.
- Click on Blind extrusion and enter the depth of 100 mm.
- Flip the orientation if required to make the feature in the opposite direction of material removal.
- Click OK to create the extrude cut feature.
- Select Extrude again from the Model tab of the ribbon interface.
- Click on the Placement tab and select Define to define the sketch.
- Select the front face as the sketch plane.
- The right plane shall be automatically taken as reference and orientation as right.
- Click on Sketch.
- The sketcher model will be activated.
- Create the sketch as shown in Figure 6.29.
- Click on the Material removal icon to make the extrude cut.
- Flip the orientation of extrusion if required.
- Click on the blind depth option and enter the value of 10 mm to depth.
- Click OK to create the extrude cut feature.
- Click on Round feature from the Model tab of the ribbon interface.

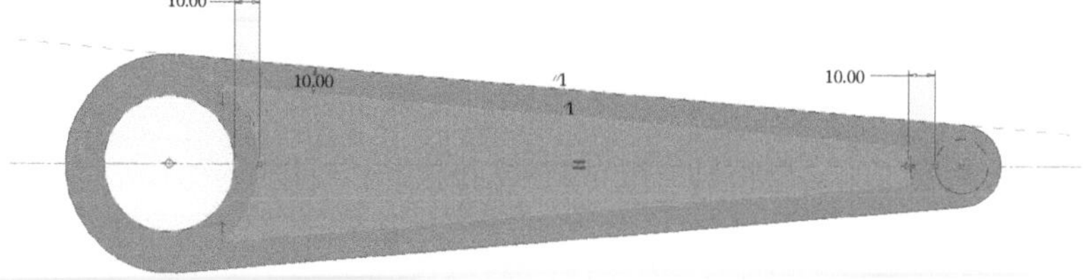

FIGURE 6.29
Constructional sketch for material removal of a pulley handle.

- Set the value of fillet to 5 mm.
- Select all four internal edges of the newly created feature.
- Click OK to create the feature.
- Select both the lastly created blind extruded cut and round feature from the model tree by pressing the Control key on the keyboard.
- Right-click and select Group to make both the features in a group.
- Select Group and select Mirror from the Model tab of the ribbon interface.
- Select the front plane as the mirror plane.
- Click on OK to create the mirrored feature.
- Select Extrude again from the Model tab of the ribbon interface.
- Click on the Placement tab and select Define to define the sketch.
- Select the front face as the sketch plane.
- The right plane shall be automatically taken as reference and orientation as right.
- Click on Sketch.
- The sketcher model will be activated.
- Create the sketch as shown in Figure 6.30.
- Click on the Material removal icon to make the extrude cut.
- Flip the orientation of extrusion if required.
- Click on the Through all option to create the cut feature through all the available material depth.
- Click OK to create the extrude cut feature.
- Click on Round feature from the Model tab of the ribbon interface.
- Set the value of fillet to 3 mm.
- Select all three internal edges of the newly created feature.
- Click OK to create the feature.
- Click on the auto round feature from the Model tab of the ribbon interface.
- Set the value of fillet to 3 mm.
- Click OK to create the feature.
- Select Extrude again from the Model tab of the ribbon interface.
- Click on the Placement tab and select Define to define the sketch.
- Select the front face as the sketch plane.
- The right plane shall be automatically taken as reference and orientation as right.
- Click on Sketch.
- The sketcher model will be activated.

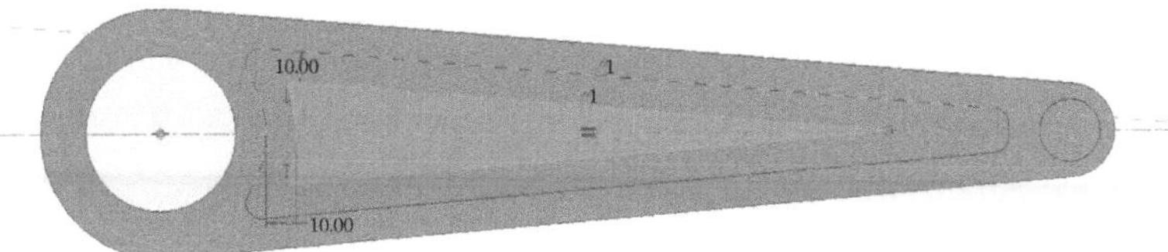

FIGURE 6.30
Constructional sketch for material removal through all features of a pulley handle.

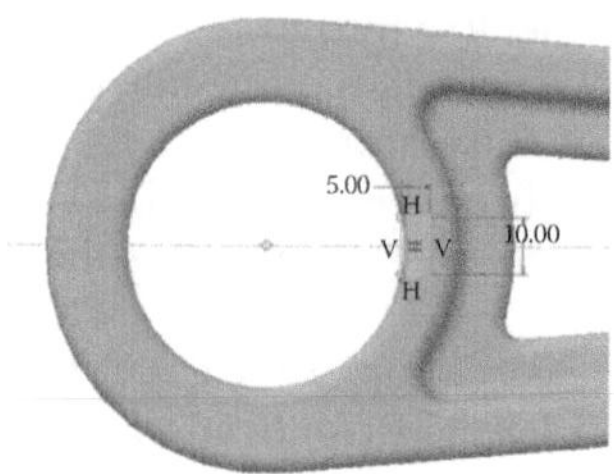

FIGURE 6.31
Constructional sketch for a keyway of a pulley handle.

- Create the sketch as shown in Figure 6.31.
- Click on the Material removal icon to make the extrude cut.
- Flip the orientation of extrusion if required.
- Click on the Through all option to create the cut feature through all the available material depth.
- Click OK to create the extrude cut feature.
- Save and exit the file.

Hands-On Exercise, Example 6.22: Creating a Fan Blade (Software Used: Creo Parametric)

Fan blade shapes are largely varied depending on the type of fan and fan manufacturer. An example of fan blade creation of an industrial fan blade is shown here.

Steps:

- Create a part file and rename it as Fan Blade Profile. (Other names may be given, but it is advisable to provide a relevant name.)
- Select Solid on the right-hand side of the box. This will activate the solid mode and environment.
- Uncheck the box of the default template and click OK to create a new solid file.
- The template selection box will appear.
- Select mmns_part_Solid. This will allow us to use the solid template file of the following units:
  - mm—millimeter for length unit
  - n—newton for force unit
  - s—second for time unit
- Once the solid part is created, go to File > Prepare > Model properties to check the units. In the upper sections, the units of the model will be displayed.
  - Note: The units usually are displayed in the summary section of any CAD software. The setting of units usually is in the tools or settings.
- The fan blade as shown in Figure 6.32 is to be created.
- A fan blade can be created by blend operation method.

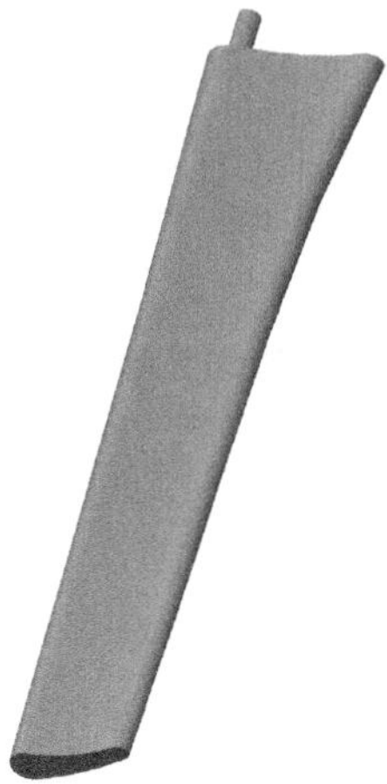

FIGURE 6.32
3-D view of a fan blade.

- Principally, it requires cross-sectional profiles.
- The sketches can be created during the blend operation or before the blend operation.
- It is preferable to create all the sketches before blend operation since any cancellations of blend operation during creation always cancel all the sketches if created during blend operation.
- Create a sketch file and rename it as Fan Blade Profile. (Other names may be given, but it is advisable to provide a relevant name.)
- Click OK to create the sketch file.
- The sketcher mode will be activated in the sketch file.
- Create the sketch as shown in Figure 6.33.
- Save and exit the sketch file.

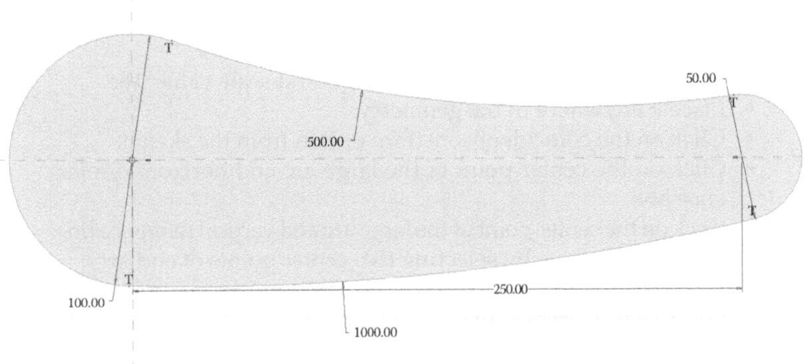

FIGURE 6.33
Constructional sketch of a fan blade.

- Select the front plane from the model tree and click on Pattern from the Model tab of the ribbon interface.
- Select direction for creation type.
- Select the z-axis of the coordinate system as the direction.
- Enter the number of patterned components as 4 and interval distance 500 mm.
- Click OK to create all patterned features.
- Click on Sketch from the Model tab of the ribbon interface.
- Select the front plane as the sketch plane.
- The right plane shall be automatically taken as reference and orientation as right.
- Click on Sketch.
- In the sketcher ribbon interface, click on the file system to get data (i.e., sketch that we created in a separate file).
- Select the newly created sketch file Fan Blade Profile from the saved folder.
- Click OK to place the sketch.
- Resize options will appear in the ribbon interface.
- Enter 0° in the field rotational angle and scale value 1.
- Place it anywhere in the geometry.
- Click on the coincident constrain option from the sketch.
- Click on the center point of the large arc and horizontal reference line.
- Click on the center point of the large arc and vertical reference line.
- Click OK to create and exit the sketch.
- Click on Sketch again from the Model tab of the ribbon interface.
- Select plane DTM1 (first component of patterned feature) as the sketch plane.
- The right plane shall be automatically taken as reference and orientation as right.
- Click on Sketch.
- In the sketcher ribbon interface, click on the file system to get data (i.e., sketch that we created in a separate file).
- Select the newly created sketch file Fan Blade Profile from the saved folder.
- Click OK to place the sketch.
- Resize options will appear in the ribbon interface.
- Enter 2.5° in the field rotational angle and scale value 0.8.
- Place it anywhere in the geometry.
- Click on the coincident constrain option from the sketch.
- Click on the center point of the large arc and horizontal reference line.
- Click on the center point of the large arc and vertical reference line.
- Draw a centerline by selecting two center points of end arcs.
- Provide an angular dimension by selecting the centerline and horizontal reference line.
- Enter the value 2.5° as the angular dimension.
- Click OK to create and exit the sketch.
- Click on Sketch again from the Model tab of the ribbon interface.

- Select plane DTM2 (first component of patterned feature) as the sketch plane.
- The right plane shall be automatically taken as reference and orientation as right.
- Click on Sketch.
- In the sketcher ribbon interface, click on the file system to get data (i.e., sketch that we created in a separate file).
- Select the newly created sketch file Fan Blade Profile from the saved folder.
- Click OK to place the sketch.
- Resize options will appear in the ribbon interface.
- Enter 5° in the field rotational angle and scale value 0.75.
- Place it anywhere in the geometry.
- Click on the coincident constrain option from the sketch.
- Click on the center point of the large arc and horizontal reference line.
- Click on the center point of the large arc and vertical reference line.
- Draw a centerline by selecting two center points of end arcs.
- Provide an angular dimension by selecting the centerline and horizontal reference line.
- Enter the value 5° as the angular dimension.
- Click OK to create and exit the sketch.
- Click on Sketch again from the Model tab of the ribbon interface.
- Select plane DTM3 (first component of patterned feature) as the sketch plane.
- The right plane shall be automatically taken as reference and orientation as right.
- Click on Sketch.
- In the sketcher ribbon interface, click on the file system to get data (i.e., sketch that we created in a separate file).
- Select the newly created sketch file Fan Blade Profile from the saved folder.
- Click OK to place the sketch.
- Resize options will appear in the ribbon interface.
- Enter 7.5° in the field rotational angle and scale value 0.7.
- Place it anywhere in the geometry.
- Click on the coincident constrain option from the sketch.
- Click on the center point of the large arc and horizontal reference line.
- Click on the center point of the large arc and vertical reference line.
- Draw a centerline by selecting two center points of end arcs.
- Provide an angular dimension by selecting the centerline and horizontal reference line.
- Enter the value 7.5° as the angular dimension.
- Click OK to create and exit the sketch.
- Click on Blend from the Model tab of the ribbon interface.
- Click on the Sections tab.
- Select the option Selected sections.
- Select the first sketch.

- Click on Insert.
- Select the second sketch.
- Click on Insert.
- Select the third sketch.
- Click on Insert.
- Select the fourth sketch.
- Click OK to create the feature.
- Click Plane from the Model tab of the ribbon interface.
- A plane creation window appears.
- Select the front plane and enter an offset value of 200 mm to create the plane opposite to the blend feature side.
- Click on Axis from the Model tab of the ribbon interface.
- Select the lastly created plane and front plane.
- Click on OK to create the axis.
- Click on Extrude from the Model tab of the ribbon interface.
- Click on the Placement tab and select Define to define the sketch.
- Select the front face as the sketch plane.
- The top plane shall be automatically taken as reference and orientation as right.
- Click on Sketch.
- The sketcher model will be activated.
- Select the lastly created axis as reference and small-end perpendicular surface of the blend feature.
- Create a circle placing the center point in the axis.
- Click on Tangent constrain from the Sketcher tab of the ribbon interface.
- Select the circle and the surface reference of the small-end perpendicular surface.
- Click OK to create and exit the sketch.
- Click on the Material removal icon to make the extrude cut.
- Flip the orientation of extrusion if required.
- Click on the Through all option to make the cut through all the available depth.
- Click on the Options tab.
- Click on the drop-down menu on the other side of the depth option and select Through all.
- Click OK to create the extrude cut feature.
- Save and exit the file.

Hands-On Exercise, Example 6.23: Spiral (Software Used: Creo Parametric)

Steps:

- Create a part file and rename it as Spiral. (Other names may be given, but it is advisable to provide a relevant name.)
- Select Solid on the right-hand side of the box. This will activate the solid mode and environment.

- Uncheck the box of the default template and click OK to create a new solid file.
- The template selection box will appear.
- Select mmns_part_Solid. This will allow us to use the solid template file of the following units:
  - mm—millimeter for length unit
  - n—newton for force unit
  - s—second for time unit
- Once the solid part is created, go to File > Prepare > Model properties to check the units. In the upper sections, the units of the model will be displayed.
  - Note: The units usually are displayed in the summary section of any CAD software. The setting of units usually is in the tools or settings.
- The spiral as shown in Figure 6.34 is to be created.
- A spiral can be created by many methods, for example, creating a horizontal helix path, but it also depends on the software.
- Click on the drop-down menu of datum from the model of the ribbon interface.
- Click on the curve from the equation.
- Select the coordinate system Cartesian.
- Click on Equation to create the equation.
- Write the following:
  - $n = 4{*}t$
  - $a = 5{*}t$
  - $x = n{*}\cos(a{*}360)$
  - $z = n{*}\sin(a{*}360)$
  - $y = 0$
- Click OK to create the equation.
- Click on the Reference tab.
- Select the available default coordinate system.
- Enter 0.20 in the From field and 1.00 in the To field.
- Click OK to create the spiral curve.
- Click on Sweep from the Model tab of the ribbon interface.
- Graphically select the sketch as a trajectory.

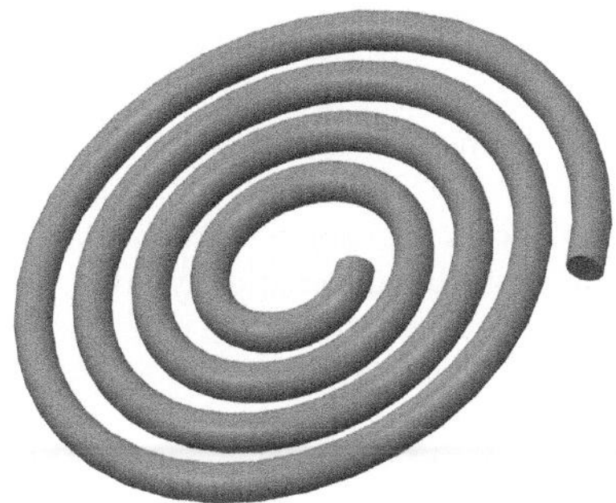

FIGURE 6.34
3-D view of a spiral.

- The system assumes the profile sketch point and shows it by displaying the directional arrow.
- Graphically click on the arrows of feature creation to draw the sweep profile in other control points.
- Click on Sketch to sketch the profile.
- The sketcher mode activates with the sketcher environment.
- Draw a circle at the intersection of references.
- Change the diameter to 0.5 mm.
- Click OK to create and exit the sketch.
- Click OK to finish the sweep.
- Save and exit the file.

Hands-On Exercise, Examples 6.24: User-Defined Parameter—Cost of Tank Shell (Software Used: Creo Parametric)

Steps:

- Create a part file and rename it as Tank_shell. (Other names may be given, but it is advisable to provide a relevant name.)
- Select Solid on the right-hand side of the box. This will activate the solid mode and environment.
- Uncheck the box of the default template and click OK to create a new solid file.
- The template selection box will appear.
- Select mmns_part_Solid. This will allow us to use the solid template file of the following units:
    - mm—millimeter for length unit
    - n—newton for force unit
    - s—second for time unit
- Once the solid part is created, go to File > Prepare > Model properties to check the units. In the upper sections, the units of the model will be displayed.
    - Note: The units usually are displayed in the summary section of any CAD software. The setting of units usually is in the tools or settings.
- The model as shown in Figure 6.35 is to be created.
- Click on Extrude from the Model tab in the ribbon interface.
- Click on the Placement tab and select Define to define the sketch.
- Select the top plane as the sketch plane.
- The right plane shall be automatically taken as reference and orientation as right.
- Click on Sketch.
- The sketcher model will be activated.
- Draw a circle of diameter 500 mm.
- Click OK to create and exit the sketch.
- Select Blind extrusion.
- Enter the value 1000 mm for extrusion depth.

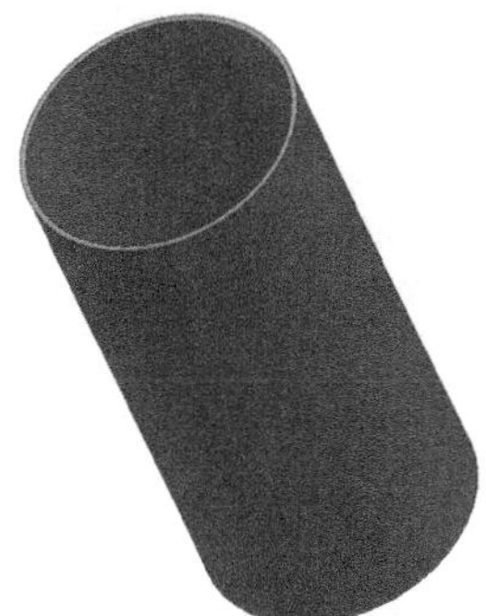

FIGURE 6.35
View of a shell of a tank.

- Click on the Thicken part creation option and enter the value of 10 mm.
- Click on OK to create the extrusion.
- Click on File > Prepare > Model properties.
- Click on Material change to change the material.
- Double-click on Steel to set steel as the material.
- Click Close at the bottom to close the model properties window.
- Click on the Analysis tab of the ribbon interface.
- Click on Mass properties.
- The mass properties analysis window will appear.
- Click on Preview in the mass properties analysis window or middle-click on the mouse button to calculate the mass properties.
- Mass properties will show in the mass properties box.
- In the bottom of the mass properties window, click on the drop-down menu and select Feature to create the analysis as a feature.
- The analyzed mass properties are saved as a feature and displayed in the model tree.
- Click on the Tools tab in the ribbon interface.
- Click on Parameters.
- The parameters window will appear.
- Add a parameter and rename it as cost_per_tonne.
- Keep the type as real number.
- Enter the cost_per_tonne parameter value as 40000.
- Click on Relation from the Tools tab of the ribbon interface.
- Click on Feature in the Look in box on the top.
- Select the feature lastly added in the analyzed mass properties.
- The parameter mass is displayed in the parameter area in the bottom of the relation window.
- Right-click on the parameter mass and select Insert to relation.
- The mass will appear in the relation window.
- Click in the front of the mass in the relation window and write down "Total_Material_Cost = cost_per_tonne*MASS."

- Click on Verify on the upper right side of the relation window.
- A notification box with a message "Relations have been successfully verified" should appear.
- Click OK to close the box.
- A parameter of Total_Material_Cost shall have been added in the parameter window displaying the total material cost.
- Click OK in the bottom section of the relation window to close the window.
- Save and exit the file.

Hands-On Exercise, Example 6.25: User-Defined Parameter—Paint of Tank Outer Surface (Software Used: Creo Parametric)

Steps:

- Create a part file and rename it as User_defined_paint. (Other names may be given, but it is advisable to provide a relevant name.)
- Select Solid on the right-hand side of the box. This will activate the solid mode and environment.
- Uncheck the box of the default template and click OK to create a new solid file.
- The template selection box will appear.
- Select mmns_part_Solid. This will allow us to use the solid template file of the following units:
  - mm—millimeter for length unit
  - n—newton for force unit
  - s—second for time unit
- Once the solid part is created, go to File > Prepare > Model properties to check the units. In the upper sections, the units of the model will be displayed.
  - Note: The units usually are displayed in the summary section of any CAD software. The setting of units usually is in the tools or settings.
- The model as shown in Figure 6.36 is to be created.
- Click on Revolve from the Model tab in the ribbon interface.
- Click on the Thicken part creation option and enter the value of 10 mm.
- Click on the Placement tab and select Define to define the sketch.
- Select the front plane as the sketch plane.
- The right plane shall be automatically taken as reference and orientation as right.
- Click on Sketch.
- The sketcher model will be activated.
- Draw the sketch as per Figure 6.37.
- Click OK to create and exit the sketch.
- Select Blind revolve.
- Enter the value 360 mm for revolving angle.

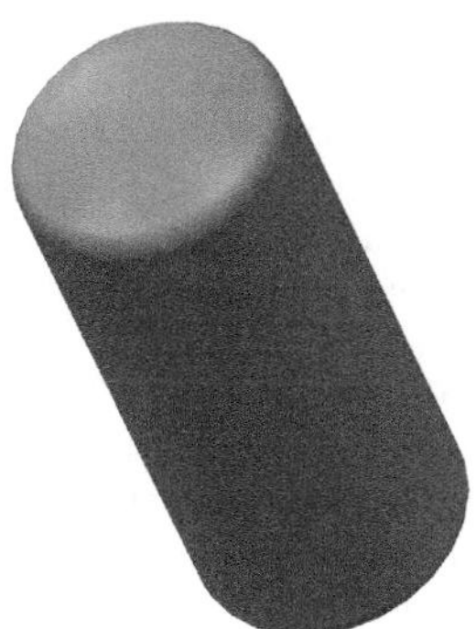

FIGURE 6.36
View of the outer surface of a tank.

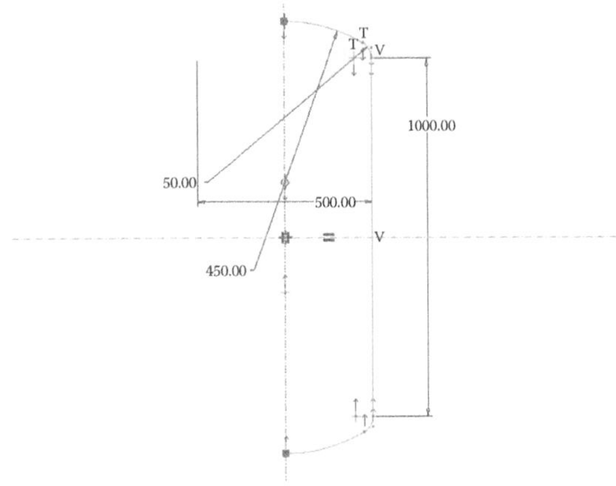

FIGURE 6.37
Constructional sketch of a tank.

- Click on OK to create the revolve feature.
- Click on the Analysis tab of the ribbon interface.
- Click on the drop-down menu of summary and select Area.
- Measure: The area analysis window will appear.
- Select all the outer surfaces one by one by pressing the Control key on the keyboard.
- The total area will be displayed in the result box.
- In the top of the measure area window, click on the drop-down menu and select Make feature to create the analysis as a feature.
- The analyzed measure of area is saved as a feature and displayed in the model tree.
- Click on Relation from the Tools tab of the ribbon interface.
- Click on Feature in the Look in box on the top.
- Select the feature lastly added in the analyzed measure area.
- The parameter Total_area is displayed in the parameter area in the bottom of the relation window.

- Right-click on the parameter mass and select Insert to relation.
- The mass will appear in the relation window.
- Click in the front of the mass in the relation window and write down "Total_Material_Cost = cost_per_tonne*MASS."
- Click on Verify on the upper right side of the relation window.
- A notification box with a message "Relations have been successfully verified" should appear.
- Click OK to close the box.
- A parameter of Total_Material_Cost shall have been added in the parameter window displaying the total material cost.
- Add a parameter and rename it as Primer_coat.
- Keep the change as integer.
- Enter the Primer_coat parameter value as 1.
- Add a parameter and rename it as Paint_coat.
- Keep the change as integer.
- Enter the Paint_coat parameter value as 3.
- Add a parameter and rename it as Primer_area_per_Liter.
- Keep the change as real number.
- Enter the Primer_area_per_Liter parameter value as 10.
- Add a parameter and rename it as Paint_area_per_Liter.
- Keep the change as real number.
- Enter the Paint_area_per_Liter parameter value as 9.
- In the relation area, write down the following:
  - Total_primer_liter = ((TOTAL_AREA/1000000)/Primer_area_per_liter)*Primer_coat
  - Total_paint_liter = ((TOTAL_AREA/1000000)/Paint_area_per_liter)*Paint_coat
  - Note: Total area is divided by 1,000,000 to obtain the area in square meters.
- Click on Verify on the upper right side of the relation window.
- A notification box with a message "Relations have been successfully verified" should appear.
- Click OK to close the box.
- Notice that two parameters, that is, Total_primer_liter and Total_paint_liter, have been added showing the requirement of primer and paint in liters.
- Click OK in the bottom section of the relation window to close the window.
- Save and exit the file.

Hands-On Exercise, Example 6.26: Creating a Knob (Software Used: Creo Parametric)

Steps:

- Create a part file and rename it as Knob. (Other names may be given, but it is advisable to provide a relevant name.)
- Select Solid on the right-hand side of the box. This will activate the solid mode and environment.

- Uncheck the box of the default template and click OK to create a new solid file.
- The template selection box will appear.
- Select mmns_part_Solid. This will allow us to use the solid template file of the following units:
  - mm—millimeter for length unit
  - n—newton for force unit
  - s—second for time unit
- Once the solid part is created, go to File > Prepare > Model properties to check the units. In the upper sections, the units of the model will be displayed.
  - Note: The units usually are displayed in the summary section of any CAD software. The setting of units usually is in the tools or settings.
- The knob as shown in Figure 6.38 is to be created.
- Click on Extrude from the Model tab in the ribbon interface.
- Click on the Placement tab and select Define to define the sketch.
- Select the front plane as the sketch plane.
- The right plane shall be automatically taken as reference and orientation as right.
- Click on Sketch.
- The sketcher model will be activated.
- Draw a circle of diameter 50 mm.
- Click OK to create and exit the sketch.
- Select mid plane extrusion.
- Enter the value 2 mm for extrusion depth.
- Click on OK to create the extrusion.
- Select Extrude again from the Model tab of the ribbon interface.
- Click on the Placement tab and select Define to define the sketch.
- Select the top face of the extruded feature as the sketch plane.
- The right plane shall be automatically taken as reference and orientation as right.
- Click on Sketch.
- The sketcher model will be activated.

FIGURE 6.38
3-D view of a knob.

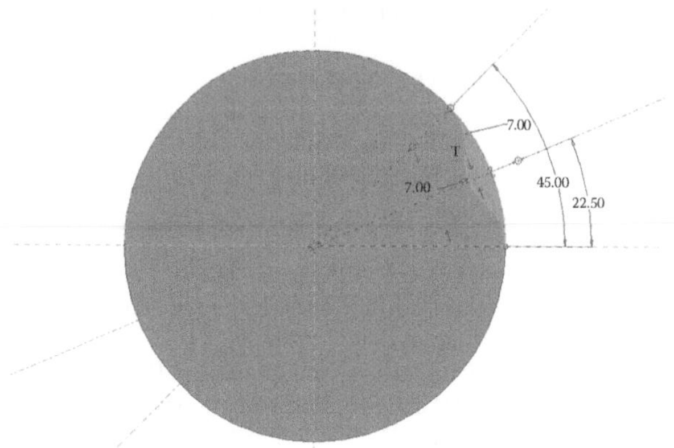

FIGURE 6.39
Constructional sketch of a knob.

- Create the sketch as shown in Figure 6.39.
- Click OK to create and exit the sketch.
- Click on the Material removal icon to make the extrude cut.
- Flip the orientation of extrusion if required.
- Click on the Through all option to make the cut through all the available depth.
- Click OK to create the extrude cut feature.
- Select Extrude again from the Model tab of the ribbon interface.
- Click on the Placement tab and select Define to define the sketch.
- Select the top face as the sketch plane.
- The right plane shall be automatically taken as reference and orientation as right.
- Click on Sketch.
- The sketcher model will be activated.
- Select the icon Project to project the entities.
- Select Loop for the projected entities creation option.
- Select the top surface; it will select the outer profile.
- Click OK to create and exit the sketch.
- Click on Blind extrusion and enter the depth of 20 mm.
- Flip the orientation if required to make the feature in the opposite direction of material removal.
- Click on the Thicken option and set the thickness as 2 mm.
- Flip the orientation outward.
- Click OK to create the extrude feature.
- Select Extrude again from the Model tab of the ribbon interface.
- Click on the Placement tab and select Define to define the sketch.
- Select the bottom face of the first extrusion as the sketch plane.
- The right plane shall be automatically taken as reference and orientation as right.
- Click on Sketch.
- The sketcher model will be activated.

- Draw a circle placing the center in the intersection of the horizontal and vertical refereces.
- Enter 20 mm as the diameter.
- Click OK to create and exit the sketch.
- Flip the orientation of extrusion if required.
- Click on the blind depth option and enter the value of 18 mm to depth.
- Click on the Thicken option and set the thickness as 2 mm.
- Flip the orientation inward.
- Select Extrude again from the Model tab of the ribbon interface.
- Click on the Placement tab and select Define to define the sketch.
- Select the bottom face of the lastly created extrusion as the sketch plane.
- The right plane shall be automatically taken as reference and orientation as right.
- Click on Sketch.
- The sketcher model will be activated.
- Create a rectangle of dimensions 20 × 5 mm keeping the horizontal hands at 20 mm.
- Place the rectangle symmetric to the horizontal and vertical reference lines.
- Click OK to create and exit the sketch.
- Select the Material removal icon to create the extrude cut feature.
- Flip the orientation of extrusion if required.
- Click on the blind depth option and enter the value of 5 mm to depth.
- Click on Round feature from the Model tab of the ribbon interface.
- Set the value of fillet to 1 mm.
- Select the internal curved edges by pressing the Control key for multiple selections.
- Click OK to create the feature.
- Click on Round feature again from the Model tab of the ribbon interface.
- Set the value of fillet to 3 mm.
- Select the external curved edge at top face of the extrusions.
- Click OK to create the feature.

Hands-On Exercise, Example 6.27: Creating a Collet (Software Used: Creo Parametric)

Steps:

- Create a part file and rename it as Collet. (Other names may be given, but it is advisable to provide a relevant name.)
- Select Solid on the right-hand side of the box. This will activate the solid mode and environment.
- Uncheck the box of the default template and click OK to create a new solid file.

- The template selection box will appear.
- Select mmns_part_Solid. This will allow us to use the solid template file of the following units:
  - mm—millimeter for length unit
  - n—newton for force unit
  - s—second for time unit
- Once the solid part is created, go to File > Prepare > Model properties to check the units. In the upper sections, the units of the model will be displayed.
  - Note: The units usually are displayed in the summary section of any CAD software. The setting of units usually is in the tools or settings.
- The collet as shown in Figure 6.40 is to be created.
- Click on Revolve from the Model tab in the ribbon interface.
- Click on the Placement tab and select Define to define the sketch.
- Select the front plane as the sketch plane.
- The right plane shall be automatically taken as reference and orientation as right.
- Click on Sketch.
- The sketcher model will be activated.
- Draw the sketch as displayed in Figure 6.41.

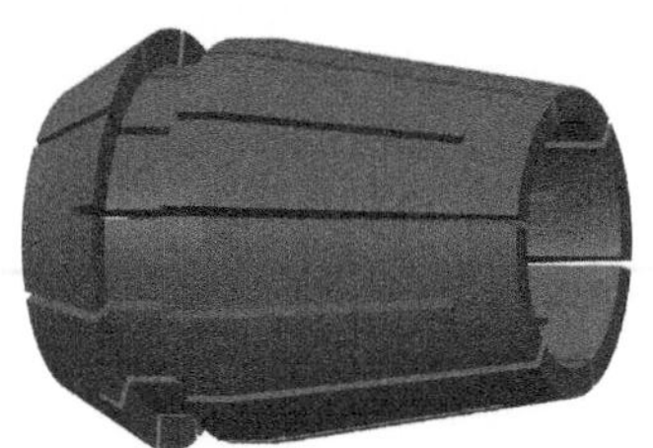

FIGURE 6.40
3-D view of a collet.

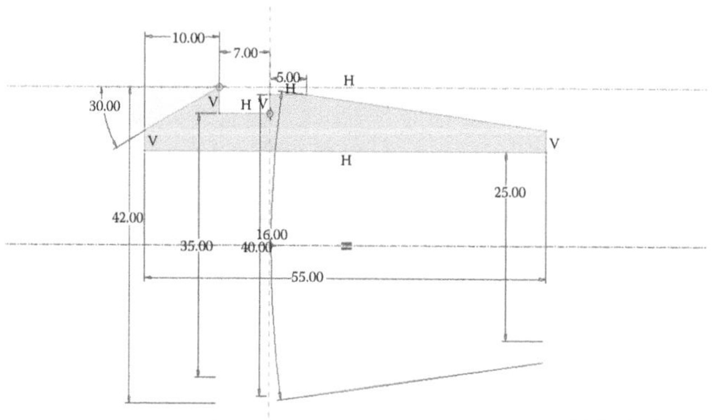

FIGURE 6.41
Constructional sketch of a collet.

- Click OK to create and exit the sketch.
- Ensure the revolve angle shall be 360°.
- Click on OK to create the revolve feature.
- Select Extrude from the Model tab of the ribbon interface.
- Click on the Placement tab and select Define to define the sketch.
- Select the left face of the extruded feature as the sketch plane.
- The right plane shall be automatically taken as reference and orientation as right.
- Click on Sketch.
- The sketcher model will be activated.
- Create a rectangle of 1 mm width starting from the horizontal reference line and ending beyond the circle.
- The rectangle shall be placed symmetrically about the vertical reference line.
- A vertical centerline on the vertical reference line is required to place the rectangle symmetrical to the vertical reference line.
- Click OK to create and exit the sketch.
- Click on the Material removal icon to make the extrude cut.
- Flip the orientation of extrusion if required.
- Enter the depth of cut as 48 mm.
- Select the lastly created extrude feature from the model tree.
- Right-click and select Pattern.
- Select Axis on the type pattern selection box.
- Select the revolved axis.
- Enter the value 6.
- Make sure to activate the equally spaced option on a 360° rotational angle.
- Click OK to create the pattern of the extrude cut feature.
- Similarly, select Extrude again from the Model tab of the ribbon interface.
- Click on the Placement tab and select Define to define the sketch.
- Select the right face of the extruded feature as the sketch plane.
- The right plane shall be automatically taken as reference and orientation as right.
- Click on Sketch.
- The sketcher model will be activated.
- Create the sketch as displayed in Figure 6.42.
- Click OK to create and exit the sketch.
- Click on the Material removal icon to make the extrude cut.
- Flip the orientation of extrusion if required.
- Enter the depth of cut as 48 mm.
- Select the lastly created extrude feature from the model tree.
- Right-click and select Pattern.
- Select Axis on the type pattern selection box.
- Select the revolved axis.
- Enter the value 6.
- Make sure to activate the equally spaced option on a 360° rotational angle.
- Click OK to create the pattern of the extrude cut feature.
- Save and exit the file.

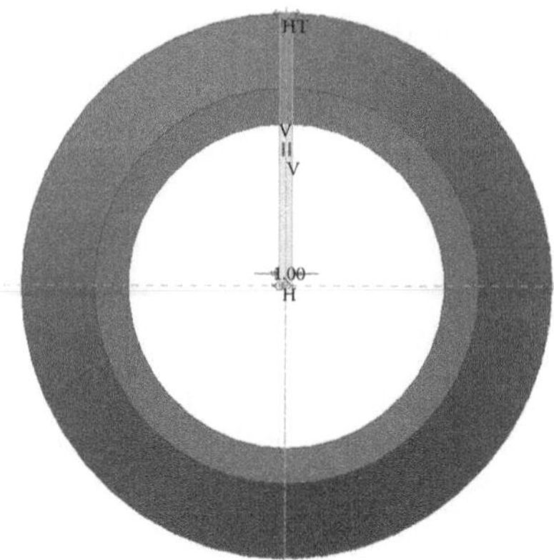

FIGURE 6.42
Constructional sketch for a material removal operation in the creation of a 3-D model of a collet.

Hands-On Exercise, Example 6.28: Creating a Cap (Software Used: Creo Parametric)

Steps:

- Create a part file and rename it as Cap. (Other names may be given, but it is advisable to provide a relevant name.)
- Select Solid on the right-hand side of the box. This will activate the solid mode and environment.
- Uncheck the box of the default template and click OK to create a new solid file.
- The template selection box will appear.
- Select mmns_part_Solid. This will allow us to use the solid template file of the following units:
    - mm—millimeter for length unit
    - n—newton for force unit
    - s—second for time unit
- Once the solid part is created, go to File > Prepare > Model properties to check the units. In the upper sections, the units of the model will be displayed.
    - Note: The units usually are displayed in the summary section of any CAD software. The setting of units usually is in the tools or settings.
- The cap as shown in Figure 6.43 is to be created.
- Click on Revolve from the Model tab in the ribbon interface.
- Click on the Placement tab and select Define to define the sketch.
- Activate the thin feature option.
- Enter the thickness as 2 mm.
- Select the front plane as the sketch plane.

FIGURE 6.43
3-D view of a cap.

- The right plane shall be automatically taken as reference and orientation as right.
- Click on Sketch.
- The sketcher model will be activated.
- Draw the sketch as displayed in Figure 6.44.
- Click OK to create and exit the sketch.
- Ensure the revolve angle shall be 360°.
- Click on OK to create the revolve feature.
- Click on Round feature from the Model tab of the ribbon interface.
- Create the fillet as shown in Figure 6.45.
- Click OK to create the feature.

- Click on Extrude from the Model tab in the ribbon interface.
- Click on the Placement tab and select Define to define the sketch.
- Select the bottom face plane as the sketch plane.
- The right plane shall be automatically taken as reference and orientation as right.
- Click on Sketch.
- The sketcher model will be activated.
- Project the outside circular edge.
- Click OK to create and exit the sketch.
- Select thin extrusion.

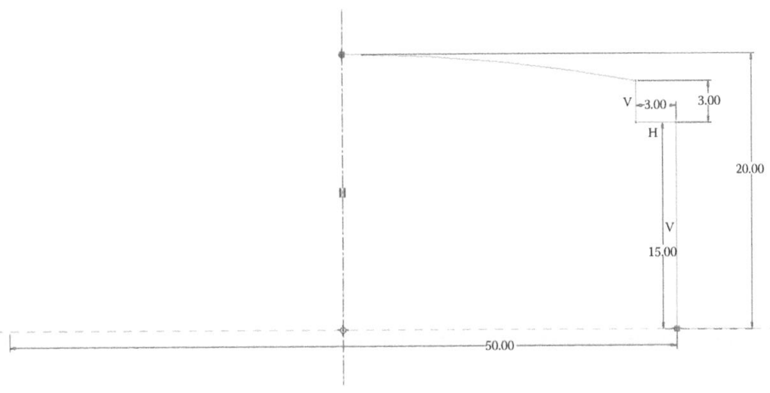

FIGURE 6.44
Constructional sketch of a cap.

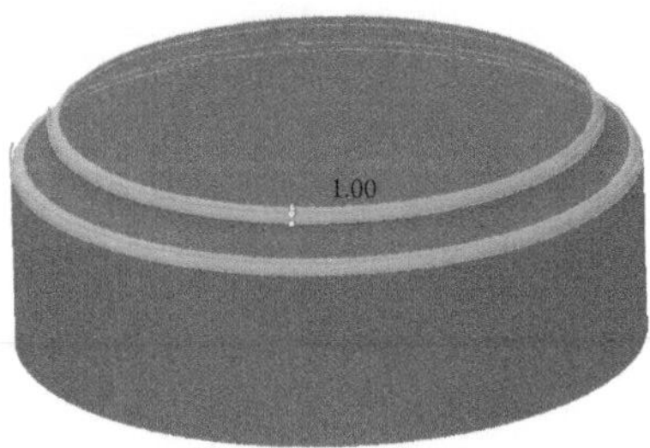

FIGURE 6.45
Constructional sketch for a material removal operation of a cap.

- Set the value as 1 mm.
- Flip the orientation of material thickness inward.
- Enter the value 2 mm for extrusion depth.
- Slip the feature creation orientation toward the body of the cap.
- Click on OK to create the extrusion.

- Click on Extrude from the Model tab in the ribbon interface.
- Click on the Placement tab and select Define to define the sketch.
- Select the inside groove face plane as the sketch plane.
- The right plane shall be automatically taken as reference and orientation as right.
- Click on Sketch.
- The sketcher model will be activated.
- Project the inside circular edge of the groove.
- Click OK to create and exit the sketch.
- Select thin extrusion.
- Set the value as 1 mm.
- Flip the orientation of material thickness outward.
- Enter the value 3 mm for extrusion depth.
- Slip the feature creation orientation downward.
- Click on OK to create the extrusion.
- Click on Round feature from the Model tab of the ribbon interface.
- Set the value of fillet to 0.5 mm.
- Select two outward edges of the newly created extrusion.
- Click OK to create the feature.

- Click on the Helical sweep option from the drop-down menu of Sweep from the Model tab of the ribbon interface.
- Click on the References tab and select Define to define the helix sweep profile.
- Select the front plane as the sketch plane.
- The right plane shall be automatically taken as reference and orientation as right.
- Click on Sketch.
- The sketcher model will be activated.

- Draw a geometry centerline on the vertical reference line in the middle.
- Draw a vertical line at the right-hand inside edge by taking the reference of the cylindrical surface. The line should start 1 mm above the bottom horizontal surface.
- Set the length of the line as 7 mm.
- Click OK to create the sketch and exit the sketcher window.
- Click Sketch to create the profile sketch above the Reference tab.
- Draw a half circle of diameter 1.5 mm placing the center at the intersection of reference lines.
- Click OK to create and exit the sketch.
- Enter the pitch value of 3 mm.
- Click on the Right-handed rule icon to create the helical spring clockwise downward.
- Click OK to create the thread.

- Click on Extrude from the Model tab in the ribbon interface.
- Click on the Placement tab and select Define to define the sketch.
- Select the top surface of the bottom extrusion as the sketch plane.
- The right plane shall be automatically taken as reference and orientation as right.
- Click on Sketch.
- The sketcher model will be activated.
- Draw the sketch as shown in Figure 6.46.
- Click OK to create and exit the sketch.
- Flip the material orientation upward.
- Enter the value 11 mm for extrusion depth.
- Click on OK to create the extrusion.

- Click on Round feature from the Model tab of the ribbon interface.
- Set the value of fillet to 0.5 mm.
- Select the side and top edges of the last extrude.
- Click OK to create the feature.

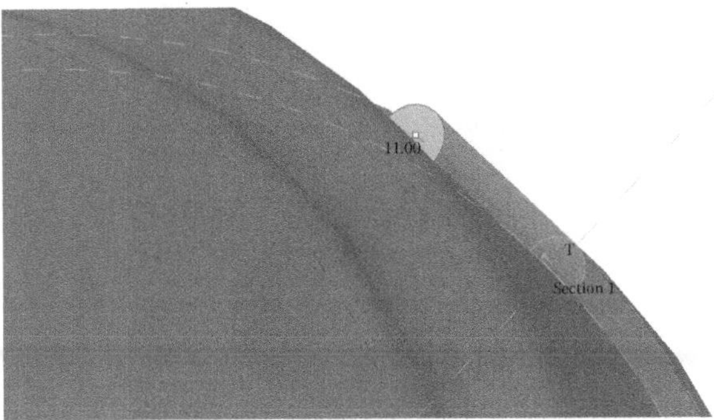

FIGURE 6.46
Constructional sketch of a feature in the creation of 3-D model of a cap.

- Select the last extrusion and round feature from the model tree by keeping the Control key pressed.
- Select the newly created group in the model tree.
- Right-click and select Pattern.
- Select Axis in the Type of pattern selection field.
- Select the central axis.
- Enter the quantity 24 for patterned feature.
- Select the option to make the pattern equally spaced within 360°.
- Click OK to create the patterned feature.
- Save and exit the file.

Hands-On Exercise, Example 6.29: Creating an Industrial Hook (Software Used: Creo Parametric)

Steps:

- Create a part file and rename it as Industrial hook. (Other names may be given, but it is advisable to provide a relevant name.)
- Select Solid on the right-hand side of the box. This will activate the solid mode and environment.
- Uncheck the box of the default template and click OK to create a new solid file.
- The template selection box will appear.
- Select mmns_part_Solid. This will allow us to use the solid template file of the following units:
  - mm—millimeter for length unit
  - n—newton for force unit
  - s—second for time unit
- Once the solid part is created, go to File > Prepare > Model properties to check the units. In the upper sections, the units of the model will be displayed.
  - Note: The units usually are displayed in the summary section of any CAD software. The setting of units usually is in the tools or settings.
- The industrial hook as shown in Figure 6.47 is to be created.

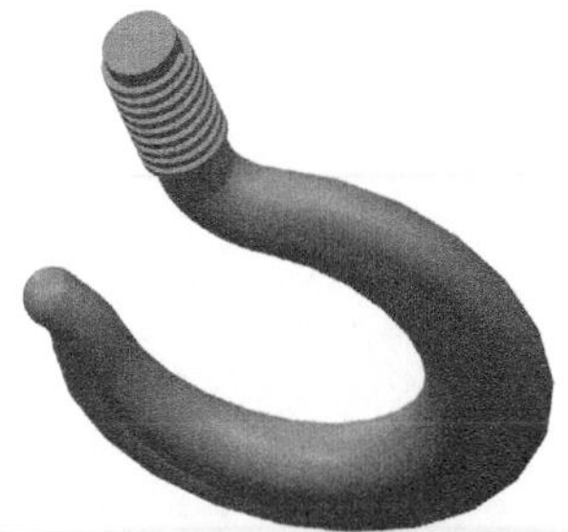

FIGURE 6.47
3-D view of an industrial hook.

- Click on Sweep from the Model tab of the ribbon interface.
- The sweep trajectory will be created first followed by a sweep operation.
- The trajectory can also be created inside the sweep operation; however, the sketch will not be available if the sweep feature is deleted.
- Select Sketch from the Model tab of the ribbon interface.
- A sketch window for selection of sketch plane will appear.
- Select the front plane as the sketch plane.
- The right plane shall automatically be selected as reference and the orientation is to be right.
- Click on Sketch.
- The sketcher window will open up and the reference selection window appears for selection of references.
- The two references (i.e., top plane as horizontal reference and right plane as vertical reference) shall be automatically taken; however, if not, please select the same.
- Create a sketch as displayed in Figure 6.48.
- Click OK to create and exit the sketch.
- Click Datum plane on the Model tab from the ribbon interface.
- Select any end point of the curve and the adjacent curve by pressing the Control key.
- Set the end point reference as Through and curve as Normal. This will allow creation of a plane perpendicular to the curve at the end point.
- Similarly, click Datum plane again on the Model tab from the ribbon interface.
- Select the other end point of the curve and the adjacent curve by pressing the Control key.
- Set the end point reference as Through and curve as Normal. This will allow creation a plane perpendicular to the curve at the end point.
- Select Sketch from the Model tab of the ribbon interface.

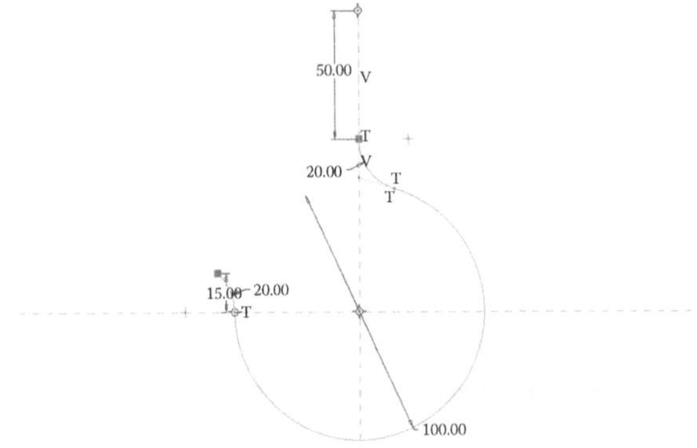

FIGURE 6.48
Constructional sketch of an industrial hook.

- A sketch window for selection of sketch plane will appear.
- Select the top plane as the sketch plane.
- The right plane shall automatically be selected as reference and the orientation is to be right.
- Click on Sketch.
- The sketcher window will open up and the reference selection window appears for selection of references.
- The two references (i.e., top plane as horizontal reference and right plane as vertical reference) shall be automatically taken; however, if not, please select the same.
- Create a sketch as displayed in Figure 6.49.
- Click OK to create and exit the sketch.
- Select Sketch again from the Model tab of the ribbon interface.
- A sketch window for selection of sketch plane will appear.
- Select the right plane as the sketch plane.
- The top plane shall automatically be selected as reference and the orientation is to be top.
- Click on Sketch.
- The sketcher window will open up and the reference selection window appears for selection of references.
- The two references (i.e., top plane as horizontal reference and right plane as vertical reference) shall be automatically taken; however, if not, please select the same.
- Create the sketch as displayed in Figure 6.50.
- Click OK to create and exit the sketch.
- Select Sketch again from the Model tab of the ribbon interface.
- A sketch window for selection of sketch plane will appear.
- Select the top plane as the sketch plane.
- The right plane shall automatically be selected as reference and the orientation is to be right.
- Click on Sketch.

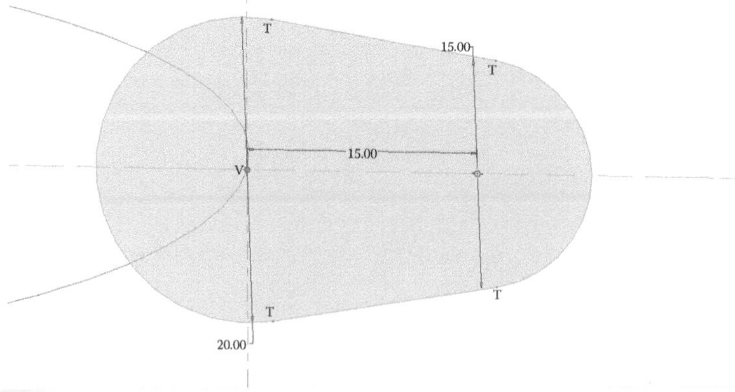

FIGURE 6.49
Constructional sketch of the profile of an industrial hook.

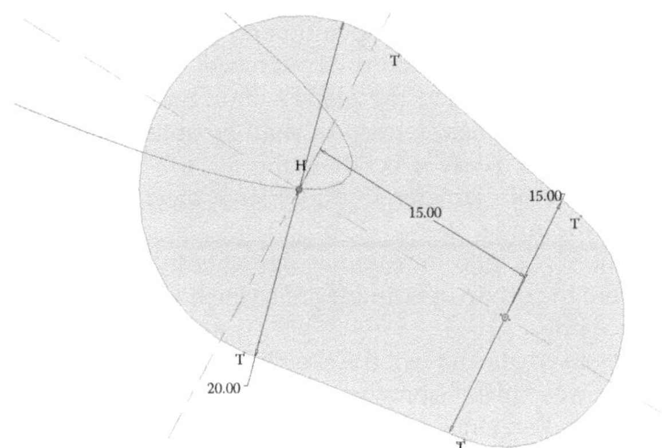

FIGURE 6.50
Constructional sketch of the profile of an industrial hook.

- The sketcher window will open up and the reference selection window appears for selection of references.
- The two references (i.e., top plane as horizontal reference and right plane as vertical reference) shall be automatically taken; however, if not, please select the same.
- Create an oblong hole of diameter 15 mm and center-to-center distance 5 mm. Pierce the origin of the top arc center to the curve.
- Click OK to create and exit the sketch.
- Select Sketch again from the Model tab of the ribbon interface.
- A sketch window for selection of sketch plane will appear.
- Select the newly created angular plane as the sketch plane.
- The right plane shall automatically be selected as reference and the orientation is to be right.
- Click on Sketch.
- The sketcher window will open up and the reference selection window appears for selection of references.
- The two references (i.e., top plane as horizontal reference and right plane as vertical reference) shall be automatically taken; however, if not, please select the same.
- Create a hole of diameter 15 mm. Pierce the origin of top arc center to the curve.
- Break the circle into four equal segments by creating centerlines from the origin at an angle 45° to the horizontal or vertical reference line.
- Click OK to create and exit the sketch.
- Select Sketch again from the Model tab of the ribbon interface.
- A sketch window for selection of sketch plane will appear.
- Select the newly created plane that is parallel to the top plane as the sketch plane.
- The right plane shall automatically be selected as reference and the orientation is to be right.
- Click on Sketch.

- The sketcher window will open up and the reference selection window appears for selection of references.
- The two references (i.e., top plane as horizontal reference and right plane as vertical reference) shall be automatically taken; however, if not, please select the same.
- Create a hole of diameter 15 mm. Pierce the origin of top arc center to the curve.
- Break the circle into four equal segments by creating centerlines from the origin at an angle 45° to the horizontal or vertical reference line.
- Click OK to create and exit the sketch.
- Click on Sweep blend from the Model tab of the ribbon interface.
- Click on the Reference tab.
- Graphically select the first sketch as a trajectory.
- The system assumes the profile sketch point and shows it by displaying the directional arrow.
- Keep the section plane control as normal to trajectory.
- Click on the Sections tab.
- Select Selected sections.
- Select all the sections one by one starting from top by clicking the inset button.
- Make sure the start point shall be in the same quadrant for all the sections.
- Click OK to create the swept blend feature.
- Click on Extrude from the Model tab in the ribbon interface.
- Click on the Placement tab and select Define to define the sketch.
- Select the top face of the swept blend feature as the sketch plane.
- The right plane shall be automatically taken as reference and orientation as right.
- Click on Sketch.
- The sketcher model will be activated.
- Draw a circle by projecting the circular edge.
- Click OK to create and exit the sketch.
- Enter the value 50 mm for extrusion depth.
- Click on OK to create the extrusion.
- Click on the Helical sweep option from the drop-down menu of Sweep from the Model tab of the ribbon interface.
- Click on References tab and select Define to define the helix sweep profile.
- Select the front plane as the sketch plane.
- The right plane shall be automatically taken as reference and orientation as right.
- Click on Sketch.
- The sketcher model will be activated.
- Draw a geometry centerline on the vertical reference line in the middle.
- Draw a vertical line at the right-hand edge by taking the reference of the cylindrical surface. The line should start 5 mm below the top horizontal surface.

- Set the length of the line as 36 mm.
- Click OK to create the sketch and exit the sketcher window.
- Click Sketch to create the profile sketch above the Reference tab.
- Create a square of 2 × 2 mm placing the left edge of the square at the vertical reference line and symmetric to the horizontal line.
- Click OK to create and exit the sketch.
- Enter the pitch value of 4 mm.
- Click on the Right-handed rule icon to create the helical spring clockwise downward.
- Click OK to create the thread.

Hands-On Exercise, Example 6.30: Creating a Multistart Thread (Software Used: Creo Parametric)

Steps:

- Create a part file and rename it as Multistart thread. (Other names may be given, but it is advisable to provide a relevant name.)
- Select Solid on the right-hand side of the box. This will activate the solid mode and environment.
- Uncheck the box of the default template and click OK to create a new solid file.
- The template selection box will appear.
- Select mmns_part_Solid. This will allow us to use the solid template file of the following units:
  - mm—millimeter for length unit
  - n—newton for force unit
  - s—second for time unit
- Once the solid part is created, go to File > Prepare > Model properties to check the units. In the upper sections, the units of the model will be displayed.
  - Note: The units usually are displayed in the summary section of any CAD software. The setting of units usually is in the tools or settings.
- The multistart thread as shown in Figure 6.51 is to be created.
- Click on Extrude from the Model tab in the ribbon interface.
- Click on the Placement tab and select Define to define the sketch.
- Select the front plane as the sketch plane.
- The right plane shall be automatically taken as reference and orientation as right.
- Click on Sketch.
- The sketcher model will be activated.
- Draw a circle of diameter 20 mm.
- Click OK to create and exit the sketch.
- Enter the value 200 mm for extrusion depth.
- Click on OK to create the extrusion.

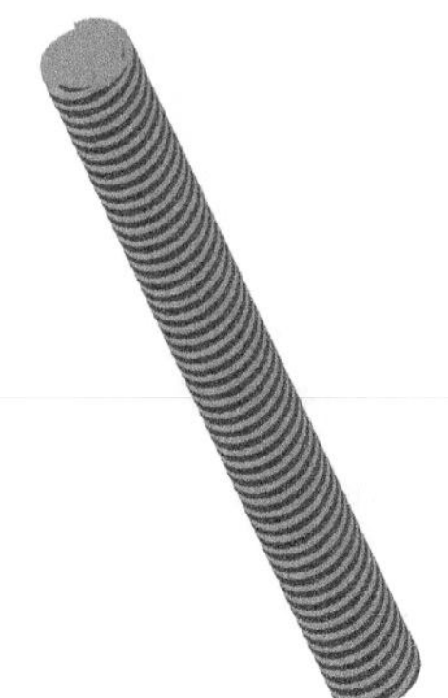

FIGURE 6.51
3-D view of a multistart thread.

- Click on the Helical sweep option from the drop-down menu of Sweep from the Model tab of the ribbon interface.
- Click on References tab and select Define to define the helix sweep profile.
- Select the front plane as the sketch plane.
- The right plane shall be automatically taken as reference and orientation as right.
- Click on Sketch.
- The sketcher model will be activated.
- Draw a geometry centerline on the vertical reference line in the middle.
- Draw a vertical line at the right-hand edge by taking the reference of the cylindrical surface. The line should start 10 mm below the bottom horizontal surface and end 10 mm above the top horizontal surface.
- Click OK to create the sketch and exit the sketcher window.
- Click Sketch to create the profile sketch above the Reference tab.
- Create a square of 2 × 2 mm placed in the second quadrant.
- Click OK to create and exit the sketch.
- Enter the pitch value of 12 mm.
- Click on the Right-handed rule icon to create the helical spring clockwise downward.
- Click OK to create the thread.
- Select the helical sweep in the model tree.
- Right-click and select Pattern.
- Select Axis in the Type of pattern selection field.
- Select the central axis.
- Enter the quantity 3 for patterned feature.
- Select the option to make the pattern equally spaced within 360°.
- Click OK to create the patterned feature.
- Save and exit the file.

Hands-On Exercise, Example 6.31: Creating a Spanner (Software Used: Creo Parametric)

Steps:

- Create a part file and rename it as Spanner. (Other names may be given, but it is advisable to provide a relevant name.)
- Select Solid on the right-hand side of the box. This will activate the solid mode and environment.
- Uncheck the box of the default template and click OK to create a new Solid file.
- The template selection box will appear.
- Select mmns_part_Solid. This will allow us to use the solid template file of the following units:
  - mm—millimeter for length unit
  - n—newton for force unit
  - s—second for time unit
- Once the solid part is created, go to File > Prepare > Model properties to check the units. In the upper sections, the units of the model will be displayed.
  - Note: The units usually are displayed in the summary section of any CAD software. The setting of units usually is in the tools or settings.
- The spanner as shown in Figure 6.52 is to be created.
- Click on Extrude from the Model tab in the ribbon interface.
- Click on the Placement tab and select Define to define the sketch.
- Select the front plane as the sketch plane.
- The right plane shall be automatically taken as reference and orientation as right.
- Click on Sketch.
- The sketcher model will be activated.
- Draw the sketch as shown in Figure 6.53.
- Click OK to create and exit the sketch.
- Select mid plane extrusion.
- Enter the value 4 mm for extrusion depth.
- Click on OK to create the extrusion.

- Click on Extrude from the Model tab in the ribbon interface.
- Click on the Placement tab and select Define to define the sketch.
- Select the front plane as the sketch plane.

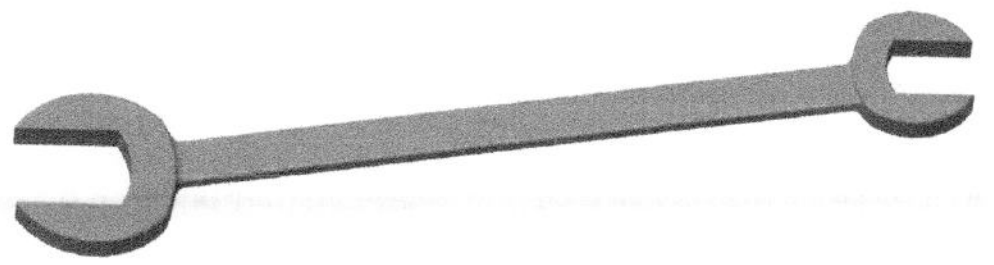

FIGURE 6.52
3-D view of a spanner.

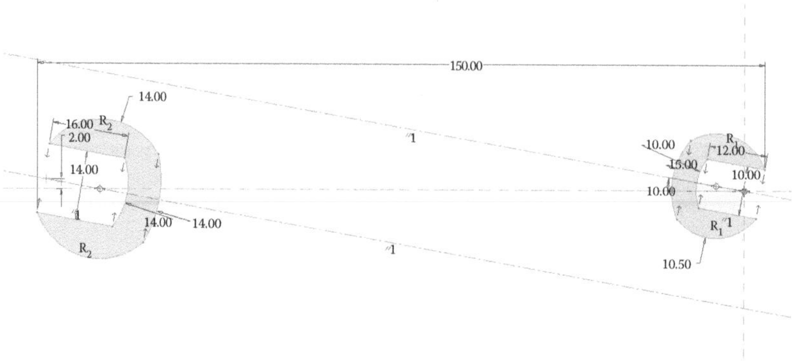

FIGURE 6.53
Constructional sketch of a spanner.

- The right plane shall be automatically taken as reference and orientation as right.
- Click on Sketch.
- The sketcher model will be activated.
- Draw the sketch as shown in Figure 6.54.
- Click OK to create and exit the sketch.
- Select mid plane extrusion.
- Enter the value 2 mm for extrusion depth.
- Click on OK to create the extrusion.

- Click on Round feature from the Model tab of the ribbon interface.
- Set the value as required for the edges shown in Figure 6.54.
- Select the edges as displayed in Figure 6.55.
- The fillets may need to be created by individual fillet operation.
- Click OK to create the feature.
- Save and exit the file.

FIGURE 6.54
Constructional sketch of features of a spanner.

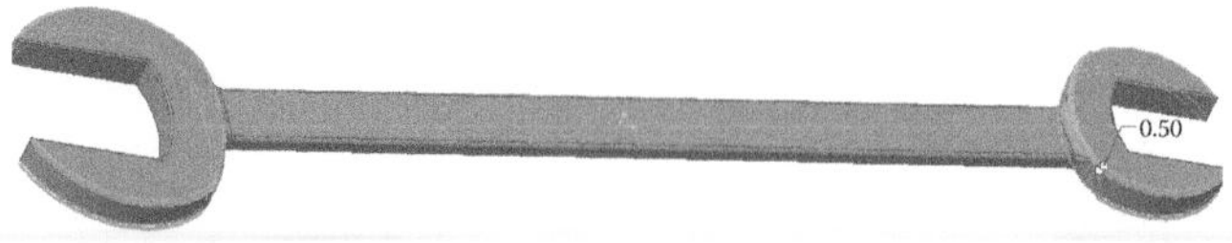

FIGURE 6.55
View of edges for round operation of a spanner.

Hands-On Exercise, Example 6.32: Creating a Screwdriver (Software Used: Creo Parametric)

Steps:

- Create a part file and rename it as Screwdriver. (Other names may be given, but it is advisable to provide a relevant name.)
- Select Solid on the right-hand side of the box. This will activate the solid mode and environment.
- Uncheck the box of the default template and click OK to create a new solid file.
- The template selection box will appear.
- Select mmns_part_Solid. This will allow us to use the solid template file of the following units:
  - mm—millimeter for length unit
  - n—newton for force unit
  - s—second for time unit
- Once the solid part is created, go to File > Prepare > Model properties to check the units. In the upper sections, the units of the model will be displayed.
  - Note: The units usually are displayed in the summary section of any CAD software. The setting of units usually is in the tools or settings.
- The screwdriver as shown in Figure 6.56 is to be created.
- Click on Revolve from the Model tab in the ribbon interface.
- Click on the Placement tab and select Define to define the sketch.
- Select the front plane as the sketch plane.

FIGURE 6.56
3-D view of a screwdriver.

- The right plane shall be automatically taken as reference and orientation as right.
- Click on Sketch.
- The sketcher model will be activated.
- Draw the sketch as displayed in Figure 6.57.
- Click OK to create and exit the sketch.
- Ensure the revolve angle shall be 360°.
- Click on OK to create the revolve feature.

- Click on Revolve from the Model tab in the ribbon interface.
- Click on the Placement tab and select Define to define the sketch.
- Select the front plane as the sketch plane.
- The right plane shall be automatically taken as reference and orientation as right.
- Click on Sketch.
- The sketcher model will be activated.
- Draw the sketch as displayed in Figure 6.58.

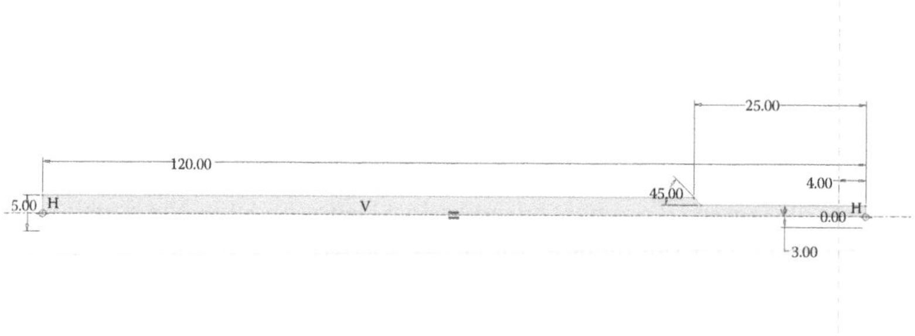

FIGURE 6.57
Constructional sketch of a screwdriver.

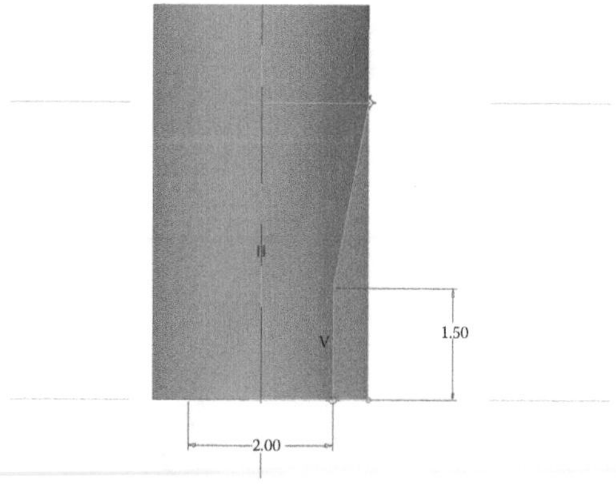

FIGURE 6.58
Constructional sketch of features of a screwdriver.

- Click OK to create and exit the sketch.
- Ensure the revolve angle shall be 360°.
- Select the material removal option.
- Click on OK to create the revolve feature.

- Select Sketch from the Model tab of the ribbon interface.
- A sketch window for selection of sketch plane will appear.
- Select the front plane as the sketch plane.
- The right plane shall automatically be selected as reference and the orientation is to be right.
- Click on Sketch.
- The sketcher window will open up and the reference selection window appears for selection of references.
- The two references (i.e., top plane as horizontal reference and right plane as vertical reference) shall be automatically taken; however, if not, please select the same.
- Create a sketch as displayed in Figure 6.59.
- Click OK to create and exit the sketch.

- Click on Round feature from the Model tab of the ribbon interface.
- Set the value of chamfer to 0.5 mm (D X D).
- Select all bottom circular edges.
- Click OK to create the feature.

- Click on Sweep from the Model tab of the ribbon interface.
- Select the sketch as sweep trajectory.
- Click on Sweep from the Model tab of the ribbon interface.
- Graphically select the sketch as a trajectory.
- The system assumes the profile sketch point and shows it by displaying the directional arrow.
- Graphically click on the arrows of feature creation to draw the sweep profile in other control points.

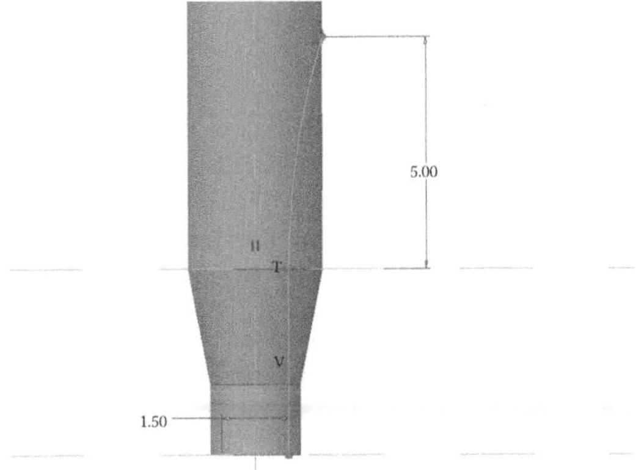

FIGURE 6.59
Constructional sketch of features of a screwdriver.

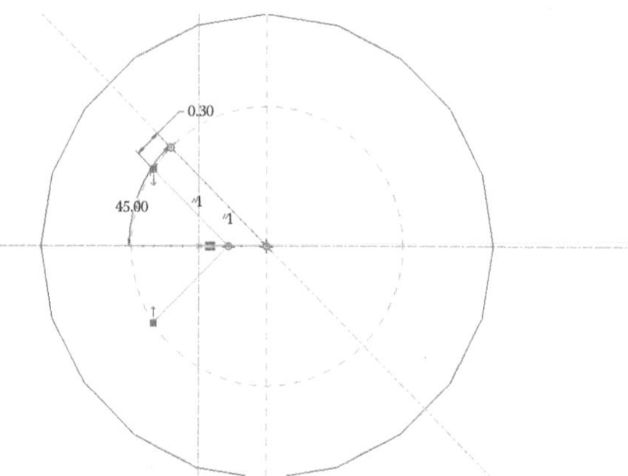

FIGURE 6.60
Constructional sketch of features of a screwdriver.

- Click on Sketch to sketch the profile.
- The sketcher mode activates with the sketcher environment.
- Draw the sketch as shown in Figure 6.60.
- Click OK to create and exit the sketch.
- Select the material removal option.
- Click OK to finish the sweep.

- Click on Round feature from the Model tab of the ribbon interface.
- Set the value of fillet to 0.5 mm.
- Select all two intermediate circular edges of the newly created feature.
- Click OK to create the feature.

- Click Ok datum plane from the Model tab of the ribbon interface.
- Select the top face as the reference plane.
- Set the offset value 10 mm downward.
- Click OK to create the datum plane.

- Click on Extrude from the Model tab in the ribbon interface.
- Click on the Placement tab and select Define to define the sketch.
- Select the newly created plane as the sketch plane.
- The right plane shall be automatically taken as reference and orientation as right.
- Click on Sketch.
- The sketcher model will be activated.
- Draw the sketch as displayed in Figure 6.61.
- Click OK to create and exit the sketch.
- Enter the value 15 mm for extrusion depth.
- If required, flip the material orientation to make the cut downward.
- Select the material removal option.
- Click on OK to create the extrusion.

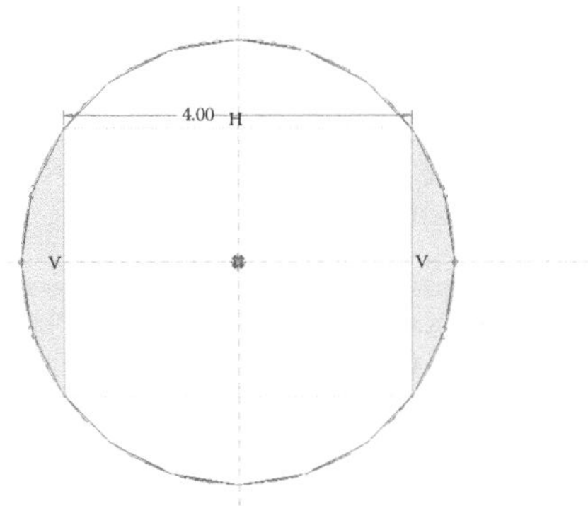

FIGURE 6.61
Constructional sketch of features of a screwdriver.

- Click on Extrude from the Model tab in the ribbon interface.
- Click on the Placement tab and select Define to define the sketch.
- Select the surface created by the last extrusion cut as the sketch plane.
- The right plane shall be automatically taken as reference and orientation as right.
- Click on Sketch.
- The sketcher model will be activated.
- Draw the sketch as displayed in Figure 6.62.
- Click OK to create and exit the sketch.
- Select the Up to surface option.
- Select the opposite cut surface of the same side.
- Click on OK to create the extrusion.

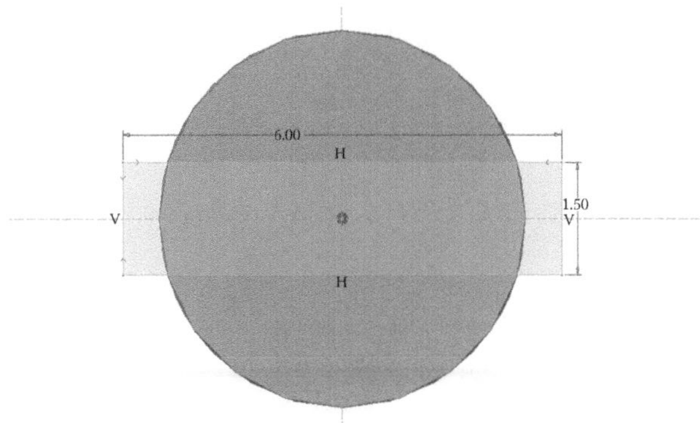

FIGURE 6.62
Constructional sketch of features of a screwdriver.

- Click on Round feature from the Model tab of the ribbon interface.
- Set the value of fillet to 0.5 mm.
- Select all two external straight edges of the lastly created extrusion.
- Click OK to create the feature.
- Save and exit the file.

Hands-On Exercise, Example 6.33: Creating a Plastic Bottle (Software Used: Creo Parametric)

Steps:

- Create a part file and rename it as Plastic bottle. (Other names may be given, but it is advisable to provide a relevant name.)
- Select Solid on the right-hand side of the box. This will activate the solid mode and environment.
- Uncheck the box of the default template and click OK to create a new solid file.
- The template selection box will appear.
- Select mmns_part_Solid. This will allow us to use the solid template file of the following units:
  - mm—millimeter for length unit
  - n—newton for force unit
  - s—second for time unit
- Once the solid part is created, go to File > Prepare > Model properties to check the units. In the upper sections, the units of the model will be displayed.
  - Note: The units usually are displayed in the summary section of any CAD software. The setting of units usually is in the tools or settings.
- The plastic bottle as shown in Figure 6.63 is to be created.

FIGURE 6.63
3-D view of a plastic bottle.

- Select the Blend feature from the ribbon at the top. Clicking on the feature opens the selection box for all other options.
- You may change the name of the feature to be created; it helps to trace back the feature in the future so that one easily can understand which feature is used where.
- Click on the References tab and select Sketched feature.
- Click on Define, select the top plane as the sketch plane, and click on Sketch.
- The horizontal and vertical references are to be taken automatically by the system.
- Draw a rectangle of dimensions 60 × 60 mm symmetric to the horizontal and vertical centerline.
- Click OK to finish the sketch.
- Click Insert to sketch another section and provide 15 as the offset from the secion1 option (offset dimension option).
- Click on Sketch to sketch a circle.
- The horizontal and vertical references are to be taken automatically by the system.
- Draw a circle placing the center point of the circle at the intersection of the horizontal and vertical reference lines.
- Make the diameter of the circle 46 mm.
- Draw two centerlines at an angle 45° to the horizontal and vertical reference lines.
- Break the circle into four parts at the intersection of centerlines.
- Click OK to finish the sketch.
- Click OK to create the blend feature.
- Click on Round feature from the Model tab of the ribbon interface.
- Set the value of fillet to 5 mm.
- Select all four inclined edges of the newly created feature.
- Click OK to create the feature.

- Click on Extrude from the Model tab in the ribbon interface.
- Click on the Placement tab and select Define to define the sketch.
- Select the bottom surface as the sketch plane.
- The right plane shall be automatically taken as reference and orientation as right.
- Click on Sketch.
- The sketcher model will be activated.
- Project the outer edges of the bottom surface.
- Click OK to create and exit the sketch.
- Select the extrusion downward.
- Enter the value 210 mm for extrusion depth.
- Click on OK to create the extrusion.

- Click on Extrude from the Model tab in the ribbon interface.
- Click on the Placement tab and select Define to define the sketch.
- Select the top surface as the sketch plane.
- The right plane shall be automatically taken as reference and orientation as right.
- Click on Sketch.

- The sketcher model will be activated.
- Project the outer edges of the top surface.
- Click OK to create and exit the sketch.
- Select the extrusion upward.
- Enter the value 20 mm for extrusion depth.
- Click on OK to create the extrusion.

- Click on Revolve from the Model tab in the ribbon interface.
- Click on the Placement tab and select Define to define the sketch.
- Select the front plane as the sketch plane.
- The right plane shall be automatically taken as reference and orientation as right.
- Click on Sketch.
- The sketcher model will be activated.
- Draw the sketch as displayed in Figure 6.64.
- Click OK to create and exit the sketch.
- Ensure the revolve angle shall be 360°.
- Click on the material removal option.
- Click on OK to create the revolve feature.

- Click Ok datum plane from the Model tab of the ribbon interface.
- Select the top plane as the reference plane.
- Set the offset value 15 mm downward.
- Click OK to create the datum plane.

- Click Shell from the Model tab of the ribbon interface.
- Set the shell thickness as 1 mm.
- Select the top circular flat surface to remove and create the bottle opening.
- Click OK to create the shell feature.

- Click on Extrude from the Model tab in the ribbon interface.
- Click on the Placement tab and select Define to define the sketch.

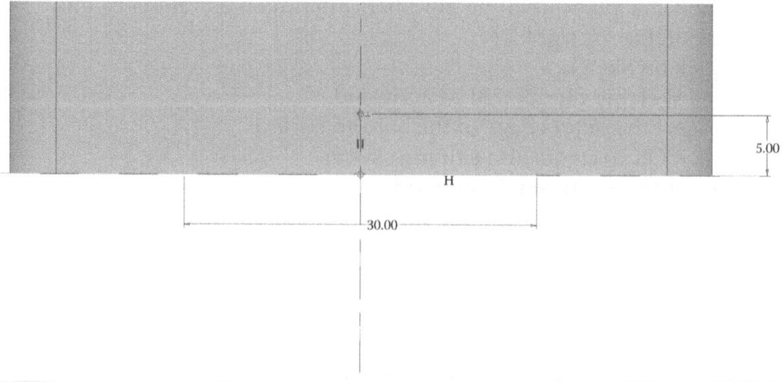

FIGURE 6.64
Constructional sketch of features of a plastic bottle.

- Select the lastly created datum plane as the sketch plane.
- The right plane shall be automatically taken as reference and orientation as right.
- Click on Sketch.
- The sketcher model will be activated.
- Draw the sketch as displayed in Figure 6.65.
- Click OK to create and exit the sketch.
- Select the extrusion downward.
- Enter the value 180 mm for extrusion depth.
- Click on OK to create the extrusion.

- Click on Round feature from the Model tab of the ribbon interface.
- Set the value of fillet to 1.5 mm.
- Select all two circular outside boundary edges of the newly created feature.
- Click OK to create the feature.

- Click on Round feature from the Model tab of the ribbon interface.
- Set the value of fillet to 2 mm.
- Select all outside boundary edges of the newly created feature.
- Click OK to create the feature.

- Select the two lastly created rounds and extrusions by pressing the Control key.
- Right-click and select Group.

- Right-click on the group and select Pattern.
- Select the direction in the pattern type selection field.
- Select the right plane for pattern direction.
- Set the value of 5 instances at 10 mm distance toward the right plane.

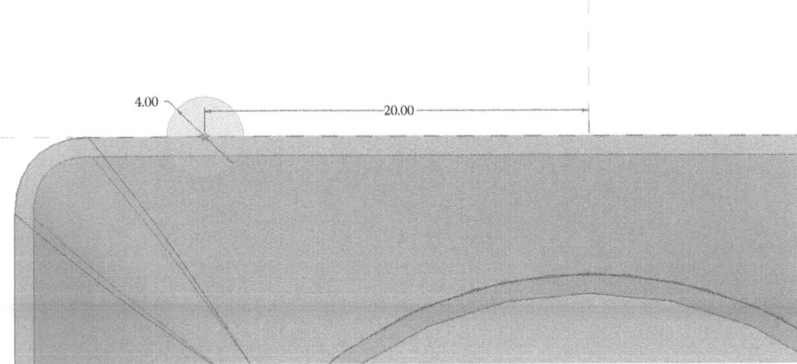

FIGURE 6.65
Constructional sketch of features of a plastic bottle.

- Click OK to create the patterned feature.
- Similarly, repeat the aforementioned steps to create the extrusion and round feature in the other three sides.

- Click on Round feature from the Model tab of the ribbon interface.
- Set the value of fillet to 3 mm.
- Select all four bottom edges and four top edges of the bottle body.
- Click OK to create the feature.
- Click on Round feature from the Model tab of the ribbon interface.
- Set the value of fillet to 7 mm.
- Select the circular edge at the bottom revolve.
- Click OK to create the feature.

- Click on Revolve from the Model tab in the ribbon interface.
- Click on the thin feature option.
- Set the value 1 mm for thin revolve.
- Click on the Placement tab and select Define to define the sketch.
- Select the front plane as the sketch plane.
- The right plane shall be automatically taken as reference and orientation as right.
- Click on Sketch.
- The sketcher model will be activated.
- Draw the sketch as displayed in Figure 6.66.
- Click OK to create and exit the sketch.
- Ensure the revolve angle shall be 360°.
- Click on OK to create the revolve feature.

- Click on Round feature from the Model tab of the ribbon interface.
- Set the value of fillet to 0.5 mm.
- Select all two external circular edges of the newly created feature.
- Click OK to create the feature.

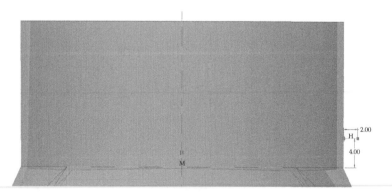

FIGURE 6.66
Constructional sketch of features of a plastic bottle.

- Click on the Helical sweep option from the drop-down menu of Sweep from the Model tab of the ribbon interface.
- Click on the References tab and select Define to define the helix sweep profile.
- Select the front plane as the sketch plane.
- The right plane shall be automatically taken as reference and orientation as right.
- Click on Sketch.
- The sketcher model will be activated.
- Draw a geometry centerline on the vertical reference line in the middle.
- Draw a vertical line at the right-hand edge by taking the reference of the cylindrical surface. The line should start 3 mm below the top horizontal surface.
- Set the length of the line as 7 mm.
- Click OK to create the sketch and exit the sketcher window.
- Click Sketch to create the profile sketch above the Reference tab.
- Draw a half circle of diameter 1.5 mm placing the center at the intersection of reference lines.
- Click OK to create and exit the sketch.
- Enter the pitch value of 3 mm.
- Click on the Right-handed rule icon to create the helical spring clockwise downward.
- Click OK to create the thread.

Hands-On Exercise, Example 6.34: Creating a Crankshaft (Software Used: Creo Parametric)

Steps:

- Create a part file and rename it as Crankshaft. (Other names may be given, but it is advisable to provide a relevant name.)
- Select Solid on the right-hand side of the box. This will activate the solid mode and environment.
- Uncheck the box of the default template and click OK to create a new solid file.
- The template selection box will appear.
- Select mmns_part_Solid. This will allow us to use the solid template file of the following units:
  - mm—millimeter for length unit
  - n—newton for force unit
  - s—second for time unit
- Once the solid part is created, go to File > Prepare > Model properties to check the units. In the upper sections, the units of the model will be displayed.
  - Note: The units usually are displayed in the summary section of any CAD software. The setting of units usually is in the tools or settings.

FIGURE 6.67
3-D view of a crankshaft.

- The crankshaft as shown in Figure 6.67 is to be created.
- Click OK datum plane from the Model tab of the ribbon interface.
- Select the right plane as the reference plane.
- Set the offset value –50 mm downward.
- Click OK to create the datum plane.
- Click on Extrude from the Model tab in the ribbon interface.
- Click on the Placement tab and select Define to define the sketch.
- Select the front plane as the sketch plane.
- The right plane shall be automatically taken as reference and orientation as right.
- Click on Sketch.
- The sketcher model will be activated.
- Draw the sketch as shown in Figure 6.68.
- Click OK to create and exit the sketch.
- Select mid plane extrusion.
- Enter the value 20 mm for extrusion depth.
- Click on OK to create the extrusion.
- Select the lastly created extrusion in the model tree.
- Right-click and select Pattern.
- Select direction as the pattern type.
- Select the right plane as the directional reference.
- Enter the quantity 2 for patterned feature.

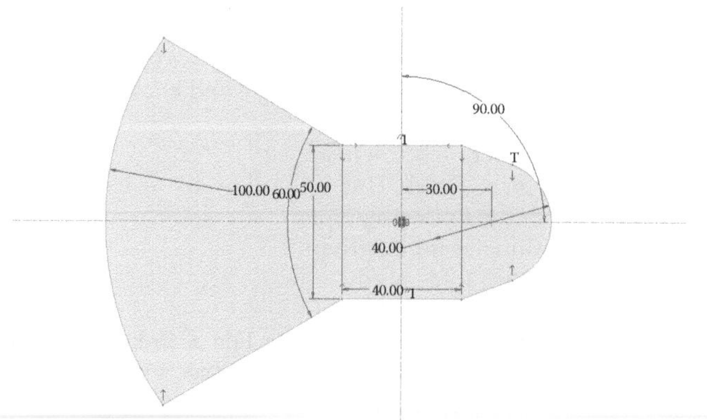

FIGURE 6.68
Constructional sketch of features of a crankshaft.

- Flip the orientation of the pattern toward the right plane.
- Click OK to create the patterned feature.

- Click on Extrude from the Model tab in the ribbon interface.
- Click on the Placement tab and select define to define the sketch.
- Select the inside surface of the first extrusion as the sketch plane.
- The right plane shall be automatically taken as reference and orientation as right.
- Click on Sketch.
- The sketcher model will be activated.
- Draw a circle of diameter 30 mm.
- Place the center of the circle on the center of the small arc with the help of taking the small circular edge reference.
- Click OK to create and exit the sketch.
- Select the Up to the entity option for depth.
- Select the inside surface of the patterned object.
- Click on OK to create the extrusion.
- Select the lastly created datum plane, pattern, and extrusion in the model tree by keeping the Control key pressed.
- Right-click and select Group to make a group with these three features.
- Select the newly created group in the model tree.
- Right-click and select Pattern.
- Select Dimension in the Type of pattern selection field.
- Select the dimensions −50 mm and the angle 90°.
- Enter the incremental value −320 for linear dimension and 90° for angular dimension.
- Enter the quantity 3 for patterned feature.
- Click OK to create the patterned feature.
- Click on Revolve from the Model tab in the ribbon interface.
- Click on the Placement tab and select Define to define the sketch.
- Select the front plane as the sketch plane.

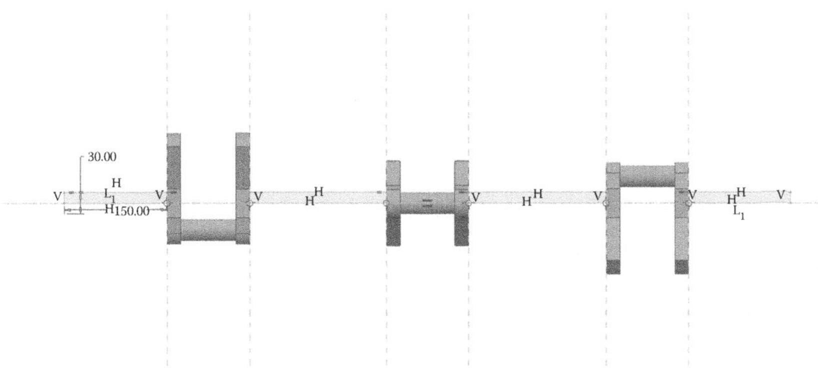

FIGURE 6.69
Constructional sketch of features of a crankshaft.

- The right plane shall be automatically taken as reference and orientation as right.
- Click on Sketch.
- The sketcher model will be activated.
- Draw the sketch as displayed in Figure 6.69.
- Click OK to create and exit the sketch.
- Ensure the revolve angle shall be 360°.
- Click on OK to create the revolve feature.
- Click on the auto round feature from the Model tab of the ribbon interface.
- Set the value of fillet to 3 mm.
- Click OK to create the feature.
- Save and exit the file.

QUESTIONS

1. What is a part?
2. Differentiate between part and feature.
3. What are the advantages of creating features in relations?
4. Describe the usability of patterns.
5. Why is the draft feature required?
6. Describe the advantages of cosmetic features.
7. Describe the uses of datum planes.
8. Differentiate the features blend sweep and swept blend.
9. Describe the different types of surfaces.
10. How are the manipulations of properties important?
11. Describe the advantages of the use of map keys.
12. What is parametric modeling?
13. Site an example of the use of parametric modeling.
14. Describe the procedure of parametric modeling.
15. What is application-based modeling?
16. How is application-based modeling advantageous?
17. Why are graphical presentations important?
18. What is a template?
19. Describe the advantage of using templates.
20. How are templates created?
21. Describe the advantages of using a customized template.

7

Assembly Methods

7.1 Introduction

Assemblies are combinations of individual parts fitted mechanically to perform a desired function. Independent parts are combined and mechanically linked in an assembly. Typically, in an assembly, all or some parts function, and the function of the assembly is achieved by these parts working together. As an example, a component consists of several links that perform a definite motion, and many links can be single parts performing actions within the assembly to achieve the function of the assembly.

Most engineering components in the real world that perform a complete solution by providing some mechanical action are assemblies. A single part can only contribute independently in static condition. A real engineering challenge mostly comprises such goals that involve components of different materials and varying properties. Several components with different materials are required to meet the design challenge, which involves the design of separate components, thus creating the assembly.

In computer-aided design (CAD), assembly is typically in a workspace with the same graphical area as in part modeling, with the main features of placing and making the constraints. Additionally, the available features in an assembly are comparatively fewer than the features in part modeling. Using assemblies in CAD requires very good understanding of the uses of all assembly features to create accurate functional assemblies, especially when large assemblies or automations are considered.

In an assembly, the set units can be different from the set units of its parts. In fact, the parts can have different units. In an assembly, the set unit affects the assembly dimensions and assembly features. When handling large assemblies, the set unit may often be confused for assembly and parts. When designing, it is recommended to set all the units in one system to avoid this kind of confusion. Here is an example of how the set units differ for an assembly.

Hands-On Exercise, Example 7.1: Changing Assembly Units (Software Used: Creo Parametric)

Steps:

- Create an assembly file with template mmns_asm_design. This will create an assembly file with set units linear, millimeter; force, newton; and time, second. Three planes are automatically added: ASM_RIGHT, ASM_TOP, and ASM_FRONT.
- Click on the Model tab, select the Assembly icon (click to add a component to the assembly), and select column 1.
- Select Default in the Placement mating condition field. This will locate the component by default. (The coordinate system of the part will be located on the coordinate system of the assembly.)
- Click again on the Assembly icon to add another part into the assembly.
- Select the assembly top plane and the part's top plane and set the mating condition to Coincide.
- Similarly, select the assembly front plate and the part's front plane and select Coincide.
- Select the assembly right plane and the part's right plane. In the Mating conditions tab, select Offset and set the value to 2000.
- Click on OK to finish the component placement.
- Now click on File > Prepare > Model properties > Units > Change.
- Select inch_pound_second. When prompted for converted or interpreted dimensions, select Interpret and click OK.
- Regenerate the model; you can see two parts (i.e., the columns are moved far apart).
- Click on the Analysis tab, select Measure, and select the two right planes of the column2 and ASM_RIGHT plane. The analyzed value will be 2000 in.
- Save the file and exit.

7.2 Assembly Workspace

Most commands in CAD are accessed in the form of an icon. Relative operational icons are grouped in the form of a toolbar. The number of such icons and toolbars is relatively large to accommodate in a single screen. Also, in most cases, not all the toolbars are necessary for every type of operation. For example, sheet metal operational toolbars are only required when working with sheet metal. However, there are some common operations that are required in any kind of modeling in CAD. For example, a dimensional analysis toolbar is required in every type of modeling. Most CAD software is programmed in such a way that only the toolbars that need to be accessible are visible and highlighted for any particular CAD modeling when selected. When selecting assembly or part modeling or sheet metal, the user interface window changes to contain only the toolbars that are required for the particular kind of modeling. This can be termed *workspace*. An assembly workspace contains the toolbars that are specially required for assembly and common operational toolbars. The toolbars are as follows:

1. Workspace-specific toolbar
    a. Component placement toolbar
    b. Component manipulation toolbar
    c. Assembly operation toolbars
2. Common toolbars
    a. Analysis toolbar
    b. Parametric and programming toolbars
    c. Datum creation toolbar
    d. View creation toolbar

The assembly workspace is shown in Figure 7.1.

Assembly structure: Every assembly has a defined structure and levels according to complexities. An assembly that has only parts can be termed a *single-level assembly*. An assembly with one subassembly, which contains only the parts, is called a *second-level assembly*.

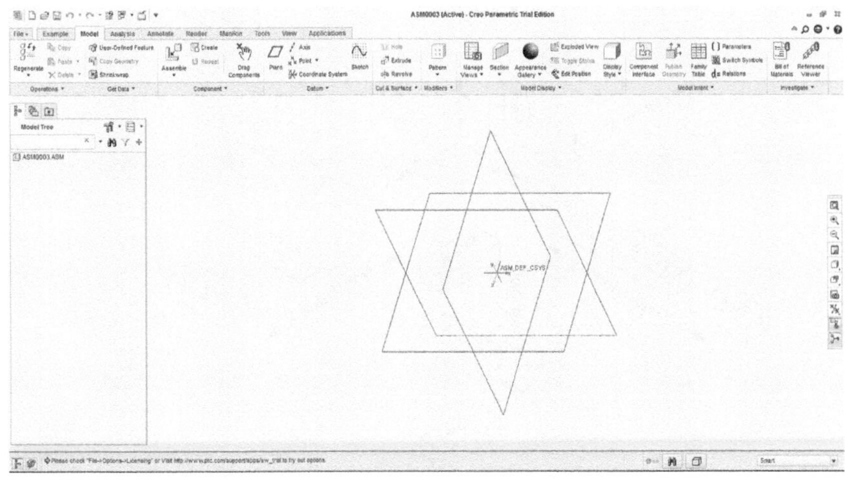

FIGURE 7.1
Assembly workspace.

7.3 Degree of Freedom

Understanding degrees of freedom is an essential requirement in order to use all the assembly tools effectively.

Degrees of freedoms are independent motion directions to which a component can alter its position. There are six degrees of freedoms in CAD (as shown in Figure 7.2):

1. Linear movement on x axis
2. Rotational movement about x axis
3. Linear movement on y axis
4. Rotational movement about y axis
5. Linear movement on z axis
6. Rotational movement about z axis

A little consideration will show that these three axes must be perpendicular to each other. The axes' orientation can be varied as per the user's choice;

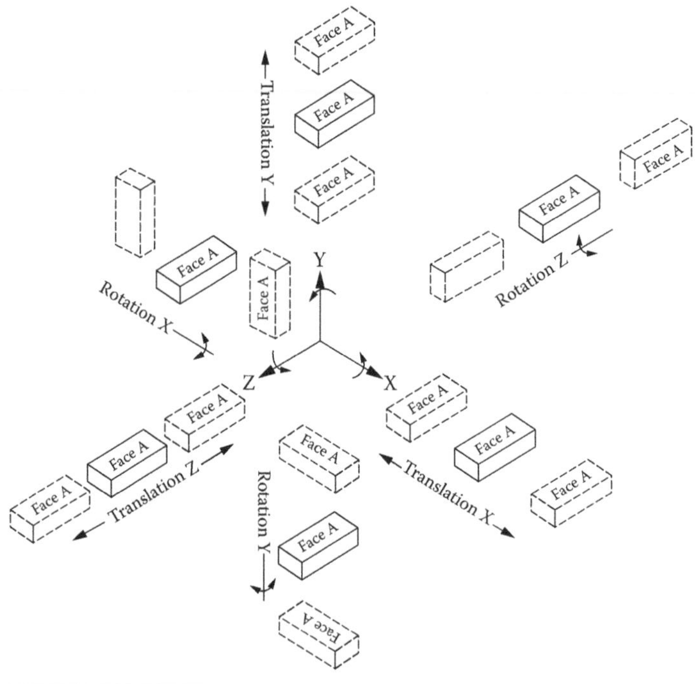

FIGURE 7.2
Degrees of freedom.

however, the relative angles between the axes cannot be changed. If these six degrees of freedom of a component are restrained, the component will be fixed to its location with respect to its mating parent component, and the component will be termed *fully constrained*. In a static assembly, all the components must be fully constrained. When some degrees of freedom are not restrained through mating to the mating parent component, they are called *partially constrained*. Except ground links, all the links in a mechanism are to be partially constrained in order to achieve the relative motion between each other. When no degrees of freedom are restrained, it is called *free component*. Free placement of components may be useful in creation of manuals or illustration work, where a proper fit of the components is not required for illustration.

7.4 Assembly Methods

Creating an assembly in CAD is typically sequential placement of components through available constrain sets. All the available constraint sets restrict one or more degrees of freedom. In any constraint set, each entity from the existing assembly and new part or subassembly will be selected, and mating conditions are applied. Nowadays, the advanced CAD software provides automatic selection for the mating conditions. The available mating conditions will automatically be displayed depending upon the selection. It is advised that the assembly constraint be applied in such a way that the assembly component can be placed only in one position in the main assembly. If the provided constrains are insufficient to define the newly assembled component in one position, the position of the component may get changed during regeneration. The mainly used available constraints are as follows:

1. *Plane constraints*: Plane constraints are the mating condition between two planes of new components to existing components. Plane surfaces can also be used for plane constraints. When planes or surfaces are selected from a component and a main assembly, there can be a linear offset or an angular offset. Much software provides various conditions like coincide, parallel, normal, and so forth. Coincide can be defined as an offset with 0 offset value. Similarly, parallel has an angular offset with 0° angles, and normal has an angular offset with a 90° angle.

   Degree of freedom in plane constraints: If the plane constraints are used with any dimension value or coincide, this restricts one translational and two rotational degrees of freedom. If the plane constraint is used for parallelism, this restricts two rotational degrees of freedom. Figure 7.3 shows an example of a plane constant.

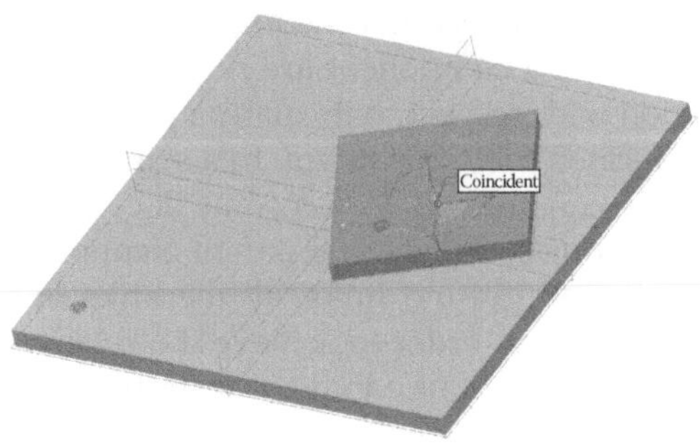

FIGURE 7.3
Example of plane constant in assembly.

2. *Axial/cylindrical constraints*: Axial/cylindrical constraints are the mating condition between two axes or cylindrical surfaces that can define an axis. When two axes or cylindrical surfaces are constrained, two translational and two rotational degrees of freedom are restricted. Figure 7.4 shows the axial or cylindrical constraints.
3. *Point constraint*: In point constraints, two points of assembly and parts are constrained. Hence, three translational degrees of freedom get restricted; however, three rotational degrees of freedom are present in the point constraint.
4. *Coordinate constraints*: A coordinate constant restricts all the degrees of freedom between an assembly and the part. Since selecting mating by a coordinate system does not allow any flexibility while assembling any part, axis orientation selection at the time of creation

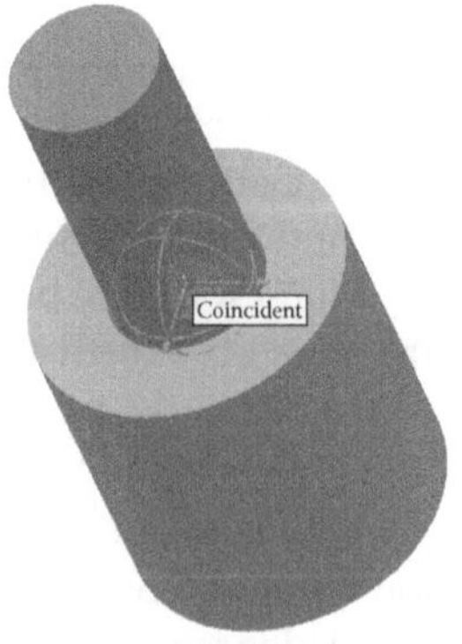

FIGURE 7.4
Example of plane axial or cylindrical constant in assembly.

TABLE 7.1

Length Unit Conversion Factor

	1 km	1 m	1 cm	1 mm	1 μm	1 in.	1 ft.	1 yd.
1 km	1	1×10^3	1×10^5	1×10^6	1×10^9	39,370.08	3280.84	1093.61
1 m	0.001	1	100	1×10^3	1×10^6	39.37	3.28	1.094
1 cm	1×10^{-5}	0.01	1	10	1×10^4	0.3937	0.0328	0.010936
1 mm	1×10^{-6}	1×10^{-3}	0.1	1	1×10^3	0.03937	0.00328	0.0010936
1 μm	1×10^{-9}	1×10^{-6}	1×10^{-4}	1×10^{-3}	1			
1 in.	2.54e–5	0.0254	2.54	25.4	25,400	1	0.08333	0.02777
1 ft.	0.0003048	0.3048	30.48	304.8	304,800	12	1	0.333
1 yd.	0.0009144	0.9144	91.44	914.4	914,400	36	3	1

TABLE 7.2

Common Assembly Constraints

	Translation *X*	Translation *Y*	Translation *Z*	Rotation *X*	Rotation *Y*	Rotation *Z*
Plane	Constrained	Free	Free	Free	Constrained	Constrained
Axis or cylindrical	Free	Constrained	Constrained	Free	Constrained	Constrained
Point	Constrained	Constrained	Constrained	Free	Free	Free

of a coordinate system is significant and carefully done. Constraint relations are shown in Table 7.1.

Some common assembly constraint sets are shown in Table 7.2.

7.5 Assembly Operations

The operations that are possible to create only in an assembly are the assembly operations. Operations can be of two types in an assembly: feature operations (i.e., creation of any operation in the assembly) and component operations (i.e., adding or manipulation of components in the assembly).

7.5.1 Feature Operations

Operations from the material point of view are basically of two types: material-adding operations and material-removing operations. Any operations that add the material, except welding, cannot be done in the assembly or assembly mode. The operations that remove materials can be done in the assembly or assembly mode. The material-removing operations in an assembly are as follows:

1. *Hole*: Creation of a hole is an essential feature in an assembly. Practically, in many cases, many holes for fastening the parts are required to be created. If these holes are created independently in the parts, there may be a possibility of mismatching. To ensure the alignment of holes, the best way to create holes is in assembly wherever possible. It ensures location alignment in all the parts and saves the time required to create holes in all the parts. CAD software also allows retaining the holes at the part level, though they are created in the assembly.

Hands-On Exercise, Example 7.2: Creating Holes in an Assembly (Software Used: Creo Parametric)

Steps:

- Create an assembly file with template mmns_asm_design. This will create an assembly file with set units linear, millimeter; force, newton; and time, second. Three planes are automatically added: ASM_RIGHT, ASM_TOP, and ASM_FRONT.
- Users are advised to create two parts independently: Plate 1 (500 × 500 × 20) and Plate 2 (200 × 200 × 20).
- Click on the Model tab, select the Assembly icon (click to add a component to the assembly), and select Plate 1.
- Select Default in the Placement mating condition field. This will locate the component by default. (The coordinate system of the part will be located on the coordinate system of the assembly.)
- Click again on the Assembly icon to add Plate 2 into the assembly.
- Select the top face of Plate 1 and bottom face of Plate 2 and select Coincide as the mating condition.
- Select one of the faces of Plate 1 (except the previous selection) and one of the faces of Plate 2 (except the previous selection), and select Coincide as the mating condition.
- Similarly, select another face, and place the component as shown in Figure 7.5.
- Click on OK to finish the component placement.
- Now click on Hole to create the hole in the assembly.
- Make 20 the diameter of the hole. Place the hole as shown in Figure 7.6.
- In the depth selection box, select the hole through all the surfaces.
- Click OK to create the hole.
- Open Plate 1. It can be seen that the created hole in the assembly does not have any effect on the part. It is logical that the hole is made in the assembly; therefore, it is not being updated in the

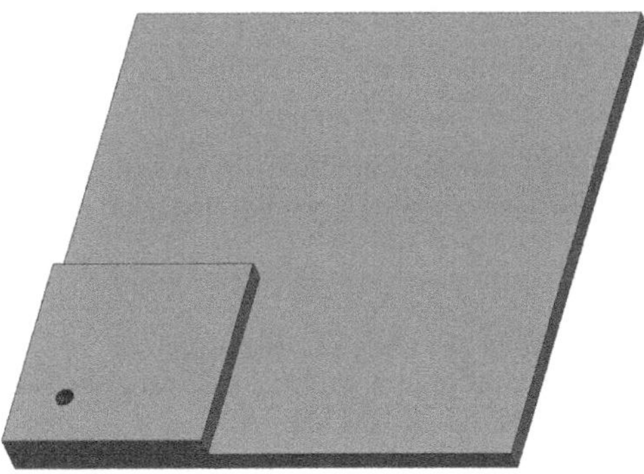

FIGURE 7.5
Example of planar constant.

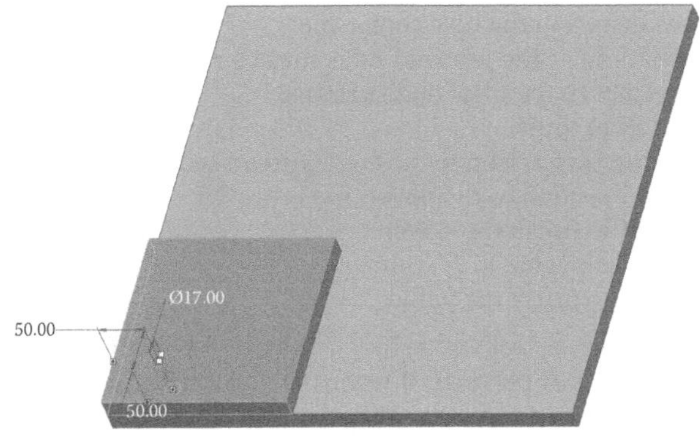

FIGURE 7.6
Example of hole feature in assembly.

part. However, in many cases, especially in top–down design, the features are required to be available in the part model, though it is created in the assembly.

- Select the hole from the model tree, select Edit definition, and click on the Interface tab.
- Turn off the automatic update; at the bottom section of the box, click on Part level. Click OK.
- Open Plate 1 or Plate 2. It can be seen that the parts are updated with the hole.
- Save the file and exit.

2. *Cut*: The cut feature is an essential feature in the assembly. Cut features are mainly used in top–down design methods, where cut sections on various parts are derived from a single section. Cuts by

extrude, revolve, sweep, helical sweep, blend, and so forth can be done simply in an assembly as an assembly feature.

Hands-On Exercise, Example 7.3: Creating an Internal Thread in an Assembly (Software Used: Creo Parametric)

Steps:

- Open the assembly as created in Example 7.2.
- Create a plane parallel to one of the side faces and through the axis of the hole. Name it ADTM1.
- Select the feature Helical sweep from the Shapes drop-down menu.
- Click on the Reference tab, and select Define to create a helix profile.
- Select the reference of the axis of the hole, the top and bottom surface of the plates, and the internal surface of the hole.
- Draw an axis in the hole centerline.
- Draw a line at the internal hole surface reference terminating between both the plate end surfaces.
- Click OK to finish.
- Select the Sketch icon to draw the thread section.
- Draw the section as displayed in Figure 7.7.
- Click OK to finish the sketch.
- Set the pitch value to 2.5 mm.
- Click OK to finish the value.

3. *Pattern of features*: A pattern of features is a very useful tool for copying features. Patterns in assembly are the same as in part mode. All the patterned features will follow exactly the same properties of the original features.

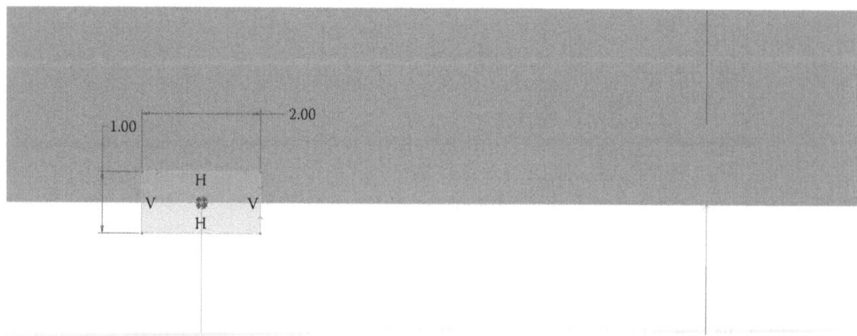

FIGURE 7.7
Constructional sketch.

Hands-On Exercise, Example 7.4: Pattern of Feature (Software Used: Creo Parametric)

Steps:

- Create a new assembly file; rename it as fill_pattern.
- Assemble a plate of size 500 × 500 × 2 mm thickness in its default location.
- Assemble another plate of the same size (a different part file is required to ensure feature intersect at the part level), as displayed in Figure 7.8.
- Create a hole of size 10 mm through and 50 mm offset from both the edges of the overlapped plate.
- Select the hole; the pattern option should be highlighted after selection of the feature.
- Click on Pattern and select Directions to create the pattern.
- Enter 50 mm for patterning in both directions.
- Enter the quantity 9 in the long side of the pattern feature creation.
- The feature will be patterned; however, the intersection or creation of a hole is at the assembly level only by default.
- Select any feature in the pattern from the model tree, and select Edit definition by right-clicking on it.
- Uncheck the box for automatic update, and change the intersection level to part level. Click on OK.
- Both the plates shall contain the holes at the part level now.
- Now again click on Edit definition by right-clicking on any of the holes in the pattern.
- Remove the intersection of any one plate and click on OK.
- The complete feature will be created on one part only.
- Change the intersection to part level and click on Automatic update; click on OK.
- Save the part and exit.

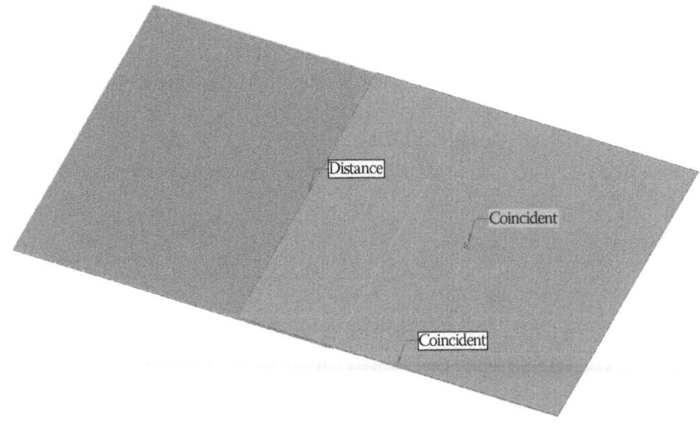

FIGURE 7.8
View during assembly of plates.

4. *Mirror of features*: Mirror of features is mirroring a feature about a plane in the assembly mode. Similar to patterns, mirrored features are also dependent on the original feature. If the position of an original feature gets changed, the position of the mirrored featured automatically gets changed as required to maintain the mirror about the preselected plane. However, the intersection level of a mirrored feature may not be dependent on the original feature.

Hands-On Exercise, Example 7.5: Mirror of Feature (Software Used: Creo Parametric)

Steps:

- Create a new assembly file; rename it as Mir_Feature.
- Assemble a C-section of 100 × 50 × 8 mm thickness of 500 mm length.
- Assemble a plate of size 500 × 50 × 10 mm thickness on the top of the C-section wide flange.
- Create a cut at the end of plate such that the cut shall intersect both the plate and the C-section.
- Select the Extrude cut feature from the model tree, and click on Mirror from the Modifiers from the drop-down menu of the ribbon.
- Select the mirror plane and click OK.
- Click on Edit definition on the original feature and select Edit definition by right-clicking on it.
- Change the intersection level to part level and click OK.
- Open any of the parts; it can be seen that only the original features are updated in the part level.
- Now in the assembly, select the mirrored feature and click on Edit definition by right-clicking on it.
- Change the intersection to part level in mirrored feature too and click on OK.
- Now the feature will be updated in both the parts in the part level.
- Now right-click on original feature and select Edit.
- Modify the dimension of the cut and click on Regenerate.
- It can be seen that the mirrored feature automatically gets updated with the original feature.
- Save the part and exit.

7.5.2 Component Operations

Component operations are mostly required and used in assembly mode. Especially in the case of large assemblies with many components or editing large assemblies, manipulation of components is often required.

The most used component operations are listed as follows:

1. *Copy/pattern*: Copy or pattern is an essential feature in an assembly to create repetitive parts with same distance or angle apart.

Hands-On Exercise, Example 7.6: Pattern of Component (Software used: Creo Parametric)

Steps:

- Create a new assembly called Flange joint.
- Assemble a component flange of 100 Nominal Bore (NB).
- Assemble another flange of 100 NB of the same class on the mating face of the existing flange, as displayed in Figure 7.9.
- Affix a bolt of suitable size to any one of the holes suitably, as displayed in the final figure.
- Right-click on the newly added bolt and select Pattern.
- The reference pattern can be automatically selected by the system; however, select Axis from the first drop-down box for axial pattern.
- Click on the Angle shaped icon for equal segment pattern.
- Enter the number of instances matched to the number of holes; for this case, it is 8.
- Similar to a hex bolt, assemble one hex nut at the opposite-site head of the bolt of flanges.
- Right-click on the nut and select Pattern; usually, the reference pattern will be activated. Click on OK to automatically get all the nuts in the respective bolts. A complete view is shown in Figure 7.10.
- Save and exit the file.

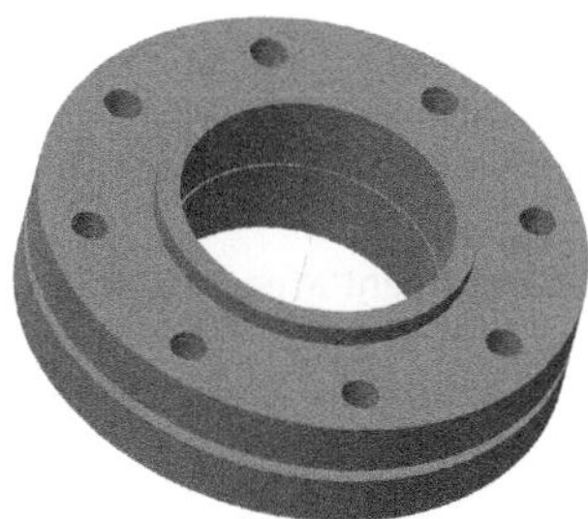

FIGURE 7.9
3-D view of assembly of flanges.

FIGURE 7.10
3-D view of assembly of flanges with hardware.

2. *Restructure*: Restructuring is manipulation of parts or subassemblies in an assembly. Restructuring is a very useful command in large assemblies where extensive component manipulation is necessary. However, extensive care should be taken while restructuring the component. It is very essential when creation of subassemblies is required by parts from the master assembly.

Hands-On Exercise, Example 7.7: Restructure (Software used: Creo Parametric)

Steps:

- Open the assembly Flange joint.
- Select the pattern group of fasteners and delete the pattern by right-clicking on it.
- Click on the Create an assembly icon in the assembly mode itself.
- Select subassembly and subtype standard and rename as Fast; click on OK.
- Creation method: locate default datums and three planes.
- Click on the first three planes of the master assembly from the model tree.
- The subassembly Fast is now added in the master assembly.
- Drag and drop the newly added assembly and place before fasteners.
- Now click on Restructure from the component drop-down menu of the ribbon interface.
- Once the Restructure menu box opens, select Hex bolt in the source component area.
- Select the target component area and select the newly created subassembly Fast.
- Click on OK. The fasteners are moved to the subassembly Fast.
- Click on Save and exit the file.

7.6 Family Table in Assembly

A family table in an assembly is used to create an array of components of the same family. An example of a family table is given as follows:

Hands-On Exercise, Example 7.8: Family Table (Software Used: Creo Parametric)

Steps:

- Open the assembly Flange joint.
- The assembly contains two mating flanges, one gasket, and required fasteners (i.e., eight hex bolts and eight hex nuts).

- Each of the components has three variations of pipe sizes: 100 NB, 150 NB, and 200 NB.
- Select Family table from the Tool tab of the ribbon interface.
- Click on the vertical arrow to select the variations.
- In the bottom section of the family items box, select Component.
- Add all the components.
- Click on OK.
- Click on the horizontal arrow and enter the table, as displayed in Figure 7.11.
- Click on the icon for verifying family tree and click on Verify.
- Verification status must appear as successful for all the components; otherwise, the problem needs to be identified and shorted out for any future use of instances.
- Select Preview to see any instances.
- Click on Save and exit the file.

7.7 Application-Based Assembly Techniques

The assembly can be created in two ways in principle: the top–down approach and bottom–up approach. Both have their advantages and limitations.

- *Relationships in assembly*: In an assembly, all the components have a relationship with only their parent assembly or any other component from higher-level assemblies. The relationships can be of two types: placement references and feature references.
- *Placement references*: When a component is assembled in an assembly, it requires a few constraints to place or define the location of the component in the assembly with respect to assembly features. These constraints are actually placement relationships of the component to its higher-level assembly. These placement relationships of any component can affect only the top-level assembly. Regardless of any change in placement references of the components, the features in the component have no effect.
- *Feature references*: References can be taken for creation of any component's features from other components' feature or features of the top-level assembly. The features that are created by taking references from other components have dependencies on the other components. If the driving feature in the top-level or sublevel assembly is modified, the dependent feature also gets modified, though it is in another component during the regeneration of the top-level assembly. The feature references are difficult to handle, and it requires good knowledge and skill in the software. Feature references are created in the top–down design method.

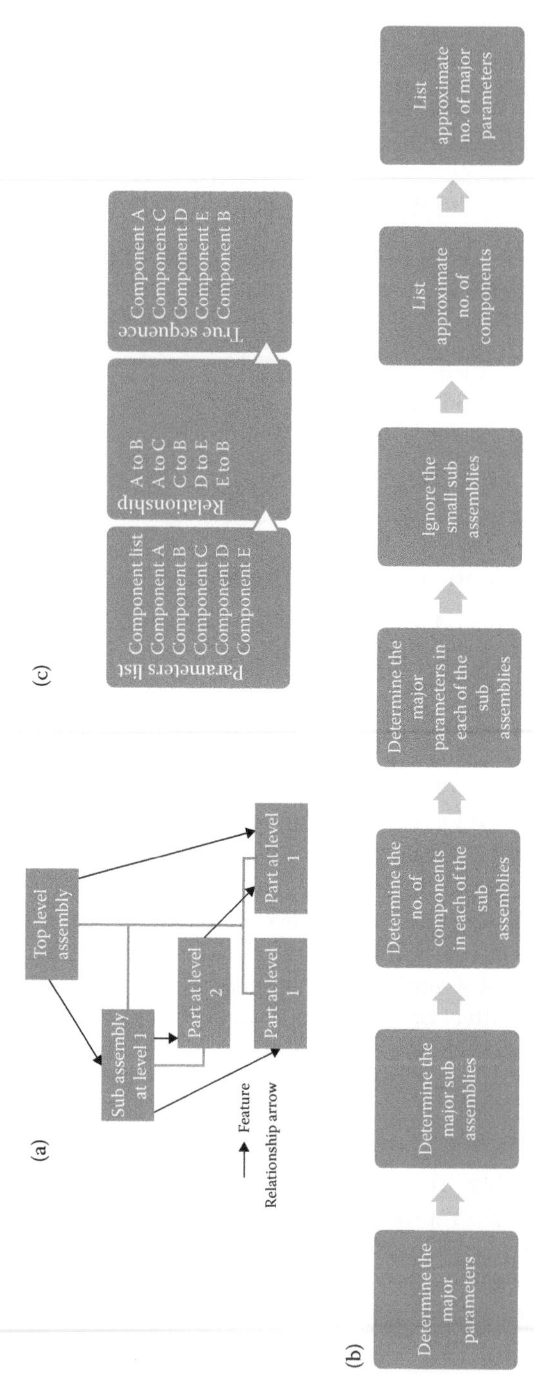

FIGURE 7.11
Family table of flanges: (a) Relationships in assembly; (b) Process of planning; (c) Sequence of modeling.

- *Top–down approach*: In the top–down approach, the main or top-level assembly is created first. The next sublevel assemblies and parts are created within the top-level assembly by taking references and dependencies from the top-level assembly or its sublevel components.

Top-level assembly relationship arrows are shown in Figure 7.11a. At the time of creation of any component into the assembly, it can take relations from the top-level assembly features or any of the sublevel components. The top–down design method is mostly used by designers who create or develop products.

Advantages:

1. The top-level assembly method can be used for creation of components that are in relation to other components in an assembly.
2. If any feature has to be modified, the subsequent dependent features of any level of component are modified. This ensures the required modification of all desired features of dependent components even if the designer forgets.
3. The mechanism is often not satisfied in the top–down approach, since feature references are present.

Disadvantages:

1. Independent feature modification in any of the components can affect other components and may generate severe errors.
2. Care is required in restructuring, which is often required for preparing final bill of materials (BOM).
3. Chances of cyclic reference errors often occur.
4. The top–down design method often requires a high level of skill on the software being used.

- *Bottom–up approach*: In the bottom–up approach, all the part-level components are made independently. Subassemblies are made when all the parts of subassemblies are available. The bottom–up approach has only placement references.

Advantages:

1. Cyclic reference error for feature creation is not available; therefore, feature generation error can be completely avoided.
2. Assembly in the bottom–up approach does not need a high level of skill in the software.
3. For creation of a mechanism, the bottom–up approach is perfect, since feature references are not present in the assembly.

Disadvantages:

1. Cyclic reference error for feature creation is not available; therefore, feature regeneration error can be completely avoided.
2. Designing is a little bit time consuming in assemblies where components are created in relation to another component's feature, since designers have to manually create or change features in part mode.
3. If any feature changes, the subsequent logically dependent feature in other parts will not be updated automatically. It can often lead to mistakes of keeping the feature of old revisions.

Designing and CAD modeling are often a combination of top–down and bottom–up approaches as per requirement to save time. Bottom–up approaches are mostly used in creating CAD models of old, fully designed jobs. However, in the case of a new design, the top–down design method is often found useful.

A top–down or bottom–up or combination requirement is based on the requirements of the design or the CAD model requirement. Every design or CAD model needs to be planned before starting, especially in the case of large assemblies or design jobs. Work in CAD can be classified as follows:

1. Design of new product
2. Modification of existing product/design of new product from existing product
3. 3-D model preparation of existing product

The approach for each type of job is different and, if planned properly, can be produced within minimum time and with minimum error.

CAD models are used for the following purposes:

1. Design and standard model preparation
2. Creation of mechanism
3. Animation
4. Visualization and presentation
5. Illustration
6. Analysis
7. Automation

Apart from each type of job, CAD models also need to use the correct approach. For example, a model created by the top–down design method may not be very useful in a mechanism; however, for static analysis, it can be perfectly useful.

7.7.1 Planning

Planning to execute each type of job is not much different in nature with respect to the top–down, bottom–up, or combined approach.

7.7.1.1 Planning for the Design of a New Product

The first kind of planning in CAD modeling is to determine the content of the work and variations. It is not possible to determine the exact values; however, an approximation with 25% tolerance can be also helpful. First, the components of the assembly are to be determined, even at a macro level. For example, for the designing of any moving equipment with certain loads, they can be determined as sliding driving units, chassis, functional driving units, and so forth even before starting any design. Assume the quantity of parts that will be required in each of the assemblies. Any subassemblies containing fewer than three components having less importance can be ignored during the counting of components. This will give a rough estimate of the total number of components required in the assembly. Assume at least 10% more components over the derived quantity. The process of planning is shown in Figure 7.11b.

After determining the approximate quantity of parts, it is essential to determine the approximate major and minor parameters. Parameters are actually the variables in the CAD that may be required to change, and in turn, the CAD model needs to be changed. Hence, it is well understood that parameters always have a great influence on the CAD models and its features. Handling of too many parameters is also not very easy and sometimes generates a lack of interest in the user or designer. Thus, it would be a good practice to classify the major and minor parameters first. The parameters that are most designated and have many dependencies and much significance can be called major parameters. For example, diameter and thickness of a tank are always major parameters. The minor parameters are mostly assumed and do not come with strict rules but are essential to provide in a manufacturing operation. For example, fillet and chamfering are common operations on the shop floor; however, the dimensions of such operations may not be so important for all the other operations and components. Since major parameters always drive many feature geometries, it would be better to classify and designate them with proper values, whereas minor parameters shall not be linked with other parameters unless otherwise logically required. The quantity of major parameters is an important criterion for selecting the correct design approaches.

Once the components and parameters are counted, the correct design approach in CAD modeling shall be decided. If the total parts exceed 50 components, a philosophy for naming all parts shall be made. This will help to quickly locate any part when it needs to be changed. This process is shown in Table 7.3.

Once the component and parameter list is created, the next approach is to decide on the procedure of design and modeling. Normally, design dominates

TABLE 7.3

Selection of Naming and Parameter Variations

Sl. No.	Component		Major Parameters	
1	Less than 50	Only meaningful name is enough for naming.	Less than 10	Preserving of parameters and variations may not be required.
2	More than 50	Philosophy of naming must be prepared, which helps users to locate and understand the model when required to be changed.	Less than 10	Preserving of parameters and variations may not be required.
3	Less than 50	Only meaningful name is enough for naming.	More than 10	Parameter list shall be prepared with variation limit.
4	More than 50	Philosophy of naming must be prepared, which helps users to locate and understand the model when required to be changed.	More than 10	Parameter list shall be prepared with variation limit.

and determines the sequence of modeling. However, it is better to prepare a sequence of assembly or drawing creation. Mostly, the major parameters of any assembly have influence on most of the components and their major parameters. Also, some component parameters or features have an influence on other components' parameters or features. Therefore, a list of prepared dependencies may be very helpful to determine the sequence of CAD modeling or even for designing. The sequence of modeling is shown in Figure 7.11c.

On preparing the true sequence, it would be easy to focus sequentially on the component that needs to be designed.

7.7.1.2 Modification of an Existing Product

Modification of an existing product is often needed with the development of subsequent products or market competition. The first task is to determine the work content for the modification. If no model is available for the existing product but all the drawings are available, then it is advisable to prepare the full 3-D models first by the bottom–up approach. This helps to visualize the need for changing the product or model. Planning of modification of an existing product is also the same as that of a new product design.

7.7.1.3 3-D Model Preparation of an Existing Product

3-D preparation of an existing product shall be done by the bottom approach if all the drawings of all components are available.

7.7.2 Assembly Parameters

Handling parameters in an assembly is a little different from handling parameters in a part model. Assembly parameters can be directly linked to preserve feature geometries. However, that can only be understood if the top-level assembly is open while editing the part. However, if one opens the part and tries to modify any of its dimensions, it would be difficult to track the assembly parameters. Assembly parameters can be used in both a placement constant and a feature constant. In the case of creation of large assemblies and top–down design, if the parameter lists are available, it is recommended to create the parameters first even before creating any features. During creation of any geometry or feature, if the parameters are not available, the chances of keeping the variables open are higher.

7.7.3 Creation of Parts or Components

Parts or components are required to be created sequentially. Even in the bottom–up approach, the sequence of component assembly shall be as per priority. Priority of assembly is also dependent on availability of a placement constant. Even in the case of the top–down design method, it is preferred that the parts that are not to be derived from the assembly be made independently so that unnecessary references and relationships can be reduced.

7.7.4 Assembly of Components

The first part that has to be assembled needs to be placed in relationship with assembly features (i.e., datum planes, axes, coordinate systems, etc.). The other parts can be placed with the feature references of the available parts. However, it is to be noted that if a part is assembled by taking references from other parts' geometry, the part will be dependent on the geometry of the previous part. Therefore, any modification or rectification of the features of previous parts may affect the assembly. Therefore, it is preferred to use assembly references wherever possible.

7.7.5 Checking

It would be a big task if someone had to check assembly after assembly of all the parts. Also, if any problem is found, the rectification would be very difficult and might affect preceding parts. Therefore, it is preferred to check all the parts immediately after placement by alteration of some dimensions. A map key or macro might be very helpful in this matter.

7.8 Presentation

1. *Graphical presentation*: A graphical presentation is a very useful tool in assembly. It is used for the following purposes:

   a. *Lightweight assembly*: A large or complicated engineering component can easily consist of more than 500 or 1000 files of parts and subassemblies. Also, every part and subassembly may include several features. Handling large assemblies needs more regeneration time and greater resources. Resources in terms of computer memory, graphics cards, and processors may be required in order to handle large assemblies. It is often seen that these resources may not be available to the users. Also, it is seen that all the parts and subassembly files are not required to be operated while opening the main assembly files. For example, there could be many components that are not dependent on certain changes of the other components but are still required to be shown in the assembly. There, it is helpful to make the assembly lightweight. The lightweight assembly means converting a master assembly to a graphical-view representation only. This would certainly exclude from the memory all the features of the components that are not being shown on the screen. However, sometimes, it is required to operate some parts or features. On-demand lightweight assembly is the best solution in this area, which allows the users to operate or create any feature in the parts or subassemblies when selected for editing.
2. *Simplified representation*: Much CAD software allows isolating of certain components to obtain different configurations. A master assembly presentation can contain all the components; however, it is sometimes necessary to display the assembly without one or more components. Simplified representation becomes very useful to obtain the displayed figure or assembly file without displaying one or more components.
3. *Symbolic presentation*: Sometimes, it is useful to show at least some indication of the actual component without having the geometries of the component. Symbolic representation becomes useful since it allows the name of the component to be shown without displaying the geometry.
4. *Exploded view*: The exploded view allows displaying of the components of an assembly in any position and also returns to its

original position in an unexploded state. This allows creating of the state to understand the assembling sequences of the original components.

5. *Sectional representation*: Sectional representation displays the sections across section planes.

The usefulness of representation is discussed in more detail in the latter part of this book.

7.9 Template Creation

An assembly template is a very useful method to reduce time and acquire more accuracy. Template creation is very useful, especially in the case of a bottom–up approach, where all the drawings can be studied before starting the 3-D creation in CAD. The common properties that are required to be included in every part easily can be customized in the assembly template itself, which has to be used when creating a new part or subassembly.

Tips for assembly:

1. Always provide a mating condition axis first during any assembly since it helps to visualize angular offset.
2. If all the components of an assembly are created in the same orientation, understanding of the location of components becomes faster when components are independently opened.

7.10 Project Work

Hands-On Exercise, Example 7.9: Creating a Paper Clip Assembly (Software Used: Creo Parametric)

PAPER CLIP (Figure 7.12)

Steps:

- Create an assembly file and rename it as Paper clip. (Other names may be given, but it is advisable to provide a relevant name.)
- Select Design on the right-hand side of the box. This will activate the standard assembly design mode and environment.
- Uncheck the box of the default template and click OK to create a new assembly design file.

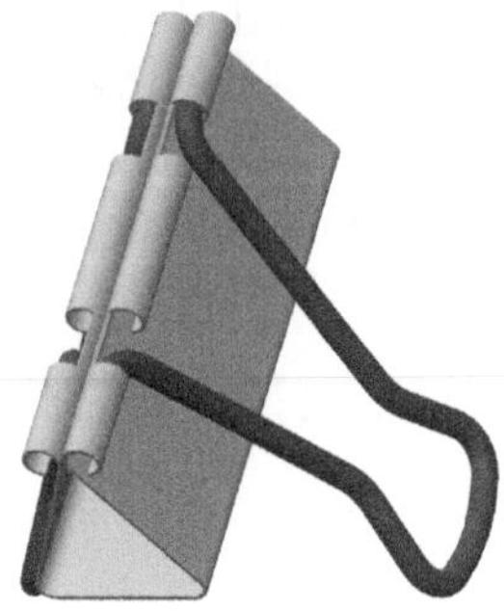

FIGURE 7.12
(See color insert.) 3-D model of a paper clip.

- The template selection box will appear.
- Select mmns_asm_Solid. This will allow use of the solid template file of the following units:
  - mm—millimeter for length unit
  - n—newton for force unit
  - s—second for time unit
- Once the assembly file is created, go to File > Prepare > Model properties to check the units. In the upper sections, the units of the model will be displayed.
  - Note: The units usually are displayed in the summary section of any CAD software. The setting of units usually is in the tools or settings.
- Click on tree filters, as shown in Figure 7.13.
- Check the box features to display in the model tree.
- Click OK to close the box of model tree items.
- The paper clip as shown in Figure 7.12 is to be created.
  - The used part files are as follows:
    - Paper_Clip_Handle in Chapter 6
    - Paper_Clip1 in Chapter 9

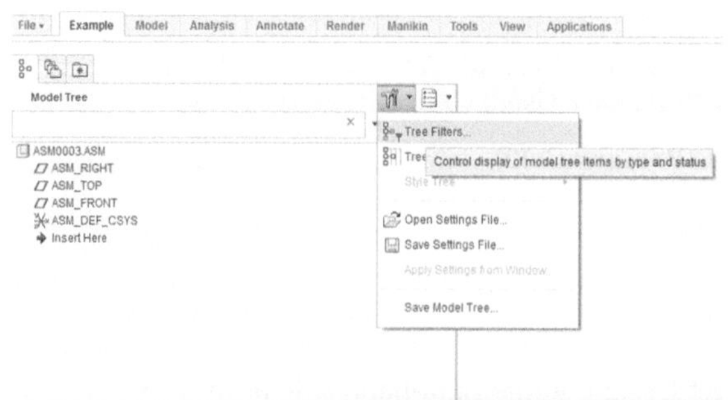

FIGURE 7.13
Model tree filter.

- Click on Assembly from the Model tab of the ribbon interface.
- Select the file Paper_Clip1 from the Chapter 9 folder.
- Click on the Placement tab.
- Select the option default in the placement option selection field.
- Click OK to assemble the component.

- Click on Assembly from the Model tab of the ribbon interface.
- Select the file Paper_Clip_Handle from the Chapter 6 folder.
- Click on the Placement tab.
- Select the extended cylindrical surface of the Paper_Clip_Handle part as a component reference, as shown in Figure 7.14.
- Select the internal cylindrical surface of Paper_Clip1 as an assembly reference, as shown in Figure 7.14.
- Select Coincident in the component reference relation field.
- Right-click on the graphical area and select New constraint.
- Select the plane RIGHT of the Paper_Clip_Handle part as a component reference.
- Select the plane FRONT of Paper_Clip1 as an assembly reference.
- Select Coincident in the component reference relation field.
- Right-click on the graphical area and select New constraint.
- Select the plane FRONT of the Paper_Clip_Handle part as a component reference.
- Select the outer large surface (the selected surface shall be on the same side of the selected internal surface) of Paper_Clip1 as an assembly reference.
- Select Parallel in the component reference relation field.

- Click on Assembly again from the Model tab of the ribbon interface.
- Select the file Paper_Clip_Handle from the Chapter 6 folder.
- Click on the Placement tab.

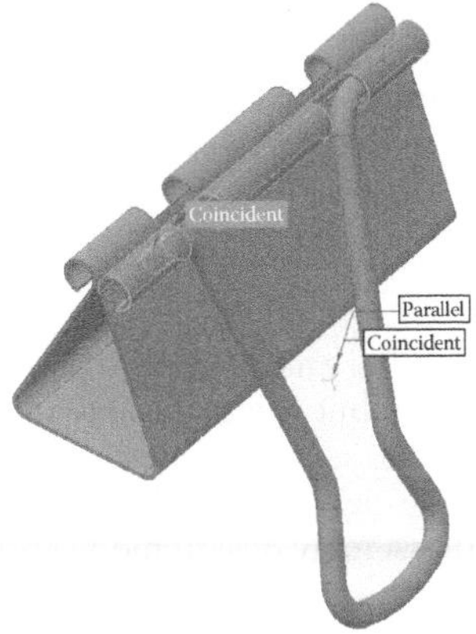

FIGURE 7.14
Cylindrical surface constraint of paper clip.

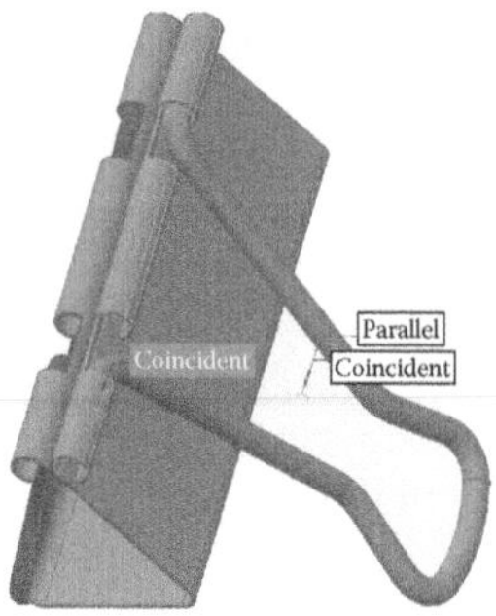

FIGURE 7.15
Cylindrical surface constraint of paper clip.

- Select the extended cylindrical surface of the Paper_Clip_Handle part as a component reference, as shown in Figure 7.15.
- Select the internal cylindrical surface of Paper_Clip1 as an assembly reference, as shown in Figure 7.15.
- Select Coincident in the component reference relation field.
- Right-click on the graphical area and select New constraint.
- Select the plane RIGHT of the Paper_Clip_Handle part as a component reference.
- Select the plane FRONT of Paper_Clip1 as an assembly reference.
- Select Coincident in the component reference relation field.
- Right-click on the graphical area and select New constraint.
- Select the plane FRONT of the Paper_Clip_Handle part as a component reference.
- Select the outer large surface (the selected surface shall be on the same side of the selected internal surface) of Paper_Clip1 as an assembly reference.
- Select Parallel in the component reference relation field.
- Save and exit the file.

Hands-on exercise, Example 7.10: Creating a Board Clip Assembly (Software Used: Creo Parametric)

BOARD CLIP

Steps:

- Create an assembly file and rename it as BOARD clip. (Other names may be given, but it is advisable to provide a relevant name.)
- Select Design on the right-hand side of the box. This will activate the standard assembly design mode and environment.
- Uncheck the box of the default template and click OK to create a new assembly design file.
- The template selection box will appear.

- Select mmns_asm_Solid. This will allow us to use the solid template file of the following units:
    - mm—millimeter for length unit
    - n—newton for force unit
    - s—second for time unit
- Once the assembly file is created, go to File > Prepare > Model properties to check the units. In the upper sections, the units of the model will be displayed.
    - Note: The units usually are displayed in the summary section of any CAD software. The setting of units usually is in the tools or settings.
- Click on tree filters, as shown in Figure 7.13.

- Check the box features to display in the model tree.
- Click OK to close the box of model tree items.
- The board clip as shown in Figure 7.16 is to be created.
    - The used part files are as follows:
        - Board_Clip_Spring in Chapter 6
        - Board_Clip1 in Chapter 6
        - Board_Clip_Pin in Chapter 6
- Click on Assembly from the Model tab of the ribbon interface.
- Select the file Board_Clip_Spring from the Chapter 6 folder.
- Click on the Placement tab.
- Select the right plane of the Board_Clip_Spring part as a component reference.
- Select the ASM_TOP plane of the main assembly as an assembly reference.
- Select Coincident in the component reference relation field.
- Right-click on the graphical area and select New constraint.
- Select the plane DTM1 of the Board_Clip_Spring part as a component reference.
- Select the ASM_RIGHT plane of the main assembly as an assembly reference.
- Select Coincident in the component reference relation field.
- Right-click on the graphical area and select New constraint.

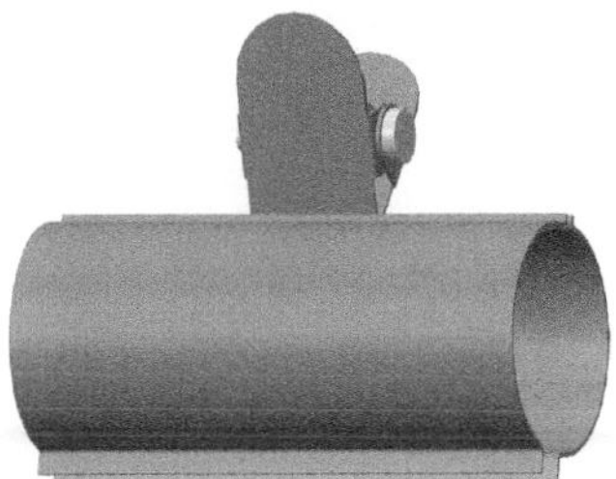

FIGURE 7.16
3-D view of a board clip.

- Select the plane FRONT of the Board_Clip_Spring part as a component reference.
- Select the ASM_FRONT plane of the main assembly as an assembly reference.
- Select Coincident in the component reference relation field.
- Flip any constraints if required.
- Click OK to assemble the component.

- Click on Assembly from the Model tab of the ribbon interface.
- Select the file Board_Clip_Clip1 from the Chapter 6 folder.
- Click on the Placement tab.
- Select the A_1 axis of the Board_Clip_Clip1 part as a component reference.
- Select the A_1 axis of Board_Clip_Spring as an assembly reference.
- Select Coincident in the component reference relation field.
- Right-click on the graphical area and select New constraint.
- Select the surface of the Board_Clip_Clip1 part, as shown in Figure 7.17, as a component reference.
- Select the top plane of Board_Clip_Spring as an assembly reference.
- Select Coincident in the component reference relation field.
- Right-click on the graphical area and select New constraint.
- Select the surface of the Board_Clip_Clip1 part as a component reference, as displayed in Figure 7.18.
- Select the vertex of the main Board_Clip_Spring as an assembly reference.
- Vertex selection filter may need to be applied if required.
- Select Coincident in the component reference relation field.

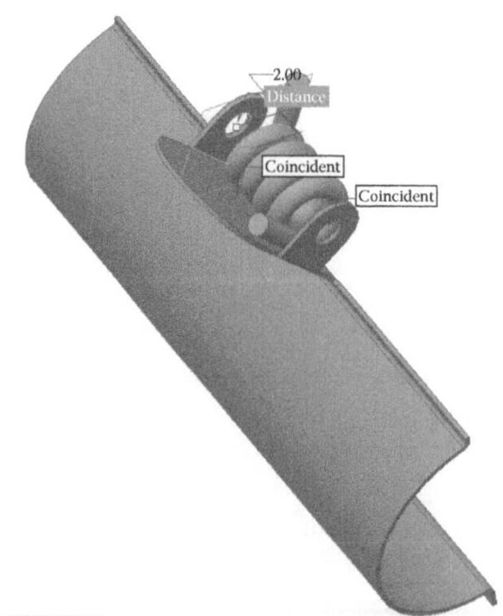

FIGURE 7.17
Surface selection of board clip during assembly.

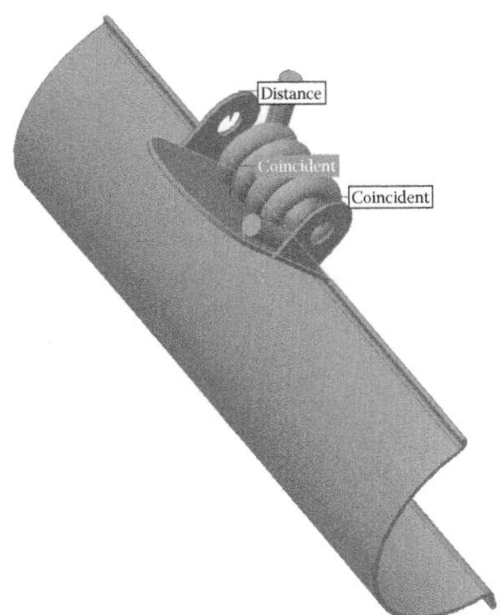

FIGURE 7.18
Surface selection of board clip during assembly.

- Flip any constraints if required.
- Click OK to assemble the component.

- Click on Assembly again from the Model tab of the ribbon interface.
- Select the file Board_Clip_Clip1 from the Chapter 6 folder.
- Click on the Placement tab.
- Select the A_2 axis of the Board_Clip_Clip1 part as a component reference.
- Select the A_1 axis of the assembled Board_Clip_Clip1 as an assembly reference.
- Select Coincident in the component reference relation field.
- Right-click on the graphical area and select New constraint.
- Select the surface of the Board_Clip_Clip1 part, as a component reference, as displayed in Figure 7.19.
- Select the surface of the assembled Board_Clip_Clip1 part, as shown in Figure 7.19, as an assembly reference.
- Select Coincident in the component reference relation field.
- Right-click on the graphical area and select New constraint.
- Select the surface of the Board_Clip_Clip1 part as a component reference, as displayed in Figure 7.20.
- Select the vertex of the main Board_Clip_Spring as an assembly reference.
- Vertex selection filter may need to be applied if required.
- Select Coincident in the component reference relation field.
- Flip any constraints if required.
- Click OK to assemble the component.
- Click on Assembly from the Model tab of the ribbon interface.
- Select the file Board_Clip_Pin from the Chapter 6 folder.
- Click on the Placement tab.

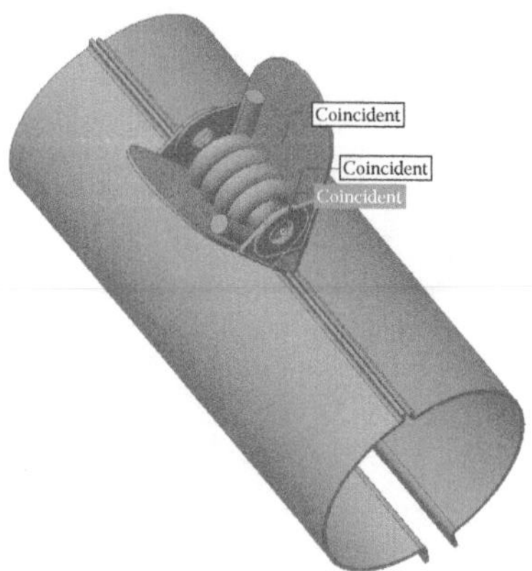

FIGURE 7.19
Surface selection of board clip during assembly.

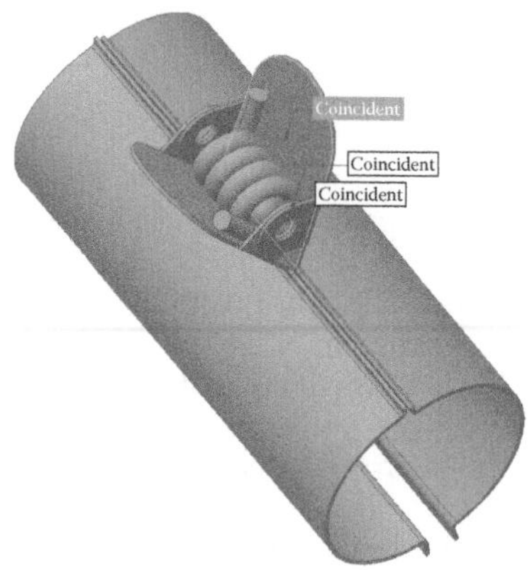

FIGURE 7.20
Surface selection of board clip during assembly.

- Select the A_1 axis of the Board_Clip_Spring part as a component reference.
- Select the A_2 axis of Board_Clip_1 as an assembly reference.
- Select Coincident in the component reference relation field.
- Right-click on the graphical area and select New constraint.
- Select the surface of the Board_Clip_Pin part, as shown in Figure 7.21, as a component reference.
- Select the surface of Board_Clip_1, as shown in Figure 7.21, as an assembly reference.
- Select Coincident in the component reference relation field.
- Right-click on the graphical area and select New constraint.
- Check the box for Allow assumption if required.
- Click OK to assemble the component.
- Double-click on the model Board_Clip_Spring.

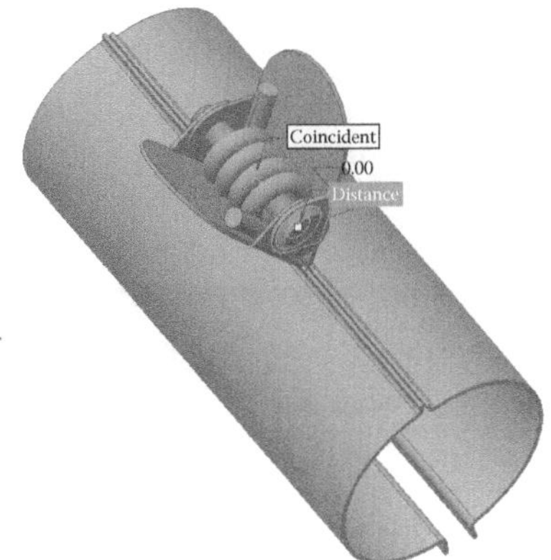

FIGURE 7.21
Surface selection of board clip during assembly.

- Right-click on the feature Extrude1 and select Edit.
- Change angular value to 143°.
- Radius to 4 mm.
- Depth to 3.85 mm.
- Click on Regenerate to regenerate the model.
- Click save and exit the file.

Hands-On Exercise, Example 7.11: Creating a Nozzle Assembly (Software Used: Creo Parametric)

NOZZLE

Steps:

- Create an assembly file and rename it as Nozzle. (Other names may be given, but it is advisable to provide a relevant name.)
- Select Design on the right-hand side of the box. This will activate the standard assembly design mode and environment.
- Uncheck the box of the default template and click OK to create a new assembly design file.
- The template selection box will appear.
- Select mmns_asm_Solid. This will allow us to use the solid template file of the following units:
  - mm—millimeter for length unit
  - n—newton for force unit
  - s—second for time unit
- Once the assembly file is created, go to File > Prepare > Model properties to check the units. In the upper sections, the units of the model will be displayed.
  - Note: The units usually are displayed in the summary section of any CAD software. The setting of units usually is in the tools or settings.
- Click on tree filters, as shown in Figure 7.13.

FIGURE 7.22
3-D view of a nozzle.

- Check the box features to display in the model tree.
- Click OK to close the box of model tree items.
- The nozzle as shown in Figure 7.22 is to be created.
- Click on the Part create on assembly mode icon from the Model tab of the ribbon interface.
- Change the name to Nozzle_Pipe.
- Select Solid as the subtype.
- Click OK.
- The creation option window will appear.
- Select the Locate default datums option.
- This will allow the creation of three default datums in the selected references.
- Select three planes in the Locate datums method area.
- Click OK and select the ASM_RIGHT, ASM_TOP, and ASM_FRONT planes.
- Notice that the part Nozzle_Pipe is added and activated.
- The ribbon interface will change to a part interface.
- Click on Extrude from the Model tab in the ribbon interface.
- Select the Thin extrusion option.
- Set the thickness as 5.2 mm.
- Click on the Placement tab and select Define to define the sketch.
- Select the top plane as the sketch plane.
- The right plane shall be automatically taken as reference and orientation as right.
- Click on Sketch.
- The sketcher model will be activated.
- Draw a circle of diameter 73 mm.
- Click OK to create and exit the sketch.
- Enter the value 150 mm for extrusion height.
- Flip upward for material direction.
- Ensure the sketch is thickening inside.
- Click on OK to create the extrusion.
- Right-click on the main assembly and select Activate.

- Click on the Part create on assembly mode icon from the Model tab of the ribbon interface.
- Change the name to R_Pad.
- Select Sheet metal as the subtype.
- Click OK.
- The creation option window will appear.
- Select the Locate default datums option.
- This will allow the creation of three default datums in the selected references.
- Select three planes in the Locate datums method area.
- Click OK and select the ASM_RIGHT, ASM_TOP, and ASM_FRONT planes.
- Notice that the part R_Pad is added and activated.
- The ribbon interface will change to a part interface.

- Click on Planar wall from the Model tab in the ribbon interface.
- Set the thickness as 10 mm.
- Click on the Placement tab and select Define to define the sketch.
- Select the top plane as the sketch plane.
- The right plane shall be automatically taken as reference and orientation as right.
- Click on Sketch.
- The sketcher model will be activated.
- Draw a circle of diameter 73 mm.
- Draw another circle of diameter 170 mm.
- Click OK to create and exit the sketch.
- Flip upward for material direction.
- Ensure the sketch is thickening inside.
- Click on OK to create the extrusion.
- Select Bend from the Model tab of the ribbon interface.
- Click on the Placement tab and select the upper face of the newly created planar wall.
- Click on the Bend line tab and select Sketch to define the bend line.
- Draw a straight line at the horizontal reference symmetric to the vertical reference and of length 1000 mm.
- Click OK to finish the sketch.
- Set the internal bend radius to 1000 mm.
- Select the option Bend to end of surface.
- Flip the bending downward if required.
- Click OK to finish the edge bend.
- Select the newly created bend in the model tree.
- Select Mirror from the editing drop-down menu of the model from the ribbon interface.
- Select the plane DTM1 as the mirror plane.
- Click OK to create the mirrored feature.
- Right-click on the main assembly and select Activate.
- Save and exit the file.

Hands-On Exercise, Example 7.12: Creating a Nozzle Assembly—Family Table Example (Software Used: Creo Parametric)

NOZZLE

Steps:

- Open the file Nozzle_Assembly.
- Right-click on the file nozzle. Select Open to open the file.
- Go to the Tools tab and click on Family table to make the other components of this family by varying the dimensions.
- The family table window will open up.
- Click on the vertical arrow icon for adding varying items.
- Click on the nozzle body.
- Select the outer diameter, thickness, and height dimension.
- Click OK on the family items window.
- Enter the data as displayed in Figure 7.23.
- Click on the Verify icon on the upper right side.
- Click on the Verify button of the family tree window.
- The verification status should be Success.
- Close the family tree window.
- Click OK to close the family table window.
- Save and close the file.
- Right-click on the file R_pad. Select Open to open the file.
- Go to the Tools tab and click on Family table to make the other components of this family by varying the dimensions.
- The family table window will open up.
- Click on the vertical arrow icon for adding varying items.
- Click on the planar wall feature.
- Select the outer diameter and inner diameter dimensions.
- Click OK on the family items window.
- Enter the data as displayed in Figure 7. 24.
- Click on the Verify icon on the upper right side.
- Click on the Verify button of the family tree window.
- The verification status should be Success.
- Close the family tree window.
- Click OK to close the family table window.
- Save and close the file.
- In Nozzle_Assembly, go to the Tools tab and click on Family table to make the other components of this family by varying the dimensions.
- The family table window will open up.

Type	Instance Name	Common Name	d1	d0	d2	F38 [EX...
	NOZZLE_PIPE	nozzle_pipe.prt	73.00	150.00	5.20	N
	NOZZLE_PIPE_500_NB	nozzle_pipe.prt_INST	510.00	200.00	8.00	Y
	NOZZLE_PIPE_65_NB	nozzle_pipe.prt_INST	73.00	150.00	5.20	*
	NOZZLE_PIPE_80_NB	nozzle_pipe.prt_INST1	88.90	150.00	5.50	*
	NOZZLE_PIPE_100_NB	nozzle_pipe.prt_INST2	114.30	150.00	6.00	*
	NOZZLE_PIPE_150_NB	nozzle_pipe.prt_INST3	168.30	150.00	7.10	*
	NOZZLE_PIPE_200_NB	nozzle_pipe.prt_INST4	219.10	200.00	6.40	*

FIGURE 7.23
Family table of nozzle parts.

Type	Instance Name	Common Name	d1	d37
	R_PAD	r_pad.prt	170.00	73.00
	R_PAD_500_NB	r_pad.prt_INST	800.00	513.00
	R_PAD_65_NB	r_pad.prt_INST	120.00	73.00
	R_PAD_80_NB	r_pad.prt_INST1	135.00	88.90
	R_PAD_100_NB	r_pad.prt_INST2	170.00	114.30
	R_PAD_150_NB	r_pad.prt_INST3	250.00	168.30
	R_PAD_200_NB	r_pad.prt_INST4	330.00	219.10

FIGURE 7.24
Family table of reinforcement pads.

Type	Instance Name	Common Name	M47 NOZZLE_PIPE	M61 R_PAD
	NOZZLE_ASSEMBLY	nozzle_assembly.asm	Y	Y
	NOZZLE_ASSEMBLY_500_NB	nozzle_assembly.asm_INST	NOZZLE_PIPE_500_NB	R_PAD_500_NB
	NOZZLE_ASSEMBLY_65_NB	nozzle_assembly.asm_INST	NOZZLE_PIPE_65_NB	N
	NOZZLE_ASSEMBLY_80_NB	nozzle_assembly.asm_INST1	NOZZLE_PIPE_80_NB	R_PAD_80_NB
	NOZZLE_ASSEMBLY_100_NB	nozzle_assembly.asm_INST2	NOZZLE_PIPE_100_NB	R_PAD_100_NB
	NOZZLE_ASSEMBLY_150_NB	nozzle_assembly.asm_INST3	NOZZLE_PIPE_150_NB	R_PAD_150_NB
	NOZZLE_ASSEMBLY_200_NB	nozzle_assembly.asm_INST4	NOZZLE_PIPE_200_NB	R_PAD_200_NB

FIGURE 7.25
Family table of nozzle assemblies.

- Click on the vertical arrow icon for adding varying items.
- Select Component in the family items window.
- Select both the component nozzle and R_Pad.
- Click OK on the family items window.
- Enter the data as displayed in Figure 7.25.
- Click on the Verify button of the family tree window.
- The verification status should be Success.
- Close the family tree window.
- Click OK to close the family table window.
- Save and exit the file.

Hands-On Exercise, Example 7.13: Creating a Manhole Top Assembly (Software Used: Creo Parametric)

MANHOLE TOP

Steps:

- Create an assembly file and rename it as Manhole top. (Other names may be given, but it is advisable to provide a relevant name.)
- Select Design on the right-hand side of the box. This will activate the standard assembly design mode and environment.
- Uncheck the box of the default template and click OK to create a new assembly design file.
- The template selection box will appear.
- Select mmns_asm_Solid. This will allow use of the solid template file of the following units:
    - mm—millimeter for length unit
    - n—newton for force unit
    - s—second for time unit

- Once the assembly file is created, go to File > Prepare > Model properties to check the units. In the upper sections, the units of the model will be displayed.
  - Note: The units usually are displayed in the summary section of any CAD software. The setting of units usually is in the tools or settings.
- Click on tree filters, as shown in Figure 7.13.

- Check the box features to display in the model tree.
- Click OK to close the box of model tree items.
- The manhole top as shown in Figure 7.26 is to be created.
- Click OK datum plane from the Model tab of the ribbon interface.
- Click on Assembly from the Model tab of the ribbon interface.
- Select the file Blind flange from the Chapter 6 folder.
- The instance selection window appears.
- Select BLIND-500NB.
- Click OK to assemble the instance of blind flange.
- Click on the Placement tab.
- Select Default in the Assembly relation drop-down box.
- The blind flange will be fixed with default plane coinciding.
- Create an assembly file and rename it as Manhole_Lifting_Assembly. (Other names may be given, but it is advisable to provide a relevant name.)
- Select Design on the right-hand side of the box. This will activate the standard assembly design mode and environment.
- Uncheck the box of the default template and click OK to create a new assembly design file.
- The template selection box will appear.
- Select mmns_asm_Solid. This will allow use of the solid template file of the following units:
  - mm—millimeter for length unit
  - n—newton for force unit
  - s—second for time unit

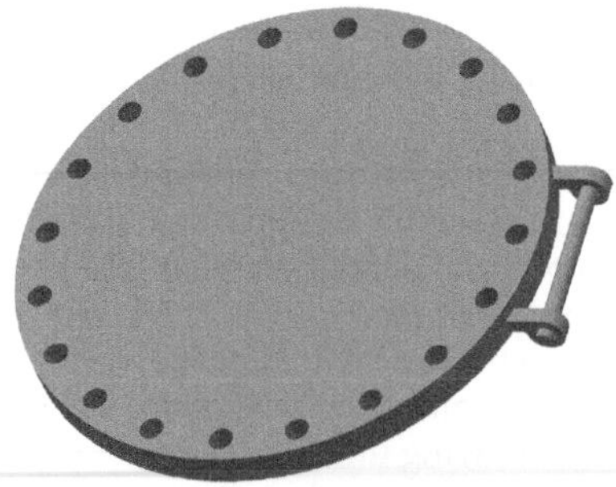

FIGURE 7.26
3-D view of a manhole cover.

- Once the assembly file is created, go to File > Prepare > Model properties to check the units. In the upper sections, the units of the model will be displayed.
  - Note: The units usually are displayed in the summary section of any CAD software. The setting of units usually is in the tools or settings.
- Click on tree filters, as shown in Figure 7.13.
- Select the front plane as the reference plane.
- Set the offset value 100 mm toward the front.
- Click OK to create the datum plane.
- The plane name will be ADTM1.

- Click on the Part create on assembly mode icon from the Model tab of the ribbon interface.
- Change the name to Clit_A.
- Select Solid as the subtype.
- Click OK.
- The creation option window will appear.
- Select the Locate default datums option.
- This will allow the creation of three default datums in the selected references.
- Select three planes in the Locate datums method area.
- Click OK and select the ASM_RIGHT, ASM_TOP, and ASM_FRONT planes.
- Notice that the part Clit_A is added and activated.
- The ribbon interface will change to a part interface.

- Click on Extrude from the Model tab in the ribbon interface.
- Click on the Placement tab and select Define to define the sketch.
- Select the ADTM1 plane as the sketch plane.
- Click on Sketch.
- The sketcher model will be activated.

- Project the sketch edges as displayed in Figure 7.27.
- Click OK to create and exit the sketch.
- Select Mid plane extrusion.
- Enter the value 20 mm for extrusion depth.
- Click on OK to create the extrude feature.
- Right-click on the main assembly and select Activate.

- Click on Round feature from the Model tab of the ribbon interface.
- Set the value of fillet to 25 mm.
- Select two edges as shown in Figure 7.28.
- Click OK to create the feature.

- Click on Round feature from the Model tab of the ribbon interface.
- Set the value of fillet to 5 mm.
- Select the edge as shown in Figure 7.29.
- Click OK to create the feature.

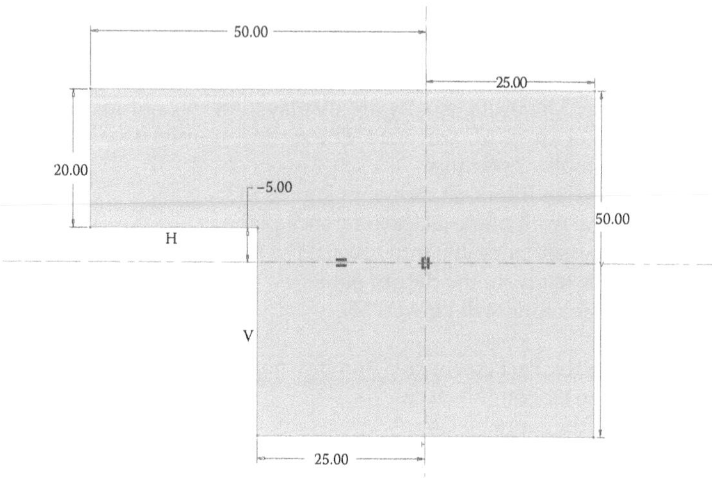

FIGURE 7.27
Constructional sketch of a hinge for a manhole.

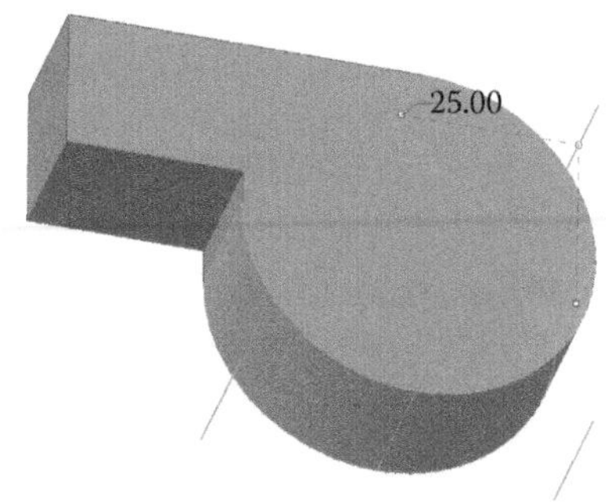

FIGURE 7.28
Edges for round operation of a hinge.

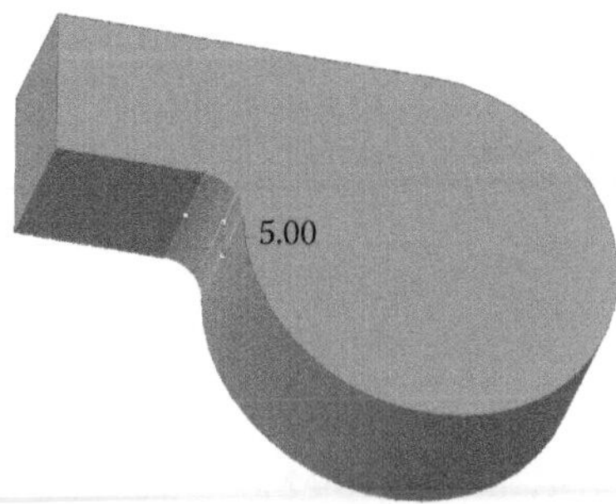

FIGURE 7.29
Edges for round operation of a hinge.

- Click on Extrude from the Model tab in the ribbon interface.
- Click on the Placement tab and select Define to define the sketch.
- Select the front flat face as the sketch plane.
- Click on Sketch.
- The sketcher model will be activated.
- Draw a circle of diameter 25 mm placing the center at the intersection of the horizontal and vertical references.
- Click OK to create and exit the sketch.
- Select through all.
- Select the Material cut option to remove the material.
- Flip the material orientation if required.
- Click on OK to create the extrusion cut.
- Right-click on the main assembly and select Activate.
- Save the assembly file.

- Click on Assembly from the Model tab of the ribbon interface.
- Select the file Clit_A from the working directory.
- Click on the Placement tab.
- Select the inside cylindrical surface of the component Clit_A part as a component reference.
- Select the inside cylindrical surface of the existing assembled component Clit_A instance as an assembly reference.
- Select Coincident in the component reference relation field.
- Right-click on the graphical area and select New constraint.
- Select the front surface of the Clit_A part as a component reference.
- Select the plane ASM_FRONT of the instance as an assembly reference.
- Select Distance in the component reference relation field.
- Enter the value 100 mm as offset distance.
- Right-click on the graphical area and select New constraint.
- Select the top surface of the component Clit_A part as a component reference.
- Select the top surface of the existing component Clit_A part as an assembly reference.
- Select Parallel in the component reference relation field.
- Flip any constraints if required.
- Click OK to assemble the component.

- Right-click on the main assembly and select Activate.

- Click on Extrude from the Model tab in the ribbon interface.
- Notice that the Material removal option is already activated and cannot be changed since in an assembly, a cut operation is only possible for joined components.
- Click on the Placement tab and select Define to define the sketch.
- Select the ASM_TOP plane as the sketch plane.
- The right plane shall be automatically taken as reference and orientation as right.
- Click on Sketch.

- The sketcher model will be activated.
- Draw a circle of diameter 698.5 mm placing the center at a distance 375 mm from the hole axis of both parts of Clit_A.
- Place the circle such that it will interfere with the smaller depth portion of Clit_A.
- Click OK to create and exit the sketch.
- Select through all extrusions.
- Flip the material direction if required to make the cut feature toward the material side.
- Click OK to exit the sketch.
- Click OK to create the feature.

- Click on the Part create on assembly mode icon from the Model tab of the ribbon interface.
- Change the name to Manhole_Hinge_Pin.
- Select Solid As the subtype.
- Click OK.
- The creation option window will appear.
- Select the Locate default datums option.
- This will allow the creation of three default datums in the selected references.
- Select three planes in the Locate datums method area.
- Click OK and select the ASM_RIGHT, ASM_TOP, and ASM_FRONT planes.
- Notice that the part Manhole_Hinge_Pin is added and activated.
- The ribbon interface will change to a part interface.
- Click on Revolve from the Model tab in the ribbon interface.
- Click on the Placement tab and select Define to define the sketch.
- Select the front plane as the sketch plane.
- The right plane shall be automatically taken as reference and orientation as right.
- Click on Sketch.
- The sketcher model will be activated.
- Draw the sketch as displayed in Figure 7.30.
- Click OK to create and exit the sketch.
- Ensure the revolve angle shall be 360°.
- Click on OK to create the revolve feature.
- Right-click on the main assembly and select Activate.
- Save and exit the file.

- Click on Assembly from the Model tab of the ribbon interface.
- Select the file Manhole_Lifting_Assembly from the working directory.
- Click on the Placement tab.
- Select the ASM_FRONT datum plane of the component Manhole_Lifting_Assembly part as a component reference.
- Select the ASM_FRONT of the instance as an assembly reference.
- Select Coincident in the component reference relation field.
- Right-click on the graphical area and select New constraint.

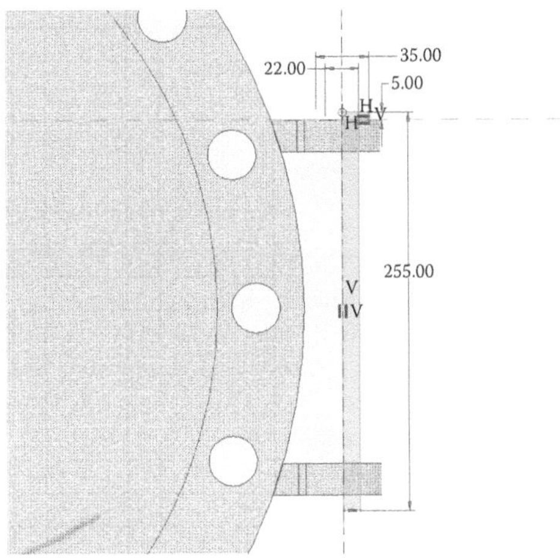

FIGURE 7.30
Construction sketch for a hinge pin.

- Select the cylindrical cut surface of the Manhole_Lifting_Assembly part as a component reference.
- Select the plane outer cylindrical surface of the blind flange as an assembly reference.
- Select Coincident in the component reference relation field.
- Right-click on the graphical area and select New constraint.
- Select the surface as displayed in Figure 7.31.
- Select Distance in the component reference relation field.
- Enter the value 5 mm.
- Flip any constraints if required.
- Click OK to assemble the component.
- Save and exit the file.

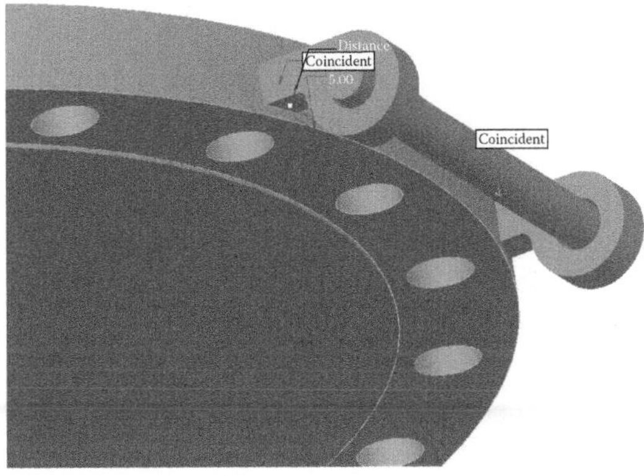

FIGURE 7.31
Mating constant in a manhole cover assembly.

Hands-On Exercise, Example 7.14: Creating a Manhole Assembly (Software Used: Creo Parametric)

MANHOLE (Figure 7.32)

Steps:

- Create an assembly file and rename it as Manhole. (Other names may be given, but it is advisable to provide a relevant name.)
- Select Design on the right-hand side of the box. This will activate the standard assembly design mode and environment.
- Uncheck the box of the default template and click OK to create a new assembly design file.
- The template selection box will appear.
- Select mmns_asm_Solid. This will allow use of the solid template file of the following units:
  - mm—millimeter for length unit
  - n—newton for force unit
  - s—second for time unit
- Once the assembly file is created, go to File > Prepare > Model properties to check the units. In the upper sections, the units of the model will be displayed.
  - Note: The units usually are displayed in the summary section of any CAD software. The setting of units usually is in the tools or settings.
- Click on tree filters, as shown Figure 7.13.
- Check the box features to display in the model tree.
- Click OK to close the box of model tree items.

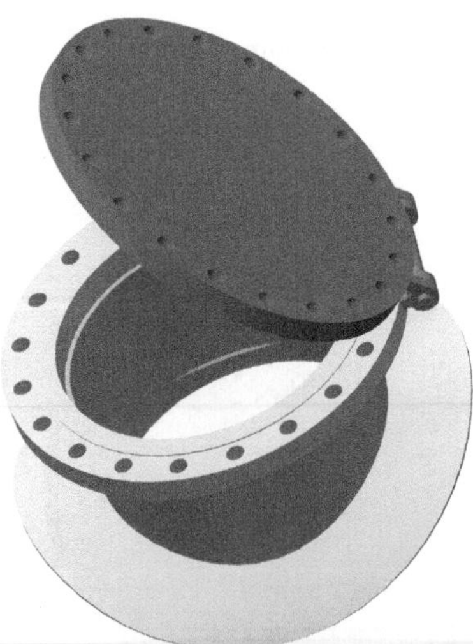

FIGURE 7.32
3-D view of a manhole assembly.

- Click on Assembly from the Model tab of the ribbon interface.
- Select the file Nozzle_Assembly from the working directory.
- Select the Nozzle_Assembly_500NB instance as the component.
- Click on the Placement tab.
- Select Default to place the component at default location.
- Click OK to assemble the component.

- Click on Assembly from the Model tab of the ribbon interface.
- Select the file Flange.
- Select the instance SORF-500NB.
- Click on the Placement tab.
- Select the bottom surface of the flange part as a component reference.
- Select the top surface of the existing nozzle as an assembly reference.
- Select Distance in the component reference relation field.
- Enter the value 20 mm as offset distance so that the nozzle is protruded inside the flange.
- Right-click on the graphical area and select New constraint.
- Select the internal cylindrical surface of the flange part as a component reference.
- Select the external cylindrical surface of the existing nozzle as an assembly reference.
- Select Coincide in the component reference relation field.
- Click OK to assemble the component.

- Click on Assembly from the Model tab of the ribbon interface.
- Select the file Clit_A from the working directory.
- Click on the Placement tab.
- Select the inside cylindrical surface of the component Clit_A part as a component reference.
- Select the outside cylindrical surface of the blind flange as an assembly reference.
- Select Coincident in the component reference relation field.
- Right-click on the graphical area and select New constraint.
- Select the front surface of the Clit_A part as a component reference.
- Select the plane ASM_FRONT of the instance as an assembly reference.
- Select Distance in the component reference relation field.
- Enter the value 100 mm as offset distance.
- Right-click on the graphical area and select New constraint.
- Select plane DTM1 of the component Clit_A part as a component reference.
- Select the front plane of the existing component SORF-Flance part as an assembly reference.
- Select Distance in the component reference relation field.
- Enter the value 375 mm as offset distance.
- Flip any constraints if required.
- Click OK to assemble the component.

- Click on Assembly from the Model tab of the ribbon interface.
- Select the file Clit_A from the working directory.
- Click on the Placement tab.
- Select the inside cylindrical surface of the component Clit_A part as a component reference.
- Select the inside cylindrical surface of the existing assembled component Clit_A instance as an assembly reference.
- Select Coincident in the component reference relation field.
- Right-click on the graphical area and select New constraint.
- Select the front surface of the Clit_A part as a component reference.
- Select the plane ASM_FRONT of the instance as an assembly reference.
- Select Distance in the component reference relation field.
- Enter the value 100 mm as offset distance.
- Right-click on the graphical area and select New constraint.
- Select the bottom surface of the component Clit_A part as a component reference.
- Select the bottom surface of the existing component Clit_A part as an assembly reference.
- Select Parallel in the component reference relation field.
- Flip any constraints if required.
- Click OK to assemble the component.

- Click on Assembly from the Model tab of the ribbon interface.
- Select the file Manhole_TOP from the working directory.
- Click on the Placement tab.
- Select the cylindrical surface of the hinge of Manhole_TOP part as a component reference.
- Select the inside cylindrical surface of the CLIT_A of the existing assembled component as an assembly reference.
- Select Coincident in the component reference relation field.
- Right-click on the graphical area and select New constraint.
- Select the ASM_FRONT plane of the Manhole_TOP part as a component reference.
- Select the plane ASM_FRONT as an assembly reference.
- Select Coincident in the component reference relation field.
- Flip any constraints if required.
- Click OK to assemble the component.
- Save and exit the file.

Hands-On exercise, Example 7.15: Creating a Tank Shell and Dish End Assembly (Software Used: Creo Parametric)

TANK SHELL AND DISH END ASSEMBLY (Figure 7.33)

Steps:

- Create an assembly file and rename it as Tank Shell and Dish End Assembly. (Other names may be given, but it is advisable to provide a relevant name.)

FIGURE 7.33
3-D view of a tank shell.

- Select Design on the right-hand side of the box. This will activate the standard assembly design mode and environment.
- Uncheck the box of the default template and click OK to create a new assembly design file.
- The template selection box will appear.
- Select mmns_asm_Solid. This will allow us to use the solid template file of the following units:
  - mm—millimeter for length unit
  - n—newton for force unit
  - s—second for time unit
- Once the assembly file is created, go to File > Prepare > Model properties to check the units. In the upper sections, the units of the model will be displayed.
  - Note: The units usually are displayed in the summary section of any CAD software. The setting of units usually is in the tools or settings.
- Click on tree filters, as shown in Figure 7.13.

- Check the box features to display in the model tree.
- Click OK to close the box of model tree items.
- The tank shell and dish end assembly as shown in Figure 7.13 is to be created.
  - The used part files are as follows:
    - Tank_Shell to be created in assembly file.
    - Dish_End to be created in assembly file.
- Click on the Part create on assembly mode icon from the Model tab of the ribbon interface.
- Change the name to Tank shell.
- Select Solid as the subtype.
- Click OK.
- The creation option window will appear.
- Select the Locate default datums option.
- This will allow the creation of three default datums in the selected references.
- Select three planes in the Locate datums method area.
- Click OK and select the ASM_RIGHT, ASM_TOP, and ASM_FRONT planes.
- Notice that the part Tank_Shell is added and activated.
- The ribbon interface will change to a part interface.

- Click on Revolve from the Model tab in the ribbon interface.
- Select the Thin option to make the feature thin.

- Set the value 10 mm as thickness of the revolved feature.
- Click on the Placement tab and select Define to define the sketch.
- Select the front plane as the sketch plane.
- The right plane shall be automatically taken as reference and orientation as right.
- Click on Sketch.
- The sketcher model will be activated.
- Draw a horizontal centerline in the horizontal reference line.
- Draw a vertical centerline in the vertical reference line.
- Draw a horizontal line symmetrical to the vertical centerline and at a distance of 500 mm from the horizontal centerline.
- Provide diameter dimension of 1000 mm by double-clicking the line twice and the horizontal centerline once.
- Set the length of the line is 5000 mm.
- Click OK to create and exit the sketch.
- Ensure the revolve angle shall be 360°.
- Click to flip on the thin feature to make 5000 mm the outer diameter.
- Click on OK to create the revolve feature.
- Right-click on the main assembly and select Activate.
- Click on the Part create on assembly mode icon from the Model tab of the ribbon interface.
- Change the name to Dish_End.
- Select Solid as the subtype.
- Click OK.
- The creation option window will appear.
- Select the Locate default datums option.
- This will allow the creation of three default datums in the selected references.
- Select three planes in the Locate datums method area.
- Click OK and select the ASM_RIGHT, ASM_TOP, and ASM_FRONT planes.
- Notice that the part Dish_End is added and activated.
- The ribbon interface will change to a part interface.
- Click on Revolve from the Model tab in the ribbon interface.
- Select the Thin option to make the feature thin.
- Set the value 12 mm as thickness of the revolved feature.
- Click on the Placement tab and select Define to define the sketch.
- Select the front plane as the sketch plane.
- The right plane shall be automatically taken as reference and orientation as right.
- Click on Sketch.
- The sketcher model will be activated.
- Draw a horizontal centerline in the horizontal reference line.
- Draw a vertical centerline in the vertical reference line.
- Draw the sketch as shown in Figure 7.34.
- Click OK to create and exit the sketch.
- Ensure the revolve angle shall be 360°.

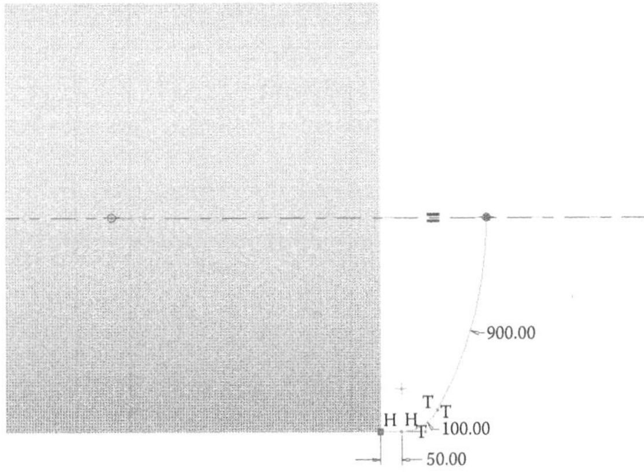

FIGURE 7.34
Constructional sketch of a dish end of a tank shell.

- Click on OK to create the revolve feature.
- Right-click on the main assembly and select Activate.
- Save and exit the file.

Hands-On Exercise, Example 7.16: Creating a Tank Saddle (Software Used: Creo Parametric)

SADDLE (Figure 7.35)

Steps:

- Create an assembly file and rename it as Saddle. (Other names may be given, but it is advisable to provide a relevant name.)
- Select Design on the right-hand side of the box. This will activate the standard assembly design mode and environment.
- Uncheck the box of the default template and click OK to create a new assembly design file.
- The template selection box will appear.

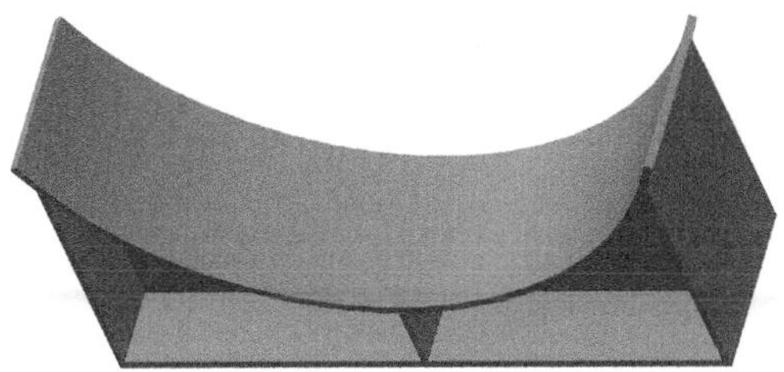

FIGURE 7.35
3-D view of a saddle assembly.

- Select mmns_asm_Solid. This will allow us to use the solid template file of the following units:
  - mm—millimeter for length unit
  - n—newton for force unit
  - s—second for time unit
- Once the assembly file is created, go to File > Prepare > Model properties to check the units. In the upper sections, the units of the model will be displayed.
  - Note: The units usually are displayed in the summary section of any CAD software. The setting of units usually is in the tools or settings.
- Click on tree filters, as shown in Figure 7.13.

- Check the box features to display in the model tree.
- Click OK to close the box of model tree items.
- The saddle as shown in Figure 7.13 is to be created.
- Select Sketch from the Model tab of the ribbon interface.
- A sketch window for selection of the sketch plane will appear.
- Select the front plane as the sketch plane.
- The right plane shall automatically be selected as reference and the orientation is to be right.
- Click on Sketch.
- The sketcher window will open up, and a reference selection window appears for selection of references.
- The two references (i.e., top plane as horizontal reference and right plane as vertical reference) shall be automatically taken; however, if not, please select the same.
- Create a sketch as displayed in Figure 7.36.
- Click OK to create and exit the sketch.

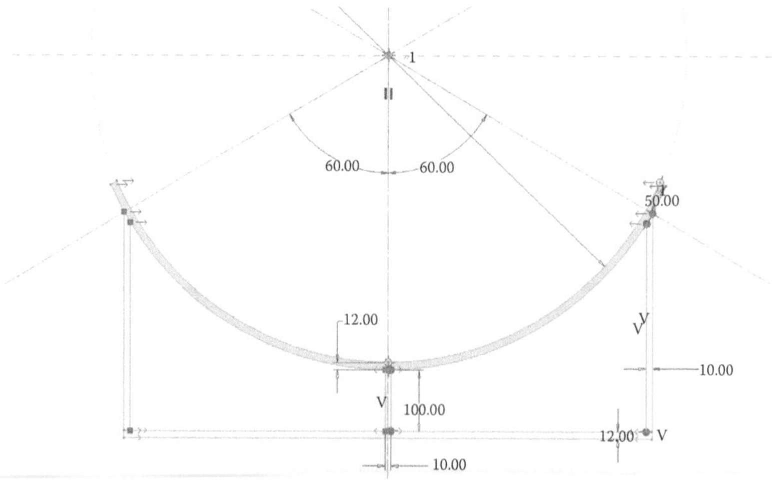

FIGURE 7.36
Constructional sketch of a saddle assembly.

- Click on the Part create on assembly mode icon from the Model tab of the ribbon interface.
- Change the name to wear_plate.
- Select Solid as the subtype.
- Click OK.
- The creation option window will appear.
- Select the Locate default datums option.
- This will allow the creation of three default datums in the selected references.
- Select three planes in the Locate datums method area.
- Click OK and select the ASM_RIGHT, ASM_TOP, and ASM_FRONT planes.
- Notice that the part wear_plate is added and activated.
- The ribbon interface will change to a part interface.

- Click on Extrude from the Model tab in the ribbon interface.
- Click on the Placement tab and select Define to define the sketch.
- Select the front plane as the sketch plane.
- The right plane shall be automatically taken as reference and orientation as right.
- Click on Sketch.
- The sketcher model will be activated.
- Project the sketch edges as displayed in Figure 7.37.
- Click OK to create and exit the sketch.
- Select Mid plane extrusion.
- Enter the value 250 mm for extrusion depth.
- Click on OK to create the extrude feature.
- Right-click on the main assembly and select Activate.
- Click on the Part create on assembly mode icon from the Model tab of the ribbon interface.
- Change the name to Base_Plate.
- Select Solid as the subtype.
- Click OK.
- The creation option window will appear.

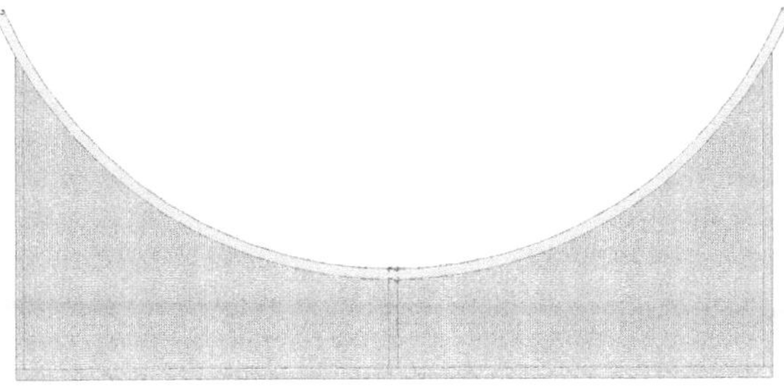

FIGURE 7.37
Projection in constructional stages of a saddle.

- Select the Locate default datums option.
- This will allow the creation of three default datums in the selected references.
- Select three planes in the Locate datums method area.
- Click OK and select the ASM_RIGHT, ASM_TOP, and ASM_FRONT planes.
- Notice that the part Base_Plate is added and activated.
- The ribbon interface will change to a part interface.

- Click on Extrude from the Model tab in the ribbon interface.
- Click on the Placement tab and select Define to define the sketch.
- Select the front plane as the sketch plane.
- The right plane shall be automatically taken as reference and orientation as right.
- Click on Sketch.
- The sketcher model will be activated.
- Project the sketch edges as displayed in Figure 7.38.
- Click OK to create and exit the sketch.
- Click on the Options tab.
- Select To selected extrusion option for both direction extrusion options.
- Select the opposite curved edges of the wear pad.
- Click on OK to create the extrude feature.
- Right-click on the main assembly and select Activate.
- Click on the Part create on assembly mode icon from the Model tab of the ribbon interface.
- Change the name to Mid_Plate.
- Select Solid as the subtype.
- Click OK.
- The creation option window will appear.
- Select the Locate default datums option.
- This will allow the creation of three default datums in the selected references.
- Select three planes in the Locate datums method area.
- Click OK and select the ASM_RIGHT, ASM_TOP, and ASM_FRONT planes.

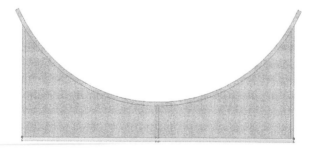

FIGURE 7.38
Projection in constructional stages of a saddle.

- Notice that the part Mid_Plate is added and activated.
- The ribbon interface will change to a part interface.

- Click on Extrude from the Model tab in the ribbon interface.
- Click on the Placement tab and select Define to define the sketch.
- Select the front plane as the sketch plane.
- The right plane shall be automatically taken as reference and orientation as right.
- Click on Sketch.
- The sketcher model will be activated.
- Project the sketch edges as displayed in Figure 7.39.
- Click OK to create and exit the sketch.
- Select Mid plane extrusion.
- Enter the value 10 mm for extrusion depth.
- Click on OK to create the extrude feature.
- Right-click on the main assembly and select Activate.
- Click on the Part create on assembly mode icon from the Model tab of the ribbon interface.
- Change the name to Side_Rib_1.
- Select Solid as the subtype.
- Click OK.
- The creation option window will appear.
- Select the Locate default datums option.
- This will allow the creation of three default datums in the selected references.
- Select three planes in the Locate datums method area.
- Click OK and select the ASM_RIGHT, ASM_TOP, and ASM_FRONT planes.
- Notice that the part Side_Rib_1 is added and activated.
- The ribbon interface will change to a part interface.

- Click on Extrude from the Model tab in the ribbon interface.
- Click on the Placement tab and select Define to define the sketch.
- Select the front plane as the sketch plane.
- The right plane shall be automatically taken as reference and orientation as right.

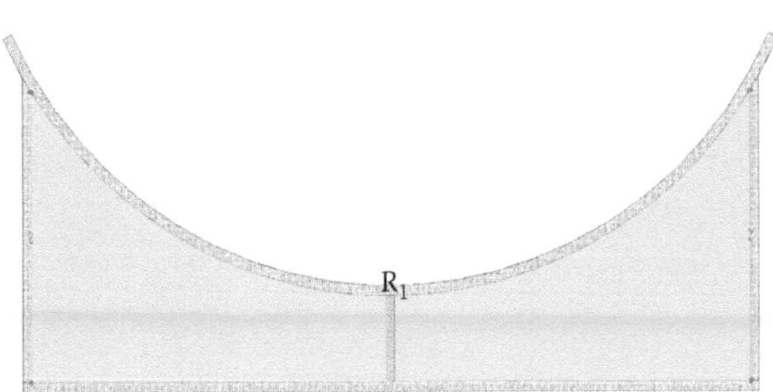

FIGURE 7.39
Projection in constructional stages of a saddle.

- Click on Sketch.
- The sketcher model will be activated.
- Project the sketch edges as displayed in Figure 7.40.
- Click OK to create and exit the sketch.
- Click on Options tab.
- Select To selected extrusion option for both direction extrusion options.
- Select the opposite curved edges of the wear pad.
- Click on OK to create the extrude feature.
- Right-click on the main assembly and select Activate.
- Click on the Part create on assembly mode icon from the Model tab of the ribbon interface.
- Change the name to Side_Rib_2.
- Select Solid as the subtype.
- Click OK.
- The creation option window will appear.
- Select the Locate default datums option.
- This will allow the creation of three default datums in the selected references.
- Select three planes in the Locate datums method area.
- Click OK and select the ASM_RIGHT, ASM_TOP, and ASM_FRONT planes.
- Notice that the part Side_Rib_2 is added and activated.
- The ribbon interface will change to a part interface.

- Click on Extrude from the Model tab in the ribbon interface.
- Click on the Placement tab and select Define to define the sketch.
- Select the front plane as the sketch plane.
- The right plane shall be automatically taken as reference and orientation as right.
- Click on Sketch.
- The sketcher model will be activated.
- Project the sketch edges as displayed in Figure 7.41.
- Click OK to create and exit the sketch.
- Click on Options tab.

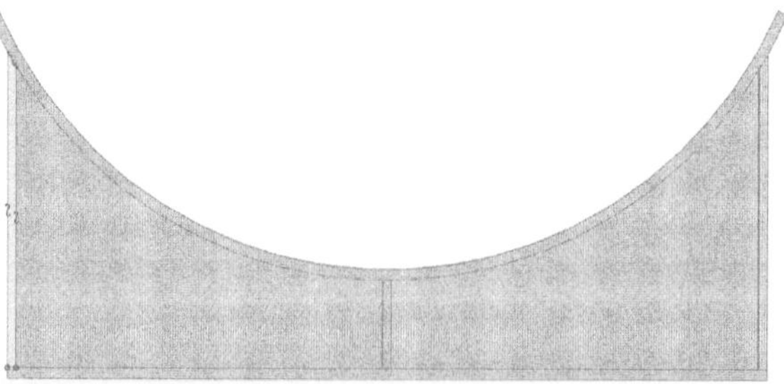

FIGURE 7.40
Projection in constructional stages of a saddle.

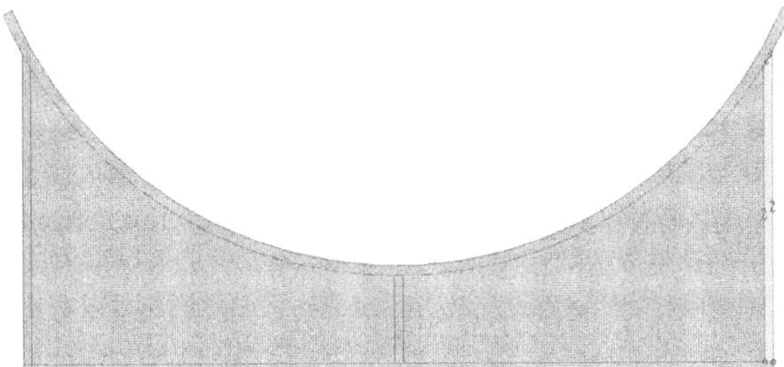

FIGURE 7.41
Projection in constructional stages of a saddle.

- Select To selected extrusion option for both direction extrusion options.
- Select the opposite curved edges of the wear pad.
- Click on OK to create the extrude feature.
- Right-click on the main assembly and select Activate.
- Click on the Part create on assembly mode icon from the Model tab of the ribbon interface.
- Change the name to Mid_Rib.
- Select Solid as the subtype.
- Click OK.
- The creation option window will appear.
- Select the Locate default datums option.
- This will allow the creation of three default datums in the selected references.
- Select three planes in the Locate datums method area.
- Click OK and select the ASM_RIGHT, ASM_TOP, and ASM_FRONT planes.
- Notice that the part Mid_Rib is added and activated.
- The ribbon interface will change to a part interface.

- Click on Extrude from the Model tab in the ribbon interface.
- Click on the Placement tab and select Define to define the sketch.
- Select the front plane as the sketch plane.
- The right plane shall be automatically taken as reference and orientation as right.
- Click on Sketch.
- The sketcher model will be activated.
- Project the sketch edges as displayed in Figure 7.42.
- Click OK to create and exit the sketch.
- Click on Options tab.
- Select To selected extrusion option.
- Select the front curved edges of the wear pad.
- Click on OK to create the extrude feature.
- Right-click on the main assembly and select Activate.
- Click on Assembly from the Model tab of the ribbon interface.
- Select the file Mid_Rib from the working directory.

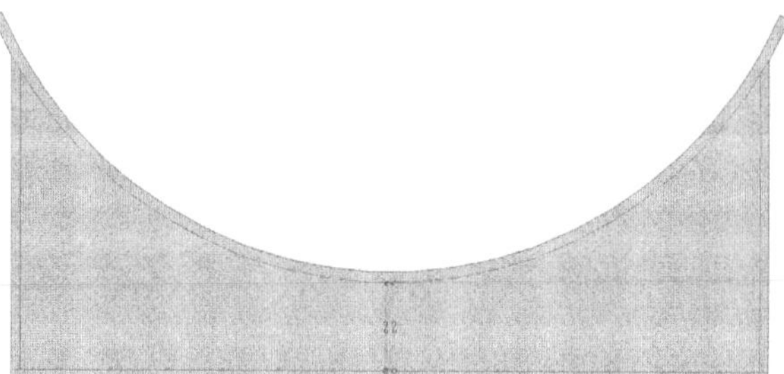

FIGURE 7.42
Projection in constructional stages of a saddle.

- Click on the Placement tab.
- Select the left bottom surface (opposite to the curved surface) of the Mid_Rib part as a component reference.
- Select the top surface of the first Base_Plate part instance as an assembly reference.
- Select Coincident in the component reference relation field.
- Right-click on the graphical area and select New constraint.
- Select the cylindrical surface of the Mid_Rib part as a component reference.
- Select the bottom surface of the first wear_plate part instance as an assembly reference.
- Select Coincident in the component reference relation field.
- Right-click on the graphical area and select New constraint.
- Select surface as shown in Figure 7.43.
- Select Coincident in the component reference relation field.
- Flip any constraints if required.
- Click OK to assemble the component.
- Save and exit the file.

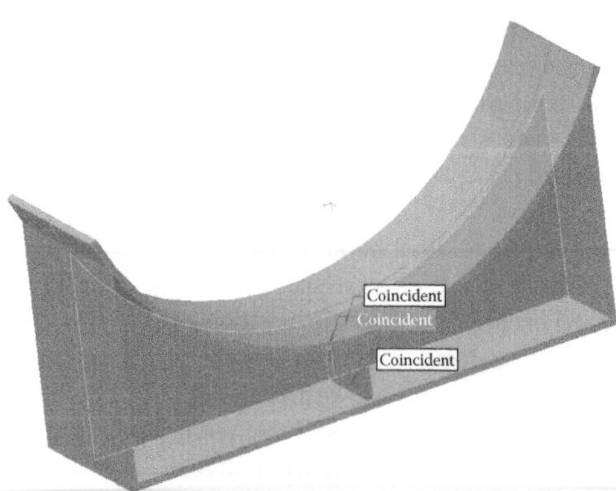

FIGURE 7.43
Surface selection during assembly of parts of a saddle.

Hands-On Exercise, Example 7.17: Creating an Expansion Joint (Software Used: Creo Parametric)

EXPANSION JOINT (Figure 7.44)

Steps:

- Create an assembly file and rename it as Expansion Joint. (Other names may be given, but it is advisable to provide a relevant name.)
- Select Design on the right-hand side of the box. This will activate the standard assembly design mode and environment.
- Uncheck the box of the default template and click OK to create a new assembly design file.
- The template selection box will appear.
- Select mmns_asm_Solid. This will allow us to use the solid template file of the following units:
    - mm—millimeter for length unit
    - n—newton for force unit
    - s—second for time unit
- Once the assembly file is created, go to File > Prepare > Model properties to check the units. In the upper sections, the units of the model will be displayed.
    - Note: The units usually are displayed in the summary section of any CAD software. The setting of units usually is in the tools or settings.
- Click on tree filters, as shown in Figure 7.13.

- Check the box features to display in the model tree.
- Click OK to close the box of model tree items.
- The expansion joint as shown in Figure 7.13 is to be created.
- Select Sketch from the Model tab of the ribbon interface.
- A sketch window for selection of the sketch plane will appear.
- Select the front plane as the sketch plane.

FIGURE 7.44
(See color insert.) 3-D view of a hinge expansion joint.

- The right plane shall automatically be selected as reference and the orientation is to be right.
- Click on Sketch.
- The sketcher window will open up, and a reference selection window appears for selection of references.
- The two references (i.e., top plane as horizontal reference and right plane as vertical reference) shall be automatically taken; however, if not, please select the same.
- Create a sketch as displayed in Figures 7.45 and 7.46.
- Click OK to create and exit the sketch.

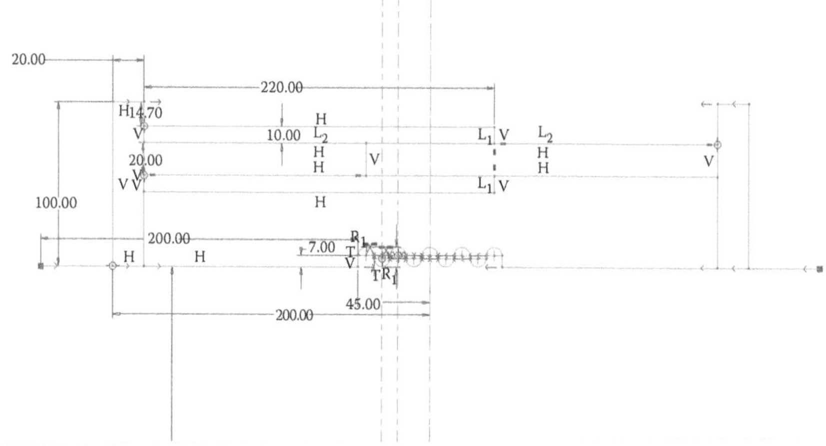

FIGURE 7.45
Constructional sketch of a hinge expansion joint.

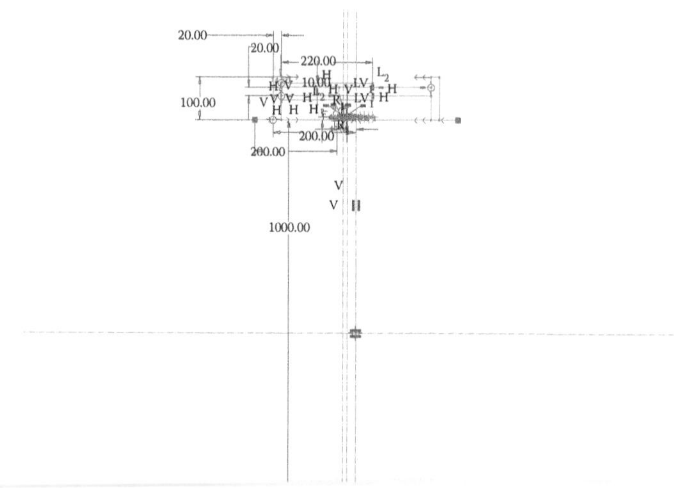

FIGURE 7.46
Constructional sketch of a hinge expansion joint.

- Click on Part create on assembly mode icon from the Model tab of the ribbon interface.
- Change the name to Pipe_1.
- Select Solid as the subtype.
- Click OK.
- The creation option window will appear.
- Select the Locate default datums option.
- This will allow the creation of three default datums in the selected references.
- Select three planes in the Locate datums method area.
- Click OK and select the ASM_RIGHT, ASM_TOP, and ASM_FRONT planes.
- Notice that the part Pipe_1 is added and activated.
- The ribbon interface will change to a part interface.

- Click on Revolve from the Model tab in the ribbon interface.
- Select the Thin option to make the feature thin.
- Set the value 10 mm as thickness of the revolved feature.
- Click on the Placement tab and select Define to define the sketch.
- Select the front plane as the sketch plane.
- The right plane shall be automatically taken as reference and orientation as right.
- Click on Sketch.
- The sketcher model will be activated.
- Draw a horizontal centerline in the horizontal reference line.
- Project the left straight edge of the sketch.
- Click OK to create and exit the sketch.
- Ensure the revolve angle shall be 360°.
- Click to flip on the thin feature to make 1000 mm the outer diameter.
- Click on OK to create the revolve feature.
- Right-click on the main assembly and select Activate.
- Click on Part create on assembly mode icon from the Model tab of the ribbon interface.
- Change the name to Pipe_2.
- Select Solid as the subtype.
- Click OK.
- The creation option window will appear.
- Select the Locate default datums option.
- This will allow the creation of three default datums in the selected references.
- Select three planes in the Locate datums method area.
- Click OK and select the ASM_RIGHT, ASM_TOP, and ASM_FRONT planes.
- Notice that the part Pipe_2 is added and activated.
- The ribbon interface will change to a part interface.

- Click on Revolve from the Model tab in the ribbon interface.
- Select the Thin option to make the feature thin.
- Set the value 10 mm as thickness of the revolved feature.
- Click on the Placement tab and select Define to define the sketch.
- Select the front plane as the sketch plane.
- The right plane shall be automatically taken as reference and orientation as right.
- Click on Sketch.
- The sketcher model will be activated.
- Draw a horizontal centerline in the horizontal reference line.
- Project the left straight edge of the sketch.
- Click OK to create and exit the sketch.
- Ensure the revolve angle shall be 360°.
- Click to flip on the thin feature to make 1000 mm the outer diameter.
- Click on OK to create the revolve feature.
- Right-click on the main assembly and select Activate.
- Click on the Component create on assembly mode icon from the Model tab of the ribbon interface.
- Change the name to Bellow.
- Select Solid as the subtype.
- Click OK.
- The creation option window will appear.
- Select the Locate default datums option.
- This will allow the creation of three default datums in the selected references.
- Select three planes in the Locate datums method area.
- Click OK and select the ASM_RIGHT, ASM_TOP, and ASM_FRONT planes.
- Notice that the part Bellow is added and activated.
- The ribbon interface will change to a part interface.

- Click on Revolve from the Model tab in the ribbon interface.
- Select the Thin option to make the feature thin.
- Set the value 1 mm as thickness of the revolved feature.
- Click on the Placement tab and select Define to define the sketch.
- Select the front plane as the sketch plane.
- The right plane shall be automatically taken as reference and orientation as right.
- Click on Sketch.
- The sketcher model will be activated.
- Draw a horizontal centerline in the horizontal reference line.
- Project the convolute portion of the sketch.
- Click OK to create and exit the sketch.
- Ensure the revolve angle shall be 360°.
- Click to flip on the thin feature to make the material orientation inward.
- Click on OK to create the revolve feature.
- Right-click on the main assembly and select Activate.

- Click on the Component create on assembly mode icon from the Model tab of the ribbon interface.
- Change the name to Hinge_Assembly.
- Select Standard as the subtype.
- Click OK.
- The creation option window will appear.
- Select the Locate default datums option.
- This will allow the creation of three default datums in the selected references.
- Select three planes in the Locate datums method area.
- Click OK and select the ASM_RIGHT, ASM_TOP, and ASM_FRONT planes.
- Right-click on the newly created assembly and select Activate.

- Click on the Component create on assembly mode icon from the Model tab of the ribbon interface.
- Change the name to Rib_1.
- Select Solid as the subtype.
- Click OK.
- The creation option window will appear.
- Select the Locate default datums option.
- This will allow the creation of three default datums in the selected references.
- Select three planes in the Locate datums method area.
- Click OK and select the ASM_RIGHT, ASM_TOP, and ASM_FRONT planes.
- Notice that the part Bellow is added and activated.
- The ribbon interface will change to a part interface.

- Click on Revolve from the Model tab in the ribbon interface.
- Click on the Placement tab and select Define to define the sketch.
- Select the front plane as the sketch plane.
- The right plane shall be automatically taken as reference and orientation as right.
- Click on Sketch.
- The sketcher model will be activated.
- Draw a horizontal centerline in the horizontal reference line.
- Draw the sketch as shown in Figure 7.47.
- Click OK to create and exit the sketch.
- Select the Mid plane option.
- Ensure the revolve angle shall be 20°.
- Click to flip on the thin feature to make the material orientation inward.
- Click on OK to create the revolve feature.
- Right-click on the main assembly and select Activate.

- Click on Extrude from the Model tab in the ribbon interface.
- Click on the Placement tab and select Define to define the sketch.
- Select the right surface of the newly created revolved feature as the sketch plane.

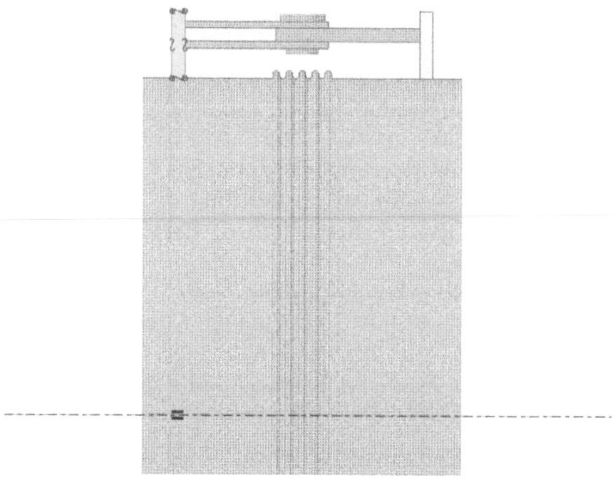

FIGURE 7.47
Constructional sketch of a hinge expansion joint.

- Click on Sketch.
- The sketcher model will be activated.
- Project the sketch as shown in Figure 7.48.
- Select Mid plane extrusion.
- Select the Through all option.
- Activate the Material removal option.
- Flip toward the material side if required.
- Click on OK to create the extrusion.
- Right-click on the main assembly and select Activate.
- Save the assembly.
- Right-click on Hinge_Assembly and select Activate.

- Click on the Component create on assembly mode icon from the Model tab of the ribbon interface.
- Change the name to Hinge_Plate1.
- Select Solid as the subtype.
- Click OK.
- The creation option window will appear.

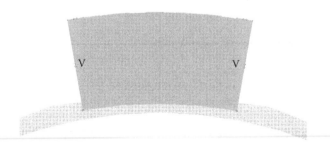

FIGURE 7.48
Constructional sketch of a hinge expansion joint.

- Select the Locate default datums option.
- This will allow the creation of three default datums in the selected references.
- Select three planes in the Locate datums method area.
- Click OK and select the ASM_RIGHT, ASM_TOP, and ASM_FRONT planes.
- Notice that the part Hinge_Plate1 is added and activated.
- The ribbon interface will change to a part interface.
- Notice that the part Hinge_Plate1 is added and activated.
- The ribbon interface will change to a part interface.
- Click on Extrude from the Model tab in the ribbon interface.
- Click on the Placement tab and select Define to define the sketch.
- Select plane DTM3 part Hinge_Plate1 as the sketch plane.
- Click on Sketch.
- The sketcher model will be activated.
- Project the sketch as shown in Figure 7.49.
- Select Mid plane extrusion.
- Set value 150 mm as the blind extrusion depth.
- Click on OK to create the extrusion.
- Right-click on the main assembly and select Activate.
- Save the assembly.
- Right-click on Hinge_Assembly and select Activate.

- Click on Assembly from the Model tab of the ribbon interface.
- Select the file Hinge_Plate1 from the working directory.
- Click on the Placement tab.
- Select the left small flat surface of the Hinge_Plate1 part as a component reference.
- Select the right surface of the first Rib_1 part instance as an assembly reference.
- Select Coincident in the component reference relation field.
- Right-click on the graphical area and select New constraint.
- Select the top surface of the Hinge_Plate1 part as a component reference.
- Select the bottom surface plane of the first Hinge_Plate1 part instance as an assembly reference.

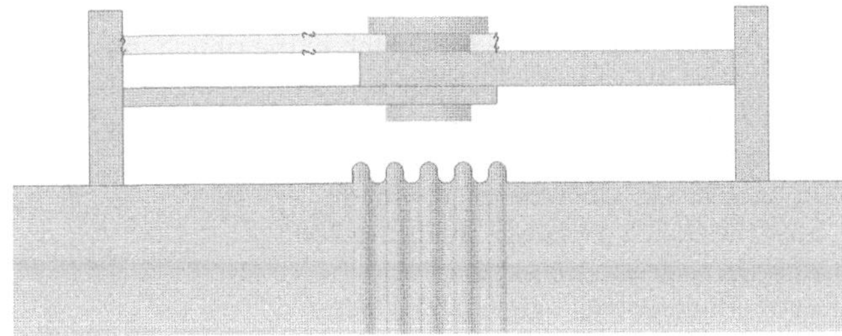

FIGURE 7.49
Constructional sketch of features of a screwdriver.

- Select Distance and set the value 20 mm in the component reference relation field.
- Right-click on the graphical area and select New constraint.
- Select the plane DTM3 of the Hinge_Plate1 part as a component reference.
- Select the ASM_FRONT plane of the main assembly as an assembly reference.
- Select Distance and set the value of 200 mm in the component reference relation field.
- Flip any constraints if required.
- Click OK to assemble the component.
- Select Hole from the Model tab of the ribbon interface.
- Select Simple hole.
- Enter 50 mm as hole diameter.
- Select the Through all option.
- Click on the Placement tab and select the top face of the first Hinge_Plate1 part instance.
- Select plane DTM1 of Rib_1 and plane ADTM3 of Hinge_Assembly as offset reference.
- Double-click on the offset and select Align in both the references.
- Click OK to create the hole feature.
- Right-click on Hinge_Assembly and select Activate.
- Click on the Component create on assembly mode icon from the Model tab of the ribbon interface.
- Change the name to Hinge_Pin.
- Select Solid as the subtype.
- Click OK.
- The creation option window will appear.
- Select the Locate default datums option.
- This will allow the creation of three default datums in the selected references.
- Select three planes in the Locate datums method area.
- Click OK and select the ASM_RIGHT, ASM_TOP, and ASM_FRONT planes.
- Notice that the part Hinge_Pin is added and activated.
- The ribbon interface will change to a part interface.
- Notice that the part Hinge_Pin is added and activated.
- The ribbon interface will change to a part interface.

- Click on Revolve from the Model tab in the ribbon interface.
- Click on the Placement tab and select Define to define the sketch.
- Select ADTM3 of Hinge_Assembly as the sketch plane.
- Click on Sketch.
- The sketcher model will be activated.
- Draw the sketch as shown in Figure 7.50.
- Click OK to create the sketch.
- Ensure the revolve angle shall be 360°.
- Click on OK to create the revolve feature.
- Right-click on the main assembly and select Activate.
- Click on Assembly from the Model tab of the ribbon interface.

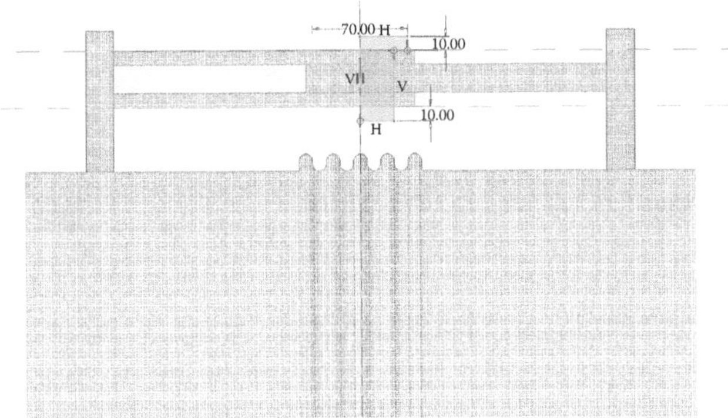

FIGURE 7.50
Constructional sketch of a hinge expansion joint.

- Select the file Hinge_Assembly from the working directory.
- Click on the Placement tab.
- Select the cylindrical surface of the Rib_1 part as a component reference.
- Select the cylindrical surface of the first Pipe_2 part as an assembly reference.
- Select Coincident in the component reference relation field.
- Right-click on the graphical area and select New constraint.
- Select the plane ADTM3 of the Hinge_Assembly part as a component reference.
- Select the ASM_FRONT plane of the main assembly as an assembly reference.
- Select Coincident in the component reference relation field.
- Right-click on the graphical area and select New constraint.
- Select the plane ADTM1 of the Hinge_Assembly part as a component reference.
- Select the ASM_RIGHT plane of the main assembly as an assembly reference.
- Select Coincident in the component reference relation field.
- Flip any constraints if required.
- Click OK to assemble the component.
- Right-click on the main assembly and select Activate.
- Save and exit the file.

Hands-On Exercise, Example 7.18: Creating a Tank Assembly (Software Used: Creo Parametric)

MANHOLE TOP (Figure 7.51)

Steps:

- Create an assembly file and rename it as Tank top. (Other names may be given, but it is advisable to provide a relevant name.)

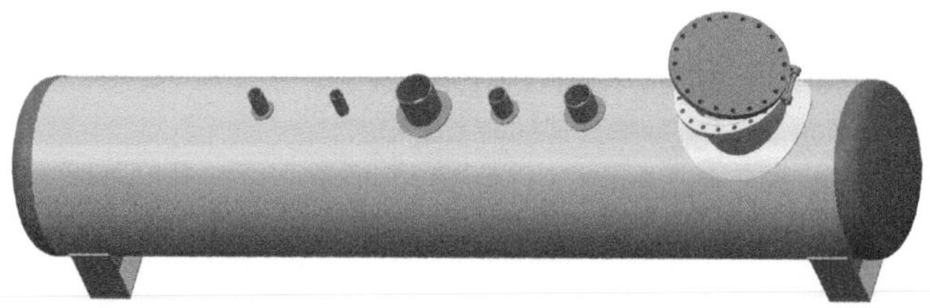

FIGURE 7.51
3-D view of a tank assembly.

- Select Design on the right-hand side of the box. This will activate the standard assembly design mode and environment.
- Uncheck the box of the default template and click OK to create a new assembly design file.
- The template selection box will appear.
- Select mmns_asm_Solid. This will allow us to use the solid template file of the following units:
  - mm—millimeter for length unit
  - n—newton for force unit
  - s—second for time unit
- Once the assembly file is created, go to File > Prepare > Model properties to check the units. In the upper sections, the units of the model will be displayed.
  - Note: The units usually are displayed in the summary section of any CAD software. The setting of units usually is in the tools or settings.
- Click on tree filters, as shown in Figure 7.13.

- Check the box features to display in the model tree.
- Click OK to close the box of model tree items.
- The tank top as shown in Figure 7.13 is to be created.
- Click OK datum plane from the Model tab of the ribbon interface.

- Click on Assembly from the Model tab of the ribbon interface.
- Select the file Tank_Shell_with_Dish_End from the Chapter 7 folder.
- Click OK to assemble the instance of Tank_Shell_with_Dish_End.
- Click on the Placement tab.
- Select Default in the Assembly relation drop-down box.
- Tank_Shell_with_Dish_End will be fixed with default plane coincide.
- Before proceeding toward assembling all the components, a few little modifications are necessary to suit the shell diameter of 1000 mm as in the dish end.
- Open the assembly Tank_Saddle.
- Right-click on sketch1 from the model tree and select Edit.
- Change the shell diameter dimension of large circle to 1000 mm.

- Click on Regenerate from the Model tab of the ribbon interface.
- Save and close the file.
- Open the file Nozzzle_Assembly generic model.
- Change the diameter of sketch1 to 1000 mm.
- Click on Regenerate from the Model tab of the ribbon interface.
- Notice that the pad diameter still has not changed.
- Double-click on R_Pad in the model tree to expand it.
- Right-click on the feature bend1 and select Edit definition.
- Change the radius value to 500 mm.
- Click OK.
- Click on Regenerate from the Model tab of the ribbon interface.
- Click on Axis from the Model tab of the ribbon interface.
- Select sketch1 graphically.
- The relation with the reference entity option shall be set to center.
- Click OK to create the datum feature.
- An axis named AA_1 will be created.
- Click on Family table from the Tools tab of the ribbon interface.
- Click on the Verify button of the family tree window.
- The verification status should be Success.
- Close the family tree window.
- Click OK to close the family table window.
- Save and close the file.

- Click on Assembly from the Model tab of the ribbon interface.
- Select the file Tank_Saddle.
- Click on the Placement tab.
- Select the internal cylindrical surface of the wear_plate part of the Tank_Saddle part as a component reference.
- Select the external cylindrical surface of the Tank_Shell part as an assembly reference.
- Select Coincide in the component reference relation field.
- Right-click on the graphical area and select New constraint.
- Select the plane ASM_FRONT of the Tank_Saddle part as a component reference.
- Select the plane ASM_RIGHT of Tank_Shell_with_Dish_End as an assembly reference.
- Enter the value 2300 mm as offset distance so that the nozzle is protruded inside the flange.
- Right-click on the graphical area and select New constraint.
- Select the plane ASM_RIGHT of the Tank_Saddle part as a component reference.
- Select the plane ASM_FRONT of Tank_Shell_with_Dish_End as an assembly reference.
- Flip the orientation suitably in each of the mating relations if required.
- Click OK to assemble the component.
- Click on Assembly again from the Model tab of the ribbon interface.
- Select the file Tank_Saddle

- Click on the Placement tab.
- Select the internal cylindrical surface of the wear_plate part of the Tank_Saddle part as a component reference.
- Select the external cylindrical surface of the Tank_Shell part as an assembly reference.
- Select Coincide in the component reference relation field.
- Right-click on the graphical area and select New constraint.
- Select the plane ASM_FRONT of the Tank_Saddle part as a component reference.
- Select the plane ASM_RIGHT of Tank_Shell_with_Dish_End as an assembly reference.
- Enter the value –2300 mm as offset distance so that the nozzle is protruded inside the flange.
- Right-click on the graphical area and select New constraint.
- Select the plane ASM_RIGHT of the Tank_Saddle part as a component reference.
- Select the plane ASM_FRONT of Tank_Shell_with_Dish_End as an assembly reference.
- Flip the orientation suitably in each of the mating relations if required.
- Click OK to assemble the component.

- Click on Assembly from the Model tab of the ribbon interface.
- Select the file Manhole_500NB.
- Click on the Placement tab.
- Select the AA_1 of Nozzle_Assembly_500NB of Manhole_500NB as a component reference.
- Select the axis A_1 of the Tank_Shell part as an assembly reference.
- Select Coincide in the component reference relation field.
- Right-click on the graphical area and select New constraint.
- Select the plane ASM_FRONT of the Manhole_500NB part as a component reference.
- Select the plane ASM_RIGHT as an assembly reference.
- Enter the value 1900 mm as offset distance so that the nozzle is protruded inside the flange.
- Right-click on the graphical area and select New constraint.
- Select the plane ASM_RIGHT of the Manhole_500NB part as a component reference.
- Select the plane ASM_FRONT of Tank_Shell_with_Dish_End as an assembly reference.
- Flip the orientation suitably in each of the mating relations if required.
- Click OK to assemble the component.

- Click on Assembly from the Model tab of the ribbon interface.
- Select the file Nozzle_Assembly_150_NB.
- Click on the Placement tab.
- Select the AA_1 of Nozzle_Assembly_150NB of Nozzle_Assembly_150_NB as a component reference.

- Select the axis A_1 of the Tank_Shell part as an assembly reference.
- Select Coincide in the component reference relation field.
- Right-click on the graphical area and select New constraint.
- Select the plane ASM_FRONT of the Nozzle_Assembly_150_NB part as a component reference.
- Select the plane ASM_FRONT of Manhole_Assembly_500NB as an assembly reference.
- Enter the value 1000 mm as offset distance so that the nozzle is protruded inside the flange.
- Right-click on the graphical area and select New constraint.
- Select the plane ASM_RIGHT of the Nozzle_Assembly_150_NB part as a component reference.
- Select the plane ASM_FRONT of Tank_Shell_with_Dish_End as an assembly reference.
- Flip the orientation suitably in each of the mating relations if required.
- Click OK to assemble the component.

- Click on Assembly from the Model tab of the ribbon interface.
- Select the file Nozzle_Assembly_150_NB.
- Click on the Placement tab.
- Select the AA_1 of Nozzle_Assembly_100NB of Nozzle_Assembly_100_NB as a component reference.
- Select the axis A_1 of the Tank_Shell part as an assembly reference.
- Select Coincide in the component reference relation field.
- Right-click on the graphical area and select New constraint.
- Select the plane ASM_FRONT of the Nozzle_Assembly_100_NB part as a component reference.
- Select the plane ASM_FRONT of Nozzle_Assembly_150_NB as an assembly reference.
- Enter the value 500 mm as offset distance so that the nozzle is protruded inside the flange.
- Right-click on the graphical area and select New constraint.
- Select the plane ASM_RIGHT of the Nozzle_Assembly_100_NB part as a component reference.
- Select the plane ASM_FRONT of Tank_Shell_with_Dish_End as an assembly reference.
- Flip the orientation suitably in each of the mating relations if required.
- Click OK to assemble the component.

- Click on Assembly from the Model tab of the ribbon interface.
- Select the file Nozzle_Assembly_200_NB.
- Click on the Placement tab.
- Select the AA_1 of Nozzle_Assembly_200NB of Nozzle_Assembly_200_NB as a component reference.
- Select the axis A_1 of the Tank_Shell part as an assembly reference.
- Select Coincide in the component reference relation field.

- Right-click on the graphical area and select New constraint.
- Select the plane ASM_FRONT of the Nozzle_Assembly_200_NB part as a component reference.
- Select the plane ASM_FRONT of Nozzle_Assembly_100_NB as an assembly reference.
- Enter the value 500 mm as offset distance so that the nozzle is protruded inside the flange.
- Right-click on the graphical area and select New constraint.
- Select the plane ASM_RIGHT of the Nozzle_Assembly_200_NB part as a component reference.
- Select the plane ASM_FRONT of Tank_Shell_with_Dish_End as an assembly reference.
- Flip the orientation suitably in each of the mating relations if required.
- Click OK to assemble the component.

- Click on Assembly from the Model tab of the ribbon interface.
- Select the file Nozzle_Assembly_65_NB.
- Click on the Placement tab.
- Select the AA_1 of Nozzle_Assembly_65NB of Nozzle_Assembly_65_NB as a component reference.
- Select the axis A_1 of the Tank_Shell part as an assembly reference.
- Select Coincide in the component reference relation field.
- Right-click on the graphical area and select New constraint.
- Select the plane ASM_FRONT of the Nozzle_Assembly_65_NB part as a component reference.
- Select the plane ASM_FRONT of Nozzle_Assembly_200_NB as an assembly reference.
- Enter the value 500 mm as offset distance so that the nozzle is protruded inside the flange.
- Right-click on the graphical area and select New constraint.
- Select the plane ASM_RIGHT of the Nozzle_Assembly_65_NB part as a component reference.
- Select the plane ASM_FRONT of Tank_Shell_with_Dish_End as an assembly reference.
- Flip the orientation suitably in each of the mating relations if required.
- Click OK to assemble the component.

- Click on Assembly from the Model tab of the ribbon interface.
- Select the file Nozzle_Assembly_80_NB.
- Click on the Placement tab.
- Select the AA_1 of Nozzle_Assembly_80NB of Nozzle_Assembly_80_NB as a component reference.
- Select the axis A_1 of the Tank_Shell part as an assembly reference.
- Select Coincide in the component reference relation field.
- Right-click on the graphical area and select New constraint.

- Select the plane ASM_FRONT of the Nozzle_Assembly_80_NB part as a component reference.
- Select the plane ASM_FRONT of Nozzle_Assembly_65_NB as an assembly reference.
- Enter the value 500 mm as offset distance so that the nozzle is protruded inside the flange.
- Right-click on the graphical area and select New constraint.
- Select the plane ASM_RIGHT of the Nozzle_Assembly_80_NB part as a component reference.
- Select the plane ASM_FRONT of Tank_Shell_with_Dish_End as an assembly reference.
- Flip the orientation suitably in each of the mating relations if required.
- Click OK to assemble the component.
- Save and exit the file.

QUESTIONS

1. What is an assembly file?
2. Describe the differences between component, part, and feature.
3. Differentiate between the bottom–up approach and top–down approach.
4. Differentiate between assembly features and part features.
5. Differentiate between a family table of assembly and parts.
6. Differentiate between feature operations and component operations.
7. Describe the advantages of application-based assembly techniques.
8. Describe the use of different types of presentations in assembly.
9. Compare different presentations of parts and assembly.
10. Describe the use of assembly templates.

8

Production Drawing Generation

8.1 Introduction

Drawings are the language of engineers to communicate with each other. Information is communicated and conveyed from one engineering discipline to another by drawings. In an engineering enterprise, a drawing is used as a master document for the entire cycle of product creation, from preparation of a bill of materials to production, inspection, sales, and so forth.

Earlier, when computer-aided design (CAD) was not available, generation from 3-D views was not possible. Therefore, draftsmen were making 2-D drawings on drawing sheets. Any dimensions that required measurement from the final component drawing had to be measured by scale. Therefore, accuracy of measurements also depended on the creation of the drawing. The accuracy of measurement is enhanced to a great extent by the use of CAD. Also, drawing of 3-D views in sheets of a complex model was very difficult before CAD and is really made easy by CAD and its 3-D features.

Once the design process is over, a component needs to be manufactured. A fully dimensionally specified drawing is to be generated from the design department with all the necessary information to produce the component. The necessary information includes different views of the component with dimensions and other manufacturing details and annotations.

CAD systems provide a customized and specially designed environment that is useful to the user for creation of the drawing. Advanced CAD software provides excellent tools to create the production drawing from the designed component.

Mostly, in advanced CAD software, the drawing file is treated as a separate file that is dependent on the respective part or assembly file. The drawing process follows the sequence shown in Figure 8.1a.

Advance CAD software automates some of the steps to reduce the effort of the users. For example, in Creo/PRO-E, if the part or assembly file is open, and the user chooses to create a new drawing file, the part or assembly file that is opened automatically gets linked with the drawing file, thus reducing the effort of the user. A step-by-step discussion of the drawing is stated as follows:

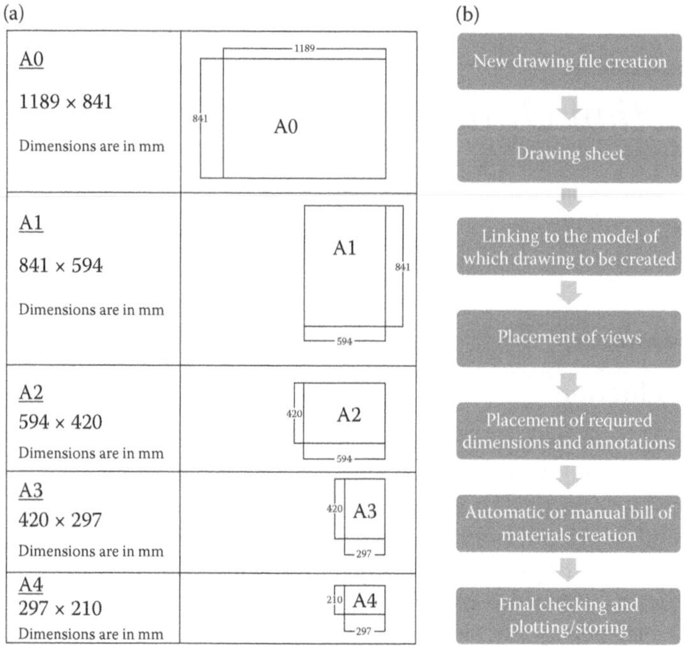

FIGURE 8.1
(a) Drawing creation process. (b) Standard drawing sheet size.

8.1.1 New Drawing File Creation

When a user chooses new file creation, the user is prompted to select the required type of file that he/she wants to create followed by the selection template file. Template files are basically pre-preserved files in the CAD software itself that provide a customized environment to the user. Users can also shorten the time and reduce the work in many cases through the template file, which is discussed in Chapter 13 on automation. In most software, users are also prompted to select the part or assembly file during the creation of a new drawing file as part of this step.

8.1.2 Drawing Sheet

The drawing sheet automatically appears if it is present in the template. However, every company has their own drawing sheet and template. Many times, drawing sheets or drawing templates need to be customized depending on the project. It is recommended to make company's customized drawing sheets and use these as drawing templates to minimize the sheet replacement work. Users always have the option in CAD to change the drawing sheet and size. The drawing sheet can be of many sizes, ranging from A to E. International Organization for Standardization (ISO) A1 to A4 sizes are used in many countries for engineering drawing printing and publications. It is understandable that the sizes are so derived that any sheet form is approximately double in width than the next lower sheet size. The basic of deriving the sheet size is that the maximum sheet size is 1 m^2, which is A0, and the side of the sheet will have a ratio of $1/\sqrt{2}$. Therefore, for the A0 sheet, the size becomes

length = 1.189 m = 1189 mm

width = 0.841 m = 841 mm

The dimensions of successive series of formats are obtained keeping the same principle of the sides of ratio $1/\sqrt{2}$ and dividing the length size by 2. The values are rounded off as shown in Figure 8.1b.

Most of the sizes of sheets are available in CAD, and also, the sizes are borrowed from the configured plotter with the CAD system so that the supported sheets are only made selectable. The sheet sizes are decided by the requirement to represent the drawing properly. It is to be remembered that a text of a height of 3 mm having a width ratio of 1 will always display 3 mm in true scale in all sizes of drawing sheets. Therefore, the requirement of an increase or decrease in text size with respect to the drawing sheet does not persist. Similarly, the dimensions and other annotations are to be decided with fixed size to display properly in all sizes of the drawing sheet. There are certain texts that need to be in larger sizes, like the drawing view's name, headings, and so forth. Hence, a detailed text and annotations of fixed sizes may be very useful, and these can be used in all drawings.

A drawing may contain much information. The quantity of information in a drawing varies with the requirement to specify the component for the required purpose. For example, a drawing of a component that is required for the erection of the component contains different information than the drawing that is required for the manufacturing of the component. Therefore, a drawing can be presented with much information and also following some rules that are globally understandable to all engineers.

Nameplates/title block: The title block should be positioned at the bottom right-hand corner, which increases ease of accessibility even if the drawing is folded. Borders are enclosed by the edges of the drawing sheet, and the drawing frame is enclosed within the borders. Grid references are positioned suitably like boxes of a chessboard, which in turn facilitates quick finding of any revisions or modifications. A title block mainly consists of the following:

- *Title of the drawing*: The title of the drawing states the contents of the drawing and its purpose, for example, general arrangement (GA) of mechanical equipment, fabrication detail of structure for a storage tank, and so forth.
- *Company information*: Information about the company that makes the component, their address, and their logo is essentially required to be

present in the title block such that any person can be informed about the manufacturer of the product.

- *Project information*: The details about the project may or may not be present in the drawing. Normally, in the turnkey project jobs, all drawings are formatted to a particular drawing format containing all the project details and a common code of practice. The project name, address, customer's name, and contractor's name are contained in the project information system.
- *Maker's name*: The names of the designer, draftsman, checker, and finally, the approval authority shall be mentioned in the title block.
- *Scale*: Scale must be mentioned along with a view of projection symbols.
- *Additional information*: There is some additional information that may be placed if required, like the purpose of the drawing, that is, for information or for approval and so forth.

8.1.3 Linking to the Model of the Drawing to Be Created

Once the drawing sheets and templates are selected and placed properly, the next task is linking the model. In CAD, more than one component can be linked and placed in the drawing. When more than one component is linked, then the model needs to be set or selected, which will immediately appear when view placement is required. It is advisable to link the required model whose views are going to be placed immediately to avoid any confusion or rework. When more than one component is required to be linked, the following are recommended and required:

1. The name of the component files shall be different.
2. Any change or renaming of the component after linking with a drawing shall not be done without proper awareness that in later stages, the drawing will no longer be associated with the linked model.
3. The location of the component in the computer shall not be changed after linking, so the "missing component" error does not appear.

8.1.4 Placement of Views

When models are linked to the drawing, the chosen views need to be placed. First, it is to be decided by the user which views will show all the required information in the drawing. The concept is that there shall be a minimum number of views that can present all the required information with ease of visualization. If more views are used, the sheet size or quantity may increase, which will not be economical and easy to handle. Also, if the views are such that they may represent the required information, however difficult to understand, then engineers who will study and work on the basis of the drawing may need more time to understand it. Therefore, a sufficient minimum number of views are to be placed in the drawing.

8.1.5 Placement of Required Dimensions and Annotations

Without dimension, no engineering drawings are specified properly. Once the views are placed, the next task is to provide the dimensions. Since, when the 3-D models are made, all the dimensions are entered, the CAD software allows retrieving those dimensions in the drawing. However, all the dimensions can also be given in the drawing mode without retrieving the dimensions from the model. If the dimensions are retrieved from the 3-D model, users are allowed to change the dimension of any model from the drawing and to regenerate the model and associated drawings for changed representation. However, in many cases, 3-D drawings are not very useful to retrieve for the following reasons:

1. The dimension provided in the 3-D models is not easy to trace, since a dimension of similar values can be present and would be difficult to identify for the required feature to be dimensioned.
2. There are many unwanted dimensions that are required to be given for making the 3-D views but may not be required to be shown in the drawing. Therefore, when the dimensions are retrieved, the entire unusual dimension may be shown in the model, which again must be removed from the drawing.
3. The dimensions may call for tolerances that are not given in the 3-D models but need to be placed in the drawing.

CAD software allows users to provide tolerances in the 3-D model itself, which in turn can be shown in the drawing. Similarly, some CAD software allows users to make the dimensions publishable when creating the 3-D models. It is suggested that users provide the dimensions in 3-D, which will be retrieved in the drawing later. The retrieved dimensions are only placed in the drawing. In this way, model accuracy will be increased.

8.1.6 Automatic or Manual Bill of Materials Creation

A bill of materials is also called a bill of quantity or item list, parts list, component list, and so forth. All the advanced CAD software allows an automatic bill of materials, which is created based on preset parameters set in the 3-D models. Materials and material properties are fetched from the assigned material of the component, and the quantities are counted for the assembly. Often, for assemblies with many stages of subassemblies, the bill of materials can be indented or flat or can contain only actual parts that need to be manufactured:

1. Indented bill of materials
2. Flat bill of materials
3. Leveled bill of materials

8.1.7 Final Checking and Plotting/Storing

The last step of drawing creation is final checking of the drawing. It is to be remembered that there are many steps involved sequentially after releasing the drawing from the designer's end. For example, for any component with many individual parts, the items or raw materials to be procured from the bill of materials of the drawing, machining, and fabrication work are the next steps, followed by erection. Therefore, any error in the drawing may generate a great penalty for the company.

The drawing shall be checked on the basis of the requirement for which the drawing is made, like production purposes or erection purposes and so forth, whether all the information for the required proposes are given or not, and so forth. Since most users prepare drawings starting from the 3-D view and annotation creation, negligence may develop in the later stages or while preparing the drawing, which in turn may produce some error due to a lack of concentration.

A drawing shall be checked for the following:

1. *Drawing's title block*: All the information in the title block like the title, project information, and so forth shall be checked.
2. *Checking the 3-D models*: Check the 3-D models for any interference and so forth.
3. *Bill of materials*: Item-wise, the bill of materials shall be checked with correspondence to drawing views and 3-D models. Mostly, if a bill of materials is checked item by item individually, the chances of error are reduced to a great extent.
4. *Annotations and cleanliness*: All the annotations and dimensions shall be checked thoroughly for any mismatches or any other kind of errors like misplacement, inadequate symbols, and so forth.

A drawing may need to be revised whenever the design changes; however, the drawing should be made so that there will be no errors that may result from any ignorance of the users. It is better to make an error-free drawing than to revise.

8.2 Structure of Drawing

A drawing has a universal structure. A typical drawing is shown in Figure 8.2 and consists of the following items:

1. *Drawing views*: The typical drawing view's positions are mentioned in Figure 8.2. The drawing can be in the proper scale or without scale.
   Piping isometrics where the aspect ratio is large are normally not drawn to scale, whereas component drawings are mostly made in the proper scale. The drawing views can be of the front, back auxiliary, or isometrics or of any kind as decided by the designer to show the required information.
2. *Nameplate*: The nameplate displays the company, project, and drawing details. Nameplates are normally made standard for every company's with respect to sheet sizes.
3. *Bill of materials*: The bill of materials consists of all the materials and quantities used in the drawing or shown in the drawing. The bill of materials shall consist of the material description, size, and special criteria of the used materials.
4. *Notes*: Notes are often used to provide information that can be shown in the form of a drawing.

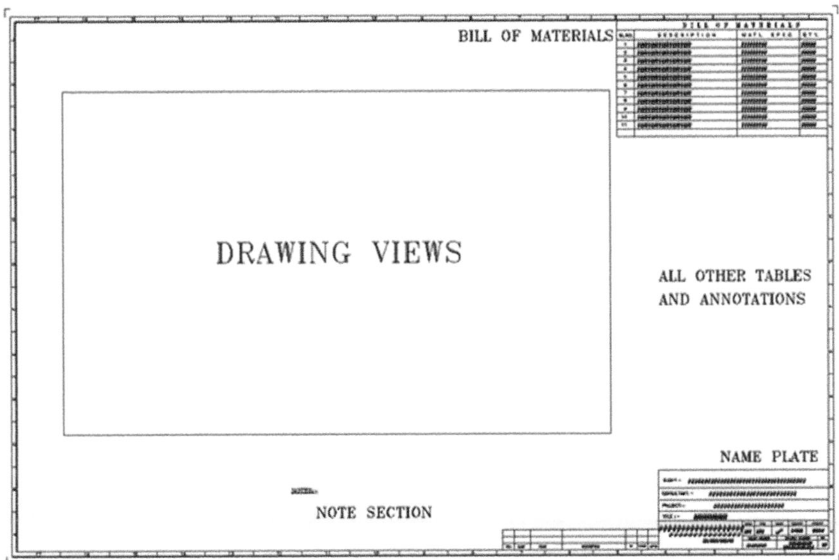

FIGURE 8.2
Standard drawing.

5. *Annotations*: Apart from the drawing, dimensions, and notes, many times it is required to have tables like design parameters, nozzle schedules, and so forth in the drawing. These kinds of information in the drawing may or may not vary in similar drawings.

8.3 Drawing View Setting and Creation

A CAD model may be a combination of several features and details. These features and details are to be completely shown and described in the drawing so that there is no confusion for the production person. Detailing of these features often leads to the addition of several orthogonal and other types of views. Before proceeding to views, we must know the method of view representation.

Designation of views:

View in direction a = view from the front
View in direction b = view from the back
View in direction c = view from the top
View in direction d = view from the bottom
View in direction e = view from the left
View in direction f = view from the right

The front view is the principal view, which needs to be placed in the drawing to represent any component. All the other views are relative to the front view and projected as per two alternative projection methods, that is, the first-angle projection method and the third-angle projection method.

If only the front view of a component is shown in the drawing, the angle of projection becomes inapplicable. However, most of the components need to be shown in more than one view to specify the component completely. Therefore, the angle of projection becomes important to understand the component's shape.

Figure 8.3 shows the front view of an object placed in two different positions or quadrants.

Imagine that the positions of views or eyes are fixed with respect to drawing screens. Now, when the object is placed in the first quadrant, we will see the object from the right side and draw it to the left of the front view. Similarly, we draw the top view at the bottom of the front view. In the case of a third-angle view, when we draw the right view at the right side of the model, similarly, we draw the top view at the top of the model. The fundamental difference between first-angle and third-angle projection is that the views alter the side in the first-angle method; however, in the case of the

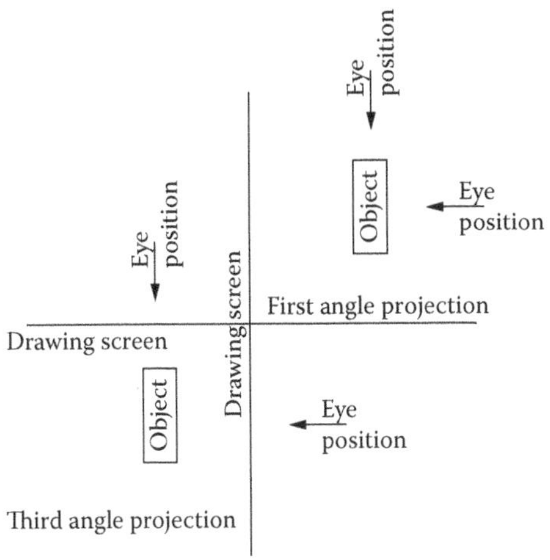

FIGURE 8.3
Projection view angle basics.

third-angle method, the view does not alter the direction. It is obvious, in the case of complex geometries and partial views, that third-angle projection becomes an efficient method of presentation. Figure 8.4 shows an example and symbols of first- and third-angle views.

Apart from the angle-of-projection method, views using reference arrows are the most advantageous method where there is confusion in understanding the drawing. The designated views can be placed irrespective of the position of the principal view. Figures 8.5 and 8.6 show an example of views using reference arrows.

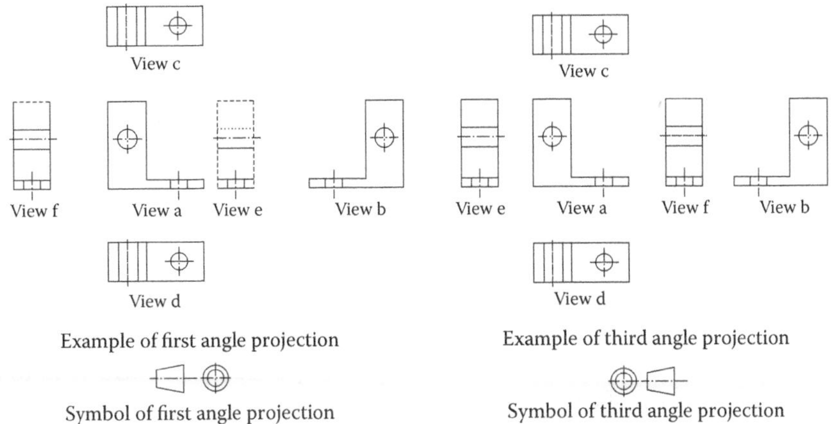

FIGURE 8.4
Views at different angles of projection.

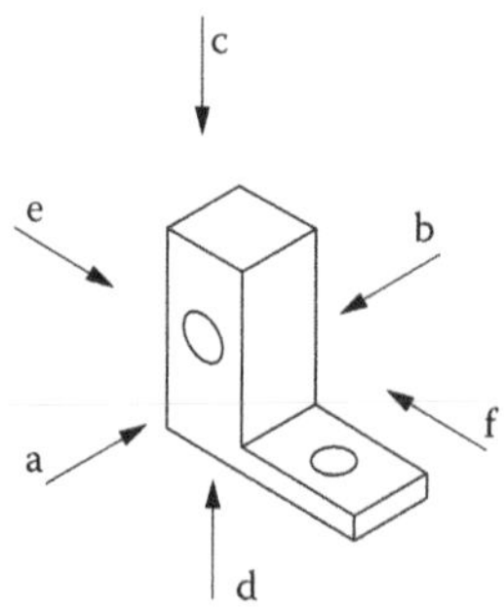

FIGURE 8.5
Example of views.

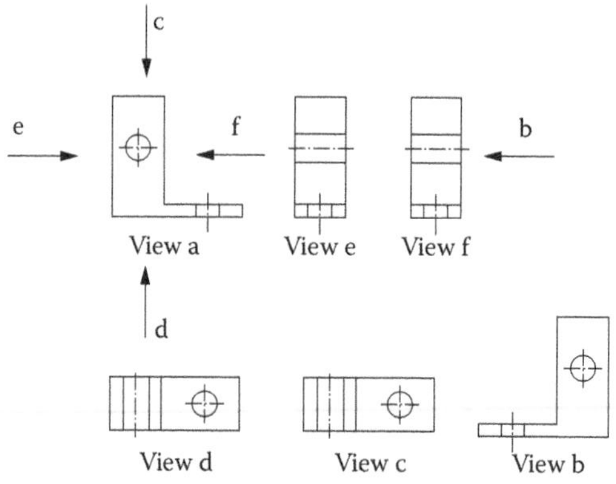

FIGURE 8.6
Example of views.

Sometimes, orthogonal views are not sufficient to show all the required information. Other views are necessary to show all the information. The most used types of views are discussed here, and they are as follows.

8.3.1 Detail Views

Detail views are needed to show the information on a particular section of any component. When a big component is shown, small geometries of the component appear too small, which can be difficult to understand. In that case, detail views are used to increase a certain portion of the drawing to a specified larger scale factor, and the detail information can be clearly shown. This saves additional space and effort to draw the full view in larger scale. Detail views are also very useful to show typical arrangements in repetitive places in a big assembly drawing. Figure 8.7 shows an example of using a

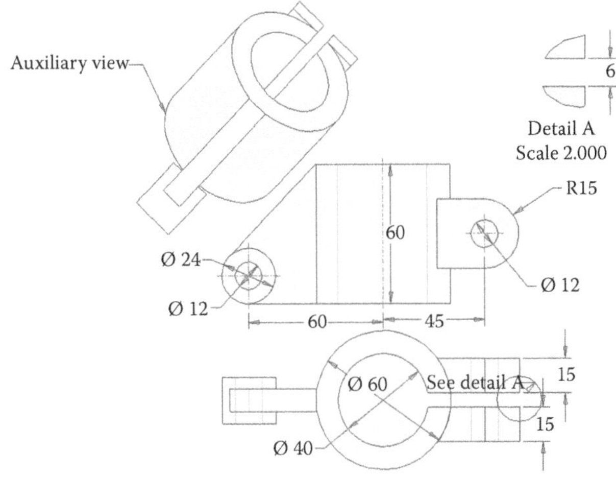

FIGURE 8.7
Example of detail view.

detail view. Here, the dimension of the gap is shown perfectly in the detail view, which is more understandable due to its increased scale factor.

8.3.2 Section Views

Section views are widely used in a drawing where a difference exists through the section of any component. Section views are the views taken in orthogonal projection if the component is cut through the imaginary section planes. Section views are particularly very useful in the case of complex casting with internal ribs and sections where the internal portion of the model cannot be viewed without sections. Section views are also used to distinguish the different materials of any assembled component by providing different types of hatch patterns. In many cases, the assembled component is displayed by half section to distinguish the internal parts from the actual view from outside. Figure 8.8a shows the sectional view of a component.

8.3.3 Partial Views

Partial views are used when a small part of any big component needs to be shown without having the full component display. Partial views are very useful to show the fit of small components over a very big component, like large pressure vessels and so forth.

8.3.4 Broken Views

Broken views are mostly used to show considerably large and uniform components. For example, in many cases, more than one journal bearing

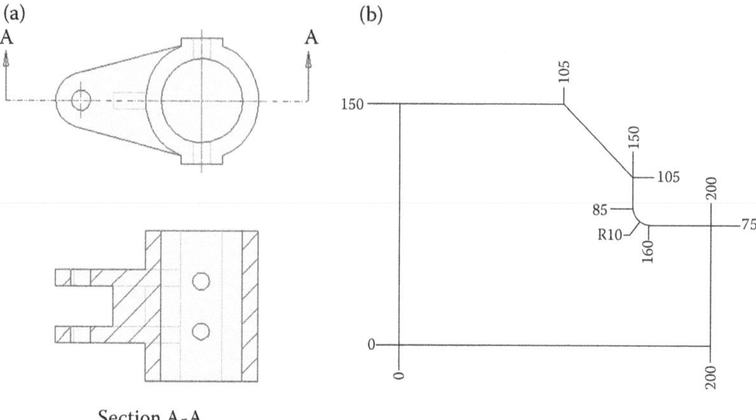

FIGURE 8.8
(a) Example of sectional view. (b) Example of ordinate dimension.

is mounted on comparatively long shafts. If the fit of each bearing is to be shown in the same drawing sheet, a broken view of shafts is essential and useful to include all the bearing fits in the same drawing.

8.3.5 Auxiliary Views

Auxiliary views are drawn when views perpendicular to any inclined edge of the component are required. Sometimes, many features need an exact perpendicular view to be understood properly. Creating a named view at all the required angles might be very time consuming. In those cases, auxiliary views are essential to show the required information properly.

8.4 Dimension

Since the component size is specified by the dimensions, they are most important. There are also different types of dimensioning that exist. Depending on the application and component geometry, each type of dimensioning has its specific advantages.

In ordinate dimensioning, all the dimensions are taken from the same reference of origin. In this way, a direct dimension would not be available from every point to point of the component; however, it facilitates marking on a flat sheet on the shop floor. Normally, angular marking of angular dimensions is not so easy compared to marking orthogonal dimensioning. Therefore, when the part has an edge that is at an angle to the horizontal or vertical axis, it would be better represented by ordinate dimensioning for work on the shop floor. Refer to Figure 8.8b; the ordinate dimensions are given.

1. On the shop floor, first, the sheet has to be selected or cut to at least the maximum size of the component; here it is 200 × 150 mm.
2. The 0,0 point must be located first.
3. Then one has to draw vertical lines at a distance of 105, 150, 160, and 200 mm from the 0,0 point towards the right-hand side.
4. Similarly, horizontal lines at a distance of 75, 85, 105, and 150 mm are to be drawn, respectively, from the 0,0 point.
5. Where the horizontal and vertical lines cross each other, the points are defined.
6. Then all the connecting lines between the points shall be drawn as required.
7. A radius of 10 mm is to be made from point 160,75 to 150,85.

This method is especially suitable for cutting of comparatively large sheets to a particular shape before bending.

Orthogonal dimensioning is the conventional method of dimensioning. The main advantage of this method is that dimensions are given as required. For example, refer to Figure 8.9.

The dimension of the vertical line associated with radius 10 can also be given from the left edge of the component; however, it is given from the right-hand edge. The reason behind this is that the dimension 50 is more

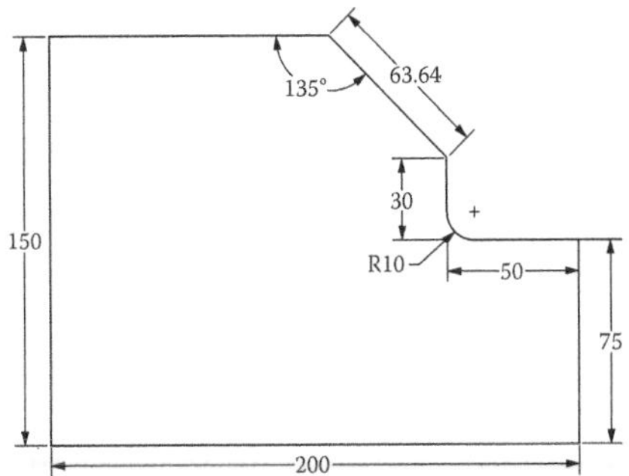

FIGURE 8.9
Example of dimensioning of a part.

important. If a dimension of 150 from the left edge is given, the tolerances of dimension 200 and 150 will apply when deriving 50 dimensions, which may result unsatisfactorily. This method of dimensioning is followed in the process of making a CAD model. Thus, when the dimensions of 3-D models are retrieved, only these dimensions are made available in the 2-D drawing.

A part of exact dimensions can be produced. There would be a variation in all the parts depending on the use of the manufacturing process for the components. Also there are some cases where accuracy of the dimensions is not necessary, and in some cases, it is. Therefore, every dimension is allowed to vary or to be produced within a range. These allowable variations of dimensions are called tolerance.

When a component requires manufacturing, its every dimension may not need to have high accuracy or close tolerance. Also, many times, costlier manufacturing operations are required for maintaining close tolerances. Therefore, first, it is to be identified which dimensions require close tolerance and how much, and then the manufacturing processes are selected to meet the requirement. There are certain terms associated with tolerance that require understanding first; they are as follows:

- *Nominal size*: The size of a part that is required from the design point of view. Any part design has to be carried out step by step. When the part is in the design stage, the tolerance need not be mentioned. After the completion of part design, the tolerances are decided. Therefore, nominal size is that which is actually required or derived in the design.
- *Basic size*: It is the size to which all the variations of the dimensions are applied. It is normally the same as nominal size.
- *Actual size*: It is the measured size of the finished part. Actual size must be kept within tolerances on basic size; otherwise, the part becomes rejected in the inspection.
- *Allowance*: It is the difference between the basic dimensions of mating parts. If a clearance exists between mating parts, the allowance is positive. If interference exists between the mating parts, the allowance is negative.
- *Limits of sizes*: When the dimensions of parts are allowed to be varied, it is obvious that there shall be two dimensions that specify the range, and the dimension of the manufactured part is to be within that range in order to get acceptance from inspection. Therefore, the dimensions that actually define the range are defining the minimum point and defining the maximum point. These minimum and maximum dimensions are called limits of sizes. The minimum dimensions are called lower limit, and the maximum dimension is called upper limit.

- *Tolerance*: It is the difference between the upper limit and the lower limit. When the upper limit and the lower limit are decided so that the basic size is within the value of the upper and lower limits, but there is no equivalent to the upper and lower limit, the tolerance is called bilateral tolerance. 20.20 ± 0.05 is an example of bilateral tolerance. When the parts are allowed to be varied in any of the directions of a basic size, either the upper limit or the lower limit exists on the basic size, and then, the tolerance is called unilateral tolerance. $20.20^{+0.05}_{-0.00}$ is an example of unilateral tolerance. Table 8.1 shows the tolerances that can be achieved by various manufacturing processes.
- *Upper deviation*: It is the difference between the upper limit and the basic size of the part.
- *Lower deviation*: It is the difference between the lower limit and the basic size of the part.
- *Actual deviation*: It is the difference between actual size and basic size.
- *Mean deviation*: It is the mean between upper deviation and lower deviation.
- *Fundamental deviation*: It is one of the two deviations chosen from the zero reference.
- *Fit*: Fit is the degree of tightness or looseness in the mating parts. The fits are of the following types:

TABLE 8.1

Manufacturing Process and Tolerances

Sl. No.	Manufacturing Process	Tolerance (IT Grade)
1	Lapping	4 and 5
2	Honing	4 and 5
3	Cylindrical grinding	5 to 7
4	Surface grinding	5 to 8
5	Broaching	5 to 8
6	Reaming	6 to 10
7	Turning	7 to 13
8	Hot rolling	8 to 10
9	Boring	8 to 13
10	Milling	10 to 13
11	Planing and shaping	10 to 13
12	Drilling	10 to 13
13	Die casting	12 to 14
14	Forging	14 to 16

- *Clearance fit*: If the limits of the dimensions between mating parts so selected that undoubtedly, a clearance must occur between the mating parts, then the fit is called clearance fit.
- *Interference fit*: If the upper and lower limits between the mating parts are so selected that the interference will occur between the mating parts, then the fit is called interference fit. Practically, when hole size is smaller than shaft size, the parts need to be pressed or forced or to be heated to fit. Sometimes, this fit is also termed *shrink fit*.
- *Transition fit*: In this type, the upper and lower limits are so selected that either clearance or interference will exist between the mating parts to a small value. Sometimes, this fit is also termed *push fit*.

Different types of limit and tolerance zone are shown in Figure 8.10.

- *Hole basis system*: In the hole basis system, the hole is kept constant. All types of fits are obtained by varying the shaft size.
- *Shaft basis system*: When the shaft's dimension is kept constant and different types of fits are obtained by the varying dimension of the hole, the system is called a shaft basis system.

The hole basis system is preferred from the manufacturing point of view, since by turning or grinding, different sizes of shafts can be produced by the same tools. However, drills and reamers are made of fixed sizes; therefore, it is difficult to get tools for different diameters of a hole easily. The hole basis system and shaft basis system are shown in Figure 8.11.

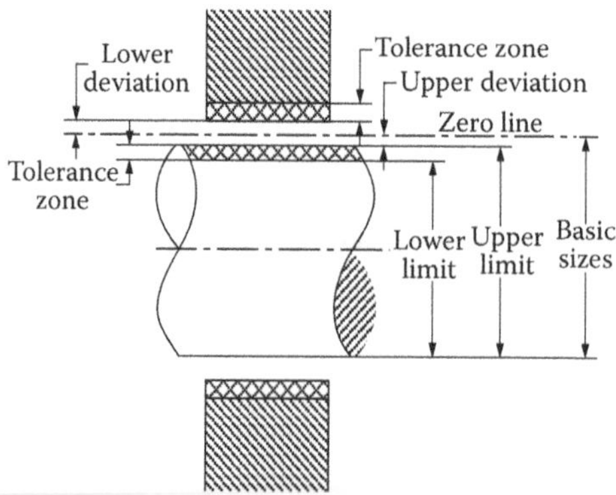

FIGURE 8.10
Limits and tolerance zone.

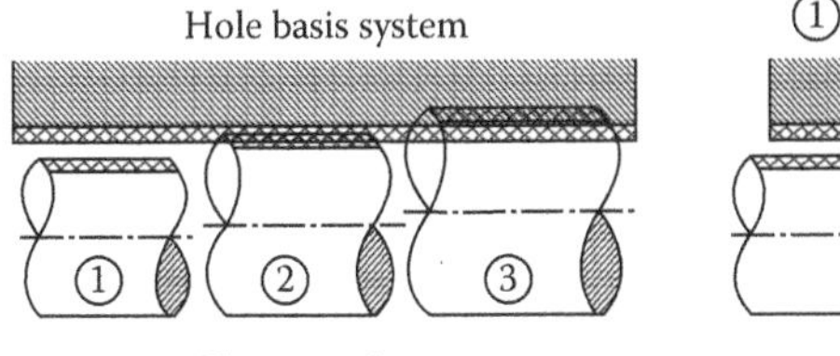

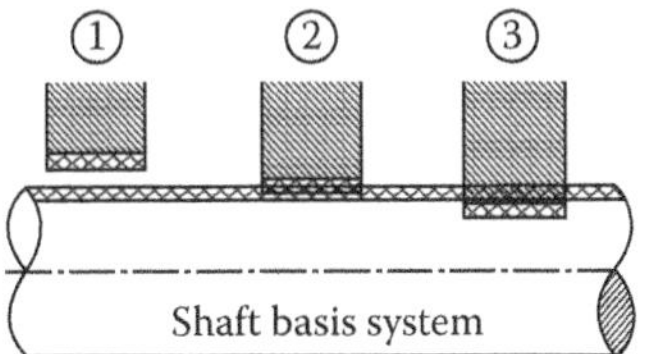

FIGURE 8.11
Hole basis system and shaft basis system.

8.5 Annotations

All the objects in the drawing except the drawing views of a component are included in the annotations. Annotations specify the component and provide all the necessary information. The important annotations are as follows.

8.5.1 Symbols

1. *Welding symbols*: If an assembled component is made by welding, its fabrication drawing needs to be prepared. The welding of parts is to be carefully shown in the drawing. The size, position, and also other important parameters are shown by the welding symbol.
   Figure 8.12 shows the method of showing welding symbols.
   Welding type is mentioned in the welding symbol. The most used welding types are mentioned in Figure 8.13.
2. *Surface texture*: Figure 8.14 shows the surface texture symbol used in the machined or unmachined surface. The first symbol shall be used with notes to display the required information, whereas the second symbol is a generalized symbol, and the third one is used to indicate an unmachined surface.

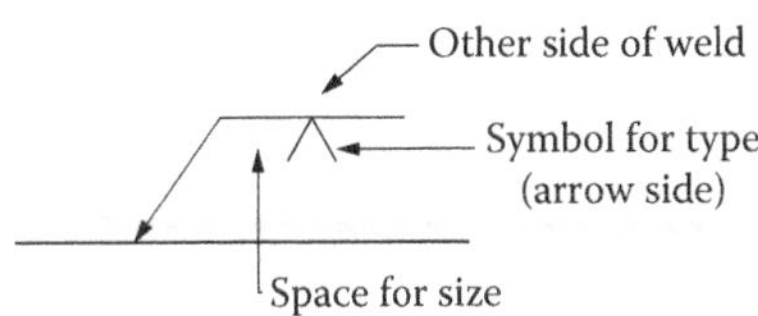

FIGURE 8.12
Method of showing welding by symbol.

Square butt weld		
Single V-butt weld		
Single-bevel butt weld		
Single V-butt weld with broad root face		
Single-bevel butt weld with broad root face		
Single U butt weld (parallel side)		
Single J butt weld		
Fillet weld		

FIGURE 8.13
Welding types and symbols.

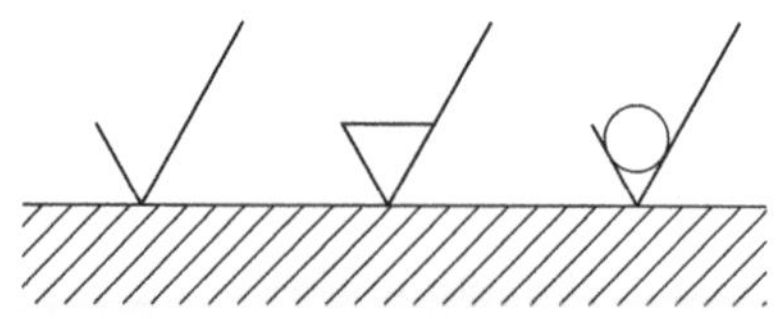

FIGURE 8.14
Types of surface texture symbols.

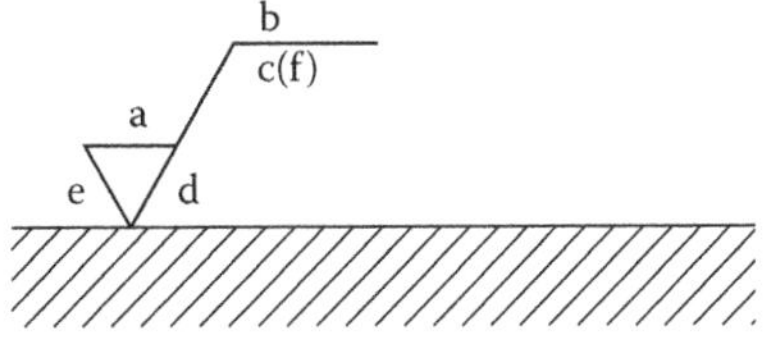

FIGURE 8.15
Specification of surface texture in the symbol. a = roughness value in microns = roughness grade number N1 to N12; b = production method, treatment or coating; c = sampling length; d = direction of lay; e = machining allowance; f = other roughness value.

Figure 8.15 shows the position of the specifications of the surface texture in the symbol. The specifications of surface texture are to be placed relative to the symbol, as shown in Figure 8.15.

The roughness value respective to the grade is shown in Table 8.2. Direction of lay is shown in Table 8.3.

TABLE 8.2
Roughness Grade Number and Roughness Values

Roughness Grade Number	Roughness Values, Ra (μm)
N1	0.025
N2	0.5
N3	0.1
N4	0.2
N5	0.4
N6	0.8
N7	1.6
N8	3.2
N9	6.3
N10	12.5
N11	25
N12	50

TABLE 8.3
Direction of Lay

Symbol	Description
=	Parallel to the plane of projection
⊥	Perpendicular to the view of projection
X	Crossed
M	Multidirectional
C	Circular
R	Radial

TABLE 8.4

Surface Roughness Symbol

▽	8 to 25 Ra (μm)	▽▽▽	0.025 to 1.6 Ra (μm)
▽▽	1.6 to 8 Ra (μm)	▽▽▽▽	<0.025 Ra (μm)

The symbols used only to show the surface roughness are in Table 8.4.

3. *Geometric dimensioning and tolerance*: Geometric dimensioning and tolerance (GD & T) is a new method for specifying tolerance. GD & T has many advantages over the conventional tolerance system. The main advantage of GD & T is that it allows the designer to specify the behavior and requirement of the component's surfaces that are used in assembly for mating. The other advantage of GD & T is the cylindrical tolerance zone. There are many good books written on this subject. The symbols and meanings are illustrated in Tables 8.5 and 8.6.

8.5.2 Notes

Notes are shown in the drawing for descriptive information that may not be suitable to show by any symbol. Notes are also sometimes referred in the drawing by special markings like an asterisk (*) mark or box markings.

TABLE 8.5

Symbol Meaning

Tolerance Type	Description	ANSI Symbol
Form	Straightness	–
Form	Flatness	c
Form	Circularity	e
Form	Cylindricity	g
Profile	Profile of a line	k
Profile	Profile of a surface	d
Orientation	Angularity	a
Orientation	Perpendicularity	b
Orientation	Parallelism	f
Location	Position	j
Location	Concentricity	r
Runout	Circular runout	h
Runout	Total runout	t

TABLE 8.6
Symbol and Modifying Term

Modifying Term	ANSI Symbol
At maximum material condition	m
At least material condition	l
Regardless of feature size	s
Project tolerance zone	p
Diameter	n
Spherical diameter	Sn
Radius	R
Spherical radius	SR
Reference	( )
Arc length	^

There some notes that are required to be placed in most drawings; they are as follows:

1. *Dimensional units*: It is not always possible to mention units in every dimension. A note for dimensions like "All dimensions are in mm and levels are in m, unless otherwise specified" may be among the most useful notes in most drawings of a plant layout.
2. *Reference drawing and documents*: A drawing may be linked with another drawing especially in the case of plant engineering, where many disciplines of engineering work together. The number of reference drawings and documents shall be mentioned in the drawing, if applicable.
3. Any information that may not be possible to show in dimensional form like thickness of paint coat, type of painting, and so forth.

Apart from these notes, there can be many other notes that are relevant and useful in studying a drawing.

8.6 Template Creation

It is obvious that in any drawing, there can be many objects and annotations that are similar and do not vary too much. Thus, effort can be reduced by increasing the repeatability of the drawing. A drawing template is a method to reduce the effort and repetitive work in a drawing. In parametric software, drawing views are generated from a 3-D model and placed as required in the drawing. If the requirements of views are determined for similar objects, then multiple views can be placed in the template to reduce the effort of view placing. However, certain alterations for view positioning and scale alteration may be required. The template shall be created as follows:

1. *Essential drawing templates*: These are the templates with a company's title block. Remember that each of the size of sheets may have similar title blocks but with different sizes. The customized sheets for every company need to be prepared such that they can be used repetitively.
2. *Template for similar drawings*: Similar drawings, like every assembly fabrication drawing, must have a bill of materials, symbol libraries, and so forth. To create a bill of materials in every drawing, an auto-region table may need to be created in the drawing, which will generate a bill of materials for every drawing. A welding symbol gallery may need to be assigned. Therefore, a template drawing with such details will save time and effort for re-creation every time.

8.7 Project Work

Hands-On Exercise, Example 8.1: Plug (Software Used, Creo Parametric; Figure 8.16)

PLUG

Steps:

- Click on Create new file.
- Click on Drawing to create a drawing file.
- Uncheck the box for using the default template.
- Change the name to Plug. (Other names may be given, but it is advisable to provide a relevant name.)

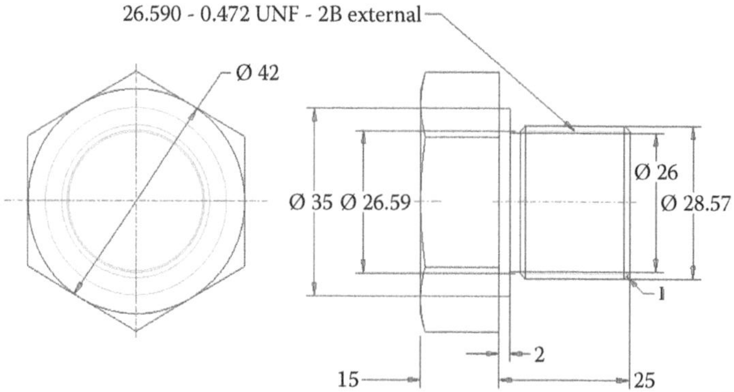

FIGURE 8.16
Drawing of a threaded plug.

- Click OK to create the drawing file.
- Select the orientation Landscape.
- Click on Empty in the Specify template field.
- Select the standard paper size A0.
- The default model field shall be empty.
- Click OK to create the drawing file with an A0 drawing sheet.

- Click on Drawing models from the Layout tab of the ribbon interface.
- A menu manager shall appear on the right side.
- Click on Add model.
- Select the file Plug from the folder Chapter 6.
- Click Open to add the file as a drawing model.
- Note that many files can be added as drawing models and that to set a model as the current model, the Set model option is used.
- Click on Done/Return to close the menu manager.

- Click on General from the Layout tab of the ribbon interface to add a general view in the drawing that may not be dependent on the other views in the drawing sheet.
- A box for selecting a combined state may appear. Select No combined state and click OK.
- Click anywhere in the drawing sheet to place the view.
- The drawing view shall appear with the Drawing view properties window.
- Double-click on the model view front and notice that the view shall change to front view as displayed in the front view in the model.
- Select Angle in the orientation method selection section.
- Enter 90 as the angle value in the respective field.
- Click on Apply.
- Click on View display in the left section of the Drawing view properties window.
- Select the display style as hidden in the right-hand side.
- Click OK to close the Drawing view properties window.
- Turn off the datum display in the top of the graphic window.

- Select the view and right-click (may need to hold longer).
- In the pop-up menu, uncheck the box for lock view movement.
- Click on the view; an arrow marking will appear.
- Move the view by dragging if required.
- Right-click on the view and select Insert projection view.
- Click on the left side of the left view to place the projected view.
- The drawing view shall appear with the Drawing view properties window.
- Click on View display in the left section of the Drawing view properties window.
- Select the display style as hidden in the right-hand side.
- Click OK to close the Drawing view properties window.

- Click on Show model annotations from the Annotate tab of the ribbon interface.
- A Show model annotations window appears.
- Click on the Show model datums tab.
- Type shall be All.
- Click on the left view.
- Click on all the check boxes to keep all the datums of the model in the drawing sheet.
- Click on the Show the model dimensions tab.
- Type shall be All.
- Click on all the check boxes to keep all the dimensions of the model in the drawing sheet.
- Click on Apply to create the annotations and click to close the model annotations window.

- Click on the Annotate tab of the ribbon interface.
- Select the 0-value dimensions or unwanted dimensions.
- Right-click and select Erase.

- Save and exit the file.

Hands-On Exercise, Example 8.2: Creating a Detail Drawing of Spanner (Software Used, Creo Parametric; Figure 8.17)

SPANNER

Steps:

- Click on Create new file.
- Click on Drawing to create a drawing file.
- Uncheck the box for using the default template.
- Change the name to Spanner. (Other names may be given, but it is advisable to provide a relevant name.)

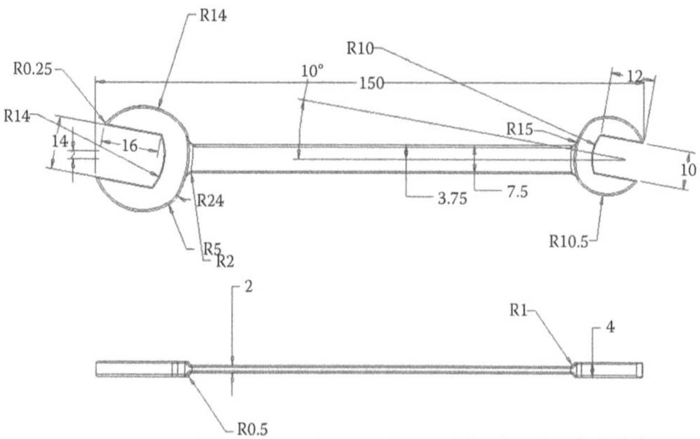

FIGURE 8.17
Detail drawing of a spanner.

- Click OK to create the drawing file.
- Select the orientation Landscape.
- Click on Empty in the Specify template field.
- Select the standard paper size A0.
- The default model field shall be empty.
- Click OK to create the drawing file with an A0 drawing sheet.

- Click on Drawing models from the Layout tab of the ribbon interface.
- A menu manager shall appear on the right side.
- Click on Add model.
- Select the file Spanner.asm from the folder Chapter 6.
- Click Open to add the file as a drawing model.
- Note that many files can be added as drawing models and that to set a model as a current model, the Set model option is used.
- Click on Done/Return to close the menu manager.

- Click on General from the Layout tab of the ribbon interface to add a general view in the drawing that may not be dependent on the other views in the drawing sheet.
- A box for selecting a combined state may appear. Select No combined state and click OK.
- Click anywhere in the drawing sheet to place the view.
- The drawing view shall appear with the Drawing view properties window.
- Double-click on the model view left and notice that the view shall change to top view as displayed in the left view in the model.
- Select Angle in the orientation method selection section.
- Enter 90 as the angle value in the respective field.
- Click on Apply.
- Click on View display in the left section of the Drawing view properties window.
- Select the display style as hidden in the right-hand side.
- Click OK to close the Drawing view properties window.
- Turn off the datum display in the top of the graphic window.

- Right-click on the view and select Insert projection view.
- Click on the bottom side of the existing view to place the projected view.
- The drawing view shall appear with the Drawing view properties window.
- Click on View display in the left section of the Drawing view properties window.
- Select the display style as hidden in the right-hand side.
- Click OK to close the Drawing view properties window.

- Click on Show model annotations from the Annotate tab of the ribbon interface.
- A Show model annotations window appears.
- Click on the Show model datums tab.

- Type shall be All.
- Click on the first placed view.
- Click on all the check boxes to keep all the datums of the model in the drawing sheet.
- Click on the Show the model dimensions tab.
- Type shall be All.
- Click on all the check boxes to keep all the dimensions of the model in the drawing sheet.
- Click on Apply to create the annotations and click to close the model annotations window.

- Click on the Annotate tab of the ribbon interface.
- Select the 0-value dimensions or unwanted dimensions.
- Right-click and select Erase.

- Save and exit the file.

Hands-On Exercise, Example 8.3: Creating a Plug Sheet Drawing (Software Used, Creo Parametric; Figure 8.18)

PLUG SHEET

Steps:

- Click on Create new file.
- Click on Drawing to create a drawing file.
- Uncheck the box for using the default template.
- Change the name to Plug Sheet. (Other names may be given, but it is advisable to provide a relevant name.)
- Click OK to create the drawing file.
- Select the orientation Landscape.
- Click on Empty in the Specify template field.
- Select the standard paper size A0.
- The default model field shall be empty.
- Click OK to create the drawing file with an A0 drawing sheet.

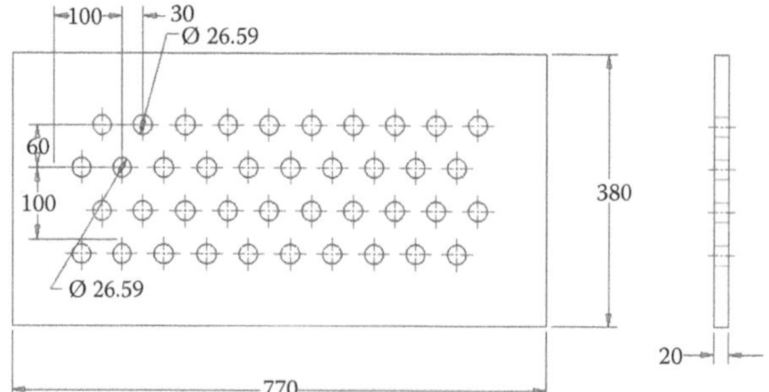

FIGURE 8.18
Drawing of a plug sheet.

- Click on Drawing models from the Layout tab of the ribbon interface.
- A menu manager shall appear on the right side.
- Click on Add model.
- Select the file Plug Sheet from the folder Chapter 6.
- Click Open to add the file as a drawing model.
- Note that many files can be added as drawing models and that to set a model as the current model, the Set model option is used.
- Click on Done/Return to close the menu manager.

- Click on General from the Layout tab of the ribbon interface to add a general view in the drawing that may not be dependent on the other views in the drawing sheet.
- A box for selecting a combined state may appear. Select No combined state and click OK.
- Click anywhere in the drawing sheet to place the view.
- The drawing view shall appear with the Drawing view properties window.
- Double-click on the model view left and notice that the view shall change to left view as displayed in the left view in the model.
- Click on View display in the left section of the Drawing view properties window.
- Select the display style as hidden in the right-hand side.
- Click OK to close the Drawing view properties window.
- Turn off the datum display in the top of the graphic window.

- Select the view and right-click (may need to hold longer).
- In the pop-up menu, uncheck the box for lock view movement.
- Click on the view; an arrow marking will appear.
- Move the view by dragging if required.
- Right-click on the view and select Insert projection view.
- Click on the right side of the front view to place the projected view.
- The drawing view shall appear with the Drawing view properties window.
- Click on View display in the left section of the Drawing view properties window.
- Select the display style as hidden in the right-hand side.
- Click OK to close the Drawing view properties window.

- Click on Show model annotations from the Annotate tab of the ribbon interface.
- A Show model annotations window appears.
- Click on the Show model datums tab.
- Type shall be All.
- Click on the left view.
- Click on all the check boxes to keep all the datums of the model in the drawing sheet.

- Click on the Show the model dimensions tab.
- Type shall be All.
- Click on all the check boxes to keep all the dimensions of the model in the drawing sheet.
- Click on Apply to create the annotations and click to close the model annotations window.

- Click on the Annotate tab of the ribbon interface.
- Select the 0-value dimensions or unwanted dimensions.
- Right-click and select Erase.

- Save and exit the file.

Hands-On Exercise, Example 8.4: Creating a Board Clip (Software Used, Creo Parametric; Figure 8.19)

BOARD CLIP

Steps:

- Click on Create new file.
- Click on Drawing to create a drawing file.
- Uncheck the box for using the default template.
- Change the name to Board_Clip. (Other names may be given, but it is advisable to provide a relevant name.)
- Click OK to create the drawing file.

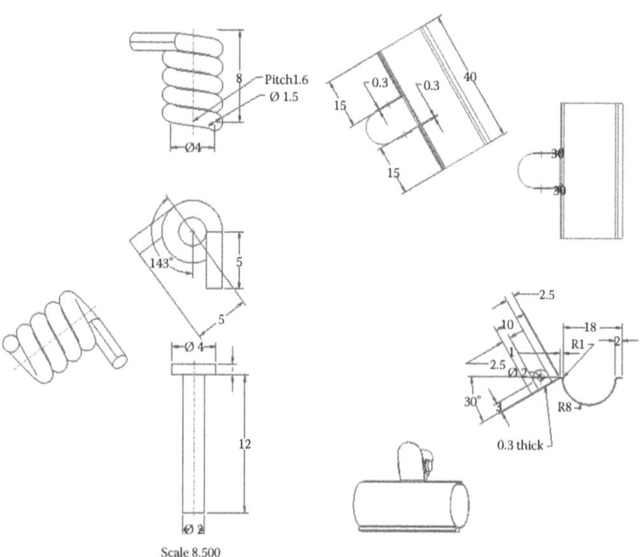

FIGURE 8.19
Drawing of a board clip.

- Select the orientation Landscape.
- Click on Empty in the Specify template field.
- Select the standard paper size A0.
- The default model field shall be empty.
- Click OK to create the drawing file with an A0 drawing sheet.

- Click on Drawing models from the Layout tab of the ribbon interface.
- A menu manager shall appear on the right side.
- Click on Add model.
- Select the file Board_Clip.asm from the folder Chapter 7.
- Click Open to add the file as a drawing model.
- Note that many files can be added as drawing models and that to set a model as the current model, the Set model option is used.
- Click on Done/Return to close the menu manager.

- Click on General from the Layout tab of the ribbon interface to add a general view in the drawing that may not be dependent on the other views in the drawing sheet.
- A box for selecting a combined state may appear. Select No combined state and click OK.
- Click anywhere in the drawing sheet to place the view.
- The drawing view shall appear with the Drawing view properties window.
- Double-click on the model view default orientation and notice that the view shall change to trimetric view as displayed in the default view in the model.
- Click on View display in the left section of the Drawing view properties window.
- Select the display style as hidden in the right-hand side.
- Click OK to close the Drawing view properties window.
- Turn off the datum display in the top of the graphic window.

- The two parts of the paper clip assembly shall be shown in the model tree, that is, BOARD_CLIP_SPRING, BOARD_CLIP1, and BOARD_CLIP_PIN.
- Select the part BOARD_CLIP1 and right-click on it.
- Select Set/add drawing model from the pop-up menu.

- Click on General from the Layout tab of the ribbon interface to add a general view in the drawing that may not be dependent on the other views in the drawing sheet.
- A box for selecting a combined state may appear. Select No combined state and click OK.
- Click anywhere in the drawing sheet to place the view.
- The drawing view shall appear with the Drawing view properties window.
- Double-click on the model view front orientation and notice that the view shall change to front view as displayed in the front view in the model.

- Click on View display in the left section of the Drawing view properties window.
- Select the display style as hidden in the right-hand side.
- Click OK to close the Drawing view properties window.
- Turn off the datum display in the top of the graphic window.

- Select the front view of the part Board_Clip1 and right-click (may need to hold longer).
- In the pop-up menu, uncheck the box for lock view movement.
- Click on the view; an arrow marking will appear.
- Move the view by dragging if required.
- Right-click on the view and select Insert projection view.
- Click on the top side of the front view to place the projected view.
- The drawing view shall appear with the Drawing view properties window.
- Click on View display in the left section of the Drawing view properties window.
- Select the display style as hidden in the right-hand side.
- Click OK to close the Drawing view properties window.

- Select the auxiliary view from the Layout tab of the ribbon interface.
- Select the right inclined edge of the front view of the part Board_Clip1 and place as shown in Figure 8.20.
- Click on the drawing models from the Layout tab of the ribbon interface.

- Click on Set model and select Board_Clip.asm as the current model.
- Right-click on the model Board_Clip_Spring and select Add/set as drawing model.

- Click on General from the Layout tab of the ribbon interface to add a general view in the drawing that may not be dependent on the other views in the drawing sheet.

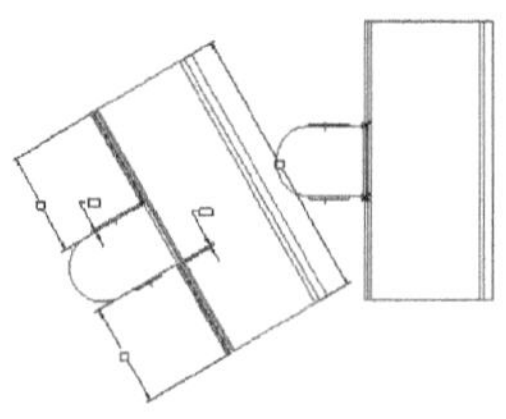

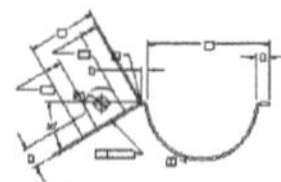

FIGURE 8.20
View creation for a drawing of a board clip.

- A box for selecting a combined state may appear. Select No combined state and click OK.
- Click anywhere in the drawing sheet to place the view.
- The drawing view shall appear with the Drawing view properties window.
- Double-click on the model view front orientation and notice that the view shall change to front view as displayed in the front view in the model.
- Click on View display in the left section of the Drawing view properties window.
- Select the display style as hidden in the right-hand side.
- Click OK to close the Drawing view properties window.
- Turn off the datum display in the top of the graphic window.

- Right-click on the view and select Insert projection view.
- Click on the bottom side of the front view to place the projected view.
- The drawing view shall appear with the Drawing view properties window.
- Click on View display in the left section of the Drawing view properties window.
- Select the display style as hidden in the right-hand side.
- Click OK to close the Drawing view properties window.

- Select the auxiliary view from the Layout tab of the ribbon interface.
- Select the right inclined edge of the front view of the part Board_Clip_Spring and place as shown in Figure 8.21.

- Click on the drawing models from the Layout tab of the ribbon interface.

- Click on Set model and select Board_Clip.asm as the current model.
- Right-click on the model Board_Clip_Pin and select Add/set as drawing model.

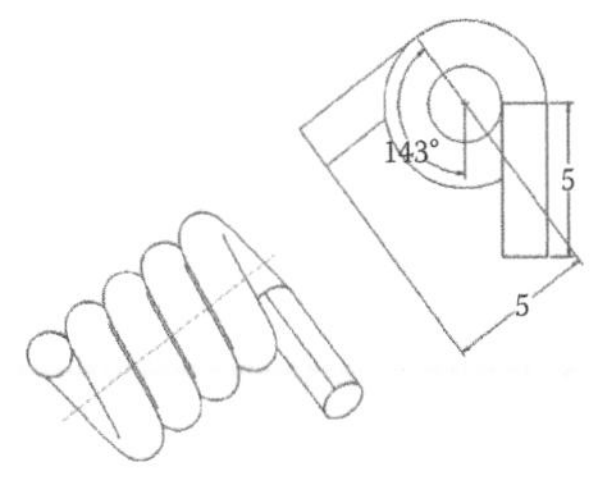

FIGURE 8.21
Drawing of a spring.

- Click on General from the Layout tab of the ribbon interface to add a general view in the drawing that may not be dependent on the other views in the drawing sheet.
- A box for selecting a combined state may appear. Select No combined state and click OK.
- Click anywhere in the drawing sheet to place the view.
- The drawing view shall appear with the Drawing view properties window.
- Double-click on the model view front orientation and notice that the view shall change to front view as displayed in the front view in the model.
- Click on View display in the left section of the Drawing view properties window.
- Select the display style as hidden in the right-hand side.
- Click OK to close the Drawing view properties window.
- Turn off the datum display in the top of the graphic window.
- Click on Show model annotations from the Annotate tab of the ribbon interface.
- A Show model annotations window appears.
- Click on the Show model datums tab.
- Type shall be All.
- Click on the views as shown in Figure 8.21.
- Click on all the check boxes to keep all the datums of the model in the drawing sheet.
- Click on the Show the model dimensions tab.
- Type shall be All.
- Click on all the check boxes to keep all the dimensions of the model in the drawing sheet.
- Click on Apply to create the annotations and click to close the model annotations window.

- Click on the Annotate tab of the ribbon interface.
- Select the 0-value dimensions or unwanted dimensions.
- Right-click and select Erase.

- Save and exit the file.

Hands-On Exercise, Example 8.5: Creating a Card Clip (Software Used, Creo Parametric; Figure 8.22)

CARD CLIP

Steps:

- Click on Create new file.
- Click on Drawing to create a drawing file.
- Uncheck the box for using the default template.
- Change the name to Card_Clip. (Other names may be given, but it is advisable to provide a relevant name.)
- Click OK to create the drawing file.

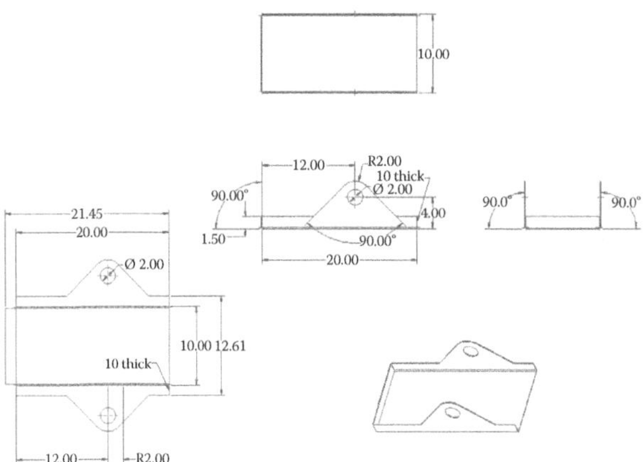

FIGURE 8.22
Drawing of a card clip.

- Select the orientation Landscape.
- Click on Empty in the Specify template field.
- Select the standard paper size A0.
- The default model field shall be empty.
- Click OK to create the drawing file with an A0 drawing sheet.

- Click on Drawing models from the Layout tab of the ribbon interface.
- A Menu manager shall appear on the right side.
- Click on Add model.
- Select the file Card Clip1 from the folder Chapter 9.
- Click Open to add the file as a drawing model.
- Note that many files can be added as drawing models and that to set a model as the current model, the Set model option is used.
- Click on Done/Return to close the menu manager.

- Click on General from the Layout tab of the ribbon interface to add a general view in the drawing that may not be dependent on the other views in the drawing sheet.
- A box for selecting a combined state may appear. Select No combined state and click OK.
- Click anywhere in the drawing sheet to place the view.
- The drawing view shall appear with the Drawing view properties window.
- Double-click on the model view default orientation and notice that the view shall change to trimetric view as displayed in the default view in the model.
- Click on View display in the left section of the Drawing view properties window.

- Select the display style as hidden in the right-hand side.
- Click OK to close the Drawing view properties window.
- Turn off the datum display in the top of the graphic window.

- Click on General from the Layout tab of the ribbon interface to add a general view in the drawing that may not be dependent on the other views in the drawing sheet.
- A box for selecting a combined state may appear. Select No combined state and click OK.
- Click anywhere in the drawing sheet to place the view.
- The drawing view shall appear with the Drawing view properties window.
- Double-click on the model view front orientation and notice that the view shall change to front view as displayed in the front view in the model.
- Click on View display in the left section of the Drawing view properties window.
- Select the display style as hidden in the right-hand side.
- Click OK to close the Drawing view properties window.
- Turn off the datum display in the top of the graphic window.

- Select the front view of the part Card_Clip1 and right-click (may need to hold longer).
- In the pop-up menu, uncheck the box for lock view movement.
- Click on the view; an arrow marking will appear.
- Move the view by dragging if required.
- Right-click on the view and select Insert projection view.
- Click on the top side of the front view to place the projected view.
- The drawing view shall appear with the Drawing view properties window.
- Click on View display in the left section of the Drawing view properties window.
- Select the display style as hidden in the right-hand side.
- Click OK to close the Drawing view properties window.

- Select the front view of the part Card_Clip1 and right-click (may need to hold longer).
- In the pop-up menu, uncheck the box for lock view movement.
- Click on the view; an arrow marking will appear.
- Move the view by dragging if required.
- Right-click on the view and select Insert projection view.
- Click on the right side of the front view to place the projected view.
- The drawing view shall appear with the Drawing view properties window.
- Click on View display in the left section of the Drawing view properties window.
- Select the display style as hidden in the right-hand side.
- Click OK to close the Drawing view properties window.

- Select the model Card_Clip1 from the model tree in the left section of the drawing window.
- The part Card_Clip1 shall be opened.
- Select View manager from the View tab of the ribbon interface.
- The view manager window will appear. Select the simplified representation (Simp Rep) tab.
- Click on New to create a new simplified representation.
- Change the name to Developed. (Other names can be given.) Press Enter.
- A menu manager shall appear.
- Select features from the menu manager.
- Select Exclude and select the feature Bend back 1 from the model tree.
- Click on Done and Done/Return in the menu manager.

- Select the tab All.
- Click on New to create a new combined state.
- Change the name to Developed. (Other names can be given.) Press Enter.
- A menu manager shall appear.
- Select features from the menu manager.
- Select the simplified representation Developed from the Simplified representation selection field.
- The orientation, cross section, and layers do not require any selection for this example.
- Click on the tick mark to create the combined state named Developed.
- Save and close the part file named Developed.

- Click on General from the Layout tab of the ribbon interface to add a general view in the drawing that may not be dependent on the other views in the drawing sheet.
- A box for selecting a combined state may appear. Select the combined state Developed and click OK.
- Click anywhere in the drawing sheet to place the view.
- The drawing view shall appear with the Drawing view properties window.
- Double-click on the model view top orientation and notice that the view shall change to top view as displayed in the top view in the model.
- Click on View display in the left section of the Drawing view properties window.
- Select the display style as hidden in the right-hand side.
- Click OK to close the Drawing view properties window.
- Turn off the datum display in the top of the graphic window.
- Click on Show model annotations from the Annotate tab of the ribbon interface.
- A Show model annotations window appears.
- Click on the Show model datums tab.

- Type shall be All.
- Click on the views as shown in Figure 8.22.
- Click on all the check boxes to keep all the datums of the model in the drawing sheet.
- Click on the Show the model dimensions tab.
- Type shall be All.
- Click on all the check boxes to keep all the dimensions of the model in the drawing sheet.
- Click on Apply to create the annotations and click to close the model annotations window.

- Click on the Annotate tab of the ribbon interface.
- Select the 0-value dimensions or unwanted dimensions.
- Right-click and select Erase.

- Save and exit the file.

Hands-On Exercise, Example 8.6: Creating a Paper Clip (Software Used, Creo Parametric; Figure 8.23)

PAPER CLIP

Steps:

- Click on Create new file.
- Click on Drawing to create a drawing file.
- Uncheck the box for using the default template.
- Change the name to Paper_Clip. (Other names may be given, but it is advisable to provide a relevant name.)
- Click OK to create the drawing file.

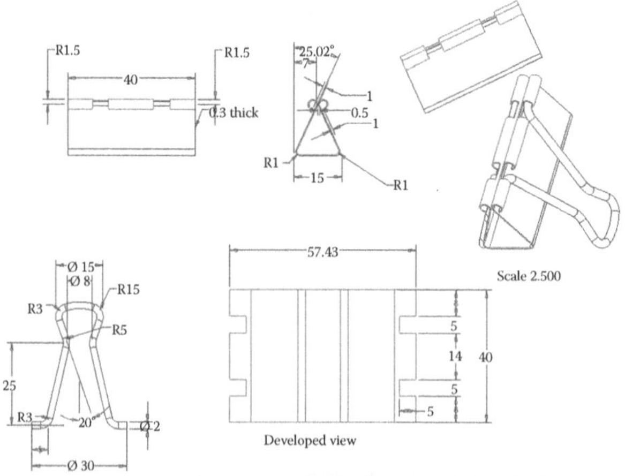

FIGURE 8.23
Drawing of a paper clip.

- Select the orientation Landscape.
- Click on Empty in the Specify template field.
- Select the standard paper size A0.
- The default model field shall be empty.
- Click OK to create the drawing file with an A0 drawing sheet.

- Click on Drawing models from the Layout tab of the ribbon interface.
- A menu manager shall appear on the right side.
- Click on Add model.
- Select the file Paper_Clip_Assembly.asm from the folder Chapter 7.
- Click Open to add the file as a drawing model.
- Note that many files can be added as drawing models and that to set a model as the current model, the Set model option is used.
- Click on Done/Return to close the menu manager.

- Click on General from the Layout tab of the ribbon interface to add a general view in the drawing that may not be dependent on the other views in the drawing sheet.
- A box for selecting a combined state may appear. Select No combined state and click OK.
- Click anywhere in the drawing sheet to place the view.
- The drawing view shall appear with the Drawing view properties window.
- Double-click on the model view default orientation and notice that the view shall change to trimetric view as displayed in the default view in the model.
- Click on View display in the left section of the Drawing view properties window.
- Select the display style as hidden in the right-hand side.
- Click OK to close the Drawing view properties window.
- Turn off the datum display in the top of the graphic window.

- The two parts of the paper clip assembly shall be shown in the model tree, that is, Paper_Clip1 and Paper_Clip_Handle.
- Select the part Paper_Clip_Handle and right-click on it.
- Select Set/add drawing model from the pop-up menu.

- Click on General from the Layout tab of the ribbon interface to add a general view in the drawing that may not be dependent on the other views in the drawing sheet.
- A box for selecting a combined state may appear. Select No combined state and click OK.
- Click anywhere in the drawing sheet to place the view.
- The drawing view shall appear with the Drawing view properties window.
- Double-click on the model view front orientation and notice that the view shall change to front view as displayed in the front view in the model.

- Click on View display in the left section of the Drawing view properties window.
- Select the display style as hidden in the right-hand side.
- Click OK to close the Drawing view properties window.
- Turn off the datum display in the top of the graphic window.

- Select the front view of the part Paper_Clip1 and right-click (may need to hold longer).
- In the pop-up menu, uncheck the box for lock view movement.
- Click on the view; an arrow marking will appear.
- Move the view by dragging if required.
- Right-click on the view and select Insert projection view.
- Click on the right side of the left view to place the projected view.
- The drawing view shall appear with the Drawing view properties window.
- Click on View display in the left section of the Drawing view properties window.
- Select the display style as hidden in the right-hand side.
- Click OK to close the Drawing view properties window.
- Select the model Paper_Clip1 from the model tree in the left section of the drawing window.
- The part Paper_Clip1 shall be opened.
- Select View manager from the View tab of the ribbon interface.
- View manager window will appear. Select the Simplified representation (Simp Rep) tab.
- Click on New to create a new simplified representation.
- Change the name to Developed. (Other names can be given.) Press Enter.
- A menu manager shall appear.
- Select features from the menu manager.
- Select Exclude and select the feature Bend back 1 from the model tree.
- Click on Done and Done/Return in the menu manager.

- Select the tab All.
- Click on New to create a new combined state.
- Change the name to Developed. (Other names can be given.) Press Enter.
- A menu manager shall appear.
- Select features from the menu manager.
- Select the simplified representation Developed from the Simplified representation selection field.
- The orientation, cross section and layers do not require any selection for this example.
- Click on the tick mark to create the combined state named Developed.
- Save and close the part file named Developed.

- Click on General from the Layout tab of the ribbon interface to add a general view in the drawing that may not be dependent on the other views in the drawing sheet.
- A box for selecting a combined state may appear. Select the combined state Developed and click OK.
- Click anywhere in the drawing sheet to place the view.
- The drawing view shall appear with the Drawing view properties window.
- Double-click on the model view top orientation and notice that the view shall change to top view as displayed in the top view in the model.
- Click on View display in the left section of the Drawing view properties window.
- Select the display style as hidden in the right-hand side.
- Click OK to close the Drawing view properties window.
- Turn off the datum display in the top of the graphic window.
- Select the auxiliary view from the Layout tab of the ribbon interface.
- Select the right inclined edge of the front view of the part Paper_Clip1 and place as shown in Figure 8.24.
- Click on the drawing models from the Layout tab of the ribbon interface.
- Click on Set model and select Paper_Clip_Assembly.asm as the current model.
- Right-click on the model Paper_Clip_Handle and select add/set as drawing model.

- Click on General from the Layout tab of the ribbon interface to add a general view in the drawing that may not be dependent on the other views in the drawing sheet.
- A box for selecting a combined state may appear. Select No combined state and click OK.
- Click anywhere in the drawing sheet to place the view.

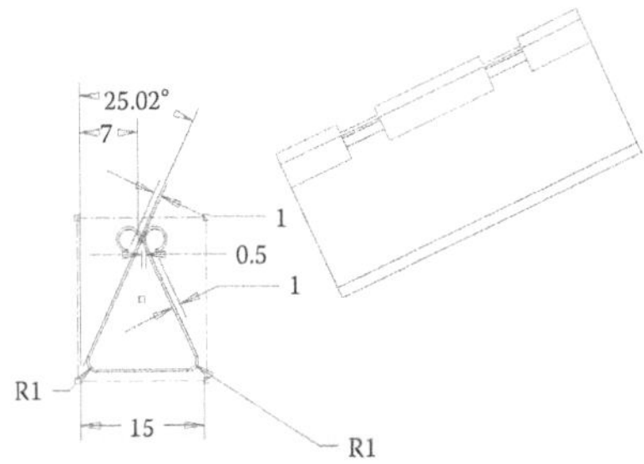

FIGURE 8.24
Drawing of the sheet metal part of a paper clip.

- The drawing view shall appear with the Drawing view properties window.
- Double-click on the model view front orientation and notice that the view shall change to front view as displayed in the front view in the model.
- Click on View display in the left section of the Drawing view properties window.
- Select the display style as hidden in the right-hand side.
- Click OK to close the Drawing view properties window.
- Turn off the datum display in the top of the graphic window.

- Click on Show model annotations from the Annotate tab of the ribbon interface.
- A Show model annotations window appears.
- Click on the Show model datums tab.
- Type shall be All.
- Click on the views as shown in Figure 8.24.
- Click on all the check boxes to keep all the datums of the model in the drawing sheet.
- Click on the Show the model dimensions tab.
- Type shall be All.
- Click on all the check boxes to keep all the dimensions of the model in the drawing sheet.
- Click on Apply to create the annotations and click to close the model annotations window.

- Click on the Annotate tab of the ribbon interface.
- Select the 0-value dimensions or unwanted dimensions.
- Right-click and select Erase.

- Save and exit the file.

Hands-On Exercise, Example 8.7: Creating Saddle Detailing (Software Used, Creo Parametric; Figure 8.25)

SADDLE DRAWING

Steps:

- Click on Create new file.
- Click on Drawing to create a drawing file.
- Uncheck the box for using the default template.
- Change the name to Tank_Saddle. (Other names may be given, but it is advisable to provide a relevant name.)
- Click OK to create the drawing file.
- Select the orientation Landscape.
- Click on Empty in the Specify template field.
- Select the standard paper size A0.
- The default model field shall be empty.
- Click OK to create the drawing file with an A0 drawing sheet.

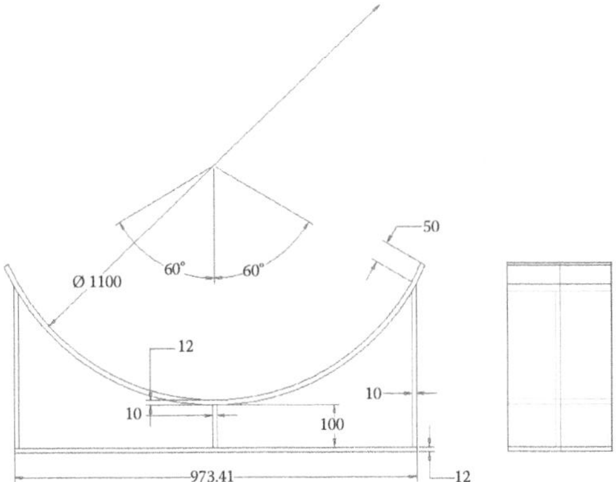

FIGURE 8.25
Drawing of a saddle.

- Click on Drawing models from the Layout tab of the ribbon interface.
- A menu manager shall appear on the right side.
- Click on Add model.
- Select the file Tanki_Saddle.asm from the folder Chapter 7.
- Click Open to add the file as a drawing model.
- Note that many files can be added as drawing models and that to set a model as the current model, the Set model option is used.
- Click on Done/Return to close the menu manager.
- Click on General from the Layout tab of the ribbon interface to add a general view in the drawing that may not be dependent on the other views in the drawing sheet.
- A box for selecting a combined state may appear. Select No combined state and click OK.
- Click anywhere in the drawing sheet to place the view.
- The drawing view shall appear with the Drawing view properties window.
- Double-click on the model view left and notice that the view shall change to front view as displayed in the left view in the model.
- Click on View display in the left section of the Drawing view properties window.
- Select the display style as hidden in the right-hand side.
- Click OK to close the Drawing view properties window.
- Turn off the datum display in the top of the graphic window.
- Click on Show model annotations from the Annotate tab of the ribbon interface.
- A Show model annotations window appears.
- Click on the Show model datums tab.
- Type shall be All.

- Click on the front view.
- Click on all the check boxes to keep all the datums of the model in the drawing sheet.
- Click on the Show the model dimensions tab.
- Type shall be All.
- Click on all the check boxes to keep all the dimensions of the model in the drawing sheet.
- Click on Apply to create the annotations and click to close the model annotations window.
- Click on the Annotate tab of the ribbon interface.
- Select the 0-value dimensions or unwanted dimensions.
- Right-click and select Erase.
- Right-click on the front view and select Projection view.
- Place the right view to the right side of the front view.
- The drawing view shall appear with the Drawing view properties window.
- Double-click on the model view left and notice that the view shall change to front view as displayed in the left view in the model.
- Click on View display in the left section of the Drawing view properties window.
- Select the display style as hidden in the right-hand side.
- Click OK to close the Drawing view properties window.
- Turn off the datum display in the top of the graphic window.

- Save and exit the file.

Hands-On Exercise, Example 8.8: Creating an Expansion Joint (Software Used, Creo Parametric; Figure 8.26)

EXPANSION JOINT

Steps:

- Click on Create new file.
- Click on Drawing to create a drawing file.
- Uncheck the box for using the default template.
- Change the name to Expansion_Joint. (Other names may be given, but it is advisable to provide a relevant name.)
- Click OK to create the drawing file.
- Select the orientation Landscape.
- Click on Empty in the Specify template field.
- Select the standard paper size A0.
- The default model field shall be empty.
- Click OK to create the drawing file with an A0 drawing sheet.

- Click on Drawing models from the Layout tab of the ribbon interface.
- A menu manager shall appear on the right side.
- Click on Add model.

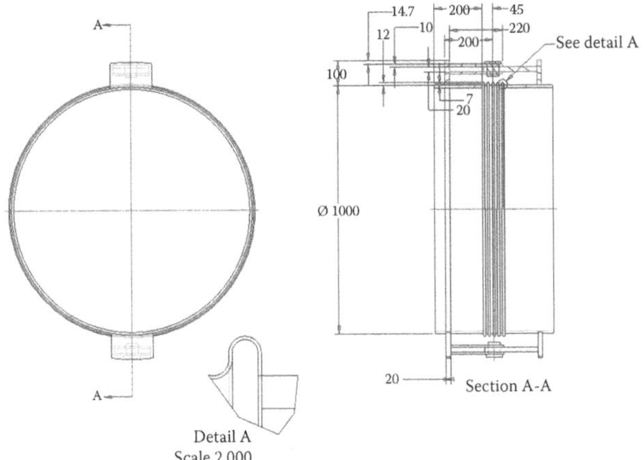

FIGURE 8.26
Drawing of a hinge expansion joint.

- Select the file Hinge_Expansion_Joint.asm from the folder Chapter 7.
- Click Open to add the file as a drawing model.
- Note that many files can be added as drawing models and that to set a model as the current model, the Set model option is used.
- Click on Done/Return to close the menu manager.

- Click on General from the Layout tab of the ribbon interface to add a general view in the drawing that may not be dependent on the other views in the drawing sheet.
- A box for selecting a combined state may appear. Select No combined state and click OK.
- Click anywhere in the drawing sheet to place the view.
- The drawing view shall appear with the Drawing view properties window.
- Double-click on the model view left and notice that the view shall change to left view as displayed in the left view in the model.
- Click on View display in the left section of the Drawing view properties window.
- Select the display style as hidden in the right-hand side.
- Click OK to close the Drawing view properties window.
- Turn off the datum display in the top of the graphic window.

- Select the view and right-click (may need to hold longer).
- In the pop-up menu, uncheck the box for lock view movement.
- Click on the view; an arrow marking will appear.
- Move the view by dragging if required.
- Right-click on the view and select Insert projection view.

- Click on the right side of the left view to place the projected view.
- The drawing view shall appear with the Drawing view properties window.
- Click on View display in the left section of the Drawing view properties window.
- Select the display style as hidden in the right-hand side.
- Click OK to close the Drawing view properties window.
- Right-click on the newly added projected view and select Properties.
- The Drawing view properties window appears.
- Click on Section in the left section of the Drawing view properties window.
- Select 2D cross section in the right side of the Drawing properties window.
- Click on Add icon to add a cross-sectional view.
- Select Create new in the first box of the View adding window.
- Select Single and click on Done.
- The view name will be asked by the system.
- Enter the view name A.
- Select the plane ASM_FRONT.
- Click on the sectioned area field and select Half.
- Graphically or from the model tree, select the plane ASM_TOP as reference.
- Click on the arrow display field.
- Select the left view to display the sectional arrow in the left view.
- Flip the orientation if required.
- Click OK to create the view.

- Click on Show model annotations from the Annotate tab of the ribbon interface.
- A Show model annotations window appears.
- Click on the Show model datums tab.
- Type shall be All.
- Click on the left view.
- Click on all the check boxes to keep all the datums of the model in the drawing sheet.
- Click on the Show the model dimensions tab.
- Type shall be All.
- Click on all the check boxes to keep all the dimensions of the model in the drawing sheet.
- Click on Apply to create the annotations and click to close the model annotations window.

- Click on the Annotate tab of the ribbon interface.
- Select the 0-value dimensions or unwanted dimensions.
- Right-click and select Erase.

- Click on the detail view from the Layout tab of the ribbon interface.
- Click on the convolution portion that is required to be shown in comparatively large scale.
- Click on the boundary of the imaginary circle around the first clicked view point to define the boundary of the detail view. Note that a minimum of three points around the view point is required to define the detail view boundary.
- Press the middle button to finish defining the view boundary.
- Click on the open space to place the detail view.
- Double-click on the detailed view to open the Drawing view properties window.
- Click on Scale in the left side of the Drawing view properties window.
- In the custom scale field, enter 2 as the view scale.
- Click OK to enlarge the view and close the Drawing properties window.
- Save and exit the file.

QUESTIONS

1. Define steps that need to be followed for creation of a drawing.
2. How are the drawing sizes classified?
3. List the different types of views and their uses.
4. What is tolerance?
5. Why does the tolerance need to be specified?
6. Define the use of the ordinate dimension.
7. What are the properties of a view that can be manipulated from the drawing without opening the 3-D file?
8. Why is the template creation required?

9

Sheet Metal

9.1 Introduction

Much computer-aided design (CAD) software provides a sheet metal module for designing the sheet metal parts. Sheet metal parts are basically modeled in the form of thin sheets, so obviously with mandatory thicknesses. In the real world, the sheet metal parts are those that are made from a sheet of single thickness. Sheets are often expressed in gauges. Table 9.1 shows sheet thicknesses corresponding to gauge number. It is also possible to model sheet metal parts in a solid modeling module by using solid modeling options. However, most of the time, for sheet metal options, it is most suitable to prepare a CAD model. Sometimes, models are also required in solid modeling options, which are then converted to sheet metal.

It is obvious that sheet metal parts are made from a single piece of raw material of sheets. The operations in sheet metals are as follows:

1. *Bending*: Bending is the operation where the sheet metal parts are bent forcefully by manual tools or applying forces in a machine through dies and punches. Bending is the most widely used operation in sheet metal. The length of the unbent part, which will exactly match the dimension of the bent part after bending, is called developed dimension. The developed length is calculated first before required bending. The flat metal sheets are cut into the developed shape and then finally placed into bending dies. After bending, the part takes its final shape. In CAD, bending is done on the sheet metal parts along the sketch bending line. When a bending operation is being done on a sheet metal part, a fixed geometry portion needs to be specified; this will remain the same after bending with respect to the base planes in the model. Figure 9.1 shows an example of bending and unbending.
2. *Punching/blanking*: Punching and blanking operations are a sheet metal cutting operation that provides the required shape to the component. When the cut piece becomes the required component that is cut from the metal sheets, this is called blanking. Similar to blanking,

TABLE 9.1

Sheet Metal Thickness Standard Chart

Gauge	Thickness (in.)	Thickness (mm)
3	0.2391	6.073
4	0.2242	5.695
5	0.2092	5.314
6	0.1943	4.935
7	0.1793	4.554
8	0.1644	4.176
9	0.1495	3.797
10	0.1345	3.416
11	0.1196	3.038
12	0.1046	2.657
13	0.0897	2.278
14	0.0747	1.897
15	0.0673	1.709
16	0.0598	1.519
17	0.0538	1.367
18	0.0478	1.214
19	0.0418	1.062
20	0.0359	0.912
21	0.0329	0.836
22	0.0299	0.759
23	0.0269	0.683
24	0.0239	0.607
25	0.0209	0.531
26	0.0179	0.455
27	0.0164	0.417
28	0.0149	0.378
29	0.0135	0.343
30	0.0120	0.305
31	0.0105	0.267
32	0.0097	0.246
33	0.0090	0.229
34	0.0082	0.208
35	0.0075	0.191
36	0.0067	0.170
37	0.0064	0.163
38	0.0060	0.152

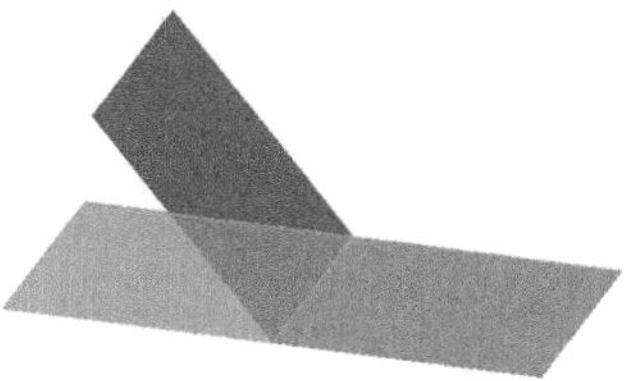

FIGURE 9.1
Example of unbending operation.

punching is a cutting operation; however, the required component is the metal sheet on which the cutting operations are done. In CAD, blanking operations are not required unless to show the process simulation, since the component creation always starts from a blank.

3. *Forming*: Forming is the operation where the material flows and takes the desired shape.

The other operations are also combinations of these three operations. Figure 9.2 shows a model of a sheet metal component. The component is made from a single sheet. It is obvious that the sheet has to be cut from the work sheet, which is called a blanking operation. After the blanking operation, the hole shall be made, which is called a punching operation. Finally, the component has to be formed in some forming die and punch combinations. Therefore, the final component is made. When we model a sheet metal

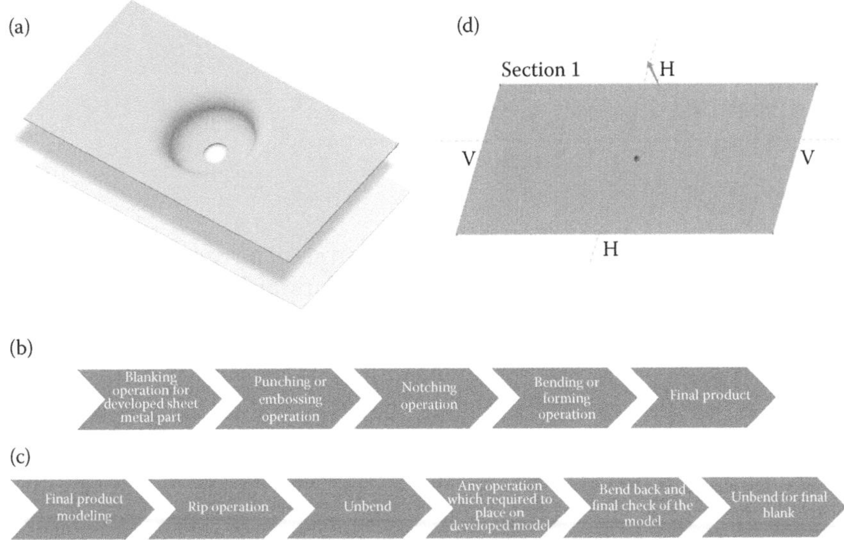

FIGURE 9.2
(a) Example of a sheet metal part. (b) Process of making a sheet metal component in real life. (c) Process of making a sheet metal component in CAD. (d) Example of a flat sheet metal wall.

part, it is obvious that the component shall be modeled as per final dimensions. To model the sheet metal part, there is a need for some more features apart from solid modeling and sheet metal operation. These are commonly called sheet metal features.

In a sheet metal module, the common operations that are performed on real sheet metal are included. The options include creation of a wall, flanges, and so forth for designing sheet metal.

During a bending operation in sheet metal parts, the fibers of the inner surface are getting compressed and under compressive loads, whereas the outer fibers are in tensile load. Hence, some elongation takes place on the outer fiber, and some compression takes place in the inner fibers. Thus, the dimension or overall length of the outer fiber gets elongated, and the inner fiber length gets reduced. The fibers from the inner surface with regard to sheet metal thickness experience reduced compressive stresses, whereas the fibers from the outer surface experience reduced tensile stresses. Thus, there must be a fiber between the outer and inner surface that experiences neither tensile nor compressive stress. The fiber that does not experience either tensile or compressive stresses in the bending is called a neutral fiber. In the developed model, the plane on which the neutral fibers are lying is called the neutral plane. The location of the neutral plane depends on various factors like mechanical properties of the material, bending allowances, and so forth. Hence, depending on the neutral plane location, the developed length may vary from model to model for the same component. Location of the neutral plane is determined by the K factor in CAD. The Y factor is also interrelated with K factor and can be derived from the K factor. The Y factor is used to calculate the developed length. The relations are as follows:

$$\text{K factor} = \delta/T$$

$$\text{Y factor} = \text{K factor} \times \pi/2$$

Calculation of the developed length:

As shown in Figure 9.3, the developed length shall be the combined length of a neutral fiber passing through the neutral plane and perpendicular to the bend edge.

$$\text{Developed length} = \text{straight length of the neutral fiber} + \text{arc length of the fiber}$$

Now, we know that the length of an arc of a given angle θ with radius R shall be

$$\Theta X\pi/180 \times R,$$

whereas ϕ is in degrees.

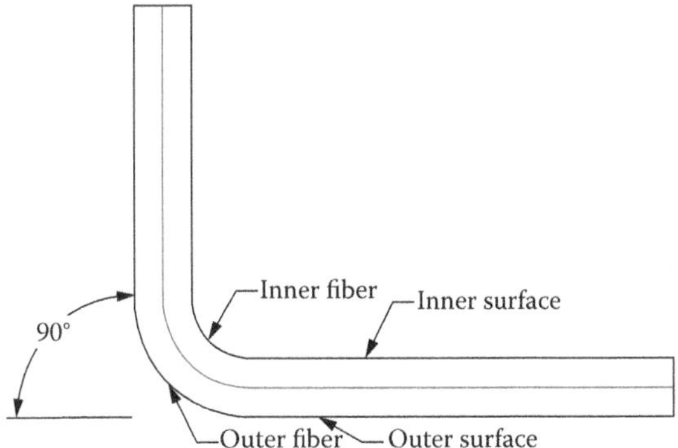

FIGURE 9.3
Example of rip relief.

If the neutral plane location is specified by K factor, the inside radius of the neutral fiber shall be $r + K \times T$, where

R = inside radius of the sheet metal part
T = thickness of the sheet metal part

Thus, the developed length shall be = $\theta X\pi/180 \times (r + K \times T)$.

Now, the developed length of the arc of the neutral plane in terms of the Y factor shall be

$$\begin{aligned}&\Theta X\pi/180\times(r+\mathrm{Y}\times T\times 2/\pi)\\&=\theta\times\pi\times(r/180+\theta \mathrm{XY}\times T/90)\\&=\theta/90(\pi\times r/90+\mathrm{Y}\times T)\end{aligned}$$

As mentioned earlier, the location of the neutral plane depends on various factors like material parts, inside radius of bend, and so forth. Thus, a multiple-bend allowance may need to be defined in the same part where feature-specific bend allowances are desirable.

Normally, when the inside radius of the bend exceeds twice the thickness, the neutral plane is likely to be in the middle; however, for inside bend radius less than twice the thickness of the sheet metal, the neutral plane lies approximately one-third of the thickness from the inner surface.

Example: Y Factor, K Factor Change

A bend table directly controls the allowances with precalculated values for respective thicknesses and radii with bend angles. A multiple-bend table can be prepared, and the same shall be applied simultaneously whenever it applies. Thus, it reduces time and effort and increases accuracy.

In practical manufacturing practice, the sequence of operations is shown in Figure 9.2a.

However, in CAD, the sequence can be followed in reverse manner, as shown in Figure 9.2b.

9.2 Features

1. *Wall*: A sheet metal component can be a combination of multiple sheet metal operations and features. The simplest feature of sheet metal is the wall creation or planar wall creation. A planar wall is the wall created with a geometry and thickness.

   Figure 9.2a shows an example of creating a flat planar wall. The wall can also be extruded from a sketch.

2. *Flange*: Flange is the operation to create a wall that is dependent on the attachment wall or first wall whose edge has been selected for the flange operation. A flange wall can be placed on a nonlinear edge. Figure 9.4 shows the flange creation method.

   Options: When creating a wall, flange, or any operation, there are many sheet metal options that the software can utilize to increase the ease of modeling. The following options are very useful when creating any sheet metal options.

   a. *Relief*: When some sections of sheet metal parts having folded edges are subject to a bend operation, there should be some provision to avoid corner overlapping after bending when the folded edges come near. This provision is created by the operation of relief. Reliefs can be provided by the time creation of folded edges by the flange creation method. The reliefs can be of the following types:

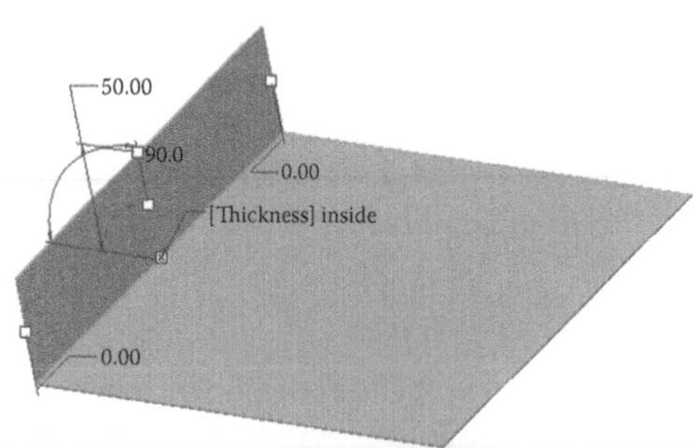

FIGURE 9.4
Example of creation of flange on the edge of a flat wall.

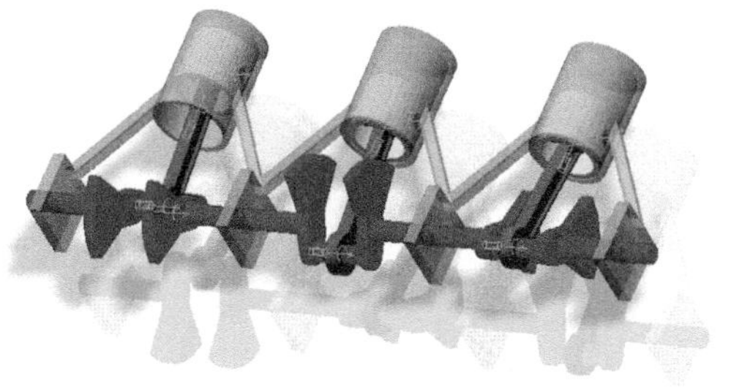

FIGURE 2.10
Piston cylinder mechanism.

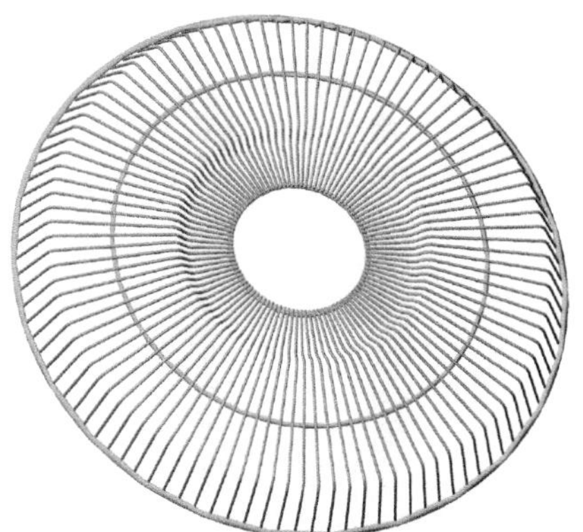

FIGURE 6.9
3-D model of a fan screen.

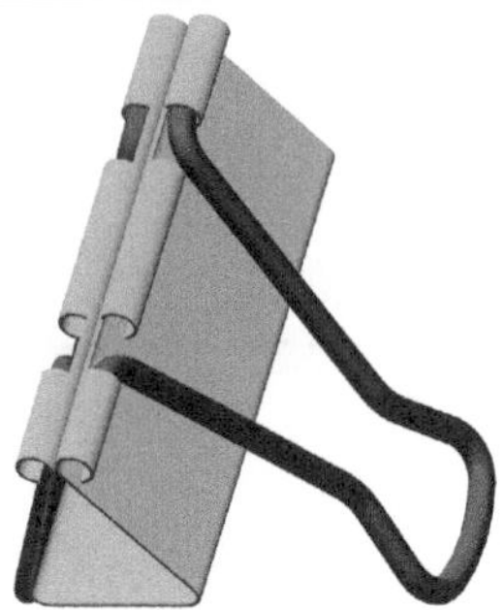

FIGURE 7.12
3-D model of a paper clip.

FIGURE 7.44
3-D view of a hinge expansion joint.

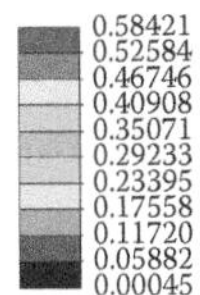

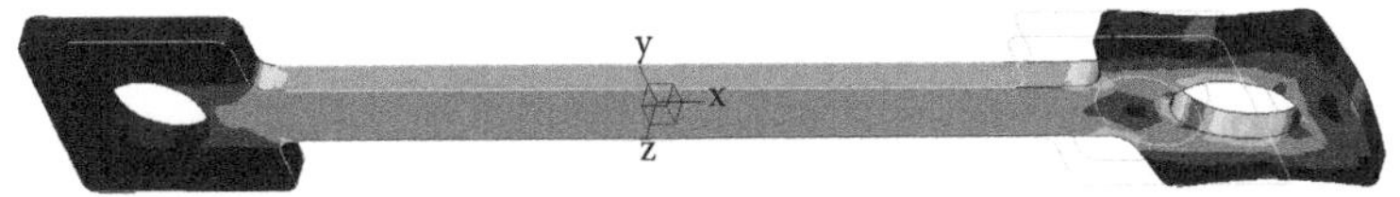

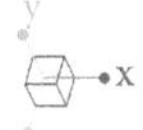

FIGURE 10.22
Results of finite element analysis.

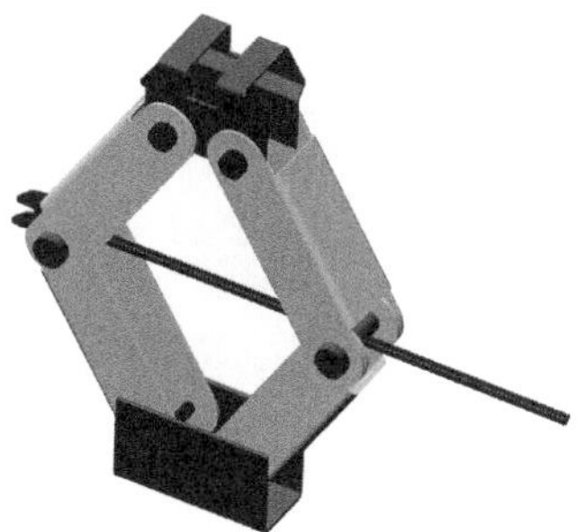

FIGURE 11.18
3-D view of screw jack.

FIGURE 11.47
3-D view of cam mechanism.

FIGURE 11.54
3-D view of universal joint.

FIGURE 12.5
Exploded view of a valve assembly.

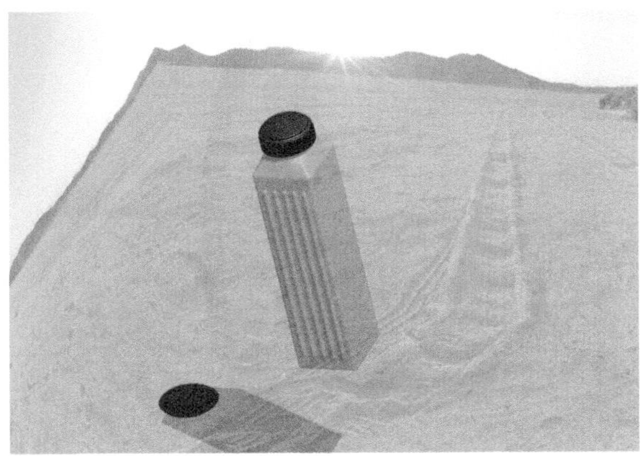

FIGURE 12.7
Example of rendering of plastic bottle.

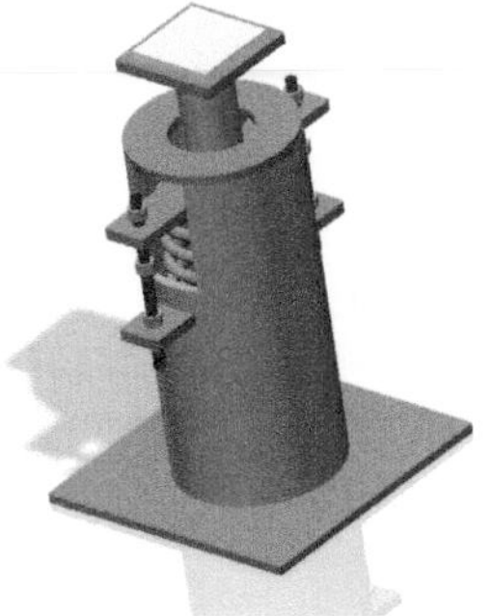

FIGURE 13.25
3-D view of spring support.

FIGURE 13.38
Assembly view of tubes in shell and tube heat exchanger.

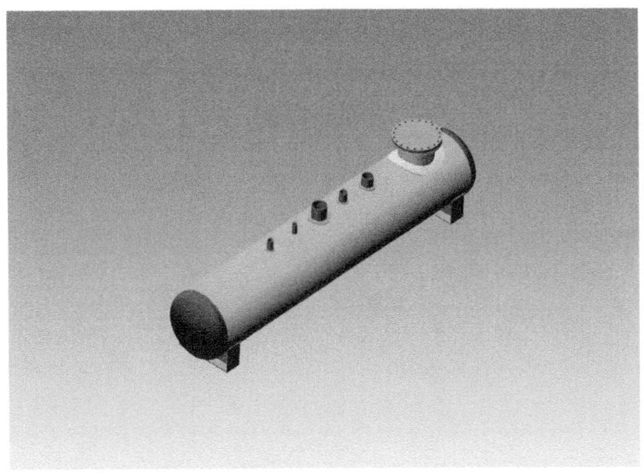

FIGURE 14.4
Example of graphical customization.

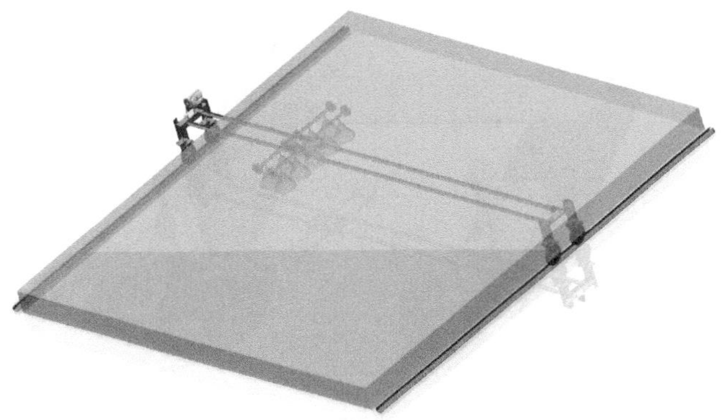

FIGURE 15.1
Industrial spray system of equipment.

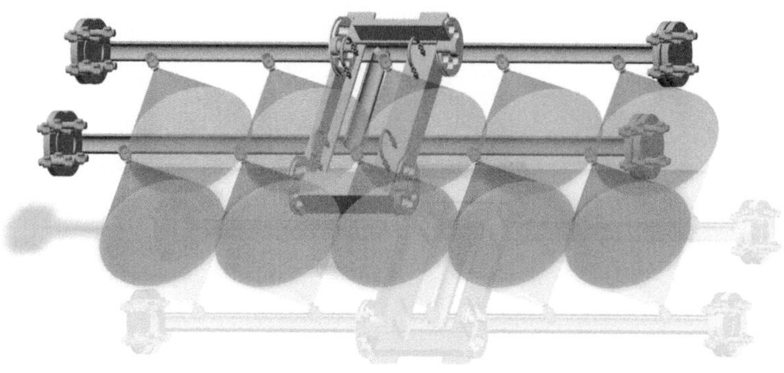

FIGURE 15.79
Three-dimensional view of a Nozzle_Assembly.

FIGURE 15.85
Three-dimensional view of the support.

i. *Rip relief*: Rip reliefs are used when tearing of the part is acceptable on the side of the edge between the edges. The rip is created up to the tangent edge of the internal radius; therefore, the dimension of a rip need not be specified. Figure 9.5 shows rip relief of a flange.

ii. *Stretch*: In stretch relief, the material in the joining section of a flange to an attached sketch is stretched. It is obvious that if the material is manufactured so that a small length of stretch segment on the existing edge is allowed, the material will be stressed more. Therefore, the amount of stretch can be controlled by the CAD software, and this can be manually entered. Figure 9.6 shows stretch relief of flange creation.

iii. *Rectangular*: Rectangular relief is similar to rip relief except for the thickness of the rip. The dimensions of rectangular rips can be manually entered. Rectangular relief can be done by none of the tools in a workshop. Figure 9.7 shows rectangular relief during flange creation.

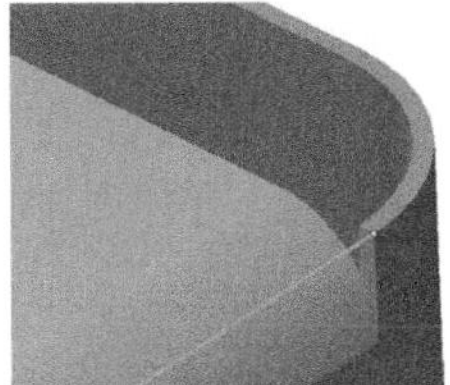

FIGURE 9.5
Example of rip relief.

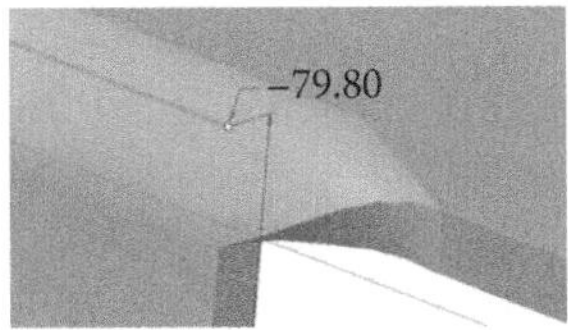

FIGURE 9.6
Example of stretch relief.

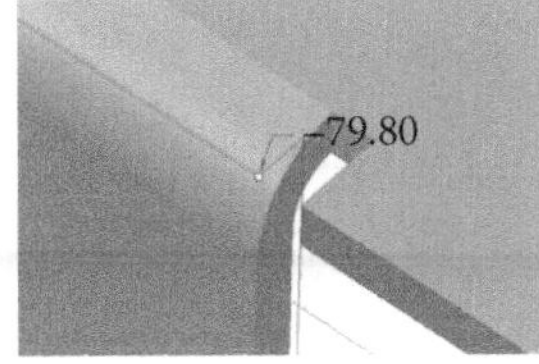

FIGURE 9.7
Example of rectangular relief.

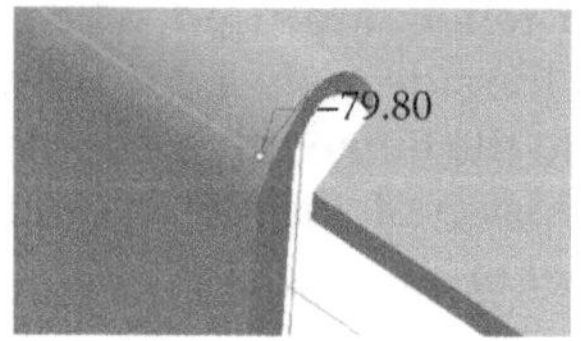

FIGURE 9.8
Example of obround relief.

   iv. *Obround*: Obround relief is similar to a rectangular rip except for the shape of relief which is controlled by the manually entered dimensions. Figure 9.8 shows rectangular relief during flange creation.

b. *Bend allowance*: CAD software allows defining of the bending allowance specific to any feature when creating the feature. It specifically undertakes the value of the Y factor, K factor, or bend table, which controls the developed length of the bend in the feature. If the bend allowance is not specified in the feature, the system considers the bend allowance that is assigned to the part and calculates developed length based on the assigned Y factor or K factor.
c. *Length adjustment*: Length adjustment is the adjustment of the length of the feature that is to be created.
d. *Shape*: Shapes define the shape of the feature. Software also allows us to define a sketched sketch other than predefined shapes.
e. *Radius*: In sheet metal, it is difficult to make a product with sharp edges in the bend portion. Most of the sheet metal product has a minimum bend radius in the inside of the bend edges. CAD also provides options to allow bend radius in the inside edges even if sharp corners exist in the sketch. This option is very useful and shaves additional operations to create fillet at the inside and outside edges of the bend. Also, by providing an automatic radius, one can keep the sketches simple, without too many constants, geometries, and dimensions.

9.3 Operation

1. *Form*: Forming is the operation to create the form in sheet metal that is used to create the drawing or embossing operation in a workshop. When a forming operation is required in a part, it is mandatory to

create a form. A forming operation is made possible by applying a punch and die between which the material will flow and take the shape inside the die and punch clearance. In a forming operation in CAD, the sheet metal can take the exterior shape of the punch or interior shape of the die. The selection of punch or die will depend on the application. Figure 9.9 shows a typical forming operation process.

a. *Punch*: In a punch form operation, the sheet metal takes the exterior shape of the punch. The punch can be modeled in the solid modeling mode. When the punch form is selected, the users are prompted to select the model that will be used as a punch. Certain selections are required to place the punch properly and the form made in the sheet metal part. Normally, punch creation of a punch form is simpler than creation of a die form for complex geometry.

Hands-On Exercise, Example 9.1: Creating a Punch Form (Software Used: Creo Parametric)

PUNCH FORM

Punch form is a very effective sheet metal operation in CAD to produce repetitive operation in various parts in comparatively lesser time. A simple example of using punch form, as illustrated in Figure 9.10, is shown in this example.

Before creating any punch form operation, it is essential to create the punch that has to be pressed on the sheet.

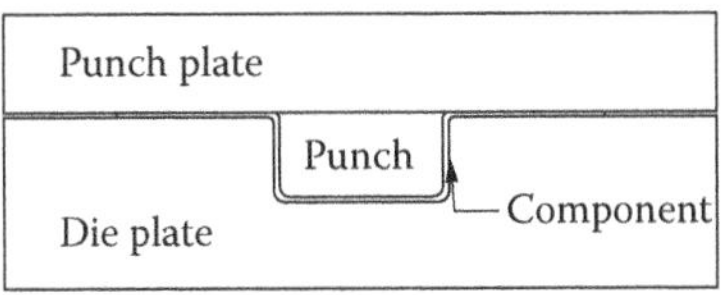

FIGURE 9.9
Technique of forming operation.

FIGURE 9.10
Example of punch form in CAD.

Steps:

- Create a part file and rename it as Punch. (Other names may be given, but it is advisable to provide a relevant name.)
- Select Solid on the right-hand side of the box. This will activate the solid mode and environment.
- Uncheck the box for default template and click OK to create a new sheet metal file.
- The template selection box will appear.
- Select mmns_part_solid. This will allow us to use the sheet metal template file of the following units:
    - mm—millimeter for length unit
    - n—newton for force unit
    - s —second for time unit
- Once the part is created, go to File > Prepare > Model properties to check the units. In the upper sections, the units of the model will be displayed.
    - Note: The units usually are displayed in the summary section of any CAD software. The setting of units usually is in the tools or settings.
- The punch, as shown in Figure 9.11, is to be created.
- Select the Extrude feature from the ribbon at the top. Clicking on the feature opens the selection box for all other options.
- You may change the name of the feature to be created; it helps to trace back the feature in the future so that one easily can understand which feature is used where.
- Click on the Placement tab, and select Define sketch; a predefined sketch also can be selected in this box. However, as no predefined sketch is available now, click on Define.
- As define is clicked, a small toolbox appears for placement of the sketch. Select the top plane for the sketch. The references for orientation (deciding orientation of the sketch plane) may need to be selected if perpendicular planes of the sketch plane are not available. Once the references are available, click on the Sketch button to define the sketch.
- The references to locate the sketch should have been automatically created; however, if the references are not taken, click on the Reference icon at the top left corner and select the references. Perpendicular planes automatically may be

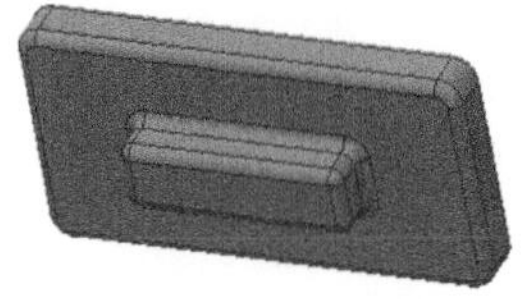

FIGURE 9.11
Example of punch creation.

taken as references. However, axes, points, or other geometries also can be taken as references. For our case, selecting the Right and Front planes for references is not automatically done.

- Select Axis from the sketch ribbon to keep the original sketch symmetrical to the axis. Draw one vertical and one horizontal axis along the references of the right and front planes.
- Once the axes are drawn, select Rectangle from the sketch ribbon, and click on the top left quadrant to locate the left corner. Then click again at the right bottom quadrant to an approximate distance as it can form a symmetrical rectangle with respect to the axes. Automatically, the symmetrical constraint is taken by the system. This practice saves time for creating constraints.
- Two weak dimensions may be displayed if the weak dimension display is activated. The dimensions to be entered are 200 for the horizontal edge dimension and 100 for the vertical edge dimension, by double-clicking the weak dimensions. Select OK.
- When creating any sketch, make sure there is no weak dimension. This ensures that sections or sketches are not overlooked.
- Once finished with the sketch, enter the value of 20 in the blind field. The flip option for changing the direction of extrusion is beside the blind depth value box.
- The blind option also can be selected via the drop-down menu beside Blind.
- Clip on Flip to create the material in the downward direction of the top plane.
- Click OK to create the feature.
- Click on Extrude again from the Model tab in the ribbon interface.
- Create an extrusion of 100 × 30 mm and 20 mm depth by following the extrude creation procedure as mentioned previously.
- The material direction shall be flipped downward.
- Click on Round from the Model tab of the ribbon interface.
- Set the value to 10 mm and select the bottom outer edges of the first extrusion.
- Select Round again.
- Set the value to 5 mm and select all the edges of the second extrusion.
- The punch is prepared now. Save and exit the file.
- Now create a sheet metal file and rename it as Punch form. (Other names may be given, but it is advisable to provide a relevant name.)
- Select Sheet metal on the right-hand side of the box. This will activate the sheet metal mode and environment.
- Uncheck the box for default template and click OK to create a new sheet metal file.
- The template selection box will appear.

- Select mmns_part_sheetmetal. This will allow us to use the sheet metal template file of the following units:
  - mm—millimeter for length unit
  - n—newton for force unit
  - s —second for time unit
- Once the sheet metal part is created, go to File > Prepare > Model properties to check the units. In the upper sections, the units of the model will be displayed.
  - Note: The units usually are displayed in the summary section of any CAD software. The setting of units usually is in the tools or settings.
- Click on Planar wall from the Model tab of the ribbon interface.
- Enter the thickness as 1 mm.
- Click on the Reference tab and select Define to define the sketch.
- The sketch creation box will appear, which allows placing of the sketch with reference and orientation.
- Click on the top plane as the selection of sketch plane.
- Automatically, the reference of the right plane should be taken.
- The orientation shall be right.
- Click on the Sketch button.
- The sketcher window opens up to create a sketch.
- Automatically, the two references of the right plane as vertical and top plane as horizontal should be present. However, if required, references shall be taken.
- Select the centerline from the Sketch tab and create one horizontal and one vertical centerline along the reference line. This will facilitate the creation of mirror geometry.
- Draw a rectangle and enter the dimension of 500 mm in both the fields of weak dimension by double-clicking on them.
- Click OK to create and exit the sketch.
- Now click on Material clip to create the feature in the downward direction.
- Now click on Form from the Model tab in the ribbon interface.
- Click on the File open icon and select Punch from the working directory.
- Assemble the punch by selecting three default planes of punch and punch form sequentially by selecting right top and front planes.
- Assemble in such a way that the top face of the punch part coincides with the top face of the punch form plate. The flip option may be useful in coinciding each of the planes if required.
- Click OK to finish the feature.
- The sheet metal part should get the outside profile of the punch.
- Save and exit the part.

b. *Die*: Die form is similar to punch form except that the material takes the shape of the interior surface of the die and forms in the sheet metal part. When control of the exterior surface in the form is essential, the die form is used.

Hands-On Exercise, Example 9.2: Creating a Die Form (Software Used: Creo Parametric)

DIE FORM

A die form is a very effective sheet metal operation in CAD to produce repetitive operations in various parts in comparatively lesser time. A simple example of using a die, as illustrated in Figure 9.12, is shown in this example.

Before creating any die form operation, it is essential to create the die that has to be pressed on the sheet.

Steps:

- Create a part file and rename it as Die. (Other names may be given, but it is advisable to provide a relevant name.)
- Select Solid on the right-hand side of the box. This will activate the solid mode and environment.
- Uncheck the box for default template and click OK to create a new sheet metal file.
- The template selection box will appear.
- Select mmns_part_solid. This will allow us to use the sheet metal template file of the following units:
    - mm—millimeter for length unit
    - n—newton for force unit
    - s—second for time unit
- Once the part is created, go to File > Prepare > Model properties to check the units. In the upper sections, the units of the model will be displayed.
    - Note: The units usually are displayed in the summary section of any CAD software. The setting of units usually is in the tools or settings.
- The die, as shown in Figure 9.13, is to be created.
- Select the Extrude feature from the ribbon at the top. Clicking on the feature opens the selection box for all other options.
- You may change the name of the feature to be created; it helps to trace back the feature in the future so that one easily can understand which feature is used where.

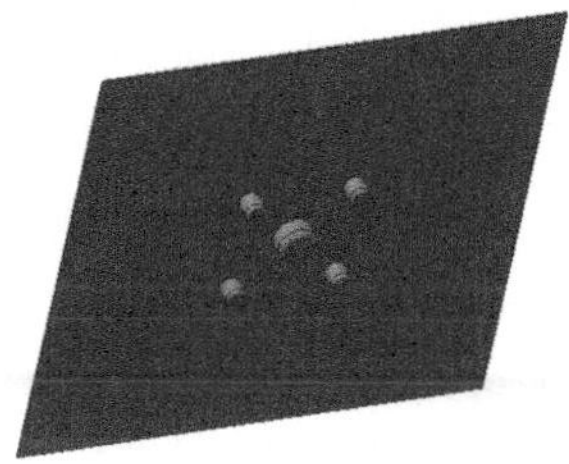

FIGURE 9.12
Example of die form.

FIGURE 9.13
Example of die creation.

- Click on the Placement tab, and select Define sketch. A predefined sketch also can be selected in this box. However, as no predefined sketch is available now, click on define.
- As Define is clicked, a small toolbox appears for placement of the sketch. Select the top plane for the sketch. The references for orientation (deciding orientation of the sketch plane) may need to be selected if perpendicular planes of the sketch plane are not available. Once the references are available, click on the Sketch button to define the sketch.
- The references to locate the sketch should have been automatically created; however, if the references are not taken, click on the Reference icon at the top left corner and select the references. Perpendicular planes automatically may be taken as references. However, axes, points, or other geometries also can be taken as references. For our case, selecting the right and front planes for references is not automatically done.
- Select Axis from the sketch ribbon to keep the original sketch symmetrical to the axis. Draw one vertical and one horizontal axis along the references of the right and front planes.
- Once the axes are drawn, select Rectangle from the sketch ribbon, and click on the top left quadrant to locate the left corner. Then click again at the right bottom quadrant to an approximate distance as it can form a symmetrical rectangle with respect to the axes. Automatically, the symmetrical constraint is taken by the system. This practice saves time for creating constraints.
- Two weak dimensions may be displayed if the weak dimension display is activated. The dimensions to be entered are 200 for the horizontal edge dimension and 200 for the vertical edge dimension, by double-clicking the weak dimensions. Select OK.
- When creating any sketch, make sure there is no weak dimension. This ensures that sections or sketches are not overlooked.
- Once finished with the sketch, enter the value of 50 in the blind field. The flip option for changing the direction of extrusion is beside the blind depth value box.
- The blind option also can be selected via the drop-down menu beside Blind.

- Clip on Flip to create the material in the downward direction of the top plane.
- Click OK to create the feature.
- Click on Extrude again from the Model tab in the ribbon interface.
- Select the top surface as the sketch plane.
- Create a circle of diameter 40 mm in the intersection of reference (front and right plane).
- The material direction shall be flipped downward.
- Click on Round from the Model tab of the ribbon interface.
- Set the value to 5 mm and select the inner circular edge of the newly created blind cut.
- Click on Extrude again from the Model tab in the ribbon interface.
- Select the top surface as the sketch plane.
- Create a circle of diameter 20 mm and 80 mm distance from the center of the previous hole at an angle of 45° with the horizontal or vertical reference line.
- The material direction shall be flipped downward.
- Click on Round from the Model tab of the ribbon interface.
- Set the value to 5 mm and select the inner circular edge of the newly created blind cut.
- Select the last hole and last fillet from the model tree.
- Right-click on them and select Group.
- Select the group, right-click on the mouse, and select Pattern.
- Select Axis in the first pattern type selection box.
- Select the central hole axis.
- Enter the number 4 for the pattern component.
- Click on the Angle icon to activate equally spaced components in 360°.
- The die is prepared now. Save and exit the file.
- Now create a sheet metal file and rename it as Die form. (Other names may be given, but it is advisable to provide a relevant name.)
- Select Sheet metal on the right-hand side of the box. This will activate the sheet metal mode and environment.
- Uncheck the box for default template and click OK to create a new sheet metal file.
- The template selection box will appear.
- Select mmns_part_sheetmetal. This will allow us to use the sheet metal template file of the following units:
    - mm—millimeter for length unit
    - n—newton for force unit
    - s—second for time unit
- Once the sheet metal part is created, go to File > Prepare > Model properties to check the units. In the upper sections, the units of the model will be displayed.
    - Note: The units usually are displayed in the summary section of any CAD software. The setting of units usually is in the tools or settings.

- Click on Planar wall from the Model tab of the ribbon interface.
- Enter the thickness as 1 mm.
- Click on the Reference tab and select Define to define the sketch.
- The sketch creation box will appear, which allows placing of the sketch with reference and orientation.
- Click on the top plane as the selection of sketch plane.
- Automatically, the reference of the right plane should be taken.
- The orientation shall be right.
- Click on the Sketch button.
- The sketcher window opens up to create a sketch.
- Automatically, the two references of the right plane as vertical and top plane as horizontal should be present. However, if required, references shall be taken.
- Select the centerline from the Sketch tab and create one horizontal and one vertical centerline along the reference line. This will facilitate the creation of mirror geometry.
- Draw a rectangle and enter the dimension of 500 mm in both the fields of weak dimension by double-clicking on them.
- Click OK to create and exit the sketch.
- Now click on Material clip to create the feature in the downward direction.
- Now click on Form from the Model tab in the ribbon interface.
- Click on the File open icon and select Die from the working directory.
- Assemble the die by selecting three default planes of the die and die form sequentially by selecting the right top and front planes.
- Assemble in such way that the top face of the die part coincides with the bottom face of the die form plate. The flip option may be useful in coinciding each of the planes if required.
- Click OK to finish the feature.
- The sheet metal part should get the outside profile of the die.
- Save and exit the part.

c. *Notches*: Notches are the cut operation for prespecified shapes with the notching tool. In CAD, notching is used with prespecified template files or user defined features (UDFs). This generates a template form cut.

d. *Rip*: Rip is the feature used to tear the sheet metal part. It provides an edge to a sheet metal part that can be used as reference for development or any other sheet metal operation. Most CAD software contains edge rip, surface rip, and sketch rip.

   i. *Edge rip*: Figure rip shows a sheet metal part having no edge across its developed direction. When the development view is required for this part, the unbend operation is commonly used. The unbend operation cannot determine which line the part is required to tear and generate the development view. In this case, the edge rip feature is used to create a rip on the selected edge. Thus, when the unbend operation is selected, the part has an edge along the edge rip operation, and this is developed as shown in Figure 9.14.

ii. *Sketch rip*: Sketch rip is used to create an edge as sketched. When the desired surface of the developed sheet metal part is such that the selection of an edge is not possible, the sketched rip becomes very useful. The edge can be straight or curved. Figure 9.15 shows an example of a sketch rip.

iii. *Surface rip*: It simply removes a particular selected surface. The surface removed by the rip can be removed by extrude cut operations; however, the surface feature saves the effort and time to make the cut operations. An application of surface rip is shown in Figure 9.16.

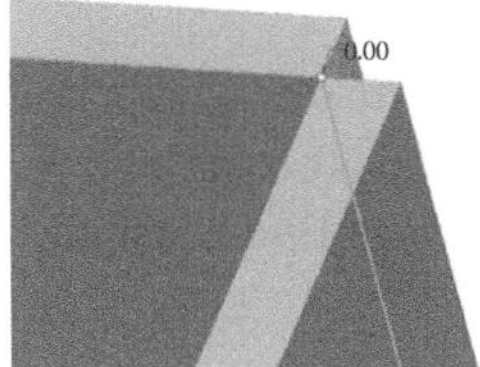

FIGURE 9.14
Example of edge rip.

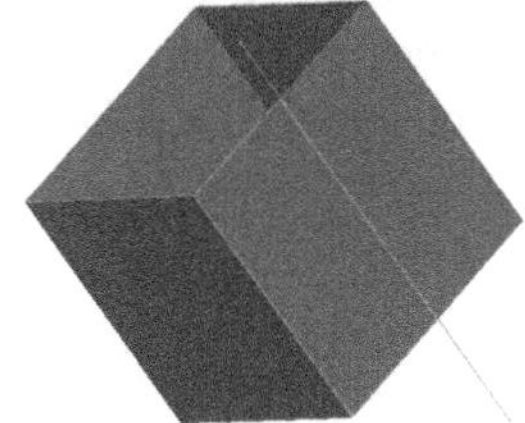

FIGURE 9.15
Example of sketch rip.

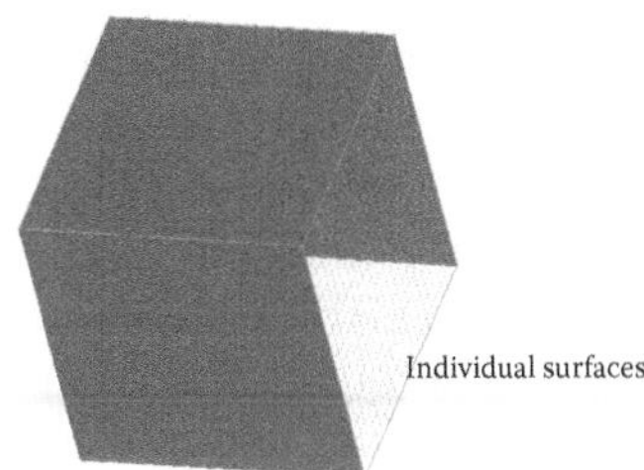

FIGURE 9.16
Example of surface rip.

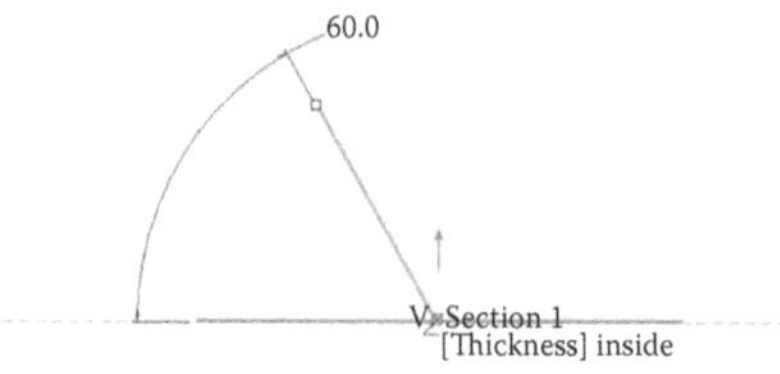

FIGURE 9.17
Example of bending.

e. *Bend/unbend*: The bend operation is done for bending of sheet metal parts along the axis of bending.

   Steps:

   i. A sheet metal model must to be created before the bending operation.
   ii. A flat surface is required where the bending is to be done.
   iii. The bending line is to be defined by the sketch or can be preselected; the bending line must cross the sheet metal boundary surface.
   iv. The required angle of bending needs to be specified along with the bending radius.
   v. Figure 9.17 shows an example of bending.

2. Many times, a bend part is created through extrusion or flange creation to model the final sheet metal part. The unbending operation is particularly useful to generate the developed model. The generated developed model is often used for a blanking operation. Figure 9.1 shows the operation of unbending.

9.4 Application-Based Modeling

Every model creation needs a planning of the sequence of features. Experienced professionals can imagine the process required to create the model of any sheet metal in lesser time; however, starting any complex sheet metal model requires a serious process plan on how to create the model step by step. In the case of solid modeling, the component can be best created by using the operations as needed to create the component in a workshop or practical life. However, in the case of sheet metal part creation, where the finished component is made first and the workshop operations follow in reverse manner to generate the model, a sequence of the process needs to be planned before starting.

1. *Planning*: The planning of feature creation for any sheet metal component depends on the quantity of features to be created and the final component shape. A sheet metal component may be used for optimization, analysis, or sheet metal manufacturing sequence creation.
2. *Model creation*: A model can be prepared by one or a combination of many operations. The designer has to decide the best method of model creation considering the needs. The model shall be created in such a way that it would be easy to change the required shape or size whenever desired.
3. *Checking*: Users many times ignore checking of the final model. Once the model is created, it is always advisable to check the final model for all the dimensions. The dimensions generated from many features may not always be the same as we think. If any dimensions remain to have an undesired value, it affects other features and final dimensions. Therefore, a final checking after preparing the model is needed.

9.5 Project Work

Hands-On Exercise, Example 9.3: Creation and Development of Eccentric Form (Software Used: Creo Parametric)

ECCENTRIC REDUCER

An eccentric reducer is often required in piping or ducting. Most eccentric reducers of small pipe bores are available in the market. However larger sizes of eccentric reducers (more than 500 mm outer diameter [OD]) are often made by plate rolling. This requires development of a reducer sheet. It is a common requirement in industry to provide the development of an eccentric reducer.

Steps:

- Create a part file and rename it as Eccentric reducer. (Other names may be given, but it is advisable to provide a relevant name.)
- Select Sheet metal on the right-hand side of the box. This will activate the sheet metal mode and environment.
- Uncheck the box for default template and click OK to create a new sheet metal file.
- The template selection box will appear.
- Select mmns_part_sheetmetal. This will allow us to use the sheet metal template file of the following units:
  - mm—millimeter for length unit
  - n—newton for force unit
  - s—second for time unit

- Once the sheet metal part is created, go to File > Prepare > Model properties to check the units. In the upper sections, the units of the model will be displayed.
  - Note: The units usually are displayed in the summary section of any CAD software. The setting of units usually is in the tools or settings.
- The eccentric reducer, as shown in Figure 9.18, is to be created.
- An eccentric reducer can be created by many methods like blend, boundary blend, and so forth. However, the most convenient method is shown here.
- To make an eccentric reducer in other CAD software, the loft feature may have to be used.
- Principally, it requires two cross sections with and without a guide curve.
- Since the reducer is of straight form, the influence of a guide curve may not be required.
- Click on Shapes and then blend to create a parallel blend.
- Set the sheet metal thickness value to 10 mm as the eccentric reducer will be made of a 10 mm sheet.
- Click on Sections tab.
- Two options will be seen there, one selected section tab and another sketched section.
  - Selected sections: Selected sections are used when sections are available in the form of a sketch before the creation of blend. This option is used when the sections are robust, critical, or difficult to draw. Since, in any case, the sketches will remain the same, even the blend feature is canceled.
  - Sketched section: This option allows creating of sketches that are to be used in the blend feature. However, even after creating one sketch, if the user decides to cancel the blend feature, the sketch will get lost and will not be available for further use.
- Select sketched sections for our current example.
- In the right-hand side, select Define to create a new sketch.
- The sketch creation box will appear, which allows placing of the sketch with reference and orientation.

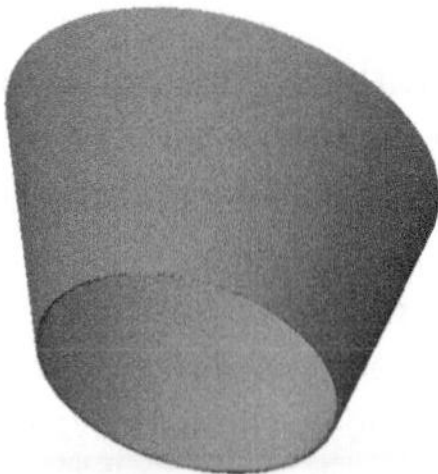

FIGURE 9.18
Example of eccentric reducer.

- Click on Front plane as the selection of sketch plane.
- Automatically, the reference of the right plane should be taken.
- The orientation shall be right.
- Click on the Sketch button.
- The sketcher window opens up to create a sketch.
- Automatically, the two references of the right plane as vertical and top plane as horizontal should be present. However, if required, references shall be taken.
- Click on Circle. Draw a circle, keeping the lower horizontal reference line as tangent in the lower portion of the circle and the center on the vertical reference line above the horizontal line.
- Click on Dimension and provide the diameter of the circle. Enter the diameter value as 1000 mm. The circle automatically changes to 1000 mm diameter keeping the provided constants.
- Click OK to finish the sketch.
- In the right-hand side of the Sections tab, click on Insert. It allows the creation of another section. Enter the value 500 mm in the Offset from section 1 box. The reducer length will be 500 mm.
- Click on the Sketch button.
- The sketcher window opens up to create a sketch.
- Automatically, the two references of the right plane as vertical and top plane as horizontal should be present. However, if required, references shall be taken.
- Click on Circle. Draw a circle, keeping the lower horizontal reference line as tangent in the lower portion of the circle and the center on the vertical reference line above the horizontal line.
- Click on Dimension and provide the diameter of the circle. Enter the diameter value as 700 mm. The circle automatically changes to 700 mm diameter keeping the provided constants.
- Click OK to finish the sketch.
- Click on the flip arrows to change the material direction if required to keep 1000 and 700 mm as outer diameters of the reducer.
- Click on OK to finish making the model of an eccentric reducer.
- Now click on Unbend; it requires a straight edge. However, the straight edge is absent in the eccentric reducer model.
- Now click on the drop-down menu of rip features and select Sketched rip.
- This will rip the material and create a straight edge.
- Click on the Placement tab and click on Define to define the sketch.
- Select the top plane for creating the sketch.
- When the reference window opens up, click on both the diametric edges as references.
- Click on Line to create a line between the two edges at the center.
- Click OK to finish the sketch.
- Now click on Flip to alter the material side.
- Click OK to finish the rip feature.
- Now click on Unbend.
- The edge should be automatically selected. If it is not selected automatically, select the edge along the rip feature.
- Click on Flip to alter the unbend direction if prompted.

- Click on OK to finish unbending.
- The eccentric reducer is developed now.
- Click on the top view to see the developed view.
- Click on the Analysis tab.
- Click on the Measure icon.
- Select the edges that are generated from circular edges of 700 mm diameter, by pressing Control. The final length will be 2158.72 mm. Now the perimeter of the 700 OD circle is 2199.11 mm. The thickness of the eccentric reducer is 10 mm. Hence, the mean diameter of the small circular side is [700 – (2 × 10/2)] = 690 mm. The perimeter of the circle of 690 mm is 2167.7 mm.
- In this type of bend with ample clearance and made by rolling, typically, the neutral plane should be in the middle of the thickness approximately.
- To adjust the neutral plane, click on File > Prepare > Model properties, and click on Y factor change.
- A new preference window shall appear. Click on the Bend allowance tab, and change the factor set to K factor from Y factor and the value to 0.5. Since it is a ratio, it does not have a unit.
- Click on Regenerate.
- Now click on the Analysis tab.
- Click on the Measure icon.
- Select the edges that are generated from circular edges of 700 mm diameter, by pressing control. The final length will be 2169.48 mm. which is approximately the same as calculated for the perimeter of mean diameter.
- Click on the Measure icon.
- Select the edges that are generated from circular edges of 1000 mm diameter, by pressing control. The final length will be 3111.95 mm, which is approximately the same as calculated for the perimeter of mean diameter of the large-diameter side, that is, 3110.17 mm for a 990 mm diameter circle.

Hands-On Exercise, Example 9.4: Creating a Transition Rectangular-to-Rectangular Section (Software Used: Creo Parametric)

TRANSITION PIECE

Steps:

- Create a part file and rename it as Transition piece. (Other names may be given, but it is advisable to provide a relevant name.)
- Select Sheet metal on the right-hand side of the box. This will activate the sheet metal mode and environment.
- Uncheck the box for default template and click OK to create a new sheet metal file.
- The template selection box will appear.

- Select mmns_part_sheetmetal. This will allow us to use the sheet metal template file of the following units:
    - mm—millimeter for length unit
    - n—newton for force unit
    - s—second for time unit
- Once the sheet metal part is created, go to File > Prepare > Model properties to check the units. In the upper sections, the units of the model will be displayed.
    - Note: The units usually are displayed in the summary section of any CAD software. The setting of units usually is in the tools or settings.
- The transition piece, as shown in Figure 9.19, is to be created.
- A transition piece can be created by many methods like blend, boundary blend, and so forth. However, the most convenient method is shown here.
- To make a transition piece in other CAD software, the loft feature may have to be used.
- Principally, it requires two cross sections with and without a guide curve.
- Since the reducer is of straight form, the influence of a guide curve may not be required.
- Click on Shapes and then blend to create a parallel blend.
- Set the sheet metal thickness value to 10 mm as the transition piece will be made of a 10 mm sheet.
- Click on the Sections tab.
- Two options will be seen there, one selected section tab and another sketched section.
    - Selected sections: Selected sections are used when sections are available in the form of a sketch before the creation of blend. This option is used when the sections are robust, critical, or difficult to draw. Since in any case the sketches will remain the same, even the blend feature is canceled.
    - Sketched section: This option allows creating of sketches that are to be used in the blend feature. However, even after creating one sketch, if the user decides to cancel the blend

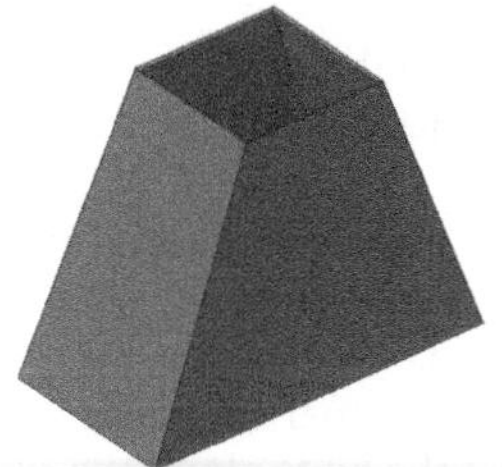

FIGURE 9.19
Example of rectangular-to-rectangular transition piece.

feature, the sketch will get lost and will not be available for further use.

- Select sketched sections for our current example.
- In the right-hand side, select Define to create a new sketch.
- The sketch creation box will appear, which allows placing of the sketch with reference and orientation.
- Click on Front plane as the selection of sketch plane.
- Automatically, the reference of the right plane should be taken.
- The orientation shall be right.
- Click on the Sketch button.
- The sketcher window opens up to create a sketch.
- Automatically, the two references of the right plane as vertical and top plane as horizontal should be present. However, if required, references shall be taken.
- Click on Rectangle. Draw an 800 × 500 rectangle mm having a top edge to horizontal reference distance of 100 mm and left edge to vertical reference distance of 100 mm.
- Right-click on the top left corner and select a start point.
- Click OK to finish and exit the sketch.
- In the right-hand side of the Sections tab, click on Insert. It allows the creation of another section. Enter the value 1000 mm in the Offset from section 1 box. The reducer length will be 500 mm.
- Click on the Sketch button.
- The sketcher window opens up to create a sketch.
- Automatically, the two references of the right plane as vertical and top plane as horizontal should be present. However, if required, references shall be taken.
- Click on Rectangle. Draw a 1000 × 1200 mm rectangle. Place the rectangle centrally with horizontal and vertical references.
- Right-click on the top left corner and select a start point.
- Click OK to finish the sketch.
- Click on the flip arrows to change the material direction so that the center section remains the outer section of the piece.
- Click on OK to finish making the model of a transition piece.
- Now click on Unbend; it requires a straight edge. However, the straight edge is absent in the transition-piece model.
- Now click on the drop-down menu of rip features and select Edge rip.
- This will rip the material along the edge.
- Select any of the transition edges (not any of the edges of any sections).
- Now click on Flip to alter the material side.
- Click OK to finish the rip feature.
- Now click on Unbend.
- The edge should be automatically selected. If it is not selected automatically, select the edge along the rip feature.
- Click on Flip to alter the unbend direction if prompted.
- Click on OK to finish unbending.
- The transition piece is developed now.
- Save and exit the file.

Hands-On Exercise, Example 9.5: Creating an Eccentric Reducer (Fixing a Long-Side Edge; Software Used: Creo Parametric)

ECCENTRIC REDUCER

An eccentric reducer is often required in piping or ducting. Most eccentric reducers of small pipe bores are available in the market. However, larger size of eccentric reducers (more than 500 mm OD) are often made by plate rolling. This requires development of a reducer sheet. It is a common requirement in industry to provide the development of an eccentric reducer.

Steps:

- Create a part file and rename it as Eccentric reducer. (Other names may be given, but it is advisable to provide a relevant name.)
- Select Sheet metal on the right-hand side of the box. This will activate the sheet metal mode and environment.
- Uncheck the box for default template and click OK to create a new sheet metal file.
- The template selection box will appear.
- Select mmns_part_sheetmetal. This will allow us to use the sheet metal template file of the following units:
  - mm—millimeter for length unit
  - n—newton for force unit
  - s—second for time unit
- Once the sheet metal part is created, go to File > Prepare > Model properties to check the units. In the upper sections, the units of the model will be displayed.
  - Note: The units usually are displayed in the summary section of any CAD software. The setting of units usually is in the tools or settings.
- The eccentric reducer, as shown in Figure 9.18, is to be created.
- An eccentric reducer can be created by many methods like blend, boundary blend, and so forth. However, the most convenient method is shown here.
- To make an eccentric reducer in other CAD software, the loft feature may have to be used.
- Principally, it requires two cross sections with and without a guide curve.
- Since the reducer is of straight form, the influence of a guide curve may not be required.
- Click on Shapes and then blend to create a parallel blend.
- Set the sheet metal thickness value to 10 mm as the eccentric reducer will be made of a 10 mm sheet.
- Click on the Sections tab.
- Two options will be seen there, one selected section tab and another sketched section.
  - Selected sections: Selected sections are used when sections are available in the form of a sketch before the creation of blend. This option is used when the sections are robust, critical, or difficult to draw. Since in any case the

sketches will remain the same, even the blend feature is canceled.
    - Sketched section: This option allows creating sketches which are to be used in the blend feature. However even after creating one sketch if user decides to cancel the blend feature, the sketch will get lost and will not be available for further use.
- Select sketched sections for our current example.
- In the right-hand side, select Define to create a new sketch.
- The sketch creation box will appear, which allows placing of the sketch with reference and orientation.
- Click on Front plane as the selection of sketch plane.
- Automatically, the reference of the right plane should be taken.
- The orientation shall be right.
- Click on the Sketch button.
- The sketcher window opens up to create a sketch.
- Automatically, the two references of the right plane as vertical and top plane as horizontal should be present. However, if required, references shall be taken.
- Click on Circle. Draw a circle, keeping the lower horizontal reference line as tangent in the lower portion of the circle and the center on the vertical reference line above the horizontal line.
- Click on Dimension and provide the diameter of the circle. Enter the diameter value as 1000 mm. The circle automatically changes to 1000 mm diameter keeping the provided constants.
- Click OK to finish the sketch.
- In the right-hand side of Sections tab, click on Insert. It allows the creation of another section. Enter the value 500 mm in the Offset from section 1 box. The reducer length will be 500 mm.
- Click on the Sketch button.
- The sketcher window opens up to create a sketch.
- Automatically, the two references of the right plane as vertical and top plane as horizontal should be present. However, if required, references shall be taken.
- Click on Circle. Draw a circle, keeping the lower horizontal reference line as tangent in the lower portion of the circle and the center on the vertical reference line above the horizontal line.
- Click on Dimension and provide the diameter of the circle. Enter the diameter value as 700 mm. The circle automatically changes to 700 mm diameter keeping the provided constants.
- Click OK to finish the sketch.
- Click on the flip arrows to change the material direction if required to keep 1000 and 700 mm as outer diameters of the reducer.
- Click on OK to finish making the model of an eccentric reducer.
- Now click on Unbend; it requires a straight edge. However, the straight edge is absent in the eccentric reducer model.
- Now click on Plane in the Model tab of the ribbon to create a datum plane.

- Select the vertex of two circular edges in the same side along the front plane and the right plane by continuing to press the Control button.
- One new plane to be created perpendicular to the right plane and two selected vertexes.
- Now click on the drop-down menu of rip features and select Sketched rip.
- This will rip the material and create a straight edge.
- Click on the Placement tab and click on Define to define the sketch.
- Select the newly created plane for creating the sketch.
- When the reference window opens up, click on both the diametric edges as references.
- Click on Line to create a line between the two edges at the center.
- Click OK to finish the sketch.
- Now click on Flip to alter the material side.
- Click OK to finish the rip feature.
- Now click on Unbend.
- The edge should be automatically selected. If it is not selected automatically, select the edge along the rip feature.
- Now click on Unbend.
- The edge should be automatically selected. If it is not selected automatically, select the edge along the rip feature.
- Click on Flip to alter the unbend direction if prompted.
- Click on OK to finish unbending.
- The eccentric reducer is developed now.
- Click on the top view to see the developed view.
- Click on the Analysis tab.
- Click on the Measure icon.
- Select the edges that are generated from circular edges of 700 mm diameter, by pressing control. The final length will be 2158.72 mm. Now the perimeter of the 700 OD circle is 2199.11 mm. The thickness of the eccentric reducer is 10 mm. Hence, the mean diameter of the small circular side is [700 – (2 × 10/2)] = 690 mm. The perimeter of the circle of 690 mm is 2167.7 mm.
- In this type of bend with ample clearance and made by rolling, typically, the neutral plane should be in the middle of the thickness approximately.
- To adjust the neutral plane, click on File > Prepare > Model properties, and click on Y factor change.
- A new preference window shall appear. Click on the Bend allowance tab, and change the factor set to K factor from Y factor and the value to 0.5. Since it is a ratio, it does not have a unit.
- Click on Regenerate.
- Now click on the Analysis tab.
- Click on the Measure icon.
- Select the edges that are generated from circular edges of 700 mm diameter, by pressing control. The final length will be 2169.48

mm, which is approximately the same as calculated for the perimeter of mean diameter.

- Click on the Measure icon.
- Select the edges that are generated from circular edges of 1000 mm diameter, by pressing control. The final length will be 3111.95 mm, which is approximately the same as calculated for the perimeter of mean diameter of the large-diameter side, that is, 3110.17 mm for a 990 mm diameter circle.

Hands-On Exercise, Example 9.6: Creating File Clip 1 (Software Used: Creo Parametric)

FILE CLIP

Steps:

- Create a part file and rename it as File_Clip_1. (Other names may be given, but it is advisable to provide a relevant name.)
- Select Sheet metal on the right-hand side of the box. This will activate the sheet metal mode and environment.
- Uncheck the box for default template and click OK to create a new sheet metal file.
- The template selection box will appear.
- Select mmns_part_sheetmetal. This will allow us to use the sheet metal template file of the following units:
    - mm—millimeter for length unit
    - n—newton for force unit
    - s—second for time unit
- Once the sheet metal part is created, go to File > Prepare > Model properties to check the units. In the upper sections, the units of the model will be displayed.
    - Note: The units usually are displayed in the summary section of any CAD software. The setting of units usually is in the tools or settings.
- The file clip, as shown in Figure 9.20, is to be created.
- The file clip will be created by the planar wall feature followed by various bends and the extrude cut feature.
- Click on the planar wall creation feature in the Model tab from the ribbon interface.
- Enter the thickness as 0.15 mm.
- Click on the Reference tab and select Define to define the sketch.

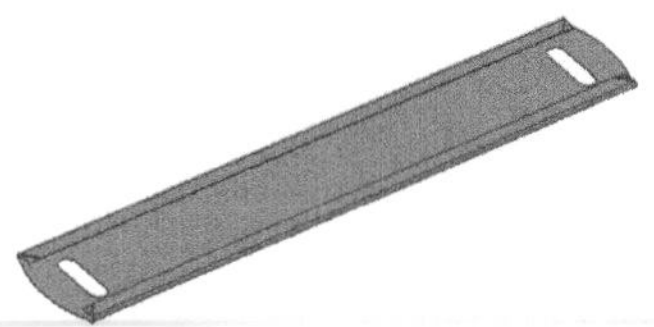

FIGURE 9.20
3-D model of a file clip's part.

- The sketch creation box will appear, which allows placing of the sketch with reference and orientation.
- Click on the top plane as the selection of sketch plane.
- Automatically, the reference of the right plane should be taken.
- The orientation shall be right.
- Click on the Sketch button.
- The sketcher window opens up to create a sketch.
- Automatically, the two references of the right plane as vertical and top plane as horizontal should be present. However, if required, references shall be taken.
- Select the centerline from the Sketch tab and create one horizontal and one vertical centerline along the reference line. This will facilitate the creation of mirror geometry.
- Draw the sketch as shown in Figure 9.21.
- Click OK to finish the sketch.
- Click on the Flip icon to alter the feature creation side; it shall be toward the bottom of the top plane.
- Click OK to finish the sketch.
- Now select Flange from the Model tab of the ribbon interface.
- Click on the Placement tab and select the long bottom edge. This will allow the creation of a feature at the top side.
- Select the shape as User defined in the first shape selection drop-down menu.
- Click on the Shape tab and click on Sketch.
- When the sketch window appears, select Sketch.
- Draw the sketch as shown in Figure 9.22.
- Once finished, click on OK to exit the sketcher window.

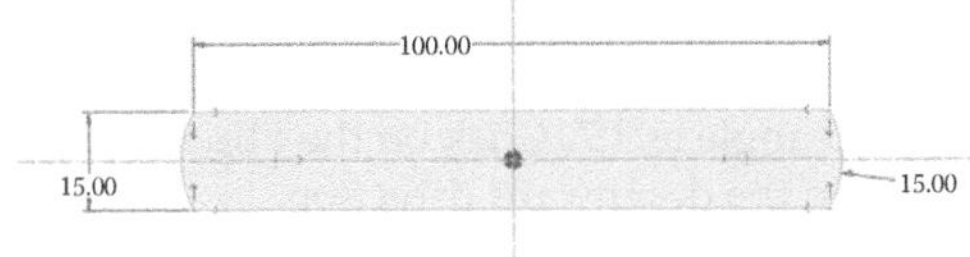

FIGURE 9.21
Constructional sketch of a file clip's part.

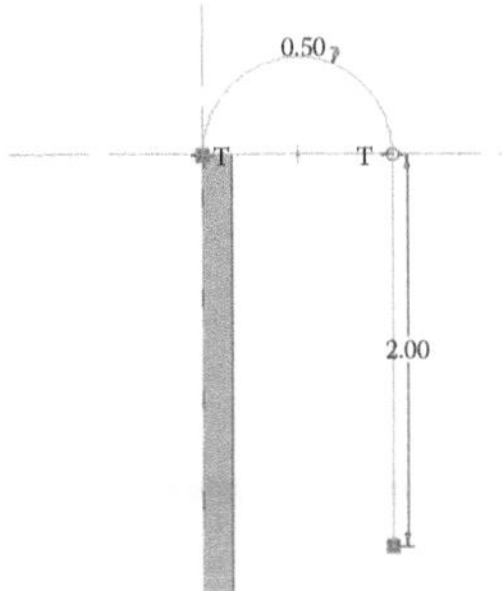

FIGURE 9.22
Constructional sketch of flange of a file clip.

- Now in the Sketch tab, click on the Save as button to save the profile. Save as "file clip flange 1."
- Click OK to create the flange as drawn.
- Now select Flange again from the Model tab of the ribbon interface.
- Click on the Placement tab and select the long bottom edge of the other side. This will allow the creation of features at the top side.
- Select the shape "file clip flange 1" in the first shape selection drop-down menu. This will create a symmetrical flange. This can also be created by mirroring the feature about the front plane.
- Select Extrude from the Model tab of the ribbon interface.
- Activate the On cut icon to make the feature for material removal.
- Click on the Placement tab and click on Define to define the sketch.
- Select the top face of the planar wall as the sketch plane to create the cut.
- Create two oblong holes of 9 × 3 mm, 90 mm apart, placed centrally.
- Click OK to finish the sketch.
- Click on the Flip icon to alter the material removal direction if required.
- Click OK to finish the extrude cut feature.
- Select Bend from the Model tab of the ribbon interface.
- Click on the Placement tab and select the upper face of new created user-defined flanges.
- Click on the Bend line tab and select Sketch to define the bend line.
- Create the sketch as shown in Figure 9.23.
- Click OK to finish the sketch.
- Enter the bend angle as 15°. Click on the Flip icon if required to make it bend in the downward direction.
- Enter the radius as 1 mm.
- Click OK to finish the edge bend.
- Similarly, create another three bends of the remaining three corners.
- Click on Unbend from the Model tab of the ribbon interface.

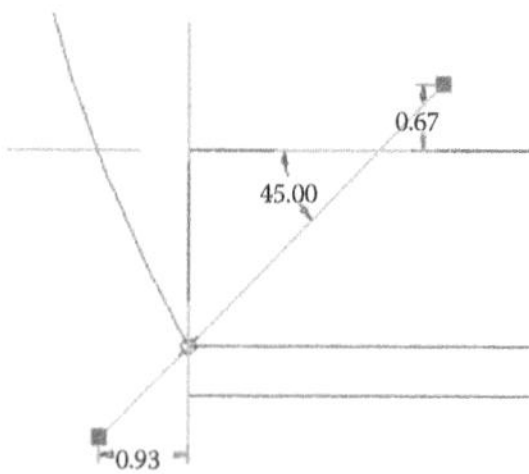

FIGURE 9.23
Constructional sketch of features of a file clip.

- Select the face of the planar wall to set the planar wall flat during unbending.
- Click OK to unbend for development view.
- Click on file, save, and exit.

Hands-On Exercise, Example 9.7: Creating a File Clip, Part 2 (Software Used: Creo Parametric)

FILE CLIP 2

Steps:

- Create a part file and rename it as File_Clip_2. (Other names may be given, but it is advisable to provide a relevant name.)
- Select Sheet metal on the right-hand side of the box. This will activate the sheet metal mode and environment.
- Uncheck the box for default template and click OK to create a new sheet metal file.
- The template selection box will appear.
- Select mmns_part_sheetmetal. This will allow us to use the sheet metal template file of the following units:
    - mm—millimeter for length unit
    - n—newton for force unit
    - s—second for time unit
- Once the sheet metal part is created, go to File > Prepare > Model properties to check the units. In the upper sections, the units of the model will be displayed.
    - Note: The units usually are displayed in the summary section of any CAD software. The setting of units usually is in the tools or settings.
- The file clip, as shown in Figure 9.24, is to be created.
- The file clip will be created by the planar wall feature followed by various bends and the extrude cut feature.
- Click on the planar wall creation feature in the Model tab from the ribbon interface.
- Enter the thickness as 0.15 mm.
- Click on the Reference tab and select Define to define the sketch.
- The sketch creation box will appear, which allows placing of the sketch with reference and orientation.

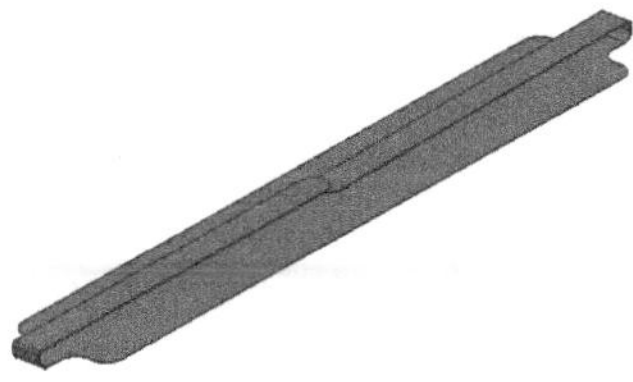

FIGURE 9.24
3-D model of a file clip's part.

- Click on the top plane as the selection of sketch plane.
- Automatically, the reference of the right plane should be taken.
- The orientation shall be right.
- Click on the Sketch button.
- The sketcher window opens up to create a sketch.
- Automatically, the two references of the right plane as vertical and top plane as horizontal should be present. However, if required, references shall be taken.
- Select the centerline from the Sketch tab and create one horizontal and one vertical centerline along the reference line. This will facilitate the creation mirror geometry.
- Draw the sketch as shown in Figure 9.25.
- Click OK to finish the sketch.
- Click on the Flip icon to alter the feature creation side; it shall be toward the bottom of the top plane.
- Click Ok to finish the sketch.
- Now click on Bend in the Model tab from the ribbon interface.
- Click on the Placement tab and select the bottom face of the part as the sketch plane.
- Click on the Bend line tab and click on Sketch.
- The sketcher interface will activate on the selected sketch plane.
- Draw a vertical line at distance of (45 – 1 – 0.15 = 43.85) from the centerline. The line should exceed the part boundaries.
- Click OK to exit the sketch.
- Similarly, create a bend on another side.
- Now click again on Bend in the Model tab of the ribbon interface.
- Click on the Placement tab and select the outer face of any of the newly bent flanges of the part as the sketch plane.
- Click on the Bend line tab and click on Sketch.
- The sketcher interface will activate on the selected sketch plane.
- Draw a horizontal line at distance of 2 mm from the bottom face of the part. The line should exceed the part boundaries.
- Click OK to exit the sketch.
- Select the outer face of any of the newly bent flanges.
- Similarly, create a similar bend in the remaining flange, except that the bend angle should be 91° in order to avoid the fouling of flanges.
- Save and exit the part.

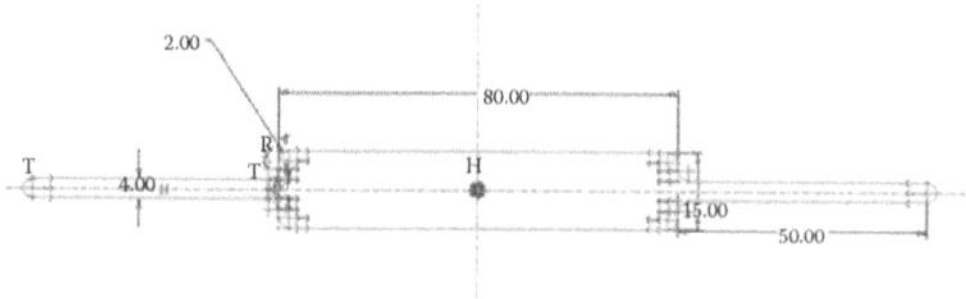

FIGURE 9.25
Constructional sketch of a file clip's part.

Hands-On Exercise, Example 9.8: Creating a Card Clip (Software Used: Creo Parametric)

CARD CLIP

Steps:

- Create a part file and rename it as card_Clip_1. (Other names may be given, but it is advisable to provide a relevant name.)
- Select Sheet metal on the right-hand side of the box. This will activate the sheet metal mode and environment.
- Uncheck the box for default template and click OK to create a new sheet metal file.
- The template selection box will appear.
- Select mmns_part_sheetmetal. This will allow us to use the sheet metal template file of the following units:
    - mm—millimeter for length unit
    - n—newton for force unit
    - s—second for time unit
- Once the sheet metal part is created, go to File > Prepare > Model properties to check the units. In the upper sections, the units of the model will be displayed.
    - Note: The units usually are displayed in the summary section of any CAD software. The setting of units usually is in the tools or settings.
- The file clip, as shown in Figure 9.26, is to be created.
- The file clip will be created by the planar wall feature followed by various bends and the extrude cut feature.
- Click on the planar wall creation feature in the Model tab from the ribbon interface.
- Enter the thickness as 0.10 mm.
- Click on the Reference tab and select Define to define the sketch.
- The sketch creation box will appear, which allows placing of the sketch with reference and orientation.
- Click on the top plane as the selection of sketch plane.
- Automatically, the reference of the right plane should be taken.
- The orientation shall be right.
- Click on the Sketch button.
- The sketcher window opens up to create a sketch.

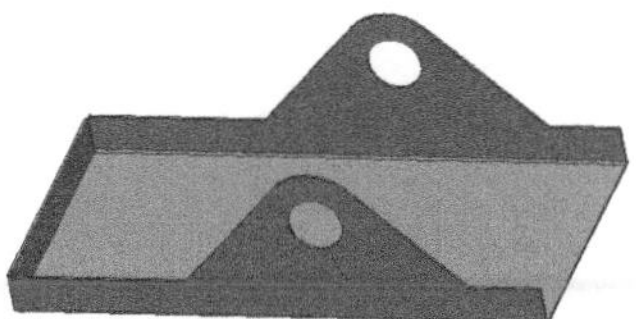

FIGURE 9.26
3-D model of a card clip's part.

- Automatically, the two references of the right plane as vertical and top plane as horizontal should be present. However, if required, references shall be taken.
- Select the centerline from the sketch tab and create one horizontal and one vertical centerline along the reference line. This will facilitate the creation of mirror geometry.
- Draw a rectangle of 20 × 10 mm.
- Click OK to finish the sketch.
- Click on the Flip icon to alter the feature creation side; it shall be toward the bottom of the top plane.
- Click OK to finish the sketch.
- Now select Flat from the Model tab of the ribbon interface.
- Click on the Placement tab and select the long bottom edge. This will allow the creation of features at the top side.
- Select the shape as User defined in the first shape selection drop-down menu.
- Click on the Shape tab and click on Sketch.
- When the sketch window appears, select Sketch.
- Draw the sketch as shown in Figure 9.27.
- Once finished, click on OK to exit the sketcher window.
- Now in the Sketch tab, click on the Save as button to save the profile. Save as "file clip flange 1."
- Click OK to create the flange as drawn.
- Now select Flange again from the Model tab of the ribbon interface.
- Click on the Placement tab and select the long bottom edge of the other side. This will allow the creation of features at the top side.
- Create a similar flange by projecting the edges of the existing flange.
- Select Extrude from the Model tab of the ribbon interface.
- Activate the On cut icon to make the feature for material removal.
- Click on the Placement tab and click on Define to define the sketch.
- Select the outer face of any newly created flat as the sketch plane to create the cut.
- Create a hole of diameter 2 mm in the center of the arc by taking the references of the arc.
- Click OK to finish the sketch.

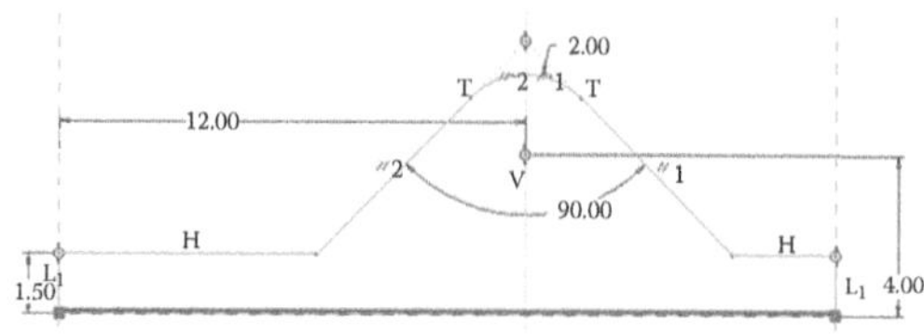

FIGURE 9.27
Constructional sketch of a card clip's part.

- Click on the Flip icon to alter the material removal direction if required.
- Select through all options in the drop-down menu of the feature creation length field.
- Click OK to finish the extrude cut feature.
- Select Flange from the Model tab of the ribbon interface.
- Select the bottom short edge of the planar wall (furthest edge from the hole center to be selected).
- Select the shape as "I" in the shape selection field.
- Click on the Shape tab and enter the value 1.5 for flange length by double-clicking on the dimension.
- Set the bend angle at 90° in the Shape tab.
- Set the internal bend radius as Thickness.
- Click OK to create the flange feature.
- Click on Unbend from the Model tab of the ribbon interface.
- Select the first planar wall top surface as fixed geometry.
- Select the automatic option for selecting all the bends.
- Click OK to finish the unbend feature.
- Save and exit the file.

Hands-On Exercise, Example 9.9: Creating a Board Clip (Software Used: Creo Parametric)

BOARD CLIP

Steps:

- Create a part file and rename it as Board_Clip_1. (Other names may be given, but it is advisable to provide a relevant name.)
- Select Sheet metal on the right-hand side of the box. This will activate the sheet metal mode and environment.
- Uncheck the box for default template and click OK to create a new sheet metal file.
- The template selection box will appear.
- Select mmns_part_sheetmetal. This will allow us to use the sheet metal template file of the following units:
    - mm—millimeter for length unit
    - n—newton for force unit
    - s—second for time unit
- Once the sheet metal part is created, go to File > Prepare > Model properties to check the units. In the upper sections, the units of the model will be displayed.
    - Note: The units usually are displayed in the summary section of any CAD software. The setting of units usually is in the tools or settings.
- The board clip, as shown in Figure 9.28, is to be created.
- The board clip will be created by the extrude wall feature followed by various bends.
- Click on the extrude feature creation feature in the Model tab from the ribbon interface.

FIGURE 9.28
3-D model of a board clip's part.

- Enter the thickness as 0.3 mm.
- Select the Mid plane extrusion option and set the value to 40 mm.
- Click on the Placement tab and select Define to define the sketch.
- The sketch creation box will appear, which allows placing of the sketch with reference and orientation.
- Click on Front plane as the selection of sketch plane.
- Automatically, the reference of the right plane should be taken.
- The orientation shall be right.
- Click on the Sketch button.
- The sketcher window opens up to create a sketch.
- Automatically, the two references of the right plane as vertical and top plane as horizontal should be present. However, if required, references shall be taken.
- Select the centerline from the sketch tab and create one horizontal and one vertical centerline along the reference line. This will facilitate the creation of mirror geometry.
- Draw the sketch as shown in Figure 9.29.
- Click OK to finish the sketch.
- Click on the Flip icon to alter the feature creation side; it shall be toward the bottom of the top plane.
- Click OK to finish the sketch.
- Click OK to create the extrude feature.
- Now select Flat from the Model tab of the ribbon interface.
- Click on the Placement tab and select the top long bottom edge. This will allow the creation of features at the top side.

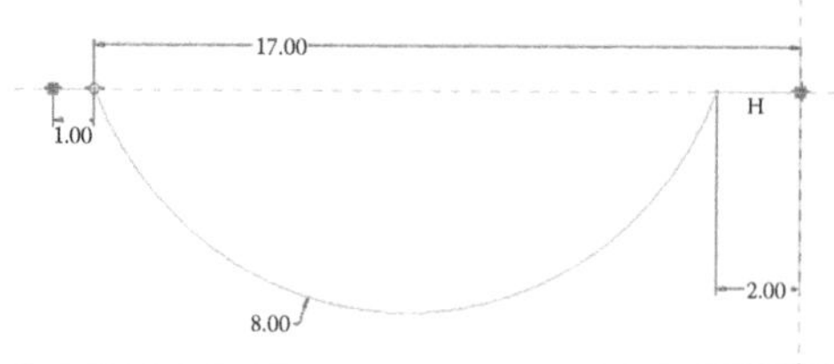

FIGURE 9.29
Constructional sketch of a board clip's part.

- Set the bend angle to 30°.
- Set the bend radius as 1 mm.
- Select the shape as User defined in the first shape selection drop-down menu.
- Click on the Shape tab and click on Sketch.
- When the sketch window appears, select Sketch.
- Draw the sketch as shown in Figure 9.30.
- Once finished, click on OK to exit the sketcher window.
- Now in the Sketch tab click, on the Save as button to save the profile. Save as "board clip flange 1."
- Click OK to create the flange as drawn.
- Now select Flat again from the Model tab of the ribbon interface.
- Click on the Placement tab and select the long edge of the newly created flat. This will allow the creation of features at the top side.
- Set the bend angle to 90°.
- Set the bend radius as Thickness.
- Select the shape as User defined in the first shape selection drop-down menu.
- Click on the Shape tab and click on Sketch.
- When the sketch window appears, select Sketch.
- Draw the sketch as shown in Figure 9.31.
- Click OK to finish the sketch.
- Click OK to create the flat.
- Similarly, create the same flat feature on the other side of the flange.
- Select Extrude cut from the Model tab in the ribbon interface.
- Select the outer face of the newly created flange as the sketch plane.

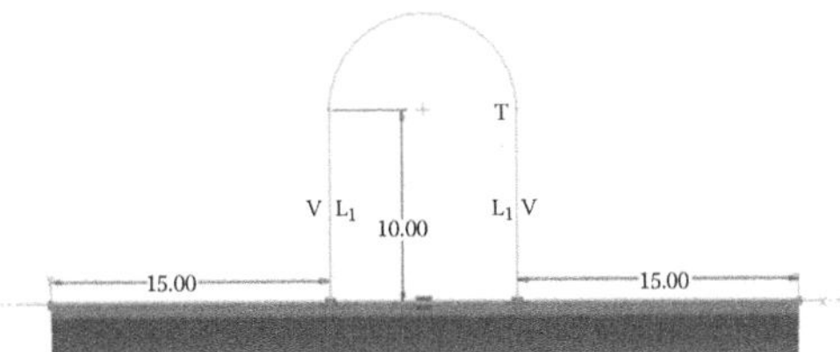

FIGURE 9.30
Constructional sketch of flange of a board clip.

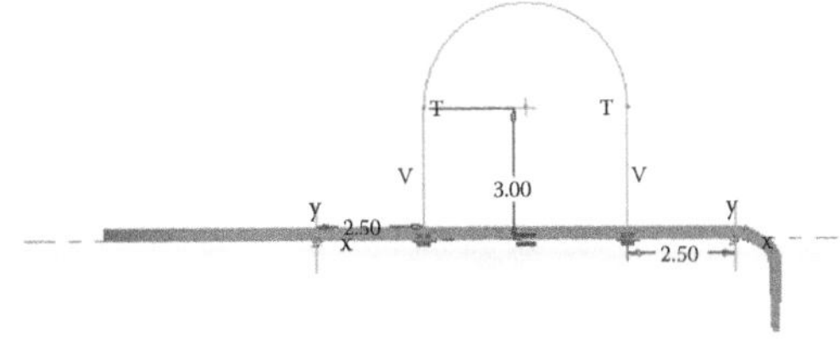

FIGURE 9.31
Constructional sketch of flange of a board clip's part.

- Draw a circle of diameter 2 mm at the center of the arc.
- Click OK to create the sketch.
- Select through all options in the depth selection field.
- Click OK to create the cut feature.

Hands-On Exercise, Example 9.10: Creating a Paper Clip (Software Used: Creo Parametric)

PAPER CLIP

Steps:

- Create a part file and rename it as Paper_Clip_1. (Other names may be given, but it is advisable to provide a relevant name.)
- Select Sheet metal on the right-hand side of the box. This will activate the sheet metal mode and environment.
- Uncheck the box for default template and click OK to create a new sheet metal file.
- The template selection box will appear.
- Select mmns_part_sheetmetal. This will allow us to use the sheet metal template file of the following units:
    - mm—millimeter for length unit
    - n—newton for force unit
    - s—second for time unit
- Once the sheet metal part is created, go to File > Prepare > Model properties to check the units. In the upper sections, the units of the model will be displayed.
    - Note: The units usually are displayed in the summary section of any CAD software. The setting of units usually is in the tools or settings.
- The paper clip, as shown in Figure 9.32, is to be created.
- The paper clip will be created by the planar wall feature followed by various bends and the extrude cut feature.
- Click on the planar wall creation feature in the Model tab from the ribbon interface.
- Enter the thickness as 0.3 mm.
- Click on the Reference tab and select Define to define the sketch.

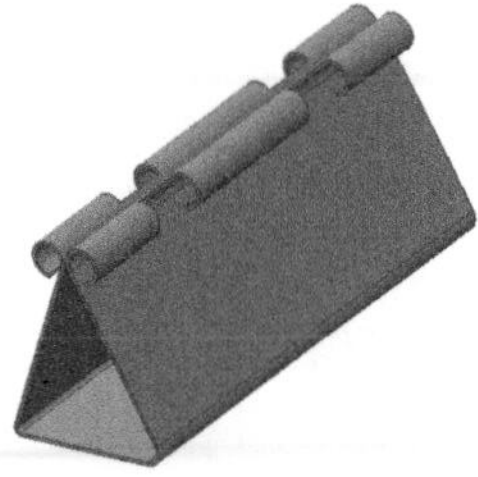

FIGURE 9.32
3-D model of a paper clip's part.

- The sketch creation box will appear, which allows placing of the sketch with reference and orientation.
- Click on the top plane as the selection of sketch plane.
- Automatically, the reference of the right plane should be taken.
- The orientation shall be right.
- Click on the Sketch button.
- The sketcher window opens up to create a sketch.
- Automatically, the two references of the right plane as vertical and top plane as horizontal should be present. However, if required, references shall be taken.
- Select the centerline from the sketch tab and create one horizontal and one vertical centerline along the reference line. This will facilitate the creation of mirror geometry.
- Draw a rectangle of 40 × 15 mm.
- Click OK to finish the sketch.
- Click on the Flip icon to alter the feature creation side; it shall be toward the bottom of the top plane.
- Click OK to finish the sketch.
- Now select Flange from the Model tab of ribbon interface.
- Click on the Placement tab and select the long bottom edge. This would allow the creation of features at the top side.
- Select the shape as User defined in the first shape selection drop-down menu.
- Click on the Shape tab and click on Sketch.
- When the sketch window appears, select Sketch.
- Draw the sketch as shown in Figure 9.33.
- Once finished, click on OK to exit the sketcher window.
- Now in the Sketch tab, click on the Save as button to save the profile. Save as "paper clip flange 1."
- Click OK to create the flange as drawn.
- Now select Flange again from the Model tab of the ribbon interface.
- Click on the Placement tab and select the long bottom edge of the other side. This will allow the creation of features at the top side.

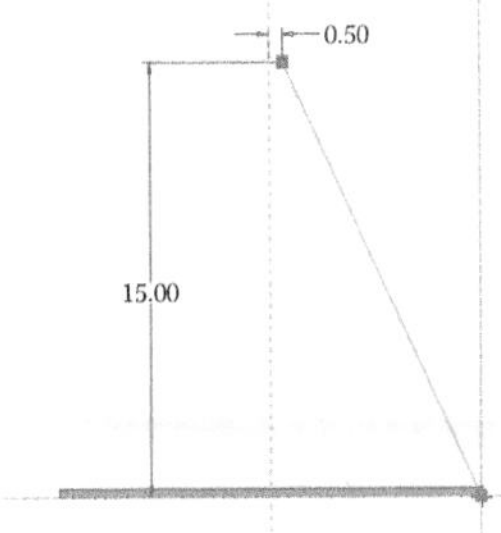

FIGURE 9.33
Constructional sketch of a paper clip's part.

- Select the shape "paper clip flange 1" in the first shape selection drop-down menu. This will create a symmetrical flange. This can also be created by mirroring the feature about the front plane.
- Click on the Flip icon to alter the material removal direction if required.
- Click OK to finish the extrude cut feature.
- Select Bend from the Model tab of the ribbon interface.
- Click on the Placement tab and select the upper face of new created user-defined flanges.
- Click on the Bend line tab and select Sketch to define the bend line.
- Create the sketch as shown in Figure 9.34.
- Click OK to finish the sketch.
- Now in the Sketch tab, click on the Save as button to save the profile. Save as "paper clip flange 2."
- Click OK to finish the edge bend.
- Similarly, create another bend on the other edge by the previously mentioned procedure and selecting "paper clip flange 2."
- Click on Unbend from the Model tab of the ribbon interface.
- Select the top face of the first planar wall to keep it fixed during the unbend operation.
- Select the auto reference to select all the bends for the unbend operation.
- Select Extrude cut from the Model tab in the ribbon interface.
- Create the sketch as per Figure 9.35.

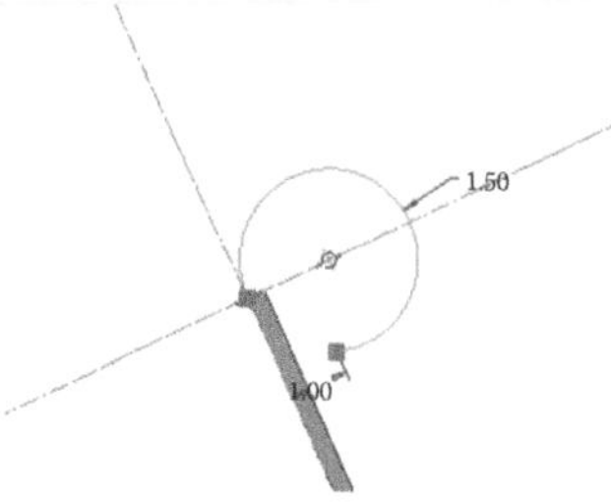

FIGURE 9.34
Constructional sketch of a paper clip's part.

FIGURE 9.35
3-D model of a concentric reducer.

Hands-On Exercise, Example 9.11: Creating a Concentric Reducer (Software Used: Creo Parametric)

CONCENTRIC REDUCER

A concentric reducer is often required in piping or ducting. Most concentric reducers of small pipe bores are available in the market. However, larger sizes of concentric reducers (more than 500 mm OD) are often made by plate rolling. This requires development of a reducer sheet. It is a common requirement in industry to provide the development of a concentric reducer.

Steps:

- Create a part file and rename it as Concentric reducer. (Other names may be given, but it is advisable to provide a relevant name.)
- Select Sheet metal on the right-hand side of the box. This will activate the sheet metal mode and environment.
- Uncheck the box for default template and click OK to create a new sheet metal file.
- The template selection box will appear.
- Select mmns_part_sheetmetal. This will allow us to use the sheet metal template file of the following units:
    - mm—millimeter for length unit
    - n—newton for force unit
    - s—second for time unit
- Once the sheet metal part is created, go to File > Prepare > Model properties to check the units. In the upper sections, the units of the model will be displayed.
    - Note: The units usually are displayed in the summary section of any CAD software. The setting of units usually in the tools or settings.
- The concentric reducer, as shown in Figure 9.35, is to be created.
- A concentric reducer can be created by many methods like blend, boundary blend, revolve, and so forth. However, the most convenient method is shown here.
- To make a concentric reducer in other CAD software, the loft feature may have to be used.
- Principally, it requires two cross sections with and without a guide curve.
- Since the reducer is of straight form, the influence of a guide curve may not be required.
- Click on Shapes and then blend to create a parallel blend.
- Set the sheet metal thickness value to 10 mm as the concentric reducer will be made of a 10 mm sheet.
- Click on the Sections tab.
- Two options will be seen there, one selected section tab and another sketched section.
    - Selected sections: Selected sections are used when sections are available in the form of a sketch before the creation of blend. This option is used when the sections are robust,

critical, or difficult to draw. Since in any case, the sketches will remain the same, even the blend feature is canceled.

  - Sketched section: This option allows creating sketches which are to be used in the blend feature. However even after creating one sketch if user decides to cancel the blend feature, the sketch will get lost and will not be available for further use.

- Select sketched sections for our current example.
- In the right-hand side, select Define to create a new sketch.
- The sketch creation box will appear, which allows placing of the sketch with reference and orientation.
- Click on Front plane as the selection of sketch plane.
- Automatically, the reference of the right plane should be taken.
- The orientation shall be right.
- Click on the Sketch button.
- The sketcher window opens up to create a sketch.
- Automatically, the two references of the right plane as vertical and top plane as horizontal should be present. However, if required, references shall be taken.
- Click on Circle. Draw a circle, keeping the center point in the intersection of the horizontal and vertical reference lines.
- Click on Dimension and provide the diameter of the circle. Enter the diameter value as 900 mm. The circle automatically changes to 900 mm diameter keeping the provided constants.
- Click OK to finish the sketch.
- In the right-hand side of the Sections tab, click on Insert. It allows the creation of another section. Enter the value 400 mm in the Offset from section 1 box. The reducer length will be 400 mm.
- Click on the Sketch button.
- The sketcher window opens up to create a sketch.
- Automatically, the two references of the right plane as vertical and top plane as horizontal should be present. However, if required, references shall be taken.
- Click on Circle. Draw a circle, keeping the center point in the intersection of the horizontal and vertical reference lines.
- Click on Dimension and provide the diameter of the circle. Enter the diameter value as 500 mm. The circle automatically changes to 500 mm diameter keeping the provided constants.
- Click OK to finish the sketch.
- Click on the flip arrows to change the material direction if required to keep 900 and 500 mm as outer diameters of the reducer.
- Click on OK to finish making the model of a concentric reducer.
- Now click on Unbend; it requires a straight edge. However, the straight edge is absent in the concentric reducer model.
- Now click on Plane in the Model tab of the ribbon to create a datum plane.
- Select the vertex of two circular edges in the same side along the front plane and the right plane by continuing to press the Control button.
- One new plane to be created perpendicular to the right plane and two selected vertexes.

- Now click on the drop-down menu of rip features and select Sketched rip.
- This will rip the material and create a straight edge.
- Click on the Placement tab and click on Define to define the sketch.
- Select the newly created plane for creating the sketch.
- When the reference window opens up, click on both the diametric edges as references.
- Click on Line to create a line between the two edges at the center.
- Click OK to finish the sketch.
- Now click on Flip to alter the material side.
- Click OK to finish the rip feature.
- Now click on Unbend.
- The edge should be automatically selected. If it is not selected automatically, select the edge along the rip feature.
- Click on Flip to alter the unbend direction if prompted.
- Click on OK to finish unbending.
- The concentric reducer is developed now.
- Click on the top view to see the developed view.
- The eccentric reducer is developed now.
- Click on the top view to see the developed view.
- Click on the Analysis tab.
- Click on the Measure icon.
- Select the edges that are generated from circular edges of 500 mm diameter, by pressing control. The final length will be 1532.49 mm. Now the perimeter of the 500 OD circle is 1570.8 mm. The thickness of the eccentric reducer is 10 mm. Hence, the mean diameter of the small circular side is [(700 – (2 × 10/2)] = 490 mm. The perimeter of the circle of 490 mm is 1539.38 mm.
- In this type of bend with ample clearance and made by rolling, typically, the neutral plane should be in the middle of the thickness approximately.
- To adjust the neutral plane, click on File > Prepare > Model properties, and click on Y factor change.
- A new preference window shall appear. Click on the Bend allowance tab, and change the factor set to K factor from Y factor and the value to 0.5. Since it is a ratio, it does not have unit.
- Click on Regenerate.
- Now click on the Analysis tab.
- Click on the Measure icon.
- Select the edges that are generated from circular edges of 500 mm diameter, by pressing control. The final length will be 1542.70 mm, which is approximately the same as calculated for the perimeter of mean diameter.
- Click on the Measure icon.
- Select the edges that are generated from circular edges of 900 mm diameter, by pressing control. The final length will be 2799.33 mm, which is approximately the same as calculated for the perimeter of mean diameter of the large-diameter side, that is, 2796.01 mm for a 890 mm diameter circle.

QUESTIONS

1. What is the sheet metal module requirement?
2. What are the advantages of the sheet metal module over solid modeling?
3. Calculate the developed length for the products specified in Examples 9.7 through 9.10.
4. Define K factor.
5. What are the uses of the bend allowance table?
6. When are punch operations and die operations required?
7. State the use of rip.
8. Explain the requirement of surface rips.
9. State the method of developing an eccentric reducer.
10. What is the difference between the flange and wall operations?
11. List different types of reliefs.
12. Describe the use of different types of reliefs.
13. Describe the use of the unbend operation.
14. Describe the use of the bend back operation.
15. Describe the advantages of application-based modeling.

10

Analysis

10.1 Introduction

Analysis is one of the most fruitful requirements in computer-aided design (CAD) after preparation or in the process of CAD modeling. Analysis can be defined as extracting and manipulation of the needed data from a CAD model. Designers need to analyze the model to ensure the conformance of the product to required standards and needs. It is discussed in the previous chapter that a CAD model comprises various features. Features might provide the direct shape of the component. Sometimes, even a material removal feature does not provide the component shape directly. In those cases, the CAD models are made in steps, and the final shape is generated through a sequence of features in order. Thus, dimensional checking of the model in every stage of the sequence becomes mandatory to ensure the final produced shape being generated in the CAD model. Sometimes, there must be limitations on the components or product that derives the model shape. In those cases, the limitations need to be measured and controlled. Mass properties like total weight, center of gravity, and so forth are some of the major limitations one may have in his/her product.

Dimensions are one of the major criteria that a user needs to know at various stages of preparation of a CAD model. CAD provides tools to measure the dimensions between various entities in a CAD model. A distance between two noncoplanar and nonintersecting lines is easily measured in CAD, whereas the same can drive out in-hand calculations with much consumed time and effort.

CAD models are often analyzed for strength and dimension criteria to ensure and verify the model in the required service condition. A manual calculation for the strength and stress values of a component may be too large and time consuming. Finite element analysis (FEA) provides an excellent solution in this type of situation, which enables engineers to validate the model with comparatively higher accuracy and in lesser time.

Kinetic analyses are among the major required analysis fields in CAD. The various positions of any point of any part in a mechanism can be traced with respect to variation of location of other parts. Advanced CAD software provides tools for simulation of mechanisms. Simulation includes the visualization of mechanisms in the running condition; thus, various positions of the parts in the running condition can be traced out easily. It also provides the simulation of mechanisms and effects of forces at various conditions, which enables us to identify the effect of forces in the mechanized parts.

In reality, we always do the optimization by analysis in a CAD model. Optimization can be defined as the process of achieving the goal of limiting some derived parameters by varying the prespecified parameters with limitations. When we create a CAD model and we check the dimensions, we always ensure the final shape by propagating the feature's dimensions. Thus, we do the optimization in our mind. CAD provides tools for optimization by selection of prespecified parameters and goals. CAD also provides the facility to visualize the optimized parts in various iterative steps and in a graph, which assists the designers in understanding the variations and limitations of the model.

There are various kinds of analysis of the engineering product. Most useful analyses are covered in this book in accordance with subject importance.

10.2 Dimensional Analysis

Dimensional analysis includes measurement and manipulation of parameters like distance, angles, length, and so forth of various entities of the CAD model. These data are required to check and improve the component design.

10.2.1 Measurements

1. *Distance*: Distance can be measured between various entities with or without project references. When no project reference is selected, the minimum distances between the two entities are shown. If the distance on a projected plane is to be ensured, a project reference is required. Figure 10.1 shows an example of dimension measurement.

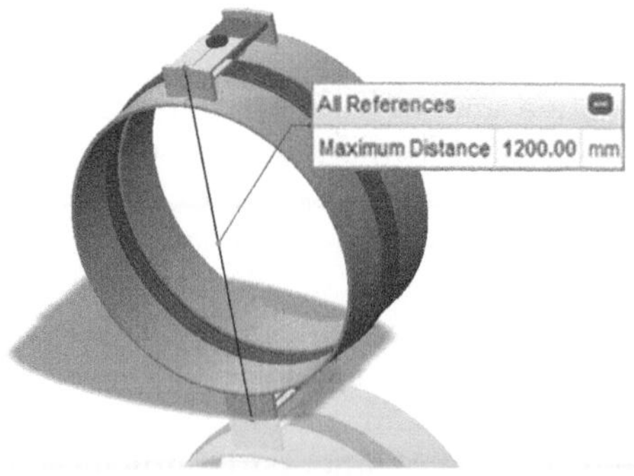

FIGURE 10.1
3-D view of expansion joint.

When two axes are selected for measurement between them, a minimum diagonal dimension is shown. However, when the projected dimension with the respective front plane is required, the project reference shall be selected.

Hands-On Exercise, Example 10.1: Linear Distance Measurement (Software Used: Creo Parametric; Figure 10.1)

LINEAR DISTANCE MEASUREMENT

Steps:

- Open the assembly file HINGE_EXPANSION_JOINT from Chapter 7.
- Click on Tree filters, as shown in Figure 10.2.

- Check the boxes of features to display in the model tree.
- Click OK to close the box of model tree items.
- The linear boundary dimension will be measured, as shown in Figure 10.3.

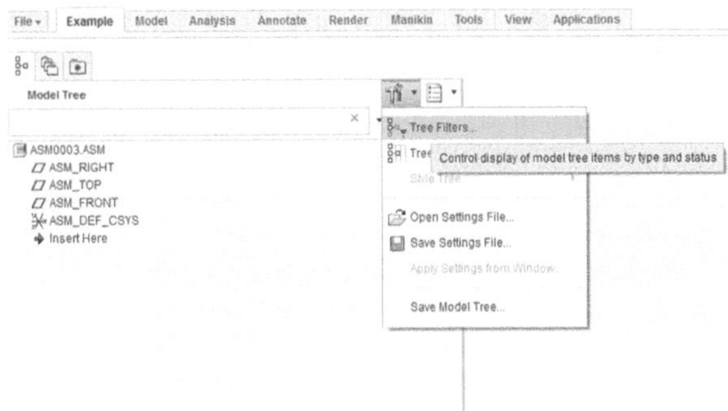

FIGURE 10.2
Model tree filter.

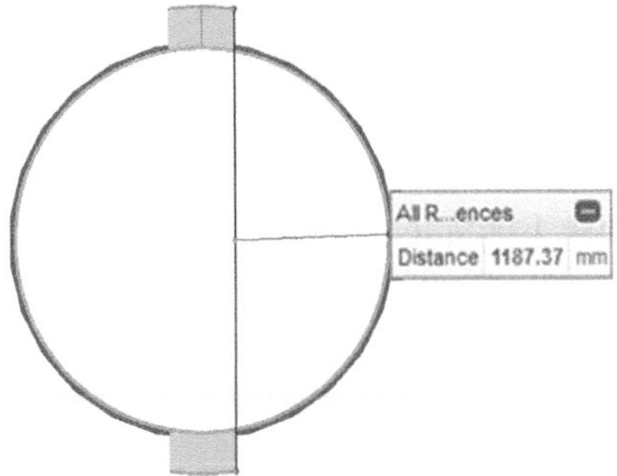

FIGURE 10.3
An example of maximum distance measurement.

- Select Summary tools from the Analysis tab of the ribbon interface.
  - Note: The distance tool also can be selected from the drop-down menu of summary tools.
- Since there is no entity in the middle of the edges, select edges by pressing the Ctrl key.
- Uncheck the box Use as Center for both the selections.
- The results shall be as shown in Figure 10.3. However, the goal of measuring the boundary dimensions of the edges is not achieved.
- Close the analysis summary toolbox by clicking the "cross" button at the top right corner of the analysis summary toolbox.

- Click OK point from the Model tab of the ribbon interface.
- Select the top circular edge of the part RIB_1 as a reference edge.
- Set the ratio value to 0.5. This will create a point at the midpoint of the edge.
- Select a new point; select the outer circular edge RIB_1, which is placed diametrically on the opposite side of the expansion joint.
- Click OK to create the points and close the point creation window.

- Select Summary tools from the Analysis tab of the ribbon interface.
  - Note: The distance tool also can be selected from the drop-down menu of summary tools.
- Select the last two created points by pressing the Ctrl key.
- The results shall be displayed as shown in Figure 10.1.
- Close the analysis summary toolbox by clicking the "cross" button at the top right corner of the analysis summary toolbox.

2. *Angles*: Angles are measured between two nonparallel intersecting planes or axes or lines. The measurement unit of an angle can be as follows:
   a. Degree
   b. Radian

Hands-On Exercise, Example 10.2: Angle Measurement (Software Used: Creo Parametric; Figure 10.4)

ANGLE MEASUREMENT

Steps:

- Open the assembly file BOARD_CLIP from Chapter 7.
- Click on Tree filters, as shown in Figure 10.2.

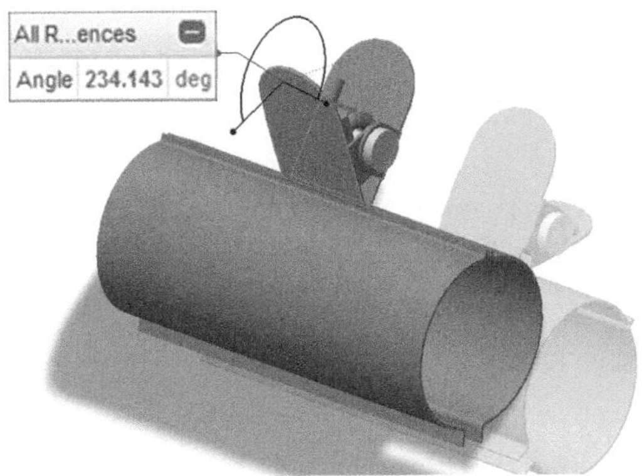

FIGURE 10.4
An example of angle measurement.

- Check the boxes of features to display in the model tree.
- Click OK to close the box of model tree items.
- The angle between the planes as a displayed dimension will be measured, as shown in Figure 10.4.
- Select Summary tools from the Analysis tab of the ribbon interface. Select the option for angle measurement.
  - Note: The angle tool also can be selected from the drop-down menu of summary tools.
- Select both the faces, as displayed in Figure 10.4.
- The measured angle shall be 125.857°. Notice that the angle option is set to Main. The angle is between the normal of the selected surfaces.
- Select the drop-down box for angle and select Supplement. The displayed value shall be 54.1428°.
- The angle is between the normal of the selected surfaces with alteration of direction of one normal.
- Select the drop-down box for angle and select Conjugate. The displayed value shall be 243.143°.
- The angle is between the normal of the selected surfaces. Change the range from 0–360° to +/–180°. The displayed value shall be –125.857°.
- Select the drop-down box for angle and select Second conjugate. The displayed value shall be –4.1428°.
- The angle is between the normal of the selected surfaces. Change the range from +/–180° to 0–360°. The displayed value shall be 305.857°.
- Click on the option button in the right top corner of the result section.
- Change the angular unit to radian from degree.
- The displayed value will change to 5.33822°.
- Save and exit the file.

3. *Curve length*: The curve length tool or option simply measures the length of any curve, that is, edges or perimeter, of any selected surface.

Hands-On Exercise, Example 10.3: Curve Length Measurement (Software Used: Creo Parametric; Figure 10.5)

CURVE LENGTH MEASUREMENT

Steps:

- Open the part file Paper_Clip_1 from Chapter 9.
- Click on Tree filters, as shown in Figure 10.2.

- Check the boxes of features to display in the model tree.
- Click OK to close the box of model tree items.
- The length of the edges as displayed will be measured, as shown in Figure 10.5.
- Select Summary tools from the Analysis tab of ribbon interface. Select the option for length measurement.
    - Note: The length tool also can be selected from the drop-down menu of summary tools.
- Select the edge, as displayed in Figure 10.6.
- If possible, multiple edges can be selected by pressing the Control key, as displayed in Figure 10.7.
- The total curve length as well as individual lengths are also displayed.
- Now select any surface of the part; notice that the perimeter of the surface or the total length of associated edges on the surface is displayed.
- Click on the right top corner of the analysis window to close it.

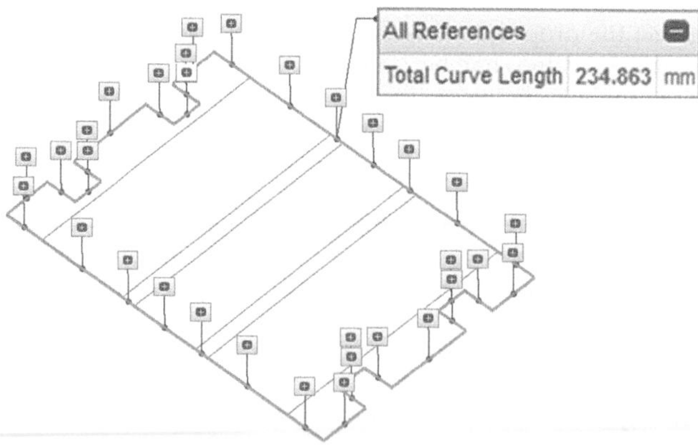

FIGURE 10.5
An example of curve length measurement.

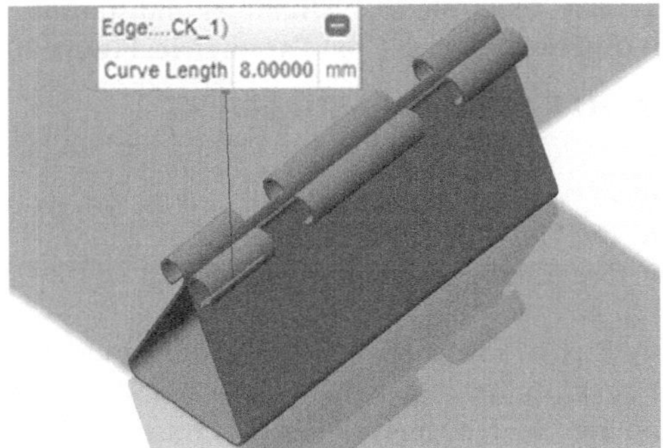

FIGURE 10.6
Selection of a single curve.

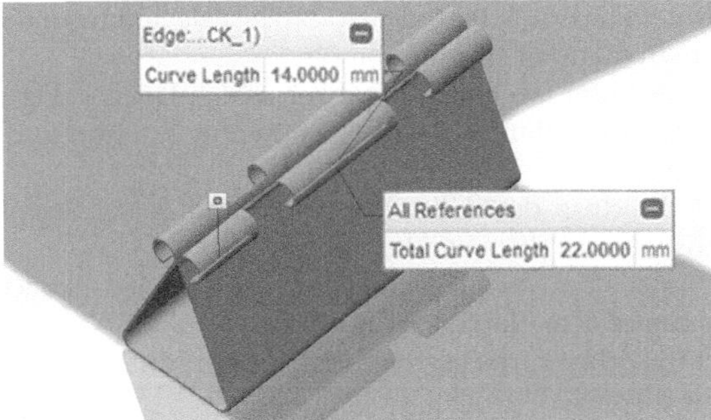

FIGURE 10.7
Selection of multiples curves.

- Drag the arrow in the model tree before the feature Bendback1, as displayed in Figure 10.8.
- The displayed model will flatten, as shown in Figure 10.5.
- Select Summary tools from the Analysis tab of the ribbon interface. Select the option for length measurement.
- Select all the individual edges by pressing the Control key, as displayed in Figure 10.5, to show the obtained total perimeter.
- Save and exit the file.

4. *Radius/diameter*

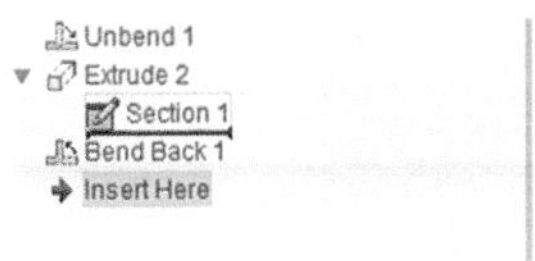

FIGURE 10.8
Alteration of features.

Hands-On Exercise, Example 10.4: Diameter and Radius Measurement (Software Used: Creo Parametric; Figure 10.9)

DIAMETER AND RADIUS MEASUREMENT

Steps:

- Open the file Flywheel from Chapter 6.
- Click on Tree filters, as shown in Figure 10.2.

- Check the boxes of features to display in the model tree.
- Click OK to close the box of model tree items.
- The diameter and radius of various round cylindrical surfaces as displayed dimensions will be measured, as shown in Figure 10.9.
- Select Summary tools from the Analysis tab of the ribbon interface. Select the option for diameter measurement.
  - Note: The diameter tool also can be selected from the drop-down menu of summary tools.
- Select the outer face of the flywheel, as displayed in Figure 10.9.
- The measured diameter will be shown as 500 mm and radius 250 mm.
- Select the drop-down box for angle and select Supplement. The displayed value shall be 54.1428°.
- The angle is between the normal of the selected surfaces with alteration of direction of one normal.
- Select the other surfaces; the diameter or radius of the surfaces will be displayed.
- Save and exit the file.

5. *Cross-sectional area*
6. *Surface area*

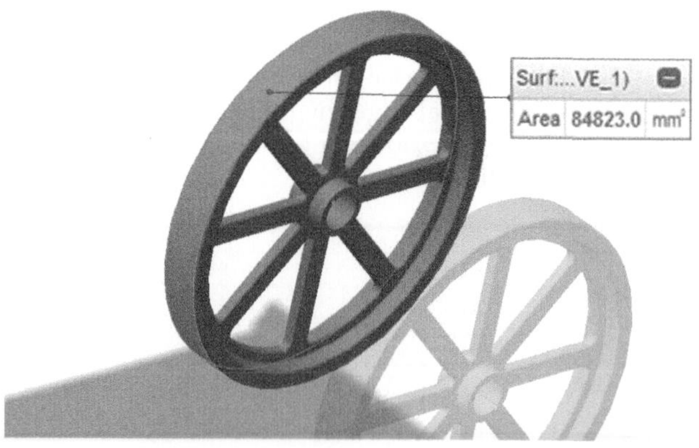

FIGURE 10.9
Example of diameter and radius measurement.

Hands-On Exercise, Example 10.5: Area Measurement (Software Used: Creo Parametric; Figure 10.10)

AREA MEASUREMENT

Steps:

- Open the assembly file Hinge_Expansion_Joint.asm from Chapter 7.
- Click on Tree filters, as shown in Figure 10.2.

- Check the boxes of features to display in the model tree.
- Click OK to close the box of model tree items.
- The area of various surfaces as a displayed dimension will be measured, as shown in Figure 10.10.
- Select Summary tools from the Analysis tab of the ribbon interface. Select the option for area measurement.
    - Note: The area tool also can be selected from the dropdown menu of summary tools.
- Select the outer surface of the pipe part, as displayed in Figure 10.11.
- The measured area is 628,319 mm^2, as shown in Figure 10.11.
- Select multiple areas by pressing the Ctrl key to obtain the total area.
- Click on the Save icon to create the measurement as a feature or save the analysis.
- Select the previously selected surface.
- Click on the project reference selection box, and select any surface that is perpendicular to the pipe surface, as displayed in Figure 10.11.
- Check: the project area will be 0 as there is no project area in the projected direction.
- Now select the area as shown in Figure 10.12.

FIGURE 10.10
Example of area measurement.

FIGURE 10.11
Example of area measurement.

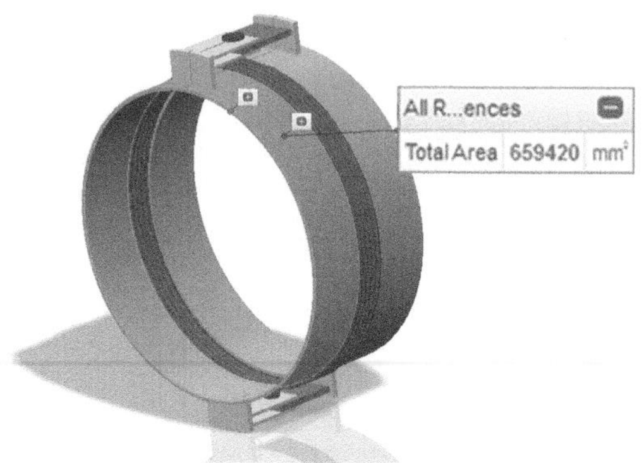

FIGURE 10.12
Example of area measurement.

- Click the Save icon in the right top corner of the analysis window to make the feature or save the analysis feature.
- Save and exit the file.

7. *Volume*

Hands-On Exercise, Example 10.6: Volume Measurement (Software Used: Creo Parametric; Figure 10.13)

ANGLE MEASUREMENT

Steps:

- Open the assembly file Tank.asm from Chapter 7.
- Click on Tree filters, as shown in Figure 10.2.
- Check the boxes of features to display in the model tree.
- Click OK to close the box of model tree items.

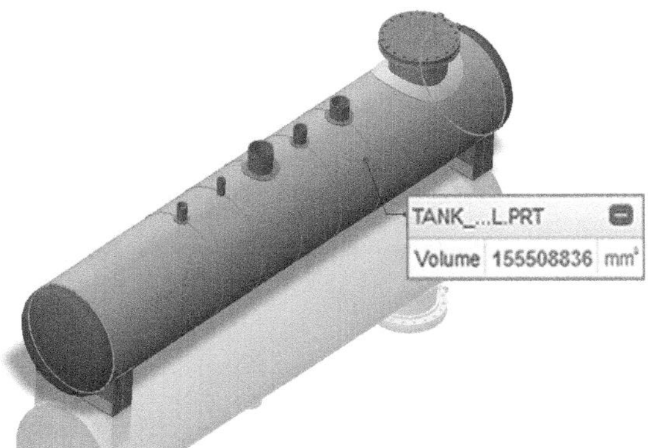

FIGURE 10.13
Example of volume measurement.

- The volume of various components as a displayed dimension will be measured, as shown in Figure 10.13.
- Select Summary tools from the Analysis tab of the ribbon interface. Select the option for angle measurement.
    - Note: The volume tool also can be selected from the drop-down menu of summary tools.
- Select the shell part, as displayed in Figure 10.13.
- The measured volume shall be 155,508,836 mm^3.
- Select the drop-down icon in the analysis window to change the unit.
- Change the length unit to m.
- The measure volume shall be displayed in m^3.
- Click the Save icon in the right top corner of the analysis window to make the feature or save the analysis feature.
- Select multiple components by pressing the Ctrl key for multi-selection to get the total volume of multiple parts.
- Save and exit the file.

8. *Interference*

Hands-On Exercise, Example 10.7: Volume Interference Measurement (Software Used: Creo Parametric; Figure 10.14)

VOLUME INTERFERENCE MEASUREMENT

Steps:

- Open the assembly file Tank.asm from Chapter 7.
- Click on Tree filters, as shown in Figure 10.2.

- Check the boxes of features to display in the model tree.
- Click OK to close the box of model tree items.
- The volume interference between the components as a displayed dimension will be measured, as shown in Figure 10.14.

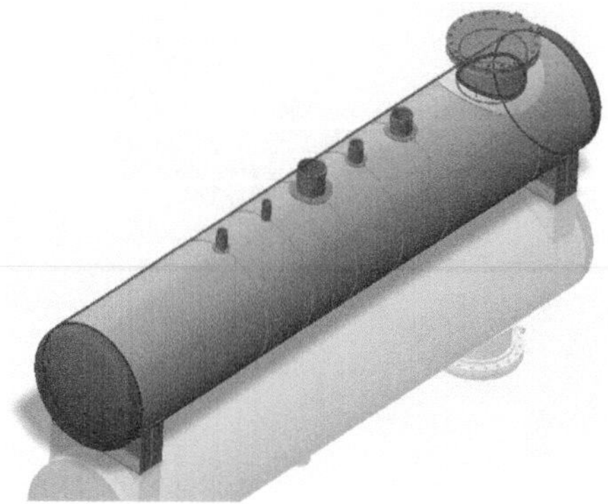

FIGURE 10.14
Example of volume interference measurement.

- Select the volume interference tool.
- All the interferences between the fitted parts are displayed in the analysis window.
- Select the subassembly option. Click on preview.
- Check that the interference between the subassemblies is displayed in the analysis window.
- Save and exit the file.

10.2.2 Mass Properties

Mass properties include determination of the following parameters:

1. Mass
2. Centroid and center of gravity
3. Moment of inertia
4. Polar moment of inertia
5. Radius of gyration
6. Section modulus
7. Torsion
8. Axis transformation

Hands-On Exercise, Example 10.8: Mass Property Measurement (Software Used: Creo Parametric; Figure 10.15)

MASS PROPERTY MEASUREMENT

Steps:

- Open the assembly file Tank.asm from Chapter 7.
- Click on Tree filters, as shown in Figure 10.2.

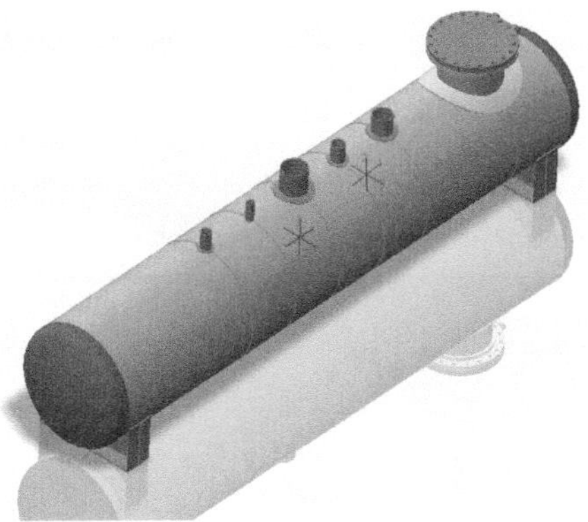

FIGURE 10.15
Example of mass property measurement.

- Check the boxes of features to display in the model tree.
- Click OK to close the box of model tree items.
- The angle between the planes as a displayed dimension will be measured, as shown in Figure 10.15.
- Select the Mass properties analysis icon from the Analysis tab of the ribbon interface.
- Click on the preview to get the mass properties.
- A density set window may appear if density of any component is not set yet.
- Leave the density window without entering the density.
- Click OK to close the density window.
- The mass properties include the following:
    - *Volume*: The total volume of the tank assembly is shown as $2.3611399e + 08$ mm^3.
    - *Surface area*: The total interior and exterior surface area is displayed as $4.3587744e + 07$ mm^2.
    - *Average density*: Average density is the average of the densities of all the parts in the assemblies and subassemblies. Average density often serves as an indication to check whether all the parts are assigned with proper densities or not.
    - *Mass*: Mass is the total mass of the assemblies including all its components. If the densities are not set properly, the calculated mass will not be accurate. Therefore, if densities of all the parts are not set properly, the best way to calculate the mass is to multiply the volume with the correct density. In this case, the tank volume is volume × density of iron = $0.2361139e$ m^3 × 7850 kg/m^3 = 1853.4941 kg, which is fairly accurate.
    - *Center of gravity*: Center of gravity is displayed as the measurement from the selected coordinate system or default. Select the coordinate system, as displayed in Figure 10.16, and uncheck the box for the default coordinate system.

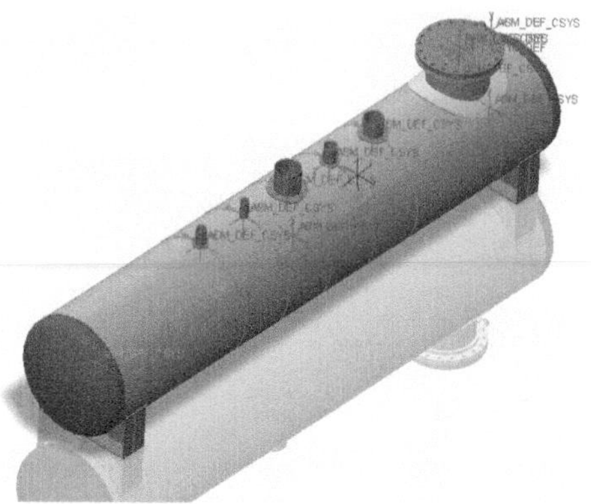

FIGURE 10.16
Example of mass property measurement using a user-defined coordination system.

- Inertia tensor.
- Principal moment of inertia.
- *Radius of gyration*: Radius of gyration is the distance squared and multiplied with the total mass of the body, giving the mass moment of inertia of the body.
- Mass properties of individual components and so forth.

- Save and exit the file.

Hands-On Exercise, Example 10.9: Section Mass Property Measurement (Software Used: Creo Parametric; Figure 10.17)

SECTION MASS PROPERTY MEASUREMENT

Steps:

- Open the assembly file HINGE_EXPANSION_JOINT from Chapter 7.
- Click on Tree filters, as shown in Figure 10.2.

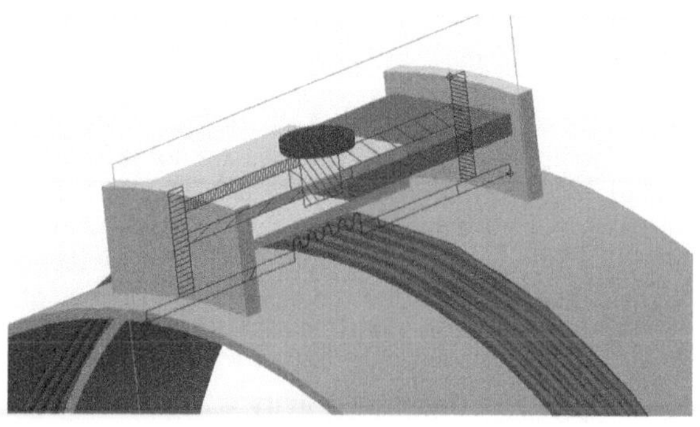

FIGURE 10.17
Example of section mass property measurement.

- Check the boxes of features to display in the model tree.
- Click OK to close the box of model tree items.
- The angle between the planes as a displayed dimension will be measured, as shown in Figure 10.17.
- Select the "x section mass properties" tool.
- Select the front plane to create the section along the front plane.
- The total area of the section will be displayed.
- Click the Save icon in the right top corner of the analysis window to make the feature or save the analysis feature.
- The measured parameters in the x section properties are as follows:
  - Area
  - Center of gravity
  - Inertia tensor
  - Polar moment of inertia and so forth
- Save and exit the file.

10.2.3 Optimization Analysis

Optimization is used to reduce the weight or cost of the model by varying certain dimensions while keeping certain dimensions or parameters constant.

10.2.4 Goal Seek Analysis

Goal seek analysis is one of the best tools to achieve the desired output dimensions by varying specific parameters or dimensions. It considers a dimension as a goal with a specified value, which has to be made by varying some parameters or dimensions within the specified limits.

10.3 Finite Element Analysis

10.3.1 Introduction

FEA is one of the most required and useful applications in CAD. Stresses on different points of different parts are determined by FEA under the prespecified load conditions. Nowadays, FEA is used in designing simple to complex engineering components in order to improve the quality and optimize the cost of the product. Interpretation of the result of FEA requires a good knowledge and understanding of engineering parameters.

FEA is used in many fields of engineering calculation and analysis such as structural analysis, fluid mechanics, heat transfer, dynamics, and so forth.

The user interface of FEA is improving day by day. However, ensuring the correct result from the FEA requires careful selection of steps in the model, especially in the case of complex geometries. Many times, FEA under real conditions with all the loads may not be economical with respect to computer capability and time. Some idealizations are required to simplify the model in order to reduce the task.

Idealization can be defined as the process of making of a simplified model through assuming of shapes and parameters in order to lighten the robust CAD model. Idealizations are extremely useful for structural components where the sections are assumed in a schematic model.

It is obvious that FEA requires skills to ensure the correct result and process. Thus, a proper understanding of FEA is needed before carrying it out in software for practical use.

10.3.2 Basic Concept

The essence of FEA is to divide the CAD model into a finite number of nonoverlapping quasi-disjoint elements and then analyze the assigning of properties of the elements.

The FEA is carried out by following this sequence: A CAD model is converted into a mesh of predefined elements of similar shape, which is called meshing. Each of the elements is connected to adjacent elements by the connecting points, called nodes. Parameters and equations are assigned to each type of element.

10.3.3 Preprocessing

Typically, preprocessing consists of conversion of a CAD model into a finite element model (FEM). A CAD model is typically a combination of solid volumes and its associated surfaces. It is natural when the analysis is carried out that a model does not need to be analyzed with the same accuracy for all the regions. Wherever the required result calls for accuracy, more emphasis has to be placed on those sections. Thus, this distinguishes the use of different elements in different regions. There are elements that comparatively provide more accurate results; however, more computer memory and time are required when analyzed. A designer or analyst has to carefully choose the elements to suit the obtained result.

Preprocessing follows these steps:

1. *Creation of or selection of elements*: Elements for FEA can be classified into three groups.
   a. *1-D elements*: 1-D elements are used for idealization and simplicity of analysis with fairly accurate results. A simple rod or beam can be analyzed faster with 1-D elements with accurate output or results. Beams, rods, and structural members are often analyzed with 1-D elements, which provide faster and accurate results. 1-D elements can be used where the entire model can be idealized in a single plane for analysis.

b. *2-D elements*: 2-D elements are surface élements used for idealization and simplicity of analysis similar to 1-D elements. An example is analysis of plates of equal thickness. 2-D elements can be used where the entire model can be imagined and idealized in a single plane.

c. *3-D elements*: 3-D elements are used where the 3-D model needs to be analyzed with all its boundary conditions, which exist in various planes in three dimensions.

2. *Assignment of properties to the elements*: The properties need to be assigned to the elements. For example, all the properties need not be assigned in the subject elements. If 1-D elements are used, bulk modulus need not be assigned, since the obtained result may not be required to have a bulk modulus effect. At the same time, one needs to be careful about the assignment of essential properties.
3. *Meshing*: Meshing is the generation of mesh in the Cad model, which converts the model to an FEA model. Meshing or mesh generation typically generates selected elements and nodal components in the model shape. A 3-D model or even 2-D geometries often have many irregular shapes, which need to be meshed carefully into small elements.
4. *Application of loads*: The following are the types of loads that can be assigned in CAD.
   a. *Force*: Force has to be specified on a point with proper directional references.
   b. *Force per unit length*: Force per unit length can be assigned on a line or curve. This allows assigning a uniformly distributed load on a line element, which is mostly used in analysis of truss.
   c. *Pressure*: Pressure load is assigned on the flat or curved surface.
   d. *Global temperature*: Global temperature is the application of ambient temperature in a model. Assignment of global temperature is often significant in analysis of models to simulate in real-life condition where ambient temperature varies.
5. *Application of boundary condition*: Boundary conditions are used to limit the boundaries under certain assumed constraints. For example, a plate is lying on a large steel surface of a base plate, and pressure is applied on the top surface. Thus, if the plate is under analysis, the bottom surface can be assumed as a fixed surface, while load can be applied on the top surface. Therefore, the surface constant with zero translation and rotational movement needs to be assigned in the bottom surface. There can also be other types of constraints that are often used in specifying boundary conditions.
   a. *Point constraints*: A point constraint fixes the selected point of a CAD model with specified translation and rotational movement.

If the translation and rotational movement are set to zero, the point will be fixed. However, if the translation is set to zero but the point is free to rotate, then the point will be in hinged condition. Translation movements are often allowed to simulate yielding conditions.

b. *Line constraints*: Line constraints are used to fix edges, which can be assumed as a straight line. Line constraints also can be allowed with translation and rotational movements to simulate various situations in real conditions.

c. *Surface constraints*: Surface constraints are widely used in analysis of 3-D models. In real life, surface constraints are widely used since most of the parts rest on a certain surface. Surface constraints also can have preallowed translation and rotational movement.

10.3.4 Solver

Solving is processing the FEA with all the supplied data. It starts with a global stiffness matrix and load matrix. Then, the element stiffness matrix is initialized. In case of numerical integration, each guess point to the stiffness matrix is found.

Shape function derivatives are found. An element stiffness matrix is found when a guess loop is completed; therefore, a global stiffness matrix with added values from an element stiffness matrix is available.

The imposed boundary conditions are also solved, and the stiffness equations become ready. When the nodal variables are available at the Gauss point of each element, outputs like strain stresses are calculated and found.

10.3.5 Postprocessing

Postprocessing is extracting the required result from the analysis.

There can be numerous results available from the analysis. The results are available in the form of nodal and element solutions. The following results often need to be checked and graphically presented:

1. *von Mises stress*: von Mises stress is the maximum tensile stress. According to the theory of maximum distortion energy, the yielding of a material occurs in a member when the distortion strain energy per unit volume reaches the limiting distortion energy.

   Mathematically, in a biaxial system,

$$(\sigma_{t1})^2 + (\sigma_{t2})^2 - 2\sigma_{t1} x \sigma_{t2} = \left(\frac{\sigma_{yt}}{F.S}\right) 2$$

   where σ_{t1}, σ_{t2} = maximum and minimum principal stress in a biaxial system; σ_{yt} = yield point stress; $F.S$ = factor of safety.

2. *Deformation*: Deformation is analysis and comparison of a deformed model with an actual model to get the deformation of various points and elements.
3. *Shear stress*: Shear stress is the stress that can be obtained by dividing the force by the tangent area of the subjective surface. Shear force measurement is often required in designing of parts under shear. Also, a shear force bending moments diagram is required in beam design.
4. *Maximum principal stress*: Maximum principal stress is the maximum normal stress that can be generated in three principal planes in a body under forces. It is observed that there would be three planes that are mutually perpendicular to each other and carry direct stress only without any shear stress. These planes are called principal planes. The direct stresses across those planes are called principal stress. The maximum stress among the principal stresses is called maximum principal stress.

Hands-On Exercise, Example 10.10: Creating FEA of a Connecting Element (Software Used: Creo Parametric; Figure 10.18)

FLYWHEEL

Steps:

- Create a part file and rename it as Connecting_Element. (Other names may be given, but it is advisable to provide a relevant name.)
- Select Solid on the right-hand side of the box. This will activate the solid mode and environment.
- Uncheck the box for the default template and click OK to create a new solid file.
- The template selection box will appear.
- Select mmns_part_Solid. This will allow us to use the solid template file of the following units:
  - mm—millimeter for length unit
  - n—newton for force unit
  - s—second for time unit
- Once the solid part is created, go to File > Prepare > Model properties to check the units. In the upper sections, the units of the model will be displayed.
  - Note: The units usually are displayed in the summary section of any CAD software. The setting of units usually is in the tools or settings.

FIGURE 10.18
Example of finite element analysis.

- The connecting element as shown in Figure 10.18 is to be created.
- Click on Extrude from the Model tab in the ribbon interface.
- Click on the Placement tab and select Define to define the sketch.
- Select the front plane as the sketch plane.
- The right plane shall be automatically taken as reference and orientation as right.
- Click on the sketch.
- The sketcher model will be activated.
- Draw the sketch, as shown in Figure 10.19.
- Click OK to create and exit the sketch.
- Select Mid plane extrusion.
- Enter the value 20 mm for extrusion depth.
- Click on OK to create the extrusion.

- Click on Extrude again from the Model tab in the ribbon interface.
- Click on the Placement tab and select Define to define the sketch.
- Select the top face as the sketch plane.
- The right plane shall be automatically taken as reference and orientation as right.
- Click on the sketch.
- The sketcher model will be activated.
- Draw the sketch, as shown in Figure 10.20.
- Set the circle diameter to 70.
- Click OK to create and exit the sketch.
- Activate the material removal option.
- Select Through all for extrusion depth.
- Flip the orientation towards the material side.
- Click on OK to create the extrusion cut.

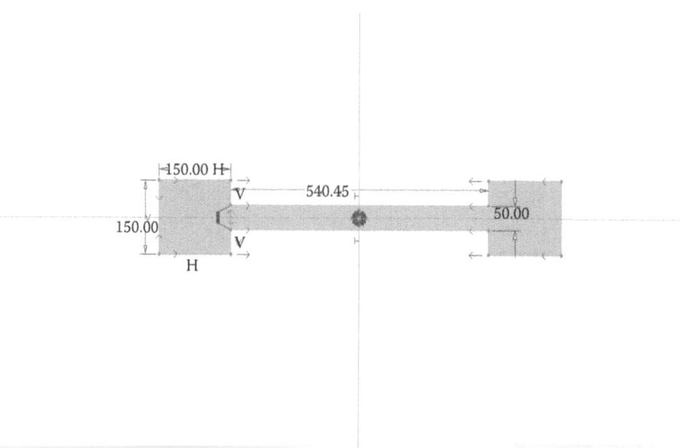

FIGURE 10.19
Constructional sketch of finite element analysis example.

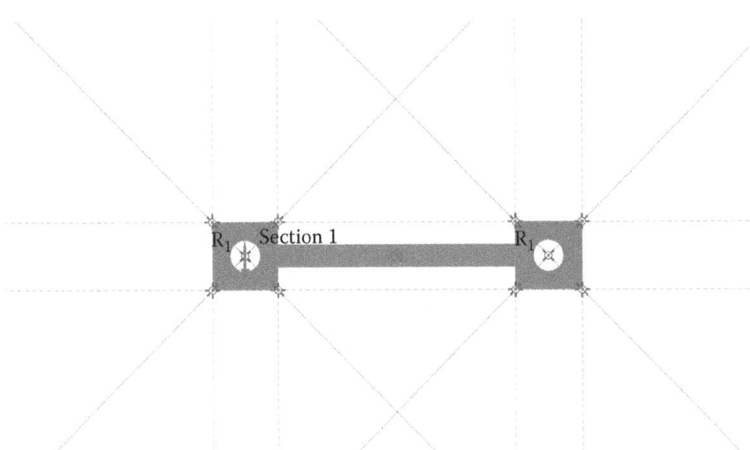

FIGURE 10.20
Constructional sketch of finite element analysis example.

- Click on Round feature from the Model tab of the ribbon interface.
- Set the value of fillet to 10 mm.
- Select all edges that are perpendicular to the front plane, except the circular edges.
- Click OK to create the feature.

- Click on Simulate from the Application tab of the ribbon interface.
- Select Simulation lite.
- Select Assign material option and click on Assign.
- Select the material steel. The material steel will be assigned to the part.
- Select Constraints and select Assign.
- Select the constraint to assign on the surface.
- Select the cylindrical surface of any of the holes.
- Click OK to create the surface constraints. Please ensure that the allowable translation and rotations are set to zero.
- Click on Loads and select Assign.
- Select Force.
- Select the flat surface of the other side of the hole.
- Enter "200 N" in the x direction vector, as shown in Figure 10.21.
- Click on Run to run the analysis.
- The run shall be completed successfully.
- Click on Results. The standards result set will be displayed with von Mises results, as shown in Figure 10.22.
- The result with displacement and maximum principal will be shown, as in Figure 10.23.

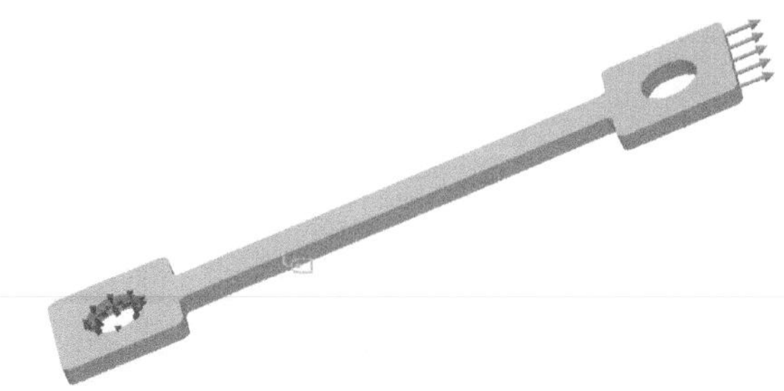

: Default bonded interface

FIGURE 10.21
Application of load and boundary condition for finite element analysis.

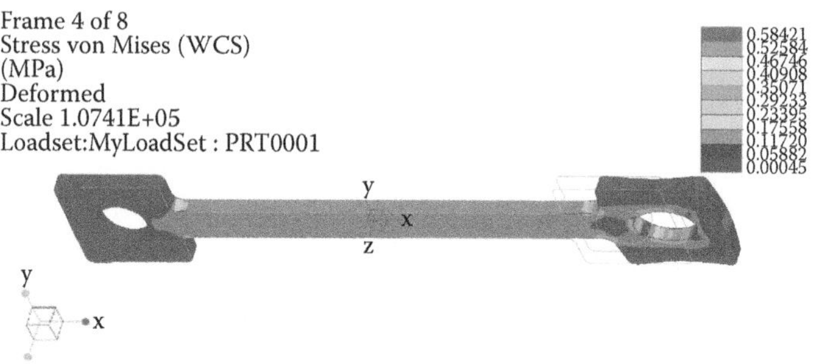

FIGURE 10.22
(See color insert.) Results of finite element analysis.

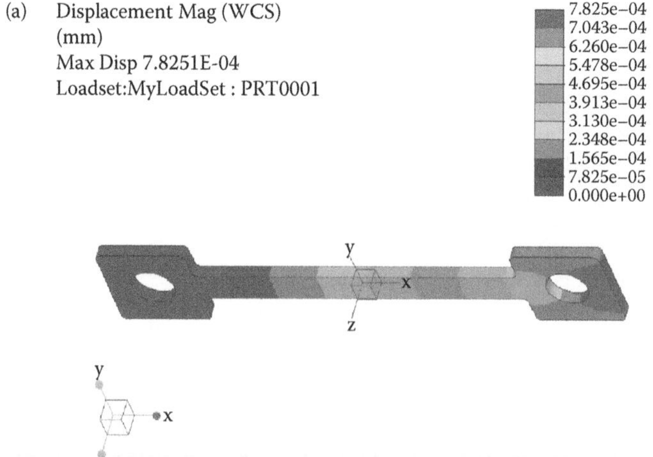

FIGURE 10.23
Results of finite element analysis.

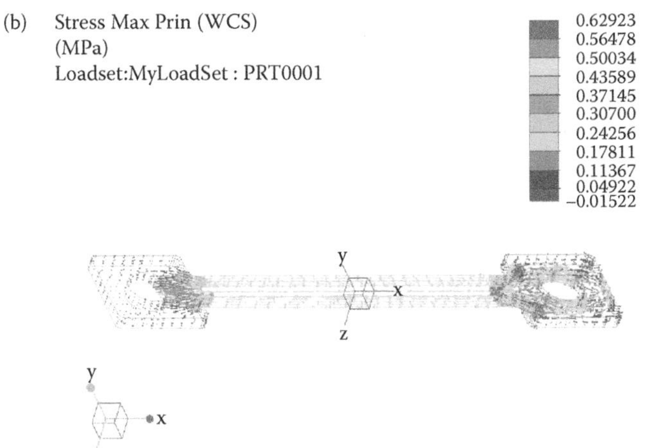

FIGURE 10.23 (Continued)
Results of finite element analysis.

10.4 Kinetic Analysis

When a machine or mechanism is simulated for actual running conditions and analyzed for the derived parameters like the position of various points, generated forces, and so forth, then the analysis is called kinetic analysis. When the analysis is carried out only for position without considering the effect of forces, then it is called kinematic analysis. However, when the forces and their effects are considered, the analyzed is called kinetic or dynamic analysis. Kinetic analysis is broadly described in the next chapter of this book.

QUESTIONS

1. What is the difference between length and distance measurement?
2. What is the significance of a projection plane in measurement? Describe with an example.
3. Why is finite element analysis so useful?
4. Describe the process of finite element analysis.
5. Describe the different types of results that can be obtained from finite element analysis.

11

Mechanism

11.1 Introduction

A mechanism is a combination of bodies that are connected in such a way that they move upon each other with definite relative motion. Every machine with metal bodies that produce work due to the movement of workpieces or tools must have one or more mechanisms within itself.

In computer-aided design (CAD), the bodies are modeled and can be connected to each other with various connection definitions and form joint configurations. Such connections provide motion relative to each other.

When the modeling of a mechanism is completed, a motor can be placed as driving equipment, which drives the mechanism and traces the motion paths of any required point or entity.

In addition, the CAD system provides the facility to analyze the dynamic and static loads on the mechanism generated by the motion.

In order to proceed with mechanisms and for better understanding, users need to be familiarized with certain terms with regard to mechanisms and CAD.

1. *Degree of freedom*: A body in 3-D can have maximum movements in six directions, three translation movements to the x-, y-, and z-axes and three rotations about each axis. These allowances on the movement are called degrees of freedom. When designing any mechanism, the required movement of any point or body is determined in terms of degree of freedom. The subsequent step in design is to produce the required allowable degree of freedom and then design accordingly. In a mechanism, every link can have a different degree of freedom except the fixed link. Degree of freedom is often used for describing pairs and joints.
2. *Planar pair*: A planar pair is a pair where two plane surfaces of two links are so constrained to each other that the links are free to move while satisfying the relation that the surfaces lie on each other. As shown in Figure 11.1a, a planar pair has three degrees of freedom, two translations and one rotational. If we consider the bottom body

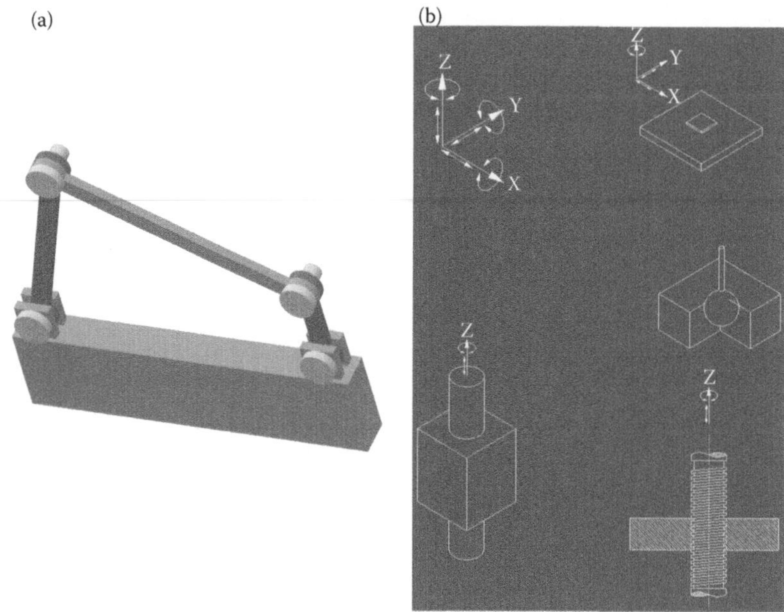

FIGURE 11.1
(a) Types of pairs. (b) 3-D view of four-bar linkage.

as a fixed link, the top body has a surface constrained to the bottom surface. There can be translation in two directions in *x* and *y* and one rotational movement about *z*. Resting bodies for their own weight such that they can be rested due to the weight to the base element are an example of a planar pair.

3. *Cylindrical pair*: A cylindrical pair has constrained cylindrical surfaces in two bodies. As shown in Figure 11.1a, a cylindrical pair has two degrees of freedom, that is, one translation and one rotation about the translation axis. Guide pins and bushes have a cylindrical pair connection.
4. *Spherical pair*: If the external surface of a sphere is constrained to another spherical surface of a spherical cut on a body, a spherical pair is formed. A spherical pair has three degrees of freedom, as shown in Figure 11.1a. Ball-and-socket joints are an example of a spherical pair.
5. *Screw pair*: Screw pairs are formed when two sliding surfaces of internal and external threads are constrained together. A screw pair has two degrees of freedom, that is, translation and rotation about the translation axis; however, these available degrees of freedom (as shown in Figure 11.1) are dependent on each other. Normally, the rotation drives the translation movement.

6. *Lower pair*: A pair that is formed by two connecting surfaces between two links is called a lower pair.
7. *Higher pair*: A higher pair is defined as a pair that is formed by connecting geometries of two pairs other than surfaces of contact. A point contact in a ball bearing or the teeth of skew-helical gears is an example of a higher pair.
8. *Wrapping pairs*.
9. *Link*: Material bodies in kinematic pairs are called links. In any connection, there shall be two links that allow movement relative to each other.
10. *Kinematic chain*: A kinematic chain is an arrangement of a series of connecting links to each other by kinematic pairs. When every link of a kinematic chain is connected to two other links, then the formed chain is called a closed chain.

11.2 Joint Modeling

In CAD software, joints are required to be defined between the links to form kinetic pairs. The different types of connections or joints are actually combinations of a few constraints. These constants are as follows:

- *Planar constraint*: A planar constraint is basically a plane or surface to a plane or surface to surface constraints. The planes can lie on each other or can be made parallel. If a planar constraint is specified, the joint has the nature of a planar pair; thus, three degrees of freedom are obtained.
- *Axis constraint or cylindrical surface constraint*: A cylindrical constraint specifies the cylindrical pair in the joint, thus obtaining two degrees of freedom.
- *Tangential surface constraint*: A tangential constraint is basically a higher pair, which has either line or point contact. It ensures at least a point that lies between two surfaces.
- *Curve-to-point constraint*: In this type of constraint, the point of any of the links of the pair always lies in any of the curves or edges of the other link of the pair. In this type of constraint, only one independent translation degree of freedom can be obtained. Various motion profiles can be obtained by this kind of constraint when used with other types of constraints.
- *Point-to-point constraint*: A point-to-point joint has three degrees of freedom, that is, rotation about three axes, x, y, and z.

11.3 Joint Classification Based on Degree of Freedom

All joints can be created based on the combination of the aforementioned constraints. In CAD, usually, the available joints are as follows:

- *Planar joint*: A planar joint uses only planar constraint. By defining a planar joint, the surfaces or planes of different links lie on each other or can be made parallel at certain distances.
- *Cylindrical joint*: A cylindrical joint uses axis or cylindrical constraint. This kind of joint can be defined by selecting axes of both the connecting links in the pair or by selecting surfaces. It is preferable to select axes because in many cases, when modification is required, surfaces need to be recreated; in that case, the internal ID of the surfaces is changed, and subsequent error occurs in the CAD system.
- *Pin joint*: A pin joint uses one planar constraint and one cylindrical constraint. In a pin joint, the planes or surfaces of the planar constraint are perpendicular to the axis of cylindrical constraint. A planar joint has three degrees of freedom, and a cylindrical joint has two degrees of freedom. The degrees of freedom of the pin joint can be found in Table 11.1.

Thus, the pin joint provides only one degree of freedom, which is rotation about the *z*-axis.

- *Slider joint*: A slider joint also uses one planar constraint and one cylindrical constraint. In a slider joint, the planes or surfaces of the planar constraint are parallel to the axis of cylindrical constraint. A planar joint has three degrees of freedom, and a cylindrical joint has two degrees of freedom. The degrees of freedom of the slider joint can be found in Table 11.2.

Thus, the slider joint provides only one translation degree of freedom, along the *x*-axis.

TABLE 11.1

Degrees of Freedom for Different Joints

Degree of Freedom	*X* Trans	*X* Rotation	*Y* Trans	*Y* Rotation	*Z* Trans	*Z* Rotation
Planar	Yes	No	Yes	No	No	Yes
Cylindrical	No	No	No	No	Yes	Yes
Pin joint	No	No	No	No	No	Yes

TABLE 11.2

Degrees of Freedom for Slider Joint

Degree of Freedom	*X* Trans	*X* Rotation	*Y* Trans	*Y* Rotation	*Z* Trans	*Z* Rotation
Planar	Yes	No	Yes	No	No	Yes
Cylindrical	Yes	Yes	No	No	No	No
Slider joint	Yes	No	No	No	No	No

- *Slot joint*: A slot joint uses curve-to-point constraint. As mentioned earlier, in the curve-to-point connection, only one independent translation degree of freedom is obtained. Screw motion can be obtained by the slot joint using a cylindrical joint between the links.
- *Ball joint*: A ball joint uses point-to-point connection. Thus, it has three rotational degrees of freedom, that is, rotation about x, y, and z.

11.3.1 Joint Selection and Use of Joints

All the aforementioned joints are used independently or in combination to define the connection between the links of any kinematic pair. The selection of a joint depends on the degree-of-freedom requirement between the links.

11.4 Connections

Apart from joints, certain software permits popular connection definitions between the links that may be difficult to create by using common joints and constraints. These connections are as follows.

Gear: Gear connections are used to establish imaginary gear relationships between gears. *Imaginary* is mentioned because the shape of the bodies does not define the gear mating parameters. These parameters are defined in the connection definition window itself. Once the connection is established, the pair behaves with and exhibits the gear relation. Before proceeding to gear, certain terms must be defined to understand gear connection. These are as follows:

1. *Pitch circle*: A pitch circle is the imaginary circle on which a gear part actually rolls over another.

2. *Pitch circle diameter*: The diameter of the pitch circle is the pitch circle diameter. It is very important since the gear ratio depends mainly on the pitch circle diameter, which needs to be specified when defining the gear connection.
3. *Pitch point*: The point of contact between two pitch circles of mating gears is called the pitch point.
4. *Pressure angle*: It is the angle between the normal of teeth of mating gears at the point of contact and the tangent at pitch point. The standard pressure angles are 14½° and 20°.
5. *Gear ratio.*
6. *Addendum*: The radial distance from the pitch circle diameter to the top of the gear teeth is called addendum.
7. *Dedendum*: The radial distance from the pitch circle diameter to the bottom of the gear teeth is called dedendum.
8. *Addendum circle*: The circle drawn through the top of the teeth is called the addendum circle.
9. *Dedendum circle*: The circle drawn through the bottom of the teeth is called the dedendum circle.
10. *Circular pitch*: It is the pitch that is measured circumferentially between points on two consecutive teeth.

    Mathematically, circular pitch = nD/T,

    where D = pitch circle diameter of the gear

    T = number of teeth of the gear

    Circular pitch must be the same for two mating gears.

    Hence if $D1$, $T1$ and $D2$, $T2$ are the pitch circle diameter and number of teeth for respective mating gears, we have circular pitch

    $$Pc = n\frac{D1}{T1} = n\frac{D2}{T2}.$$

    Hence, we have $\frac{D1}{T1} = \frac{D2}{T2}$.
11. *Diametral pitch*: It is defined as the ratio of the number of teeth to the pitch circle diameter.

    Mathematically, diametral pitch = $T/D = n/Pc$,

    where T = number of teeth

    D = pitch circle diameter

    Pc = circular pitch
12. *Module*: It is the ratio of pitch circle diameter to the number of teeth. It is denoted by m.

    Mathematically, module = D/T.

The following types of gear connections are the most used in CAD mechanism simulation.

Spurs: Spur gears are widely used in industry. Most advanced CAD software that allows mechanisms has this connection definition. To define a spur gear connection, the following have to be identified:

Gear 1 or pinion

Gear 2 or carrier

Pitch circle diameter

Bevel

Worm

Rack and pinion

Cam

Belt

Motors: Motors are defined in a connection axis or motion axis, which is created by defining connections. CAD software provides a wide range of motion profiles to define.

Servomotors: A servomotor can be defined in the motion axis, which enables all the kinematic and dynamic analysis. Required outputs are defined in the servomotors. The output can be in terms of position, velocity, or acceleration. These parameters are set with respect to time.

Force motors: A force motor basically is a motor on which the input can be a force that is measured with respect to time, or any existing measures can be taken as variables. The force motor is used for dynamic analysis with variation in forces.

Forces: Though most of the automatic movement is defined by using motors in any mechanism, the use of forces enables users to simulate a mechanism's behavior under dynamic analysis when forces on a particular link are known.

11.5 Curve Tracing

The examples are carried out in Creo. It is suggested that users carry out and check the mechanisms on their own in any software that allows mechanisms.

Use of mechanism

Four-bar linkage: A four-bar linkage is shown in Figure 11.1b.

Hands-On Exercise, Example 11.1: Four-Bar Linkage (Software Used: Creo Parametric; Figure 11.1b)

FOUR-BAR LINKAGE

Steps:

- Create a part file and rename it as Ground_Body. (Other names may be given, but it is advisable to provide a relevant name.)
- Select Solid on the right-hand side of the box. This will activate the solid mode and environment.
- Uncheck the box of the default template and click OK to create a new solid file.
- The template selection box will appear.
- Select mmns_part_Solid. This will allow us to use the solid template file of the following units:
  - mm—millimeter for length unit
  - n—newton for force unit
  - s—second for time unit
- Once the solid part is created, go to File > Prepare > Model properties to check the units. In the upper sections, the units of the model will be displayed.
  - Note: The units usually are displayed in the summary section of any CAD software. The setting of units usually is in the tools or settings.
- Click on Extrude from the Model tab in the ribbon interface.
- Click on the Placement tab and select Define to define the sketch.
- Select the top plane as the sketch plane.
- The right plane shall be automatically taken as reference and orientation as right.
- Click on Sketch.
- The sketcher model will be activated.
- Draw a rectangle of 2000 × 200 symmetrically about the horizontal and vertical references.
- A centerline may need to be drawn along the reference to draw the aforementioned rectangle.
- Draw a circle of diameter 40 placing the center at the intersection of the horizontal and vertical references.
- Click OK to create and exit the sketch.
- Flip the material orientation downward.
- Enter the value 500 mm for extrusion depth.
- Click on OK to create the extrusion.

- Click on Extrude from the Model tab in the ribbon interface.
- Click on the Placement tab and select Define to define the sketch.
- Select the top plane as the sketch plane.
- The right plane shall be automatically taken as reference and orientation as right.
- Click on Sketch.

- The sketcher model will be activated.
- Draw the sketch as shown in Figure 11.2.
- Click OK to create and exit the sketch.
- Flip the material orientation upward.
- Enter the value 200 mm for extrusion depth.
- Click on OK to create the extrusion.

- Click on Extrude from the model tab in the ribbon interface.
- Click on the Placement tab and select Define to define the sketch.
- Select the front surface of the lastly created feature extrude_2 as the sketch plane.
- The top surface shall be automatically taken as reference and orientation as top.
- Click on Sketch.
- The sketcher model will be activated.
- Draw the sketch as shown in Figure 11.3.
- Click OK to create and exit the sketch.
- Flip the orientation toward the material side.
- Select Through all for depth option.
- Activate the material removal option.
- Click on OK to create the extrusion.

- Save and exit the file.

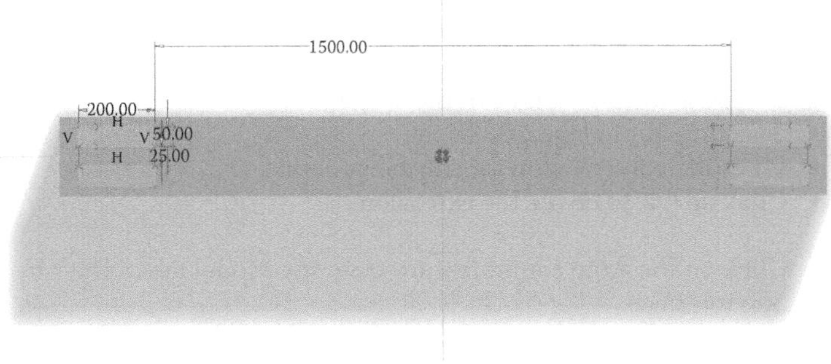

FIGURE 11.2
Constructional sketch of four-bar linkage.

FIGURE 11.3
Constructional sketch of four-bar linkage.

- Create a part file and rename it as Linkage1. (Other names may be given, but it is advisable to provide a relevant name.)
- Select Solid on the right-hand side of the box. This will activate the solid mode and environment.
- Uncheck the box of the default template and click OK to create a new solid file.
- The template selection box will appear.
- Select mmns_part_Solid. This will allow us to use the solid template file of the following units:
    - mm—millimeter for length unit
    - n—newton for force unit
    - s—second for time unit
- Once the solid part is created, go to File > Prepare > Model properties to check the units. In the upper sections, the units of the model will be displayed.
    - Note: The units usually are displayed in the summary section of any CAD software. The setting of units usually is in the tools or settings.
- Click on Extrude from the Model tab in the ribbon interface.
- Click on the Placement tab and select Define to define the sketch.
- Select the front plane as the sketch plane.
- The right plane shall be automatically taken as reference and orientation as right.
- Click on Sketch.
- The sketcher model will be activated.
- Draw the sketch as displayed in Figure 11.4.
- Click OK to create and exit the sketch.
- Select the mid plane extrusion.
- Enter the value 50 mm for extrusion depth.
- Click on OK to create the extrusion.

- Click on the Auto round feature from the Model tab of the ribbon interface.
- Set the value of fillet to 3 mm.

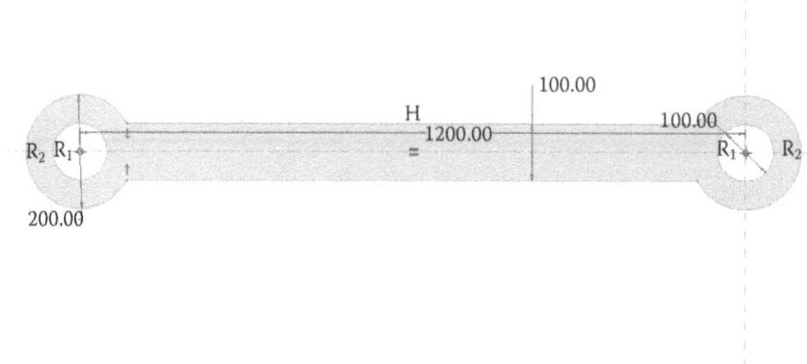

FIGURE 11.4
Constructional sketch of four-bar linkage.

- Select the edges that are not required to be round.
- Click OK to create the feature.

- Save and exit the file.

- Create a part file and rename it as Linkage2. (Other names may be given, but it is advisable to provide a relevant name.)
- Select Solid on the right-hand side of the box. This will activate the solid mode and environment.
- Uncheck the box of the default template and click OK to create a new solid file.
- The template selection box will appear.
- Select mmns_part_Solid. This will allow us to use the solid template file of the following units:
  - mm—millimeter for length unit
  - n—newton for force unit
  - s—second for time unit
- Once the solid part is created, go to File > Prepare > Model properties to check the units. In the upper sections, the units of the model will be displayed.
  - Note: The units usually are displayed in the summary section of any CAD software. The setting of units usually is in the tools or settings.
- Click on Extrude from the Model tab in the ribbon interface.
- Click on the Placement tab and select Define to define the sketch.
- Select the front plane as the sketch plane.
- The right plane shall be automatically taken as reference and orientation as right.
- Click on Sketch.
- The sketcher model will be activated.
- Draw the sketch as displayed in Figure 11.5.
- Click OK to create and exit the sketch.
- Select the mid plane extrusion.

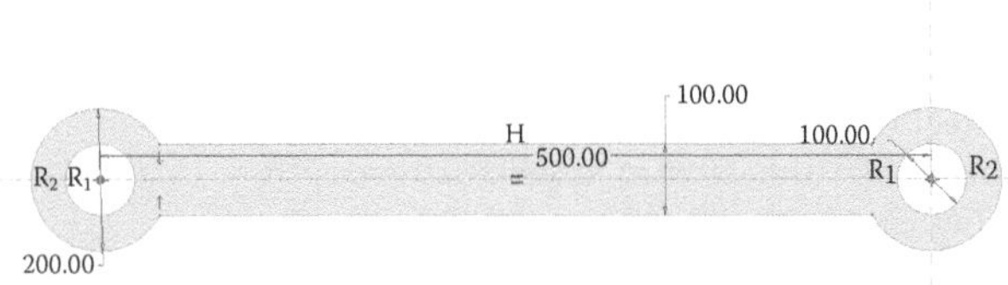

FIGURE 11.5
Constructional sketch of four-bar linkage.

- Enter the value 50 mm for extrusion depth.
- Click on OK to create the extrusion.

- Click on the Auto round feature from the Model tab of the ribbon interface.
- Set the value of fillet to 3 mm.
- Select the edges that are not required to be round.
- Click OK to create the feature.

- Save and exit the file.

- Create a part file and rename it as Linkage3. (Other names may be given, but it is advisable to provide a relevant name.)
- Select Solid on the right-hand side of the box. This will activate the solid mode and environment.
- Uncheck the box of the default template and click OK to create a new solid file.
- The template selection box will appear.
- Select mmns_part_Solid. This will allow us to use the solid template file of the following units:
    - mm—millimeter for length unit
    - n—newton for force unit
    - s—second for time unit
- Once the solid part is created, go to File > Prepare > Model properties to check the units. In the upper sections, the units of the model will be displayed.
    - Note: The units usually are displayed in the summary section of any CAD software. The setting of units usually is in the tools or settings.
- Click on Extrude from the Model tab in the ribbon interface.
- Click on the Placement tab and select Define to define the sketch.
- Select the front plane as the sketch plane.
- The right plane shall be automatically taken as reference and orientation as right.
- Click on Sketch.
- The sketcher model will be activated.
- Draw the sketch as displayed in Figure 11.6.
- Click OK to create and exit the sketch.
- Select the mid plane extrusion.
- Enter the value 50 mm for extrusion depth.
- Click on OK to create the extrusion.

- Click on the Auto round feature from the Model tab of the ribbon interface.
- Set the value of fillet to 3 mm.
- Select the edges that are not required to be round.
- Click OK to create the feature.

- Save and exit the file.

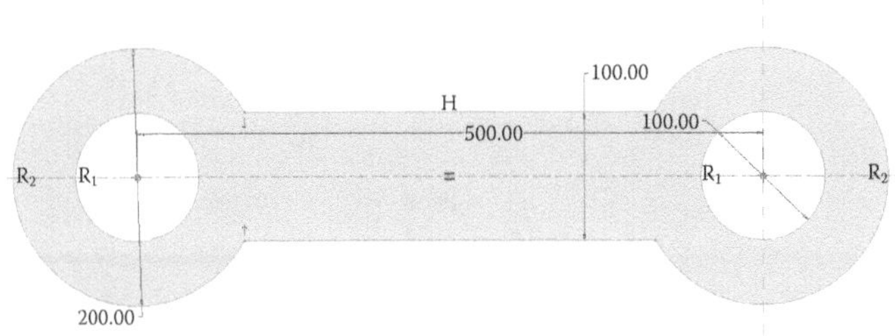

FIGURE 11.6
Constructional sketch of four-bar linkage.

- Create a part file and rename it as Pin. (Other names may be given, but it is advisable to provide a relevant name.)
- Select Solid on the right-hand side of the box. This will activate the solid mode and environment.
- Uncheck the box of the default template and click OK to create a new solid file.
- The template selection box will appear.
- Select mmns_part_Solid. This will allow us to use the solid template file of the following units:
  - mm—millimeter for length unit
  - n—newton for force unit
  - s—second for time unit
- Once the solid part is created, go to File > Prepare > Model properties to check the units. In the upper sections, the units of the model will be displayed.
  - Note: The units usually are displayed in the summary section of any CAD software. The setting of units usually is in the tools or settings.

- Click on Revolve from the Model tab in the ribbon interface.
- Click on the Placement tab and select Define to define the sketch.
- Select the top plane as the sketch plane.
- The right plane shall be automatically taken as reference and orientation as right.
- Click on Sketch.
- The sketcher model will be activated.
- Draw a centerline at the vertical reference as the axis of revolution.
- Draw the sketch as displayed in Figure 11.7.
- Click OK to create and exit the sketch.
- Ensure the revolve angle shall be 360°.
- Click on OK to create the revolve feature.
- Save and exit the file.

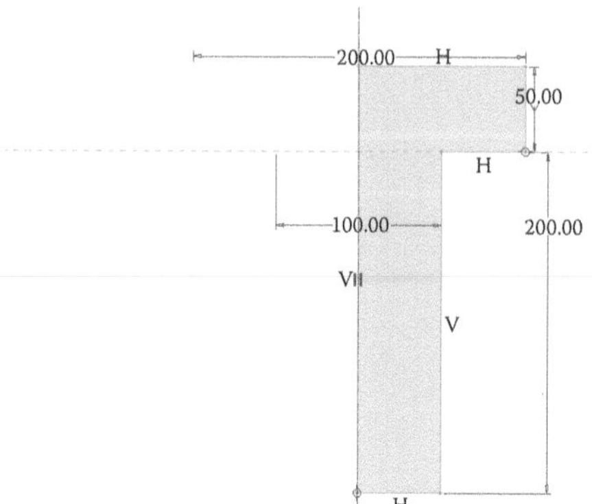

FIGURE 11.7
Constructional sketch of four-bar linkage.

- Create an assembly file and rename it as 4_bar_Linkage.asm. (Other names may be given, but it is advisable to provide a relevant name.)
- Select Design on the right-hand side of the box. This will activate the standard assembly design mode and environment.
- Uncheck the box of the default template and click OK to create a new assembly design file.
- The template selection box will appear.
- Select mmns_asm_Solid. This will allow us to use the solid template file of the following units:
  - mm—millimeter for length unit
  - n—newton for force unit
  - s—second for time unit
- Once the assembly file is created, go to File > Prepare > Model properties to check the units. In the upper sections, the units of the model will be displayed.
  - Note: The units usually are displayed in the summary section of any CAD software. The setting of units usually is in the tools or settings.
- Click on tree filters as shown in Figure 11.8.
- Check the box features to display in the model tree.
- Click OK to close the box of model tree items.
- Click on Apply. Notice that the features are displayed in the model tree.

- Click on Assembly from the Model tab of the ribbon interface.
- Select the file Ground_body.
- Click on the Placement tab.
- Select Default in the constraint type selection option box.
- Click OK to assemble the component.

- Click on Assembly from the Model tab of the ribbon interface.
- Select the file Linkage1.

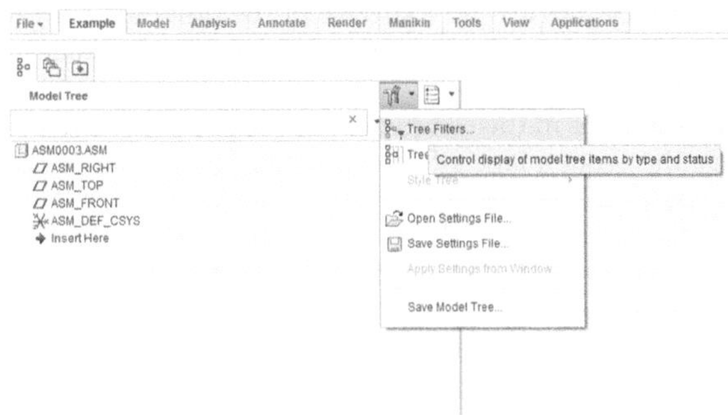

FIGURE 11.8
Model tree filter.

- Click on the Placement tab.
- Select Pin joint in the connection selection field.
- Select any cylindrical hole surface of the Linkage1 part as a component reference.
- Select the cylindrical hole surface of feature extrude_3 of the Ground_body as an assembly reference.
- Select Coincide in the component reference relation field.
- Select the front plane of the Linkage1 part as a component reference.
- Select the front plane of Ground_body as an assembly reference.
- Select Coincide in the component reference relation field.
- Click OK to assemble the component.

- Click on Assembly from the Model tab of the ribbon interface.
- Select the file Linkage2.
- Click on the Placement tab.
- Select Pin joint in the connection selection field.
- Select any cylindrical hole surface of the Linkage2 part as a component reference.
- Select the cylindrical hole surface of the other side (where linkage1 is not attached) of feature extrude_3 of Ground_body as an assembly reference.
- Select Coincide in the component reference relation field.
- Select the front plane of the Linkage2 part as a component reference.
- Select the front plane of Ground_body as an assembly reference.
- Select Coincide in the component reference relation field.
- Click OK to assemble the component.

- Click on Assembly from the Model tab of the ribbon interface.
- Select the file Linkage3.
- Click on the Placement tab.
- Select Pin joint in the connection selection field.
- Select any cylindrical hole surface of the Linkage3 part as a component reference.

- Select the cylindrical hole surface of the other side of the hole (where the ground body is not attached) of Linkage1 as an assembly reference.
- Select Coincide in the component reference relation field.
- Select the front plane of the Linkage3 part as a component reference.
- Select the front plane of Linkage1 as an assembly reference.
- Select Coincide in the component reference relation field.
- Click on Add sets.
- Select Cylindrical joint in the connection selection field.
- Select the free cylindrical hole surface of the Linkage3 part as a component reference.
- Select the free cylindrical hole surface of Linkage2 as an assembly reference.

- Click OK to assemble the component.

A servomotor is placed on the motion axis between link 3 and link 4. The output profile is specified in terms of position with respect to time, and the cosine profile is selected with the position variation of ±30° and the offset reference 90°.

Analysis:

1. Set the analysis for 10 s with a frame rate of 10 at 0.1 interval.
2. Run the analysis for position, which is purely a kinematic analysis, that is, without considering the effect of forces.

Result:

1. Select the measure window and create a measure that is the position of the motion axis between link 1 and link 3. Select the analysis and measure, and then plot it to graph. The position of the motion axis is well plotted against each time step.
2. Playback of analysis can be created for animation and further storage into video files for presentation.

Slider crank mechanism

Procedure of making a CAD model:

Hands-On Exercise, Example 11.2: Crank Cylinder (Software Used: Creo Parametric; Figure 11.9)

CRANK CYLINDER

Steps:

- Create a part file and rename it as Cylinder. (Other names may be given, but it is advisable to provide a relevant name.)

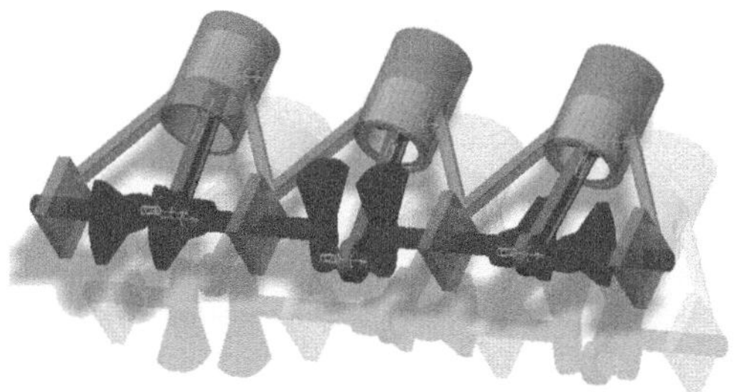

FIGURE 11.9
3-D view of crank cylinder mechanism.

- Select Sheet metal on the right-hand side of the box. This will activate the sheet metal mode and environment.
- Uncheck the box of the default template and click OK to create a new solid file.
- The template selection box will appear.
- Select mmns_part_Solid. This will allow us to use the solid template file of the following units:
  - mm—millimeter for length unit
  - n—newton for force unit
  - s—second for time unit
- Once the solid part is created, go to File > Prepare > Model properties to check the units. In the upper sections, the units of the model will be displayed.
  - Note: The units usually are displayed in the summary section of any CAD software. The setting of units usually is in the tools or settings.
- Click on Revolve from the Model tab in the ribbon interface.
- Click on the Placement tab and select Define to define the sketch.
- Select the top plane as the sketch plane.
- The right plane shall be automatically taken as reference and orientation as right.
- Click on Sketch.
- Activate the Thicken option.
- The sketcher model will be activated.
- Draw the sketch as displayed in Figure 11.10.
- Click OK to create and exit the sketch.
- Ensure the revolve angle shall be 360°.
- Enter 10 as the thickness.
- Flip the material orientation outward.
- Click on OK to create the revolve feature.

- Click on Extrude from the Model tab in the ribbon interface.
- Click on the Placement tab and select Define to define the sketch.
- Select the front plane as the sketch plane.

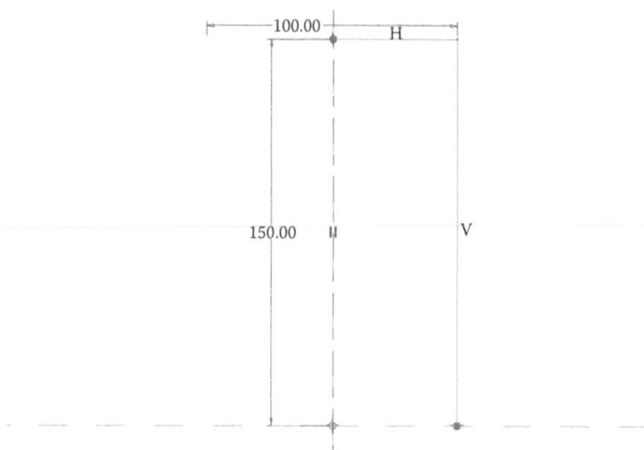

FIGURE 11.10
Constructional sketch of crank cylinder mechanism.

- The right plane shall be automatically taken as reference and orientation as right.
- Click on Sketch.
- The sketcher model will be activated.
- Draw the sketch as shown in displayed Figure 11.11.
- Click OK to create and exit the sketch.
- Select the mid plane extrusion.
- Enter the value 20 mm for extrusion depth.
- Click on OK to create the extrusion.

- Click on Extrude from the Model tab in the ribbon interface.
- Click on the Placement tab and select Define to define the sketch.
- Select the left surface of extrude_2 as the sketch plane.

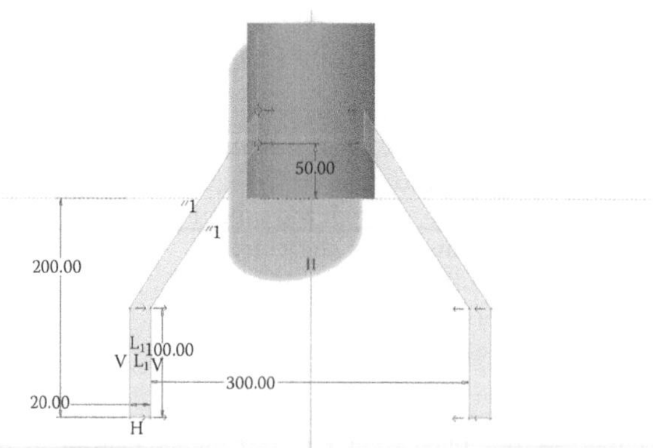

FIGURE 11.11
Constructional sketch of crank cylinder mechanism.

- The top plane shall be automatically taken as reference and orientation as right.
- Click on Sketch.
- The sketcher model will be activated.
- Draw the sketch as shown in displayed Figure 11.12.
- Click OK to create and exit the sketch.
- Flip the orientation toward the material side.
- Enter the value 20 mm for extrusion depth.
- Click on OK to create the extrusion.

- Select the features revolve_1, extrude_1, and extrude_2.
- Right-click and select Group.
- Select the group.
- Select Pattern from the Model tab of the ribbon interface.
- Select the direction.
- Select the right plane.
- Enter the value 320 mm.
- Enter the number of instances 3.
- Flip the material side.

- Click on Extrude from the Model tab in the ribbon interface.
- Click on the Placement tab and select Define to define the sketch.
- Select the right surface of the last extrusion.
- The top plane shall be automatically taken as reference and orientation as right.
- Click on Sketch.
- The sketcher model will be activated.
- Project the edges of extrude_2.
- Click OK to create and exit the sketch.
- Flip the orientation toward the material side.
- Enter the value 20 mm for extrusion depth.
- Click on OK to create the extrusion.

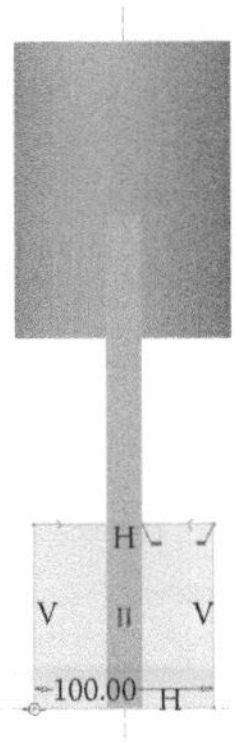

FIGURE 11.12
Constructional sketch of crank cylinder mechanism.

- Click on Extrude from the Model tab in the ribbon interface.
- Click on the Placement tab and select Define to define the sketch.
- Select the right surface of extrude_2 of the last pattern instance as the sketch plane.
- The top plane shall be automatically taken as reference and orientation as right.
- Click on Sketch.
- The sketcher model will be activated.
- Draw the sketch as displayed in Figure 11.13.
- Click OK to create and exit the sketch.
- Select Through all option.
- Flip the orientation toward the material side.
- Activate the material removal option.
- Click on OK to create the extrusion.

- Save and exit the file.

- Create a part file and rename it as Piston. (Other names may be given, but it is advisable to provide a relevant name.)
- Select Sheet metal on the right-hand side of the box. This will activate the sheet metal mode and environment.
- Uncheck the box of the default template and click OK to create a new solid file.
- The template selection box will appear.
- Select mmns_part_Solid. This will allow us to use the solid template file of the following units:
  - mm—millimeter for length unit
  - n—newton for force unit
  - s—second for time unit

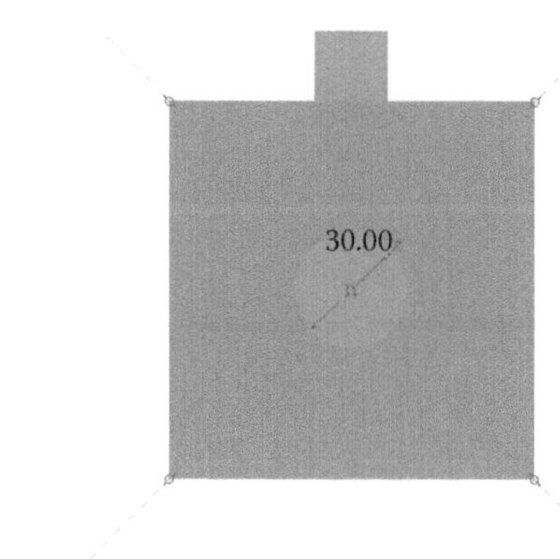

FIGURE 11.13
Constructional sketch of crank cylinder mechanism.

- Once the solid part is created, go to File > Prepare > Model properties to check the units. In the upper sections, the units of the model will be displayed.
    - Note: The units usually are displayed in the summary section of any CAD software. The setting of units usually is in the tools or settings.
- Click on Revolve from the Model tab in the ribbon interface.
- Click on the Placement tab and select Define to define the sketch.
- Select the front plane as the sketch plane.
- The right plane shall be automatically taken as reference and orientation as right.
- Click on Sketch.
- Activate the Thicken option.
- The sketcher model will be activated.
- Draw the sketch as displayed in Figure 11.14.
- Click OK to create and exit the sketch.
- Ensure the revolve angle shall be 360°.
- Enter 10 as the thickness.
- Flip the material orientation inward.
- Click on OK to create the revolve feature.

- Click on Extrude from the Model tab in the ribbon interface.
- Click on the Placement tab and select Define to define the sketch.
- Select the right plane as the sketch plane.
- The top plane shall be automatically taken as reference and orientation as left.
- Click on Sketch.
- The sketcher model will be activated.
- Draw a circle of diameter 25 placing the center at the intersection of the horizontal and vertical references.

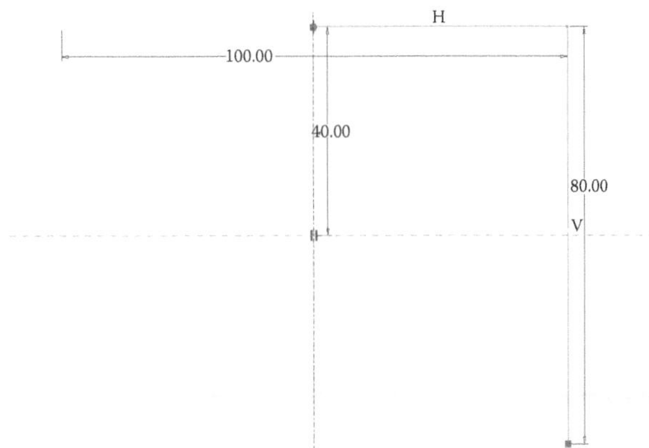

FIGURE 11.14
Constructional sketch of crank cylinder mechanism.

- Click OK to create and exit the sketch.
- Select Through all option.
- Flip the orientation toward the material side.
- Activate the material removal option.
- Select Through all for both the sides.
- Click on OK to create the extrusion.

- Save and exit the file.

- Create a part file and rename it as Connecting_Pin. (Other names may be given, but it is advisable to provide a relevant name.)
- Select Sheet metal on the right-hand side of the box. This will activate the sheet metal mode and environment.
- Uncheck the box of the default template and click OK to create a new solid file.
- The template selection box will appear.
- Select mmns_part_Solid. This will allow us to use the solid template file of the following units:
  - mm—millimeter for length unit
  - n—newton for force unit
  - s—second for time unit
- Once the solid part is created, go to File > Prepare > Model properties to check the units. In the upper sections, the units of the model will be displayed.
  - Note: The units usually are displayed in the summary section of any CAD software. The setting of units usually is in the tools or settings.
- Click on Extrude from the Model tab in the ribbon interface.
- Click on the Placement tab and select Define to define the sketch.
- Select the right plane as the sketch plane.
- The top plane shall be automatically taken as reference and orientation as left.
- Click on Sketch.
- The sketcher model will be activated.
- Draw a circle of diameter 25 placing the center at the intersection of the horizontal and vertical references.
- Click OK to create and exit the sketch.
- Select the mid plane extrusion.
- Enter the value 90 mm for extrusion depth.
- Click on OK to create the extrusion.

- Create a part file and rename it as Connecting_Rod. (Other names may be given, but it is advisable to provide a relevant name.)
- Select Sheet metal on the right-hand side of the box. This will activate the sheet metal mode and environment.
- Uncheck the box of the default template and click OK to create a new solid file.

- The template selection box will appear.
- Select mmns_part_Solid. This will allow us to use the solid template file of the following units:
  - mm—millimeter for length unit
  - n—newton for force unit
  - s—second for time unit
- Once the solid part is created, go to File > Prepare > Model properties to check the units. In the upper sections, the units of the model will be displayed.
  - Note: The units usually are displayed in the summary section of any CAD software. The setting of units usually is in the tools or settings.
- Click on Extrude from the Model tab in the ribbon interface.
- Click on the Placement tab and select Define to define the sketch.
- Select the top plane as the sketch plane.
- The right plane shall be automatically taken as reference and orientation as right.
- Click on Sketch.
- The sketcher model will be activated.
- Draw the sketch as shown in Figure 11.15.
- Click OK to create and exit the sketch.
- Select the mid plane extrusion.
- Enter the value 20 mm for extrusion depth.
- Click on OK to create the extrusion.

- Click on Extrude from the Model tab in the ribbon interface.
- Click on the Placement tab and select Define to define the sketch.
- Select the top flat face as the sketch plane.
- The right plane shall be automatically taken as reference and orientation as right.
- Click on Sketch.
- The sketcher model will be activated.

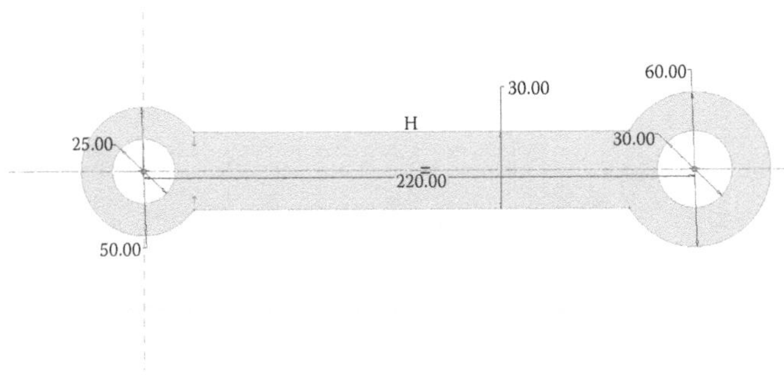

FIGURE 11.15
Constructional sketch of crank cylinder mechanism.

- Draw the sketch as shown in Figure 11.16.
- Click OK to create and exit the sketch.
- Flip the orientation toward the material side.
- Enter the value 5 mm for extrusion depth.
- Activate the material removal option.
- Click on OK to create the extrusion.

- Select the feature extrude_2.
- Click on Mirror from the Model tab of the ribbon interface.
- Select the top plane as the reference plane.
- Click OK to create the mirror feature.
- Save and exit the file.

- Create an assembly file and rename it as Piston Assembly. (Other names may be given, but it is advisable to provide a relevant name.)
- Select Design on the right-hand side of the box. This will activate the standard assembly design mode and environment.
- Uncheck the box of the default template and click OK to create a new assembly design file.
- The template selection box will appear.
- Select mmns_asm_Solid. This will allow us to use the solid template file of the following units:
  - mm—millimeter for length unit
  - n—newton for force unit
  - s—second for time unit
- Once the assembly file is created, go to File > Prepare > Model properties to check the units. In the upper sections, the units of the model will be displayed.
  - Note: The units usually are displayed in the summary section of any CAD software. The setting of units usually is in the tools or settings.
- Click on tree filters as shown in Figure 11.8.
- Click on the check box for displaying features.
- Click on Apply. Notice that the features are displayed in the model tree.

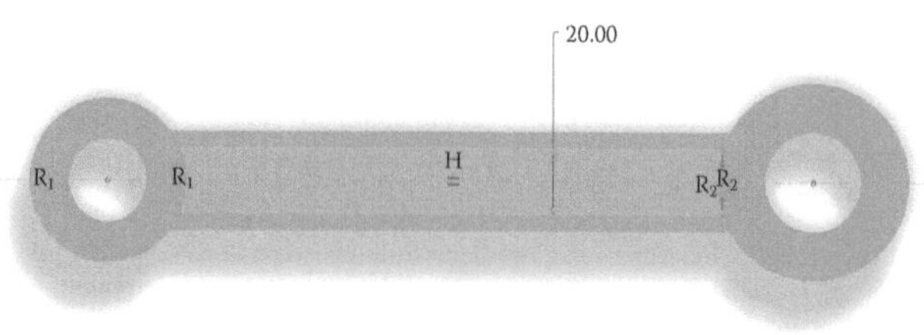

FIGURE 11.16
Constructional sketch of crank cylinder mechanism.

- Click on Assembly from the Model tab of the ribbon interface.
- Select the file Piston.
- Click on the Placement tab.
- Select Default in the constraint type selection option box.
- Click OK to assemble the component.

- Click on Assembly from the Model tab of the ribbon interface.
- Select the file Connecting_Pin.
- Click on the Placement tab.
- Select Pin joint in the connection selection field.
- Select the cylindrical surface of the Connecting_Pin part as a component reference.
- Select the cylindrical hole surface of the piston part as an assembly reference.
- Select the plane right of the Connecting_Pin part as a component reference.
- Select the right plane of the piston part as an assembly reference.
- Select Coincide in the component reference relation field.
- Click OK to assemble the component.

- Click on Assembly from the Model tab of the ribbon interface.
- Select the file Connecting_Rod.
- Click on the Placement tab.
- Select Pin joint in the connection selection field.
- Select the cylindrical hole surface of the Connecting_Rod part as a component reference.
- Select the cylindrical surface of Connecting_Pin as an assembly reference.
- Select the top right of the Connecting_Rod part as a component reference.
- Select the right plane of the Piston part as an assembly reference.
- Select Coincide in the component reference relation field.
- Click OK to assemble the component.

- Create an assembly file and rename it as Mechanism Assembly. (Other names may be given, but it is advisable to provide a relevant name.)
- Select Design on the right-hand side of the box. This will activate the standard assembly design mode and environment.
- Uncheck the box of the default template and click OK to create a new assembly design file.
- The template selection box will appear.
- Select mmns_asm_Solid. This will allow us to use the solid template file of the following units:
    - mm—millimeter for length unit
    - n—newton for force unit
    - s—second for time unit
- Once the assembly file is created, go to File > Prepare > Model properties to check the units. In the upper sections, the units of the model will be displayed.

    - Note: The units usually are displayed in the summary section of any CAD software. The setting of units usually is in the tools or settings.
- Click on tree filters as shown in Figure 11.8.

- Click on Apply. Notice that the features are displayed in the model tree.

- Click on Assembly from the Model tab of the ribbon interface.
- Select the file Piston_Assembly.
- Click on the Placement tab.
- Select Default in the constraint type selection option box.
- Click OK to assemble the component.

- Click on Assembly from the Model tab of the ribbon interface.
- Select the file Crank_Shaft.
- Click on the Placement tab.
- Select Pin joint in the connection selection field.
- Select the cylindrical surface of the Crank_Shaft part as a component reference.
- Select the cylindrical hole surface of the Cylinder part as an assembly reference.
- Select the plane right of the Crank_Shaft part as a component reference.
- Select the right plane of the Cylinder part as an assembly reference.
- Select Coincide in the component reference relation field.
- Click OK to assemble the component.

- Click on Assembly from the Model tab of the ribbon interface.
- Select the file Piston_Assembly.
- Click on the Placement tab.
- Select Cylinder joint in the connection selection field.
- Select the cylindrical surface of the piston part as a component reference.
- Select the inner cylindrical surface of the Cylinder part as an assembly reference.
- Select the references as shown in Figure 11.17.
- Click OK to assemble the component.

- Similarly assemble two other pistons in two other cylinder shells by following the aforementioned method.
- Click on Drag components from the Model tab of the ribbon interface.
- Click on the part Crank_Shaft and rotate. Notice the result.
- Save and exit the file.

1. Assign a servomotor on the motion axis of the journal to crank. In real life, though the force is transferred from the piston itself, here, we are assigning a motor to the crank for kinematic analysis.

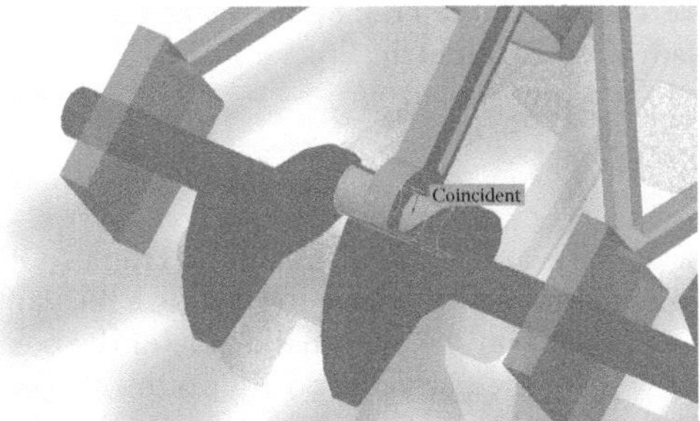

FIGURE 11.17
Constructional view of crank cylinder mechanism.

2. Assign a constant velocity profile 10 of rpm. Hence, the velocity in degrees per second will be

$$\text{rpm}/60 \times 360° = 60°/\text{s}$$

Analysis:

1. Set the analysis for 10 s with a frame rate of 10 at 0.1 interval.
2. Run the analysis for position, which is purely a kinematic analysis, that is, without considering the effect of forces.

Result:

1. Select the measure window and create a measure by selecting any point on the top surface of the pistons.
2. Playback of analysis can be created for animation and further storage into video files for presentation.

Curve tracing: Generation of the path of any point on the moving link of a mechanism is called curve tracing. Curve tracing generates a 3-D curve, which can be used further for design and analysis of the mechanism and its associative components. Analytical curves can also be generated for curve tracing, which is useful to equate with the analysis. Here, some examples of curve tracing are presented.

Epicycloid and hypocycloid: When a circle rotates on a straight line, the path of any point on the circle is called *cycloid*. When the circle externally rotates on another circle, the path of the point on the rotating circle is called *epicycloid*. Similarly, when the circle rotates internally on any circle, the path of the point of the rotating circle is called *hypocycloid*.

Involutes

Example of mechanism:

Screw Jack system

Procedure of making a CAD model:

Hands-On Exercise, Example 11.3: Screw Jack (Software Used: Creo Parametric; Figure 11.18)

SCREW JACK

Steps:

- Create a part file and rename it as Base_Frame. (Other names may be given, but it is advisable to provide a relevant name.)
- Select sheet metal on the right-hand side of the box. This will activate the sheet metal mode and environment.
- Uncheck the box of the default template and click OK to create a new solid file.
- The template selection box will appear.
- Select mmns_part_Solid. This will allow us to use the solid template file of the following units:
    - mm—millimeter for length unit
    - n—newton for force unit
    - s—second for time unit
- Once the solid part is created, go to File > Prepare > Model properties to check the units. In the upper sections, the units of the model will be displayed.
    - Note: The units usually are displayed in the summary section of any CAD software. The setting of units usually is in the tools or settings.
- Click on Extrude from the Model tab in the ribbon interface.
- Click on the Placement tab and select Define to define the sketch.
- Select the front plane as the sketch plane.
- The right plane shall be automatically taken as reference and orientation as right.
- Click on Sketch.
- The sketcher model will be activated.
- Draw the sketch as shown in Figure 11.19.
- Click OK to create and exit the sketch.
- Select the mid plane extrusion and enter 150 as the extrusion depth.

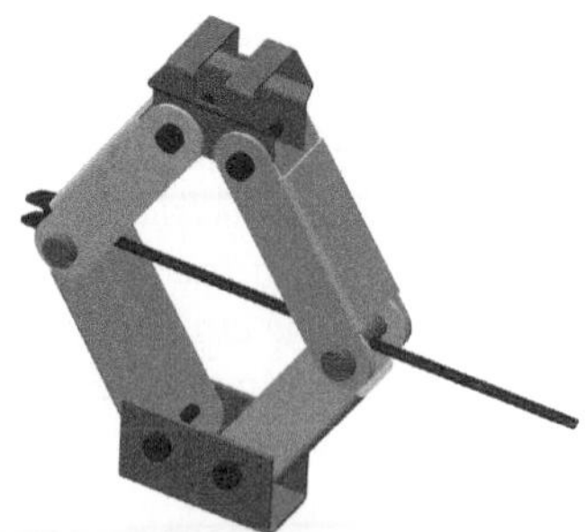

FIGURE 11.18
(See color insert.) 3-D view of screw jack.

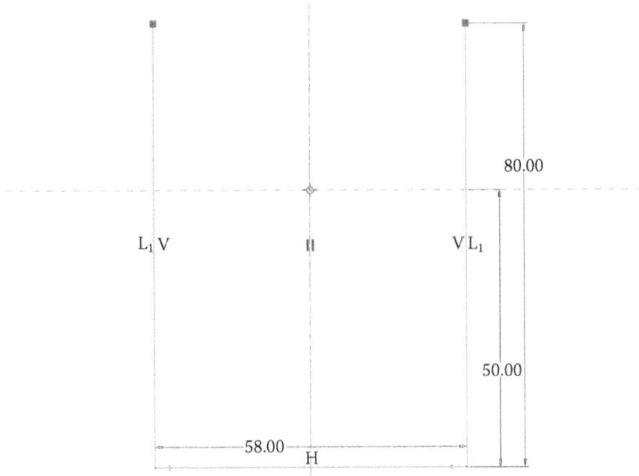

FIGURE 11.19
Constructional sketch of screw jack.

- Enter the value 4 mm for thickness.
- Flip the material orientation outward.
- Click on OK to create the extrusion.

- Click on Extrude from the Model tab in the ribbon interface.
- Click on the Placement tab and select Define to define the sketch.
- Select the front face of the lastly created extrusion as the sketch plane.
- The front plane shall be automatically taken as reference and orientation as left.
- Click on Sketch.
- The sketcher model will be activated.
- Draw the sketch as shown in Figure 11.20.
- Click OK to create and exit the sketch.
- The material removal option shall be automatically activated.
- Select Through all option.
- Flip the orientation toward the material side.
- Click on the Options tab.
- Check the box to add bends on sharp edges.
- Select thickness as Bend radius as inside.
- Click on OK to create the extrusion.

- Save and exit the file.

- Create a part file and rename it as Link. (Other names may be given, but it is advisable to provide a relevant name.)
- Select Sheet metal on the right-hand side of the box. This will activate the sheet metal mode and environment.
- Uncheck the box of the default template and click OK to create a new solid file.

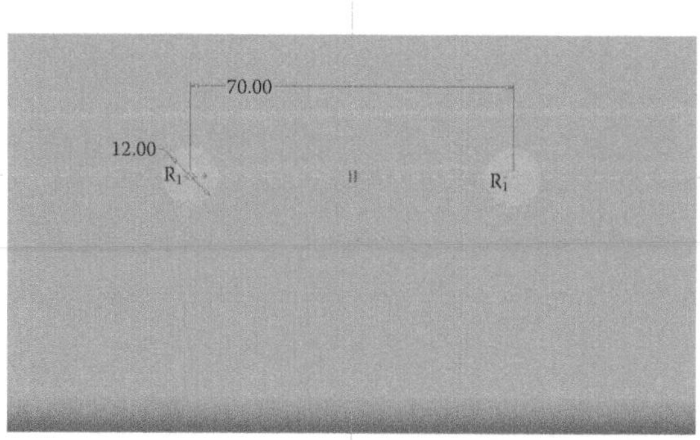

FIGURE 11.20
Constructional sketch of screw jack.

- The template selection box will appear.
- Select mmns_part_Solid. This will allow us to use the solid template file of the following units:
  - mm—millimeter for length unit
  - n—newton for force unit
  - s—second for time unit
- Once the solid part is created, go to File > Prepare > Model properties to check the units. In the upper sections, the units of the model will be displayed.
  - Note: The units usually are displayed in the summary section of any CAD software. The setting of units usually is in the tools or settings.
- Click on Extrude from the Model tab in the ribbon interface.
- Click on the Placement tab and select Define to define the sketch.
- Select the front plane as the sketch plane.
- The right plane shall be automatically taken as reference and orientation as right.
- Click on Sketch.
- The sketcher model will be activated.
- Draw the sketch as shown in Figure 11.21.
- Click OK to create and exit the sketch.
- Flip the extrusion upward.
- Enter the value 150 mm for extrusion depth.
- Select the Thicken part option.
- Enter 4 as the thickness value.
- Flip the material side outward.
- Check the box to add bends on sharp edges.
- Select thickness as Bend radius as inside.
- Click on OK to create the extrusion.

- Click on flat from the model tab of the ribbon interface.
- Select the edge as displayed in Figure 11.22.

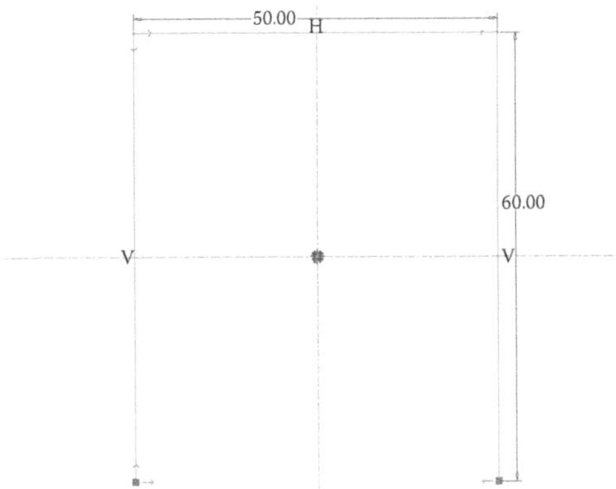

FIGURE 11.21
Constructional sketch of screw jack.

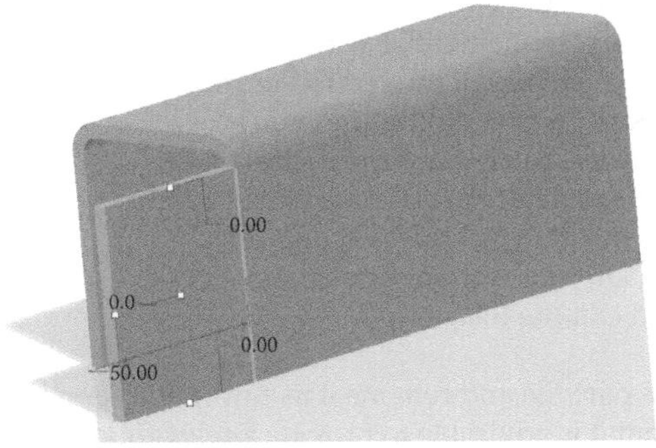

FIGURE 11.22
Constructional sketch of screw jack.

- Enter 0° as the flat bend angle.
- Click on the Shape tab.
- Double-click the flat dimension and enter 50.
- Click OK to create the flat.
- Similarly, select four other similar outside edges and create four flats.

- Click on the Round feature from the Model tab of the ribbon interface.
- Set the value of fillet to 28 mm.
- Select the edges as displayed in Figure 11.23.
- Click OK to create the feature.

- Click on Extrude from the Model tab in the ribbon interface.
- Click on the Placement tab and select Define to define the sketch.
- Select the front flat surface as the sketch plane.

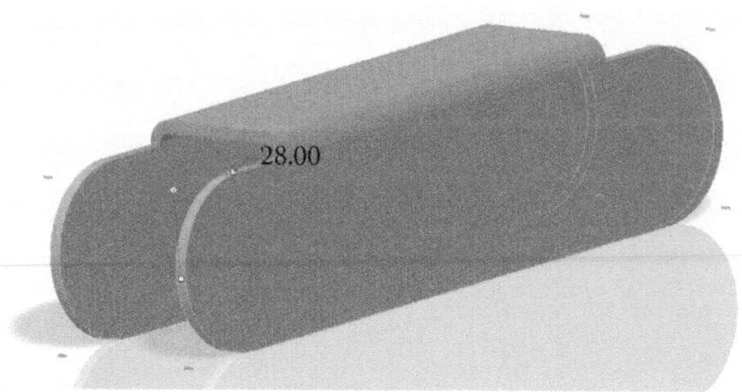

FIGURE 11.23
Constructional sketch of screw jack.

- The top plane shall be automatically taken as reference and orientation as top.
- Click on Sketch.
- The sketcher model will be activated.
- Draw the sketch as displayed in Figure 11.24.
- Click OK to create and exit the sketch.
- Flip the extrusion toward the material side.
- Select Through all option for extrusion depth.
- Activate the material removal option.
- Click on OK to create the extrusion.
- Save and exit the file.

- Create a part file and rename it as Link1. (Other names may be given, but it is advisable to provide a relevant name.)
- Select Sheet metal on the right-hand side of the box. This will activate the sheet metal mode and environment.
- Uncheck the box of the default template and click OK to create a new solid file.
- The template selection box will appear.
- Select mmns_part_Solid. This will allow us to use the solid template file of the following units:
  - mm—millimeter for length unit
  - n—newton for force unit
  - s—second for time unit

FIGURE 11.24
Constructional sketch of screw jack.

- Once the solid part is created, go to File > Prepare > Model properties to check the units. In the upper sections, the units of the model will be displayed.
    - Note: The units usually are displayed in the summary section of any CAD software. The setting of units usually is in the tools or settings.
- Click on Extrude from the Model tab in the ribbon interface.
- Click on the Placement tab and select Define to define the sketch.
- Select the front plane as the sketch plane.
- The right plane shall be automatically taken as reference and orientation as right.
- Click on Sketch.
- The sketcher model will be activated.
- Draw the sketch as shown in Figure 11.25.
- Click OK to create and exit the sketch.
- Flip the extrusion upward.
- Enter the value 150 mm for extrusion depth.
- Select the Thicken part option.
- Enter 4 as the thickness value.
- Flip the material side outward.
- Check the box to add bends on sharp edges.
- Select thickness as Bend radius as inside.
- Click on OK to create the extrusion.

- Click on Flat from the Model tab of the ribbon interface.
- Select the edge as displayed in Figure 11.26.
- Enter 0° as the flat bend angle.
- Click on the Shape tab.
- Double-click the flat dimension and enter 50.
- Click OK to create the flat.
- Similarly, select four other similar outside edges and create four flats.

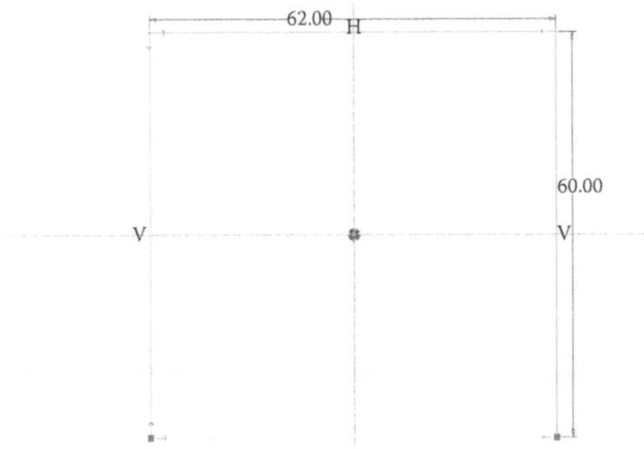

FIGURE 11.25
Constructional sketch of screw jack.

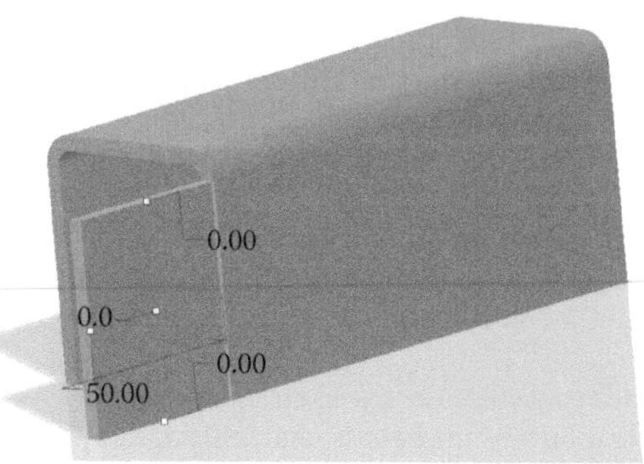

FIGURE 11.26
Constructional sketch of screw jack.

- Click on the Round feature from the Model tab of the ribbon interface.
- Set the value of fillet to 28 mm.
- Select the edges as displayed in Figure 11.27.
- Click OK to create the feature.

- Click on Extrude from the Model tab in the ribbon interface.
- Click on the Placement tab and select Define to define the sketch.
- Select the front flat surface as the sketch plane.
- The top plane shall be automatically taken as reference and orientation as top.
- Click on Sketch.
- The sketcher model will be activated.
- Draw the sketch as displayed in Figure 11.28.
- Click OK to create and exit the sketch.
- Flip the extrusion toward the material side.
- Select Through all option for extrusion depth.
- Activate the material removal option.

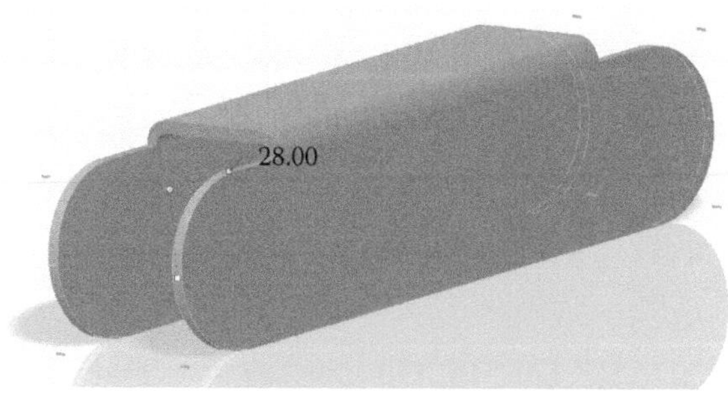

FIGURE 11.27
Constructional sketch of screw jack.

FIGURE 11.28
Constructional sketch of screw jack.

- Click on OK to create the extrusion.
- Save and exit the file.

- Create a part file and rename it as Top_Frame. (Other names may be given, but it is advisable to provide a relevant name.)
- Select Sheet metal on the right-hand side of the box. This will activate the sheet metal mode and environment.
- Uncheck the box of the default template and click OK to create a new solid file.
- The template selection box will appear.
- Select mmns_part_Solid. This will allow us to use the solid template file of the following units:
  - mm—millimeter for length unit
  - n—newton for force unit
  - s—second for time unit
- Once the solid part is created, go to File > Prepare > Model properties to check the units. In the upper sections, the units of the model will be displayed.
  - Note: The units usually are displayed in the summary section of any CAD software. The setting of units usually is in the tools or settings.
- Click on Extrude from the Model tab in the ribbon interface.
- Click on the Placement tab and select Define to define the sketch.
- Select the front plane as the sketch plane.
- The right plane shall be automatically taken as reference and orientation as right.
- Click on Sketch.
- The sketcher model will be activated.
- Draw the sketch as shown in Figure 11.29.
- Click OK to create and exit the sketch.
- Flip the extrusion upward.
- Enter the value 120 mm for extrusion depth.
- Select the Thicken part option.
- Enter 3 as the thickness value.
- Flip the material side outward.
- Check the box to add bends on sharp edges.
- Select thickness as Bend radius as inside.
- Click on OK to create the extrusion.

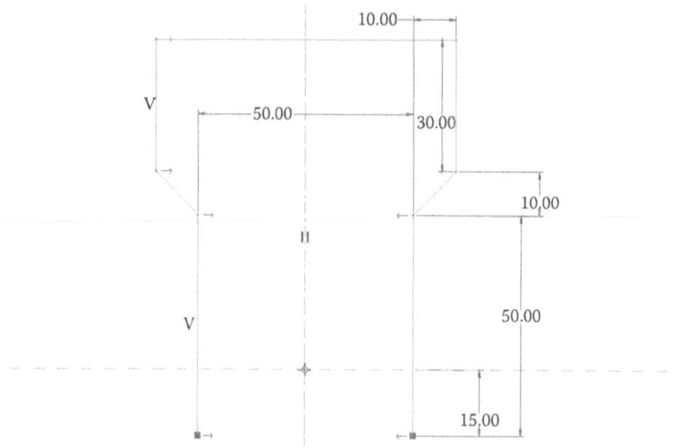

FIGURE 11.29
Constructional sketch of screw jack.

- Click on Extrude from the Model tab in the ribbon interface.
- Click on the Placement tab and select Define to define the sketch.
- Select the flat surface as displayed in Figure 11.30 as the sketch plane.
- The top plane shall be automatically taken as reference and orientation as top.
- Click on Sketch.
- The sketcher model will be activated.
- Draw the sketch as displayed in Figure 11.30.
- Click OK to create and exit the sketch.
- Flip the extrusion toward the material side.
- Select Through all option for extrusion depth.
- Activate the material removal option.
- Click on OK to create the extrusion.
- Save and exit the file.

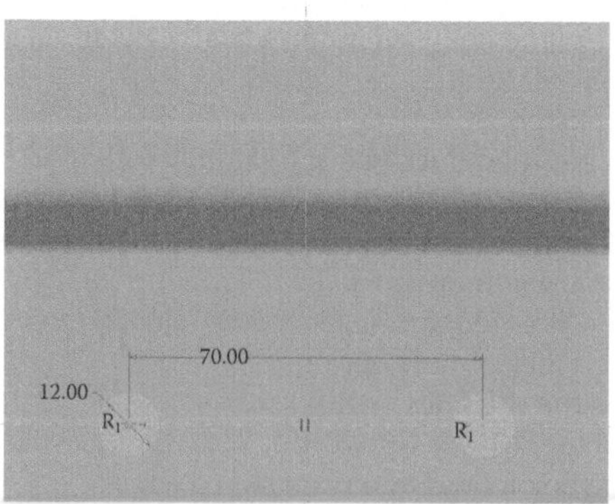

FIGURE 11.30
Constructional sketch of screw jack.

- Click on Extrude from the Model tab in the ribbon interface.
- Click on the Placement tab and select Define to define the sketch.
- Select the flat surface as displayed in Figure 11.31 as the sketch plane.
- The top plane shall be automatically taken as reference and orientation as top.
- Click on Sketch.
- The sketcher model will be activated.
- Draw the sketch as displayed in Figure 11.31.
- Click OK to create and exit the sketch.
- Flip the extrusion toward the material side.
- Select Through all option for extrusion depth.
- Activate the material removal option.
- Click on OK to create the extrusion.
- Save and exit the file.

- Create a part file and rename it as Pin1. (Other names may be given, but it is advisable to provide a relevant name.)
- Select Solid on the right-hand side of the box. This will activate the solid mode and environment.
- Uncheck the box of the default template and click OK to create a new solid file.
- The template selection box will appear.
- Select mmns_part_Solid. This will allow us to use the solid template file of the following units:
    - mm—millimeter for length unit
    - n—newton for force unit
    - s—second for time unit
- Once the solid part is created, go to File > Prepare > Model properties to check the units. In the upper sections, the units of the model will be displayed.

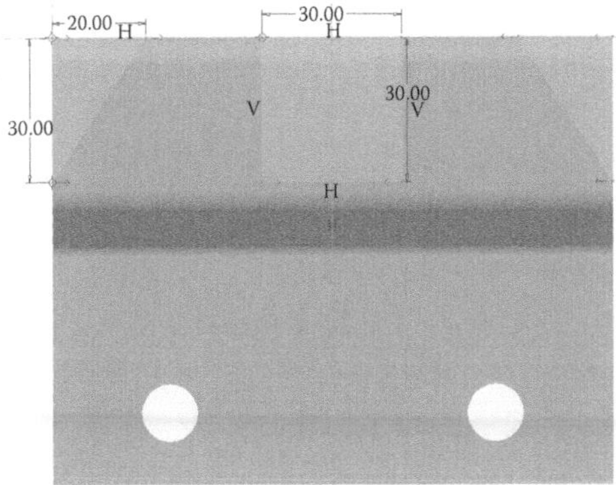

FIGURE 11.31
Constructional sketch of screw jack.

- Note: The units usually are displayed in the summary section of any CAD software. The setting of units usually is in the tools or settings.

- Click on Revolve from the Model tab in the ribbon interface.
- Click on the Placement tab and select Define to define the sketch.
- Select the front plane as the sketch plane.
- The right plane shall be automatically taken as reference and orientation as right.
- Click on Sketch.
- The sketcher model will be activated.
- Draw a centerline at the vertical reference as the axis of revolution.
- Draw the sketch as displayed in Figure 11.32.
- Click OK to create and exit the sketch.
- Ensure the revolve angle shall be 360°.
- Click on OK to create the revolve feature.

- Click on Extrude from the Model tab in the ribbon interface.
- Click on the Placement tab and select Define to define the sketch.
- Select the front plane as the sketch plane.
- The right plane shall be automatically taken as reference and orientation as right.
- Click on Sketch.
- The sketcher model will be activated.
- Take the references of the top and right planes as horizontal and vertical references.
- Draw a circle of diameter 6 placing the center at the intersection of the horizontal and vertical references.
- Click OK to create and exit the sketch.
- Activate the material removal option.

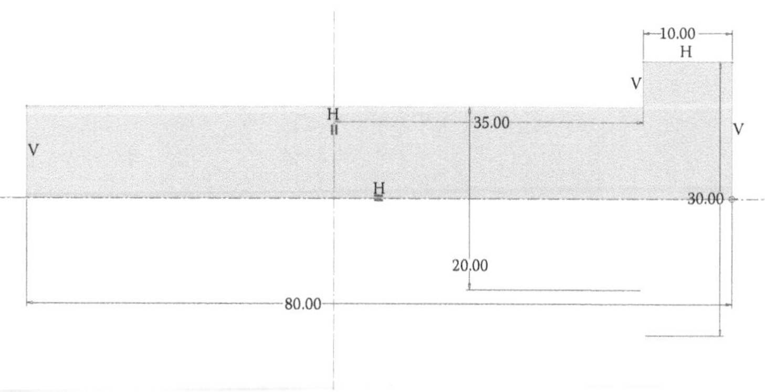

FIGURE 11.32
Constructional sketch of screw jack.

- Select Through all option for both sides.
- Click on OK to create the extrusion.

- Click on the Helical sweep option from the drop-down menu for sweep from the Model tab of the ribbon interface.
- Click on the References tab and select Define to define the helix sweep profile.
- Select the top plane as the sketch plane.
- The right plane shall be automatically taken as reference and orientation as right.
- Click on Sketch.
- The sketcher model will be activated.
- Draw the sketch as shown in Figure 11.33.
- Click OK to create the sketch and exit the sketcher window.
- Click Sketch to create the profile sketch above the Reference tab.
- Create the sketch as shown in Figure 11.34.
- Click OK to create and exit the sketch.
- Enter the pitch value of 4 mm.
- Activate the material removal option.
- Click on the left-handed rule icon to create the thread clockwise downward.
- Click OK to create the thread.

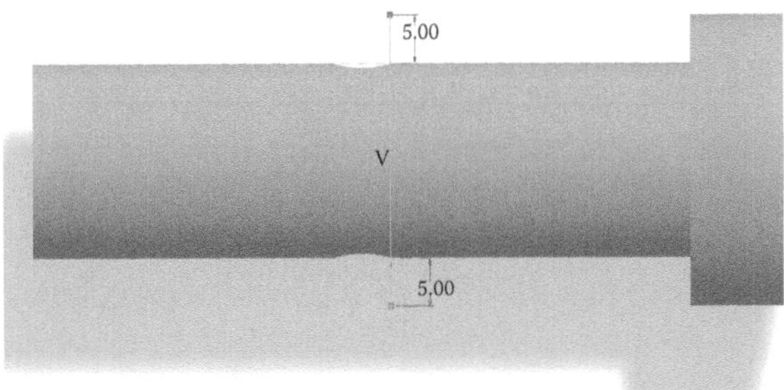

FIGURE 11.33
Constructional sketch of screw jack.

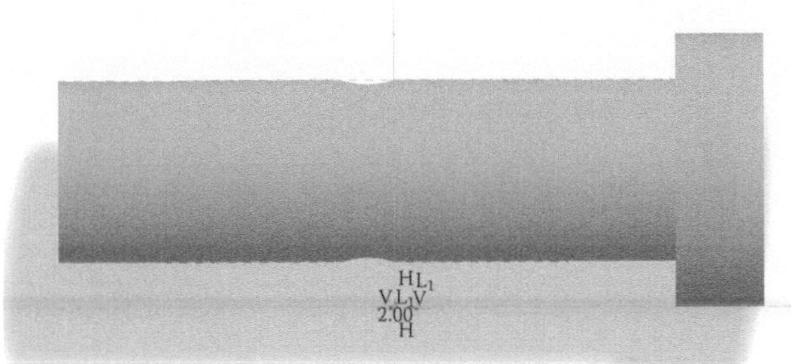

FIGURE 11.34
Constructional sketch of screw jack.

- Click on Extrude from the Model tab in the ribbon interface.
- Click on the Placement tab and select Define to define the sketch.
- Select the top plane as the sketch plane.
- The right plane shall be automatically taken as reference and orientation as right.
- Click on Sketch.
- The sketcher model will be activated.
- Draw the sketch as displayed in Figure 11.35.
- Click OK to create and exit the sketch.
- Select Through all option for extrusion depth for both the sides.
- Activate the material removal option.
- Click on OK to create the extrusion.

- Click on the datum point from the Model tab of the ribbon interface.
- Select the references as shown in Figure 11.36.
- Click OK to create the datum point.
- Save and exit the file.

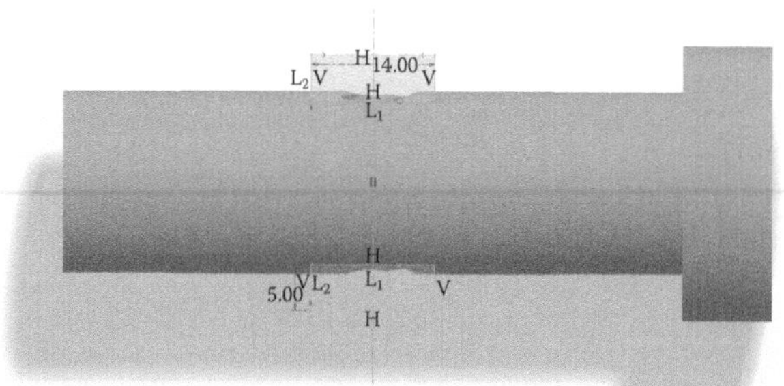

FIGURE 11.35
Constructional sketch of screw jack.

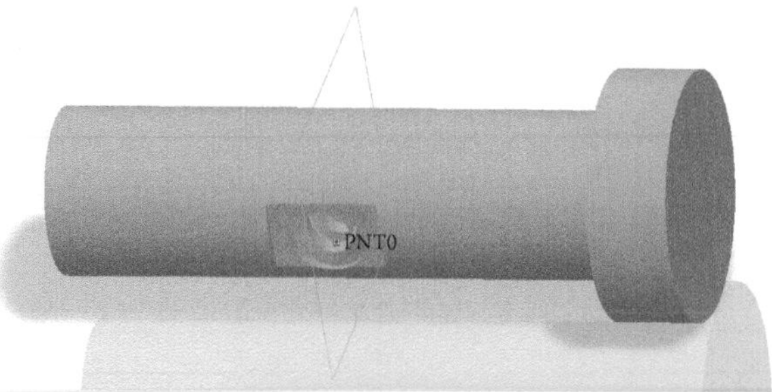

FIGURE 11.36
Constructional sketch of screw jack.

- Create a part file and rename it as Pin. (Other names may be given, but it is advisable to provide a relevant name.)
- Select Solid on the right-hand side of the box. This will activate the solid mode and environment.
- Uncheck the box of the default template and click OK to create a new solid file.
- The template selection box will appear.
- Select mmns_part_Solid. This will allow us to use the solid template file of the following units:
  - mm—millimeter for length unit
  - n—newton for force unit
  - s—second for time unit
- Once the solid part is created, go to File > Prepare > Model properties to check the units. In the upper sections, the units of the model will be displayed.
  - Note: The units usually are displayed in the summary section of any CAD software. The setting of units usually is in the tools or settings.

- Click on Revolve from the Model tab in the ribbon interface.
- Click on the Placement tab and select Define to define the sketch.
- Select the front plane as the sketch plane.
- The right plane shall be automatically taken as reference and orientation as right.
- Click on Sketch.
- The sketcher model will be activated.
- Draw a centerline at the vertical reference as the axis of revolution.
- Draw the sketch as displayed in Figure 11.37.
- Click OK to create and exit the sketch.
- Ensure the revolve angle shall be 360°.
- Click on OK to create the revolve feature.

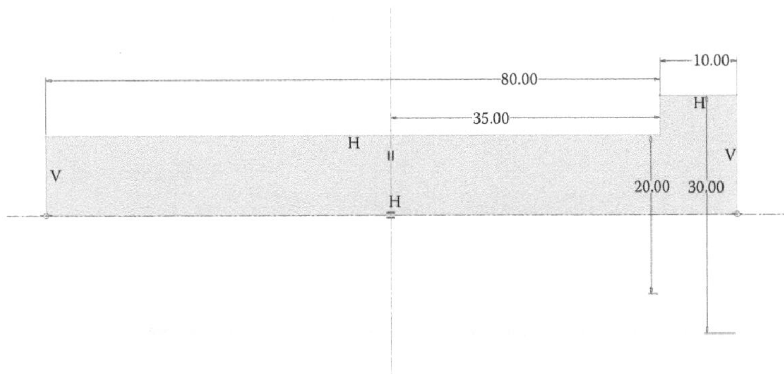

FIGURE 11.37
Constructional sketch of screw jack.

- Click on Extrude from the Model tab in the ribbon interface.
- Click on the Placement tab and select Define to define the sketch.
- Select the front plane as the sketch plane.
- The right plane shall be automatically taken as reference and orientation as right.
- Click on Sketch.
- The sketcher model will be activated.
- Take the references of the top and right planes as horizontal and vertical references.
- Draw a circle of diameter 6 placing the center at the intersection of the horizontal and vertical references.
- Click OK to create and exit the sketch.
- Activate the material removal option.
- Select Through all option for both sides.
- Click on OK to create the extrusion.

- Click on Extrude from the Model tab in the ribbon interface.
- Click on the Placement tab and select Define to define the sketch.
- Select the top plane as the sketch plane.
- The right plane shall be automatically taken as reference and orientation as right.
- Click on Sketch.
- The sketcher model will be activated.
- Draw the sketch as displayed in Figure 11.38.
- Click OK to create and exit the sketch.
- Select Through all option for extrusion depth for both the sides.
- Activate the material removal option.
- Click on OK to create the extrusion.

- Save and exit the file.

- Create a part file and rename it as Screw. (Other names may be given, but it is advisable to provide a relevant name.)
- Select Solid on the right-hand side of the box. This will activate the solid mode and environment.
- Uncheck the box of the default template and click OK to create a new solid file.

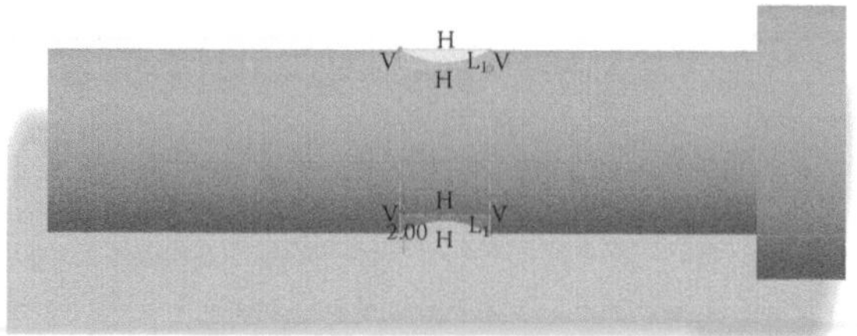

FIGURE 11.38
Constructional sketch of screw jack.

- The template selection box will appear.
- Select mmns_part_Solid. This will allow us to use the solid template file of the following units:
  - mm—millimeter for length unit
  - n—newton for force unit
  - s—second for time unit
- Once the solid part is created, go to File > Prepare > Model properties to check the units. In the upper sections, the units of the model will be displayed.
  - Note: The units usually are displayed in the summary section of any CAD software. The setting of units usually is in the tools or settings.

- Click on Revolve from the Model tab in the ribbon interface.
- Click on the Placement tab and select Define to define the sketch.
- Select the front plane as the sketch plane.
- The right plane shall be automatically taken as reference and orientation as right.
- Click on Sketch.
- The sketcher model will be activated.
- Draw a centerline at the vertical reference as the axis of revolution.
- Draw the sketch as displayed in Figure 11.39.
- Click OK to create and exit the sketch.
- Ensure the revolve angle shall be 360°.
- Click on OK to create the revolve feature.

- Click on Extrude from the Model tab in the ribbon interface.
- Activate the Material thicken option.
- Enter 3 as the thickness.
- Click on the Placement tab and select Define to define the sketch.
- Select the top plane as the sketch plane.

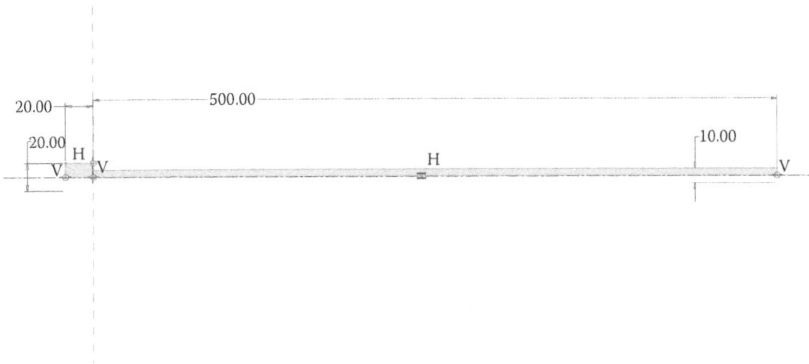

FIGURE 11.39
Constructional sketch of screw jack.

- The right plane shall be automatically taken as reference and orientation as right.
- Click on Sketch.
- The sketcher model will be activated.
- Take the references of the top and right planes as horizontal and vertical references.
- Draw the sketch as shown in Figure 11.40.
- Click OK to create and exit the sketch.
- Select the mid plane option.
- Enter 40 as the extrusion depth.
- Flip the material thickness inside.
- Click on OK to create the extrusion.

- Click on the Round feature from the Model tab of the ribbon interface.
- Set the value of fillet to 3 mm.
- Select two inside edges of the lastly created feature extrude_1.
- Click on the Sets tab.
- Click on add new sets.
- Enter the radius as 6 mm.
- Select two outside edges of the lastly created feature extrude_1.
- Click OK to create the feature.

- Click on Extrude from the Model tab in the ribbon interface.
- Click on the Placement tab and select Define to define the sketch.
- Select the flat surface as displayed in Figure 11.40 as the sketch plane.
- The top plane shall be automatically taken as reference and orientation as top.
- Click on Sketch.
- The sketcher model will be activated.
- Draw the sketch as displayed in Figure 11.41.

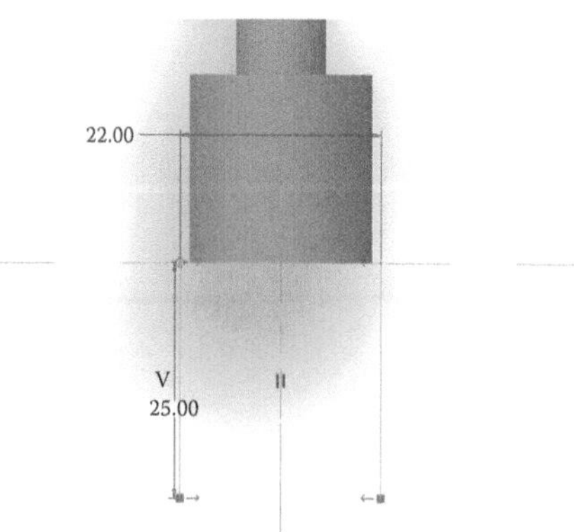

FIGURE 11.40
Constructional sketch of screw jack.

- Click OK to create and exit the sketch.
- Flip the extrusion toward the material side.
- Select Through all option for extrusion depth.
- Activate the material removal option.
- Click on OK to create the extrusion.

- Click on the Round feature from the Model tab of the ribbon interface.
- Set the value of fillet to 10 mm.
- Select all four edges as shown in Figure 11.42.
- Click OK to create the feature.

- Click on the Helical sweep option from the drop-down menu for sweep from the Model tab of the ribbon interface.
- Click on the References tab and select Define to define the helix sweep profile.
- Select the front plane as the sketch plane.

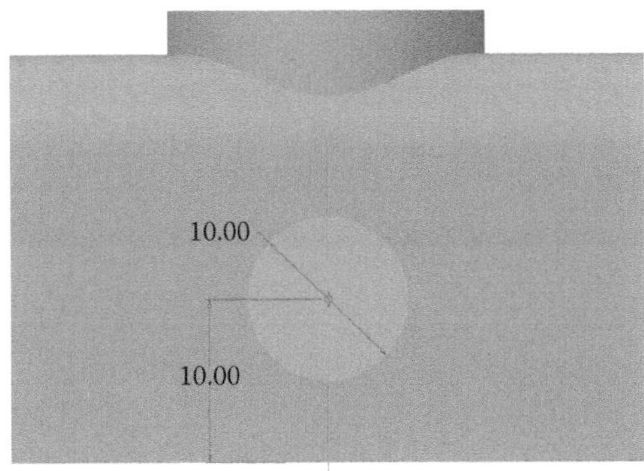

FIGURE 11.41
Constructional sketch of screw jack.

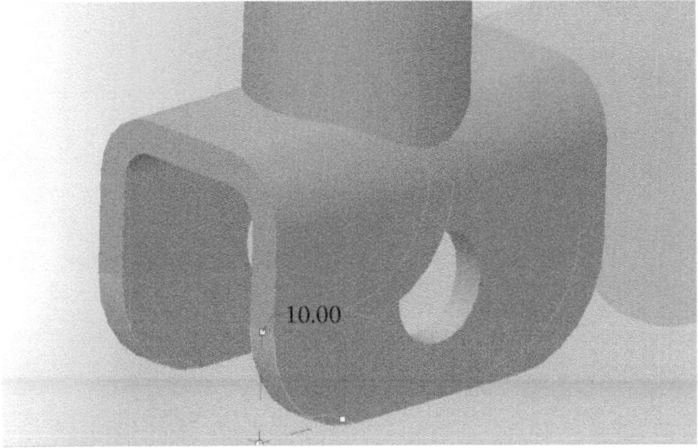

FIGURE 11.42
Constructional view of screw jack.

- The right plane shall be automatically taken as reference and orientation as right.
- Click on Sketch.
- The sketcher model will be activated.
- Draw the sketch as shown in Figure 11.43.
- Click OK to create the sketch and exit the sketcher window.
- Click Sketch to create the profile sketch above the Reference tab.
- Create the sketch as shown in Figure 11.44.
- Click OK to create and exit the sketch.
- Enter the pitch value of 4 mm.
- Activate the material removal option.
- Click on the left-handed rule icon to create the thread clockwise downward.
- Click OK to create the thread.

- Save and exit the file.

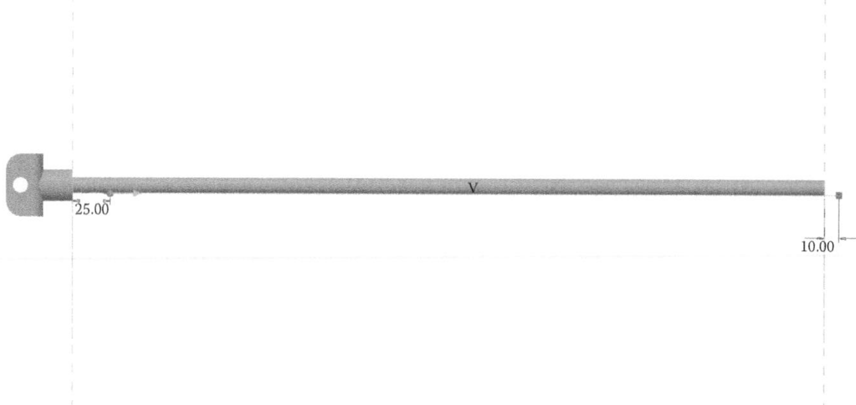

FIGURE 11.43
Constructional sketch of screw jack.

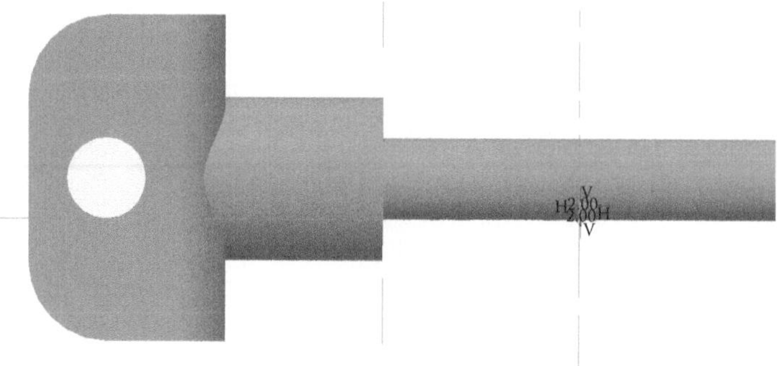

FIGURE 11.44
Constructional sketch of screw jack.

- Create a part file and rename it as Pin12. (Other names may be given, but it is advisable to provide a relevant name.)
- Select Solid on the right-hand side of the box. This will activate the solid mode and environment.
- Uncheck the box of the default template and click OK to create a new solid file.
- The template selection box will appear.
- Select mmns_part_Solid. This will allow us to use the solid template file of the following units:
  - mm—millimeter for length unit
  - n—newton for force unit
  - s—second for time unit
- Once the solid part is created, go to File > Prepare > Model properties to check the units. In the upper sections, the units of the model will be displayed.
  - Note: The units usually are displayed in the summary section of any CAD software. The setting of units usually is in the tools or settings.

- Click on Revolve from the Model tab in the ribbon interface.
- Click on the Placement tab and select Define to define the sketch.
- Select the front plane as the sketch plane.
- The right plane shall be automatically taken as reference and orientation as right.
- Click on Sketch.
- The sketcher model will be activated.
- Draw a centerline at the vertical reference as the axis of revolution.
- Draw the sketch as displayed in Figure 11.45.
- Click OK to create and exit the sketch.

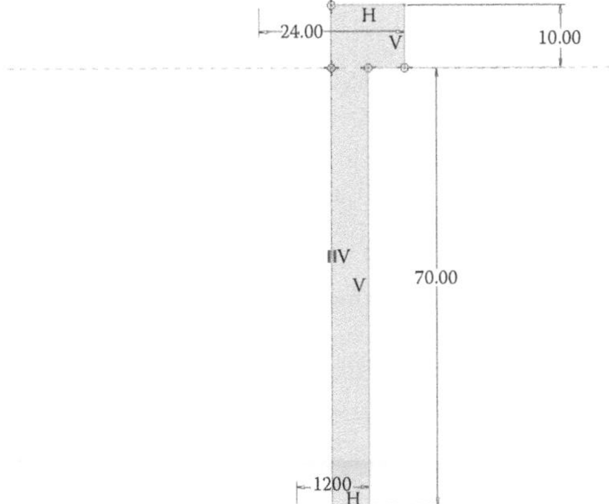

FIGURE 11.45
Constructional sketch of screw jack.

- Ensure the revolve angle shall be 360°.
- Click on OK to create the revolve feature.

- Save and exit the file.

- Create an assembly file and rename it as Screw_Jack.asm. (Other names may be given, but it is advisable to provide a relevant name.)
- Select Design on the right-hand side of the box. This will activate the standard assembly design mode and environment.
- Uncheck the box of the default template and click OK to create a new assembly design file.
- The template selection box will appear.
- Select mmns_asm_Solid. This will allow us to use the solid template file of the following units:
    - mm—millimeter for length unit
    - n—newton for force unit
    - s—second for time unit
- Once the assembly file is created, go to File > Prepare > Model properties to check the units. In the upper sections, the units of the model will be displayed.
    - Note: The units usually are displayed in the summary section of any CAD software. The setting of units usually is in the tools or settings.
- Click on tree filters as shown in Figure 11.8.
- Check the box features to display in the model tree.
- Click on Apply. Notice that the features are displayed in the model tree.

- Click on Assembly from the Model tab of the ribbon interface.
- Select the file Base_Frame.
- Click on the Placement tab.
- Select Default in the constraint type selection option box.
- Click OK to assemble the component.

- Click on Assembly from the Model tab of the ribbon interface.
- Select the file Link.
- Click on the Placement tab.
- Select Pin joint in the connection selection field.
- Select any cylindrical hole surface of the Link part as a component reference.
- Select any cylindrical hole surface of feature extrude_3 of Base_Frame as an assembly reference.
- Select Coincide in the component reference relation field.
- Select the right plane of the link part as a component reference.
- Select the right plane of Base_Frame as an assembly reference.
- Select Coincide in the component reference relation field.
- Click OK to assemble the component.

- Click on Assembly from the Model tab of the ribbon interface.
- Select the file Link.

- Click on the Placement tab.
- Select Pin joint in the connection selection field.
- Select any cylindrical hole surface of the Link part as a component reference.
- Select the other cylindrical hole surface of feature extrude_3 of Base_Frame as an assembly reference.
- Select Coincide in the component reference relation field.
- Select the right plane of the link part as a component reference.
- Select the right plane of Base_Frame as an assembly reference.
- Select Coincide in the component reference relation field.
- Click OK to assemble the component.

- Click on Assembly from the Model tab of the ribbon interface.
- Select the file Link1.
- Click on the Placement tab.
- Select Pin joint in the connection selection field.
- Select any cylindrical hole surface of the Link1 part as a component reference.
- Select the cylindrical hole surface of the first Link as an assembly reference.
- Select Coincide in the component reference relation field.
- Select the right plane of the Link1 part as a component reference.
- Select the right plane of Link as an assembly reference.
- Select Coincide in the component reference relation field.
- Click OK to assemble the component.

- Click on Assembly from the Model tab of the ribbon interface.
- Select the file Link1.
- Click on the Placement tab.
- Select Pin joint in the connection selection field.
- Select any cylindrical hole surface of the Link1 part as a component reference.
- Select the cylindrical hole surface of the second Link as an assembly reference.
- Select Coincide in the component reference relation field.
- Select the right plane of the Link1 part as a component reference.
- Select the right plane of Link as an assembly reference.
- Select Coincide in the component reference relation field.
- Click OK to assemble the component.

- Click on Assembly from the Model tab of the ribbon interface.
- Select the file Top_Frame.
- Click on the Placement tab.
- Select Pin joint in the connection selection field.
- Select any cylindrical hole surface of the Top_Frame part as a component reference.
- Select the cylindrical hole surface of the first Link1 as an assembly reference.
- Select Coincide in the component reference relation field.

- Select the right plane of the Top_Frame part as a component reference.
- Select the right plane of Link1 as an assembly reference.
- Select Coincide in the component reference relation field.
- Click on Add new set.
- Select Cylinder joint in the connection selection field.
- Select the other cylindrical hole surface of the Top_Frame part as a component reference.
- Select the cylindrical hole surface of the second Link1 as an assembly reference.
- Click on Add new set.
- Select Planar joint in the connection selection field.
- Select the front plane of the Top_Frame part as a component reference.
- Select the plane ASM_FRONT as an assembly reference.
- Select Coincide in the component reference relation field.

- Click OK to assemble the component.

- Click on Assembly from the Model tab of the ribbon interface.
- Select the file PIN1.
- Click on the Placement tab.
- Select Pin joint in the connection selection field.
- Select the cylindrical surface of the PIN1 part as a component reference.
- Select the large cylindrical hole surface of the first Link1 as an assembly reference.
- Select Coincide in the component reference relation field.
- Select the sitting flat face of the PIN1 part as a component reference.
- Select the outside flat face of the first Link1 as an assembly reference.
- Select Coincide in the component reference relation field.

- Click OK to assemble the component.

- Click on Assembly from the Model tab of the ribbon interface.
- Select the file PIN1.
- Click on the Placement tab.
- Select Pin joint in the connection selection field.
- Select the cylindrical surface of the PIN1 part as a component reference.
- Select the large cylindrical hole surface of the second Link1 as an assembly reference.
- Select Coincide in the component reference relation field.
- Select the sitting flat face of the PIN1 part as a component reference.
- Select the outside flat face of the second Link1 as an assembly reference.
- Select Coincide in the component reference relation field.

- Click OK to assemble the component.

- Click on Assembly from the Model tab of the ribbon interface.
- Select the file Screw.
- Click on the Placement tab.
- Select Pin joint in the connection selection field.
- Select the cylindrical surface of the Screw part as a component reference.
- Select the large cylindrical surface of PIN2 as an assembly reference.
- Select Coincide in the component reference relation field.
- Select the sitting flat face of the Screw part as a component reference.
- Select the outside sitting flat face of PIN2 as an assembly reference.
- Select Coincide in the component reference relation field.
- Click on Add new set.
- Select Cylinder joint in the connection selection field.
- Select the outside cylindrical surface of the screw part as a component reference.
- Select the cylindrical hole surface of PIN1 as an assembly reference.
- Click on Add new set.
- Select Slot joint in the connection selection field.
- Select the datum point of PIN1 as an assembly reference.
- Select the sweep curve as displayed in Figure 11.46. (Note: All the curves have to be selected by pressing the Control key.)
- Click OK to assemble the component.

- Click on Assembly from the Model tab of the ribbon interface.
- Select the file PIN12.
- Click on the Placement tab.

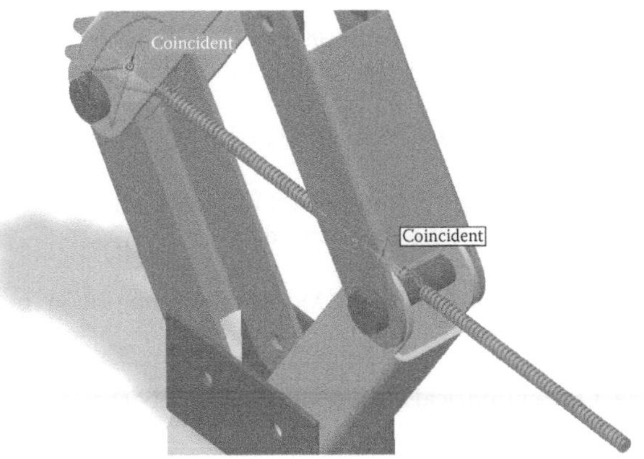

FIGURE 11.46
Constructional view of screw jack.

- Select the cylindrical surface of the PIN12 part as a component reference.
- Select the small cylindrical hole surface of the first Link1 as an assembly reference.
- Select Coincide in the component reference relation field.
- Select the sitting flat face of the PIN12 part as a component reference.
- Select the outside flat face of the first Link1 as an assembly reference.
- Select Coincide in the component reference relation field.
- Click OK to assemble the component.

- Click on Assembly from the Model tab of the ribbon interface.
- Select the file PIN12.
- Click on the Placement tab.
- Select the cylindrical surface of the PIN12 part as a component reference.
- Select the small cylindrical hole surface of the first Link1 as an assembly reference.
- Select Coincide in the component reference relation field.
- Select the sitting flat face of the PIN12 part as a component reference.
- Select the outside flat face of the second Link1 as an assembly reference.
- Select Coincide in the component reference relation field.
- Click OK to assemble the component.

- Click on Assembly from the Model tab of the ribbon interface.
- Select the file PIN12.
- Click on the Placement tab.
- Select the cylindrical surface of the PIN12 part as a component reference.
- Select any cylindrical hole surface of Base_Frame as an assembly reference.
- Select Coincide in the component reference relation field.
- Select the sitting flat face of the PIN12 part as a component reference.
- Select the outside flat face of the first Base_Frame as an assembly reference.
- Select Coincide in the component reference relation field.
- Click OK to assemble the component.

- Click on Assembly from the Model tab of the ribbon interface.
- Select the file PIN12.
- Click on the Placement tab.
- Select the cylindrical surface of the PIN12 part as a component reference.
- Select the other cylindrical hole surface of Base_Frame as an assembly reference.
- Select Coincide in the component reference relation field.

- Select the sitting flat face of the PIN12 part as a component reference.
- Select the outside flat face of the first Base_Frame as an assembly reference.
- Select Coincide in the component reference relation field.
- Click OK to assemble the component.

- Click on the drag components from the Model tab of the ribbon interface.
- Click on the screw part and rotate graphically. Notice that the jack will be lifted with the rotation of the screw.

Playback of analysis can be created for animation and further storage into video files for presentation.

Cam follower mechanism

Procedure of making a CAD model:

Hands-On Exercise, Example 11.4: Cam Mechanism (Software Used: Creo Parametric; Figure 11.47)

CAM MECHANISM

Steps:

- Create a part file and rename it as Fixed_Frame. (Other names may be given, but it is advisable to provide a relevant name.)
- Select Solid on the right-hand side of the box. This will activate the solid mode and environment.
- Uncheck the box of the default template and click OK to create a new solid file.
- The template selection box will appear.
- Select mmns_part_Solid. This will allow us to use the solid template file of the following units:
  - mm—millimeter for length unit
  - n—newton for force unit
  - s—second for time unit

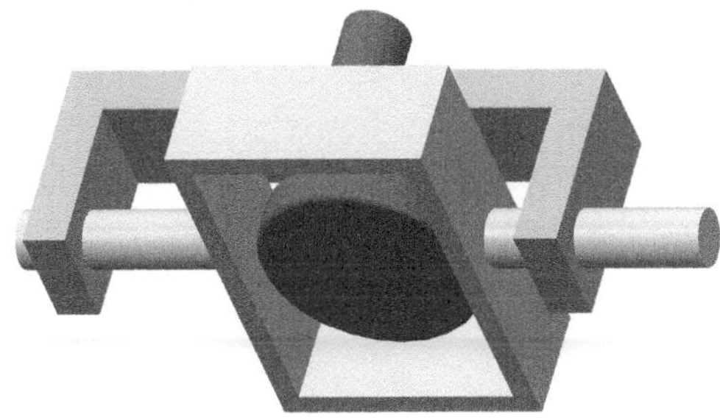

FIGURE 11.47
(See color insert.) 3-D view of cam mechanism.

- Once the solid part is created, go to File > Prepare > Model properties to check the units. In the upper sections, the units of the model will be displayed.
    - Note: The units usually are displayed in the summary section of any CAD software. The setting of units usually is in the tools or settings.
- Click on Extrude from the Model tab in the ribbon interface.
- Click on the Placement tab and select Define to define the sketch.
- Select the front plane as the sketch plane.
- The right plane shall be automatically taken as reference and orientation as right.
- Click on Sketch.
- The sketcher model will be activated.
- Draw a rectangle of 250 × 50 symmetrically about the horizontal and vertical references.
- A centerline may need to be drawn along the reference to draw the aforementioned rectangle.
- Click OK to create and exit the sketch.
- Enter the value 20 mm for extrusion depth.
- Click on OK to create the extrusion.

- Click on Extrude from the Model tab in the ribbon interface.
- Click on the Placement tab and select Define to define the sketch.
- Select the front plane as the sketch plane.
- The right plane shall be automatically taken as reference and orientation as right.
- Click on Sketch.
- The sketcher model will be activated.
- Draw the sketch as shown in Figure 11.48.
- Click OK to create and exit the sketch.
- Flip the material orientation upward.
- Enter the value 70 mm for extrusion depth.
- Click on OK to create the extrusion.

FIGURE 11.48
Constructional sketch of cam mechanism.

- Click on Extrude from the Model tab in the ribbon interface.
- Click on the Placement tab and select Define to define the sketch.
- Select the side surface of the lastly created feature extrude_2 as the sketch plane.
- The top surface shall be automatically taken as reference and orientation as top.
- Click on Sketch.
- The sketcher model will be activated.
- Draw the sketch as shown in Figure 11.49.
- Click OK to create and exit the sketch.
- Flip the orientation toward the material side.
- Select Through all for depth option.
- Activate the material removal option.
- Click on OK to create the extrusion.

- Click on Extrude from the Model tab in the ribbon interface.
- Click on the Placement tab and select Define to define the sketch.
- Select the front plane as the sketch plane.
- The top surface shall be automatically taken as reference and orientation as top.
- Click on Sketch.
- The sketcher model will be activated.
- Draw a circle of diameter 25 placing the center at the intersection of the horizontal and vertical references.
- Click OK to create and exit the sketch.
- Flip the orientation toward the material side.
- Select Through all for depth option.
- Activate the material removal option.
- Click on OK to create the extrusion.

- Save and exit the file.

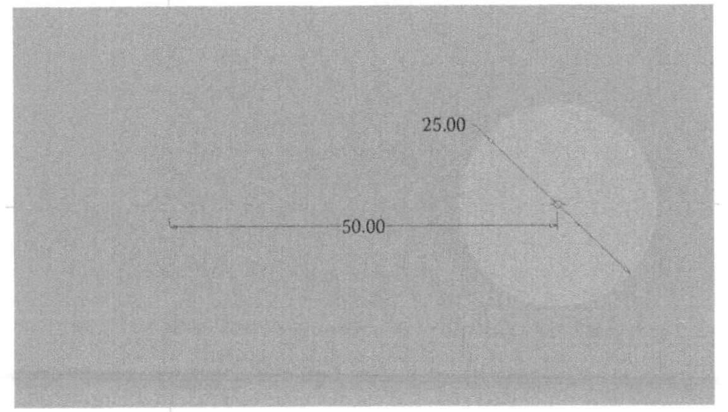

FIGURE 11.49
Constructional sketch of cam mechanism.

- Create a part file and rename it as Frame. (Other names may be given, but it is advisable to provide a relevant name.)
- Select Solid on the right-hand side of the box. This will activate the solid mode and environment.
- Uncheck the box of the default template and click OK to create a new solid file.
- The template selection box will appear.
- Select mmns_part_Solid. This will allow us to use the solid template file of the following units:
    - mm—millimeter for length unit
    - n—newton for force unit
    - s—second for time unit
- Once the solid part is created, go to File > Prepare > Model properties to check the units. In the upper sections, the units of the model will be displayed.
    - Note: The units usually are displayed in the summary section of any CAD software. The setting of units usually is in the tools or settings.
- Click on Extrude from the Model tab in the ribbon interface.
- Click on the Placement tab and select Define to define the sketch.
- Select the front plane as the sketch plane.
- The right plane shall be automatically taken as reference and orientation as right.
- Click on Sketch.
- The sketcher model will be activated.
- Draw the sketch as displayed in Figure 11.50.
- Click OK to create and exit the sketch.
- Enter the value 20 mm for extrusion depth.
- Activate the Thicken option.

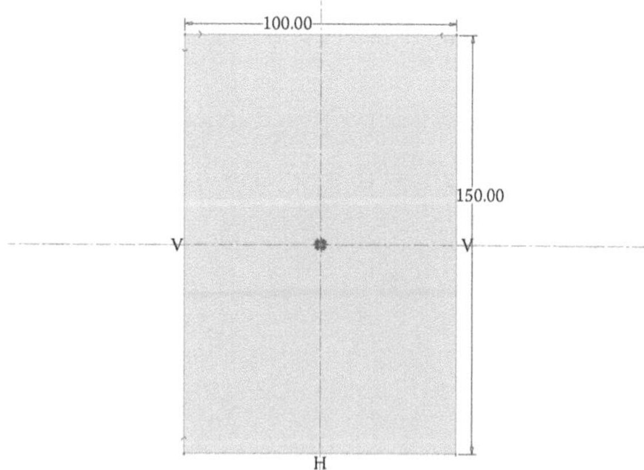

FIGURE 11.50
Constructional sketch of cam mechanism.

- Enter 10 as the thickness.
- Flip the material side outward.
- Click on OK to create the extrusion.

- Click on Extrude from the Model tab in the ribbon interface.
- Click on the Placement tab and select Define to define the sketch.
- Select the outer-side larger surface of extrude_1 as the sketch plane.
- The front plane shall be automatically taken as reference and orientation as left.
- Click on Sketch.
- The sketcher model will be activated.
- Take the references of the top and front planes as horizontal and vertical references.
- Draw a circle of diameter 25 placing the center at the intersection of the horizontal and vertical references.
- Click OK to create and exit the sketch.
- Flip the material orientation upward.
- Enter the value 100 mm for extrusion depth.
- Click on OK to create the extrusion.
- Select the lastly created feature extrude_2.
- Click on Mirror from the Model tab of the ribbon interface.
- Select the right plane and click OK to create the mirrored feature.

- Save and exit the file.

- Create a part file and rename it as Cam. (Other names may be given, but it is advisable to provide a relevant name.)
- Select Solid on the right-hand side of the box. This will activate the solid mode and environment.
- Uncheck the box of the default template and click OK to create a new solid file.
- The template selection box will appear.
- Select mmns_part_Solid. This will allow us to use the solid template file of the following units:
  - mm—millimeter for length unit
  - n—newton for force unit
  - s—second for time unit
- Once the solid part is created, go to File > Prepare > Model properties to check the units. In the upper sections, the units of the model will be displayed.
  - Note: The units usually are displayed in the summary section of any CAD software. The setting of units usually is in the tools or settings.
- Click on Extrude from the Model tab in the ribbon interface.
- Click on the Placement tab and select Define to define the sketch.

- Select the front plane as the sketch plane.
- The right plane shall be automatically taken as reference and orientation as right.
- Click on Sketch.
- The sketcher model will be activated.
- Draw the sketch as displayed in Figure 11.51.
- Click OK to create and exit the sketch.
- Enter the value 20 mm for extrusion depth.
- Select the mid plane extrusion.
- Click on OK to create the extrusion.

- Click on Extrude from the Model tab in the ribbon interface.
- Click on the Placement tab and select Define to define the sketch.
- Select the flat surface of extrude_1 as the sketch plane.
- The front plane shall be automatically taken as reference and orientation as left.
- Click on Sketch.
- The sketcher model will be activated.
- Take the references of top and front planes as horizontal and vertical references.
- Draw a circle of diameter 30 placing the center at the intersection of the horizontal and vertical references.
- Click OK to create and exit the sketch.
- Flip the material orientation upward.
- Enter the value 100 mm for extrusion depth.
- Click on OK to create the extrusion.

- Click on the Round feature from the Model tab of the ribbon interface.
- Set the value of fillet to 3 mm.

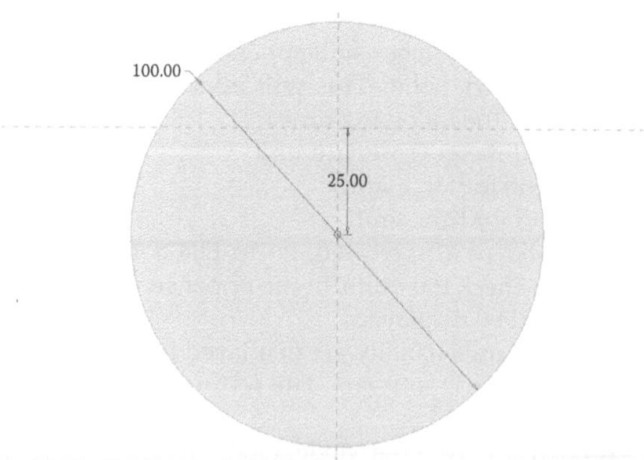

FIGURE 11.51
Constructional sketch of cam mechanism.

- Select all circular edges.
- Click OK to create the feature.

- Save and exit the file.

- Create an assembly file and rename it as Cam_Mechanism. (Other names may be given, but it is advisable to provide a relevant name.)
- Select Design on the right-hand side of the box. This will activate the standard assembly design mode and environment.
- Uncheck the box of the default template and click OK to create a new assembly design file.
- The template selection box will appear.
- Select mmns_asm_Solid. This will allow us to use the solid template file of the following units:
  - mm—millimeter for length unit
  - n—newton for force unit
  - s—second for time unit
- Once the assembly file is created, go to File > Prepare > Model properties to check the units. In the upper sections, the units of the model will be displayed.
  - Note: The units usually are displayed in the summary section of any CAD software. The setting of units usually is in the tools or settings.
- Click on tree filters as shown in Figure 11.52.
- Click on Apply. Notice that the features are displayed in the model tree.

- Click on Assembly from the Model tab of the ribbon interface.
- Select the file Fixed_Frame.
- Click on the Placement tab.
- Select Default in the constraint type selection option box.
- Click OK to assemble the component.

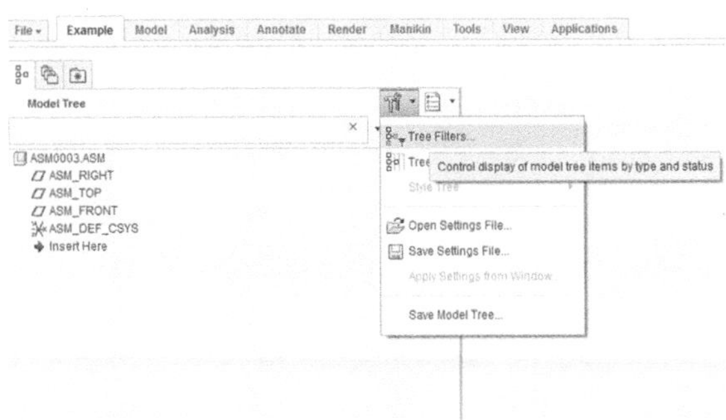

FIGURE 11.52
Model tree filter.

- Click on Assembly from the Model tab of the ribbon interface.
- Select the file Frame.
- Click on the Placement tab.
- Select Pin joint in the connection selection field.
- Select the cylindrical surface of the Connecting_Pin part as a component reference.
- Select the cylindrical hole surface of the piston as an assembly reference.
- Select Coincide in the component reference relation field.
- Select the front plane of the frame part as a component reference.
- Select the plane ASM_FRONT as an assembly reference.
- Select Coincide in the component reference relation field.
- Click OK to assemble the component.

- Click on Assembly from the Model tab of the ribbon interface.
- Select the file Cam.
- Click on the Placement tab.
- Select Pin joint in the connection selection field.
- Select the smaller cylindrical surface of the Cam part as a component reference.
- Select the cylindrical hole surface of feature extrude_4 of Fixed_Frame as an assembly reference.
- Select Coincide in the component reference relation field.
- Click OK to assemble the component.

- Click on Mechanism from the Application tab of the ribbon interface.
- Click on the cams on the connections group.
- For CAM 1, select the surfaces of the cam, and for CAM@, select surfaces of the frame as displayed in Figure 11.53.
- For CAM@, boundary vertices may need to be selected.
- Click OK to create the cam connection.
- Close the mechanism bar from the ribbon interface.

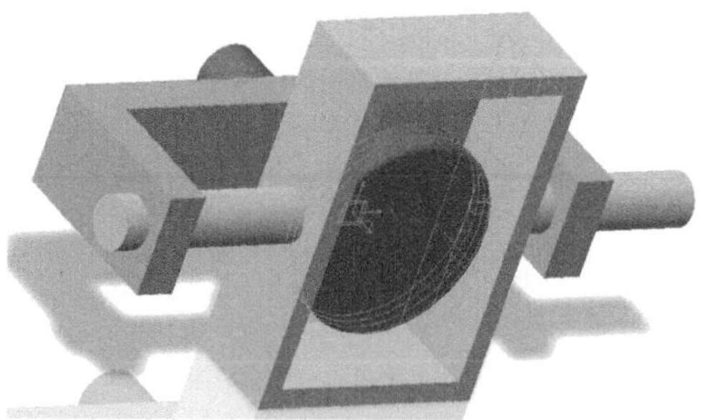

FIGURE 11.53
Constructional view of cam mechanism.

- Click on Drag components from the Model tab of the ribbon interface.
- Click on the part Cam and rotate it. The oscillating movement of the frame can be noticed.

1. Assign a servo motor on the motion axis of the journal to the pin joint.
2. Assign a constant velocity profile of 10 rpm. Hence, the velocity in degrees per second will be

$$\text{rpm}/60 \times 360° = 60°/\text{s}$$

Analysis:

1. Set the analysis for 10 s with a frame rate of 10 at 0.1 interval.
2. Run the analysis for position, which is purely a kinematic analysis, that is, without considering the effect of forces.

Result:

1. Select the measure window and create a measure by selecting any point on the top surface of the pistons.
2. Playback of analysis can be created for animation and further storage into video files for presentation.

11.6 Project Work

Hands-On Exercise, Example 11.5: Universal Joint (Software Used: Creo Parametric; Figure 11.54)

UNIVERSAL JOINT

Steps:

- Create a part file and rename it as Frame. (Other names may be given, but it is advisable to provide a relevant name.)
- Select Solid on the right-hand side of the box. This will activate the solid mode and environment.
- Uncheck the box of the default template and click OK to create a new solid file.
- The template selection box will appear.
- Select mmns_part_Solid. This will allow us to use the solid template file of the following units:
    - mm—millimeter for length unit
    - n—newton for force unit
    - s—second for time unit

FIGURE 11.54
(See color insert.) 3-D view of universal joint.

- Once the solid part is created, go to File > Prepare > Model properties to check the units. In the upper sections, the units of the model will be displayed.
  - Note: The units usually are displayed in the summary section of any CAD software. The setting of units usually is in the tools or settings.
- Click on Extrude from the Model tab in the ribbon interface.
- Click on the Placement tab and select Define to define the sketch.
- Select the top plane as the sketch plane.
- The right plane shall be automatically taken as reference and orientation as right.
- Click on Sketch.
- The sketcher model will be activated.
- Draw a rectangle of 100 × 100 symmetrically about the horizontal and vertical references.
- A centerline may need to be drawn along the reference to draw the aforementioned rectangle.
- Draw a circle of diameter 40 placing the center at the intersection of the horizontal and vertical references.
- Click OK to create and exit the sketch.
- Select the mid plane extrusion.
- Enter the value 20 mm for extrusion depth.
- Click on OK to create the extrusion.
- Save and exit the file.

- Create a part file and rename it as UJ_Link. (Other names may be given, but it is advisable to provide a relevant name.)
- Select Solid on the right-hand side of the box. This will activate the solid mode and environment.
- Uncheck the box of the default template and click OK to create a new solid file.
- The template selection box will appear.
- Select mmns_part_Solid. This will allow us to use the solid template file of the following units:

    - mm—millimeter for length unit
    - n—newton for force unit
    - s—second for time unit
- Once the solid part is created, go to File > Prepare > Model properties to check the units. In the upper sections, the units of the model will be displayed.
    - Note: The units usually are displayed in the summary section of any CAD software. The setting of units usually is in the tools or settings.
- Click on Extrude from the Model tab in the ribbon interface.
- Click on the Placement tab and select Define to define the sketch.
- Select the top plane as the sketch plane.
- The right plane shall be automatically taken as reference and orientation as right.
- Click on Sketch.
- The sketcher model will be activated.
- Draw the sketch as displayed in Figure 11.55.
- Click OK to create and exit the sketch.
- Select the mid plane extrusion.
- Enter the value 40 mm for extrusion depth.
- Click on OK to create the extrusion.

- Click on Revolve from the Model tab in the ribbon interface.
- Click on the Placement tab and select Define to define the sketch.
- Select the top plane as the sketch plane.
- The right plane shall be automatically taken as reference and orientation as right.
- Click on Sketch.
- The sketcher model will be activated.

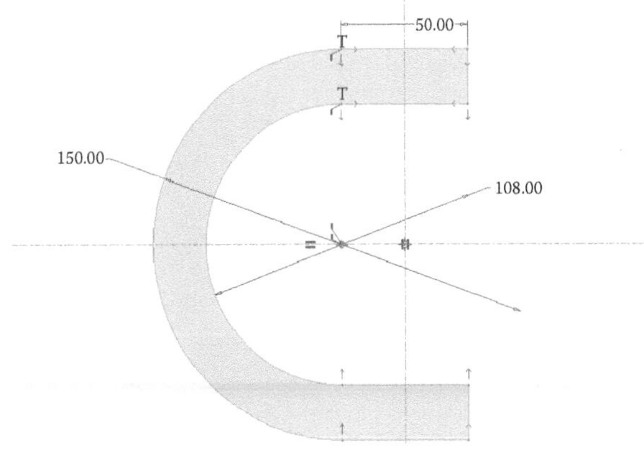

FIGURE 11.55
Constructional sketch of universal joint.

- Draw the sketch as displayed in Figure 11.56.
- Click OK to create and exit the sketch.
- Ensure the revolve angle shall be 360°.
- Click on OK to create the revolve feature.

- Click on Extrude again from the Model tab in the ribbon interface.
- Click on the Placement tab and select Define to define the sketch.
- Select the front flat surface as the sketch plane.
- The right plane shall be automatically taken as reference and orientation as right.
- Click on Sketch.
- The sketcher model will be activated.
- Take the references of the right and top planes.
- Draw a circle of diameter 25 placing the center at the intersection of the horizontal and vertical references.
- Click OK to create and exit the sketch.
- Flip the extrusion toward the material side.
- Select Through all option for extrusion depth.
- Flip the material toward the bottom.
- Activate the material removal option.
- Click on OK to create the extrusion.
- Save and exit the file.

- Create a part file and rename it as Universal. (Other names may be given, but it is advisable to provide a relevant name.)
- Select Solid on the right-hand side of the box. This will activate the solid mode and environment.
- Uncheck the box of the default template and click OK to create a new solid file.
- The template selection box will appear.

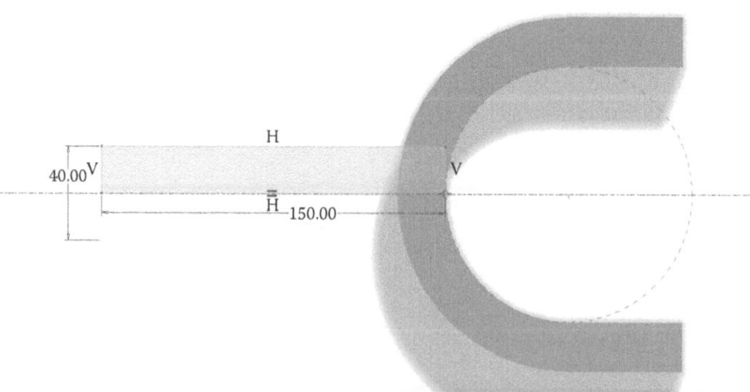

FIGURE 11.56
Constructional sketch of universal joint.

- Select mmns_part_Solid. This will allow us to use the solid template file of the following units:
  - mm—millimeter for length unit
  - n—newton for force unit
  - s—second for time unit
- Once the solid part is created, go to File > Prepare > Model properties to check the units. In the upper sections, the units of the model will be displayed.
  - Note: The units usually are displayed in the summary section of any CAD software. The setting of units usually is in the tools or settings.
- Click on Revolve from the Model tab in the ribbon interface.
- Click on the Placement tab and select Define to define the sketch.
- Select the top plane as the sketch plane.
- The right plane shall be automatically taken as reference and orientation as right.
- Click on Sketch.
- The sketcher model will be activated.
- Draw a centerline at the vertical reference as the axis of revolution.
- Draw a half circle of diameter 100 placing the center at the intersection of the horizontal and vertical references as displayed in Figure 11.57.
- Click OK to create and exit the sketch.
- Ensure the revolve angle shall be 360°.
- Click on OK to create the revolve feature.

- Click on Extrude again from the Model tab in the ribbon interface.
- Click on the Placement tab and select Define to define the sketch.

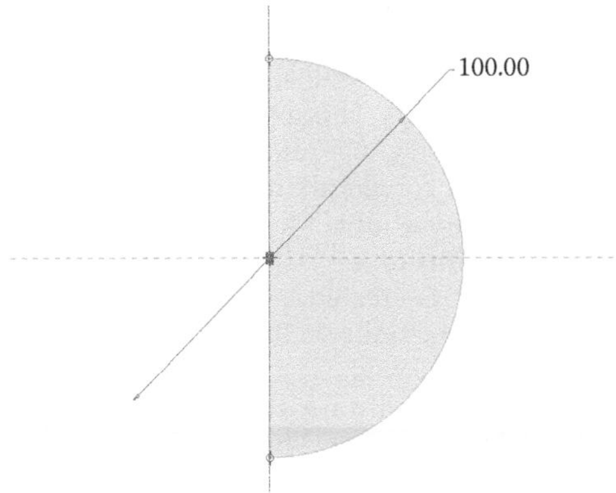

FIGURE 11.57
Constructional sketch of universal joint.

- Select the right plane as the sketch plane.
- The top plane shall be automatically taken as reference and orientation as top.
- Click on Sketch.
- The sketcher model will be activated.
- Draw a circle of diameter 25 placing the center at the intersection of the horizontal and vertical references.
- Click OK to create and exit the sketch.
- Select Through all option for both sides.
- Activate the material removal option.
- Click on OK to create the extrusion.

- Click on Extrude again from the Model tab in the ribbon interface.
- Click on the Placement tab and select Define to define the sketch.
- Select the front plane as the sketch plane.
- The right plane shall be automatically taken as reference and orientation as right.
- Click on Sketch.
- The sketcher model will be activated.
- Draw a circle of diameter 25 placing the center at the intersection of the horizontal and vertical references.
- Click OK to create and exit the sketch.
- Select Through all option for both sides.
- Activate the material removal option.
- Click on OK to create the extrusion.

- Save and exit the file.

- Create a part file and rename it as Pin. (Other names may be given, but it is advisable to provide a relevant name.)
- Select Solid on the right-hand side of the box. This will activate the solid mode and environment.
- Uncheck the box of the default template and click OK to create a new solid file.
- The template selection box will appear.
- Select mmns_part_Solid. This will allow us to use the solid template file of the following units:
  - mm—millimeter for length unit
  - n—newton for force unit
  - s—second for time unit
- Once the solid part is created, go to File > Prepare > Model properties to check the units. In the upper sections, the units of the model will be displayed.
  - Note: The units usually are displayed in the summary section of any CAD software. The setting of units usually is in the tools or settings.

- Click on Revolve from the Model tab in the ribbon interface.
- Click on the Placement tab and select Define to define the sketch.

- Select the top plane as the sketch plane.
- The right plane shall be automatically taken as reference and orientation as right.
- Click on Sketch.
- The sketcher model will be activated.
- Draw a centerline at the vertical reference as the axis of revolution.
- Draw the sketch as displayed in Figure 11.58.
- Click OK to create and exit the sketch.
- Ensure the revolve angle shall be 360°.
- Click on OK to create the revolve feature.
- Save and exit the file.

- Create an assembly file and rename it as Universal_Joint.asm. (Other names may be given, but it is advisable to provide a relevant name.)
- Select Design on the right-hand side of the box. This will activate the standard assembly design mode and environment.
- Uncheck the box of the default template and click OK to create a new assembly design file.
- The template selection box will appear.
- Select mmns_asm_Solid. This will allow us to use the solid template file of the following units:
    - mm—millimeter for length unit
    - n—newton for force unit
    - s—second for time unit
- Once the assembly file is created, go to File > Prepare > Model properties to check the units. In the upper sections, the units of the model will be displayed.
    - Note: The units usually are displayed in the summary section of any CAD software. The setting of units usually is in the tools or settings.

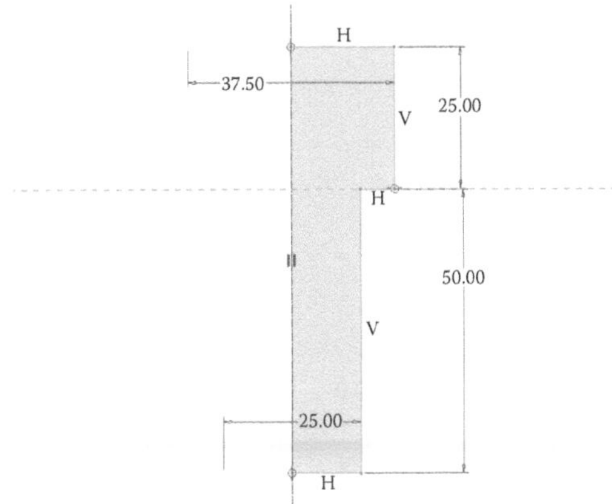

FIGURE 11.58
Constructional sketch of universal joint.

- Click on tree filters as shown in Figure 11.59.
- Check the box features to display in the model tree.
- Click on Apply. Notice that the features are displayed in the model tree.

- Click on Assembly from the Model tab of the ribbon interface.
- Select the file Frame.
- Click on the Placement tab.
- Select Default in the constraint type selection option box.
- Click OK to assemble the component.

- Click OK datum plane from the Model tab of the ribbon interface.
- Select the references as shown in Figure 11.60.
- Click OK to create the datum plane.

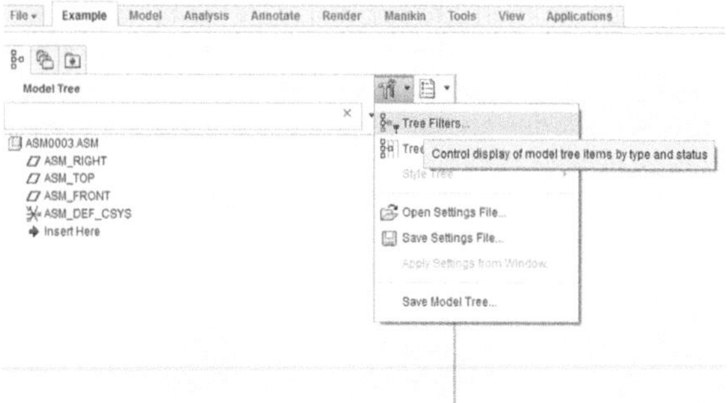

FIGURE 11.59
Model tree filter.

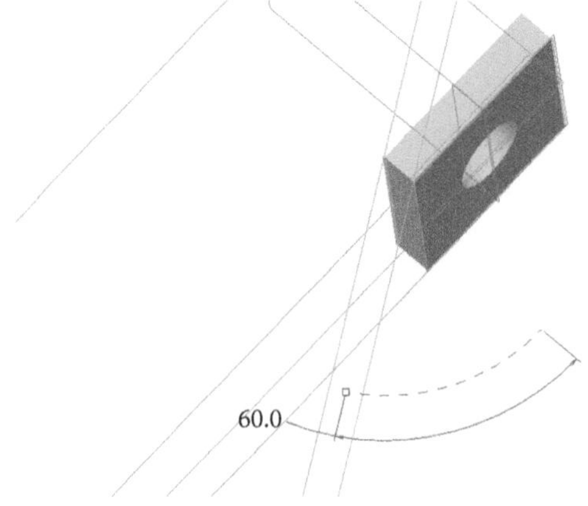

FIGURE 11.60
Constructional sketch of universal joint.

- Click OK datum plane from the Model tab of the ribbon interface.
- Select the lastly created plane ADTM1 as the reference plane.
- Set the offset value 300 mm downward.
- Click OK to create the datum plane.

- Click on Assembly from the Model tab of the ribbon interface.
- Select the file UJ-Link.
- Click on the Placement tab.
- Select Pin joint in the connection selection field.
- Select the cylindrical surface of revolve_1 of the UJ-link part as a component reference.
- Select the cylindrical hole surface of the frame as an assembly reference.
- Select Coincide in the component reference relation field.
- Select the bottom face of the UJ-link part as a component reference.
- Select the bottom surface of the frame as an assembly reference, as displayed in Figure 11.61.
- Enter the value 60 mm as the offset distance.
- Right-click on the graphical area and select New constraint.
- Click OK to assemble the component.

- Click on Assembly from the Model tab of the ribbon interface.
- Select the file Universal again.
- Click on the Placement tab.
- Select Pin joint in the connection selection field.
- Select the cylindrical hole surface of the Universal part as a component reference.

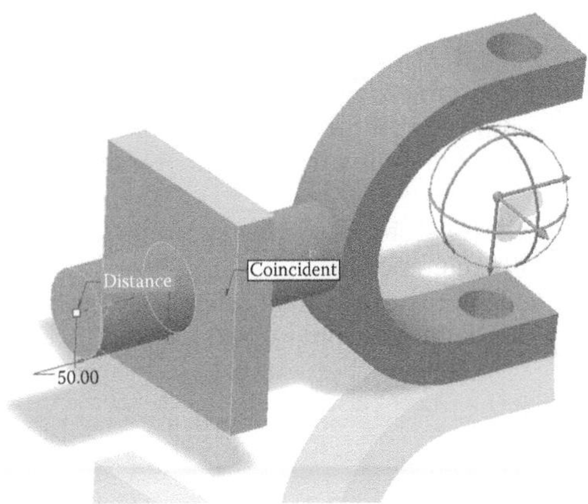

FIGURE 11.61
Constructional view of universal joint.

- Select the cylindrical hole surface of UJ-Link as an assembly reference.
- Select Coincide in the component reference relation field.
- Select the perpendicular plane of the selected hole surface of the Universal part as a component reference.
- Select the front plane of UJ-Link as an assembly reference.
- Select Coincide in the component reference relation field.
- Click OK to assemble the component.

- Click on Assembly from the Model tab of the ribbon interface.
- Select the file Universal again.
- Click on the Placement tab.
- Select Pin joint in the connection selection field.
- Select the cylindrical hole surface of the UJ-Link part as a component reference.
- Select the cylindrical hole surface of Universal as an assembly reference.
- Select Coincide in the component reference relation field.
- Select the front plane of UJ-Link as a component reference.
- Select the perpendicular plane of the selected hole surface of the Universal part as an assembly reference.
- Select Coincide in the component reference relation field.
- Click OK to assemble the component.

- Click on Assembly from the Model tab of the ribbon interface.
- Select the file Frame.
- Click on the Placement tab.
- Select the top plane of the frame part as a component reference.
- Select the plane ASM_TOP as an assembly reference.
- Select Coincide in the component reference relation field.
- Right-click on the graphical area and select New constraint.
- Select the front larger face of the frame part as a component reference.
- Select the plane ADTM2 as an assembly reference.
- Select Coincide in the component reference relation field.
- Right-click on the graphical area and select New constraint.
- Select the cylindrical hole surface of the frame part as a component reference.
- Select the cylindrical surface of the feature revolve_1 of UJ-Link as an assembly reference.
- Select Coincide in the component reference relation field.
- Click OK to assemble the component.

- Save and exit the file.

QUESTIONS

1. Define degree of freedom.
2. How many degrees of freedom are there?

3. Classify the different types of joints.
4. Compare between higher pair and lower pair.
5. Describe different types of connections.
6. Describe the use of curve tracing.
7. Site an example of curve tracing.

12

Graphical Presentation and Animation

12.1 Introduction

Designers create computer-aided design (CAD) models of simple to complex engineering components in CAD through various operations and features. When creating a single component or part, which can have numerous operations, designers conceptualize and determine the final product requirement and build the CAD model with manipulations of CAD features. In reality, the product is also made from several manufacturing operations like casting, machining, and so forth. An assembled component may have operations like casting of some components followed by machining and assembly. If there are numerous dimensions and annotations in the model, it would be preferable to the manufacturers to have a specific set of data displayed on their own process requirement. This creates a display of visualization of the same model in a different manner. Similarly, plant designers create a CAD model of the entire plant and its equipment that is used with much discipline by engineers as per their requirement. Civil engineers would be interested to make the detailed support and structural items and outlined dimensions of any other equipment that is not in their scope of work, like piping, detailed pressure vessels, and so forth, whereas piping engineers are more concerned with piping detailing than detailed structural or any other type of items. This creates a need for representation of the same CAD model in different manners.

Graphical visualization of alteration of the display of the same component in different manners reduces the effort of creating a similar component for every discipline of different levels of users and also enhances understanding of the product manufacturing sequences.

12.2 Presentations and Classifications

A CAD model can be presented with various levels of detail by simplified representation. Most of the CAD software provides this tool; the name can vary from software to software. In some software, it is called lightweight geometry. Lightweight geometry is used to make the light weight. Light weight is, in terms of memory, used for a particular model or component. Advanced CAD software like Creo has extended the functionality by allowing view hide and suppressed functionality in simplified representation. The advantages of simplified representations are as follows:

1. Display of external surfaces to reduce computer memory
2. Display of model according to the user's need
3. Generating different views in the drawing as per need
4. Display of the process simulation

Users are also allowed to create various representations by creating combinations of various representation states of components.

Simplified representations are of the following types:

1. *Master representation*: Master representation is simply a representation without using any simplified representation functionality. Therefore, master representation will be always present, and it represents the original components or features in assembly or part.
2. Graphical representation
3. Symbolic representation
4. Geometry representation
5. User-driven representation

12.3 Application of Presentation

Various presentations are needed in a CAD model. Often, customers or technicians understand the functionality of any component by seeing the CAD model on the screen. Apart from viewing, CAD models are also used in various works like illustration, drawing or process representations, and so forth. The popular uses of representations are as follows:

1. *Illustration*: Illustration of CAD models is needed mainly in the creation of manuals. Types of manuals also vary, like parts manuals, operation manuals, service manuals, and so forth. Therefore, the illustration requirement also varies with the type of manual. Parts manuals often require the exploded view with part marking, whereas operation manuals need illustration for displaying functionality.

Hands-On Exercise, Example 12.1: Ball Valve Illustration (Software Used: Creo Parametric; Figure 12.1)

BALL VALVE

Steps:

- Open the assembly file MECHANISED_3_PIECE_BALL_VALVES from Chapter 7.
- Click on tree filters as shown in Figure 12.2.

- Check the box features to display in the model tree.
- Click OK to close the box of model tree items.

- Click on the drop-down arrow to manage views and select View manager.
- A view manager window appears; click on the Explode tab.
- Click on New. A default exploded view named Exp0001 will be added. Rename it to Complete and press Enter.
- Click on Edit and select Edit position.

FIGURE 12.1
Exploded view of a valve assembly.

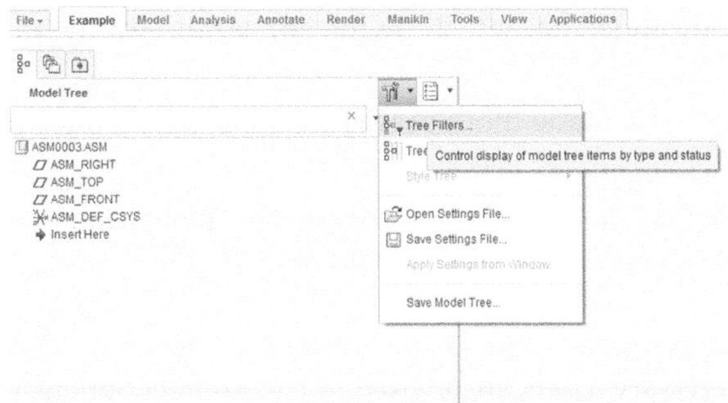

FIGURE 12.2
Model tree filter.

- Select Translate.
- Click on the Reference tab.
- Select the component End_Piece and move leftward graphically.
- Similarly, position all the components as displayed in Figure 12.1.
- The Save display elements window will appear.
- The exploded view check box shall be automatically checked.
- Click OK to save the view.

2. *Drawing presentation*: The drawings are made to serve various purposes. A drawing used for production shall have all the symbols, tolerance and manufacturing data needed for the production whereas an assembly drawing shall have the needed data for assembly of sub-assemblies and its components. A CAD model may need to be shown in different drawings with different data to serve different purposes. Thus a CAD model needs to be presented differently in different drawing presentations.
3. *Process presentation*: A process is represented sequentially by displaying various stages of the creation process of any component. Hide features or components in different stages can display the component in the actual form at various stages of its creation process.

12.4 Rendering

Rendering or photo rendering is creating a picture presentation with graphical texture or prespecified colors on the model and background.

The purposes of rendering are as follows:

1. A proposed model can be shown with a graphical aid for marketing an advertisement.
2. Real-world rendering will show the appearance of the model in the real world.
3. A model can be made more presentable through rendering.

Hands-On Exercise, Example 12.2: Paper Clip Rendering (Software Used: Creo Parametric; Figure 12.3)

Steps:

Before proceeding to creation, please note that a little bit of adjustment may be necessary. Users are advised to try different options to understand the effects of the process.

- Open the assembly file Paper_Clip.asm from Chapter 7.
- Click on tree filters as shown in Figure 12.2.

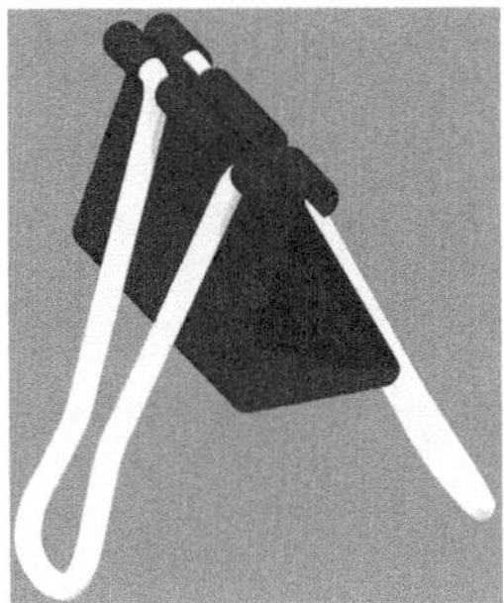

FIGURE 12.3
Example of rendering of a paper clip.

- Check the box features to display in the model tree.
- Click OK to close the box of model tree items.

- Click on the drop-down arrow for the appearance gallery from the View tab of the ribbon interface.
- Right-click on the model section and select New.
- Enter the name of the color as 3.
- Click on Color shade in the Basic tab.
- Click anywhere on the color wheel to select the color or enter the following value in the RGB field.
  - R = 248.8
  - G = 250.0
  - B = 247.6
- Click OK to close the color editor window.
- Set the following:
  - Color intensity: 100
  - Ambient: 100
  - Shine: 100
  - Highlight: 30
  - Reflection: 28
  - Transparency: 0
- Click Close to close the model appearance window.
- Click on the drop-down arrow for the appearance gallery from the View tab of the ribbon interface.
- Click on the color 3 from the model section.
- Select the part Paper_Clip_Handle from the model tree or graphically.
- Press the middle button to assign the color.

- Similarly, click on the drop-down arrow for the appearance gallery from the View tab of the ribbon interface.
- Right-click on the model section and select New.
- Enter the name of the color as 4.
- Click on Color shade in the Basic tab.

- Click anywhere on the color wheel to select the color or enter the following value in the RGB field.
  - R = 64.4
  - G = 64.1
  - B = 63.6
- Click OK to close the color editor window.
- Set the following:
  - Color intensity: 69
  - Ambient: 55
  - Shine: 100
  - Highlight: 30
  - Reflection: 0
  - Transparency: 0

- Click Close to close the model appearance window.
- Click on the drop-down arrow for the appearance gallery from the View tab of the ribbon interface.
- Click on the color 4 from the model section.
- Click on the part Paper_Clip1 from the model tree or graphically.
- Press the middle button to assign the color.

- Click on the Render tab of the ribbon interface.
- Click on Scene.
- Click on the Room tab.
- In the room appearance palette, click on any palette.
- In the room appearances, select the default PTC-STD-SILVER-POLISHED.
- Close the room appearance editor window.
- Similarly, double-click on other palettes and set PTC-STD-SILVER-POLISHED.
- Close the scene window.
- Click on the render window.
- The render model shall be displayed as shown in Figure 12.3.
- Save and exit the file.

12.5 Animation

Animation is a very useful application of CAD. A functionality of a model can be best understood by animation of the function.

1. *Application of animation*: Animations are applied for various purposes. The applications are as follows:
   a. Animation of the installation or dismantling process
   b. Simulation of functionality

    c. Process of creating the component
    d. Simulation of service requirement
2. Exploding view animation
    a. Valve animation

Hands-On Exercise, Example 12.3: Ball Valve Animation (Software Used: Creo Parametric; Figure 12.4)

BALL VALVE

Steps:

- Open the assembly file Ball Valve from Chapter 7.
- Click on tree filters as shown in Figure 12.2.

- Check the box features to display in the model tree.
- Click OK to close the box of model tree items.

- Click on the drop-down arrow to manage views and select View manager.
- A view manager window appears; click on the Explode tab.
- Click on New. A default exploded view named Exp0001 will be added. Press Enter.
- Click on Edit and select Rename. Rename it as Complete.
- Click on Edit and select Edit position.
- Select Translate.
- Click on the Reference tab.
- Select all the components from the model tree except Body.
- Click on the movement reference area.
- Select the plane ASM_RIGHT.
- Drag the arrow in the graphics area and place all the components as displayed in Figure 12.5.
- Click on OK to create the exploded view.

FIGURE 12.4
3-D view of a valve assembly.

FIGURE 12.5
(See color insert.) Exploded view of a valve assembly.

- In the view manager, click Edit and save to create the exploded view.
- The Save display elements window will appear.
- The exploded view check box shall be automatically checked.
- Click OK to save the view.

- Click on Animation from the Tools tab of the ribbon interface.
- Click on Key frame sequence from the ribbon interface of the animation tab.
- The key frame sequence window shall appear.
- In the key frame sequence tab, select the exploded view complete at time 0. Click on the plus icon to add.
- Similarly, select Unexploded, enter the time 30, and add it again as displayed in Figure 12.6.
- Check the box for Follow explode sequence.
- Click OK to close the key frame sequence window.
- Notice that the partial animation window is being shown in the upper section of the graphical area.
- Double-click on the timeline.

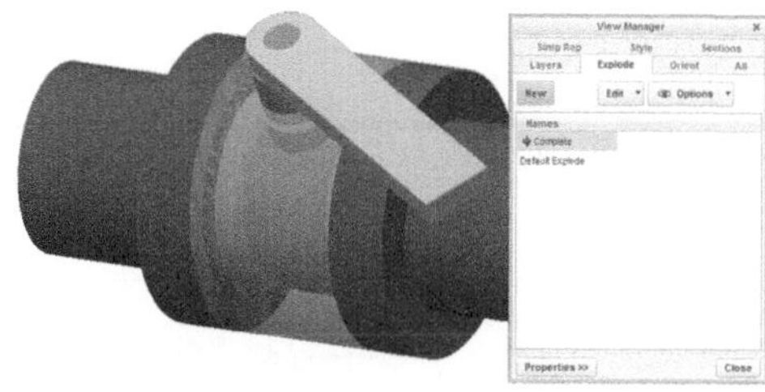

FIGURE 12.6
View animation setup window in a CAD system.

- The animation time domain will appear.
- Change the end time to 30.
- Click OK to close the window.

3. *Motion animation*: Mechanism
4. Process animation
   a. Valve and tank animation

12.6 Project Work

Hands-On Exercise, Example 12.4: Plastic Bottle Rendering (Software Used: Creo Parametric; Figure 12.7)

PLASTIC BOTTLE RENDERING

Steps:

Before proceeding to creation, please note that a little bit of adjustment may be necessary. Users are advised to try different options to understand the effects of the process.

- Open the assembly file Plastic_Bottle.asm from Chapter 7.
- Click on tree filters as shown in Figure 12.2.

- Check the box features to display in the model tree.
- Click OK to close the box of model tree items.

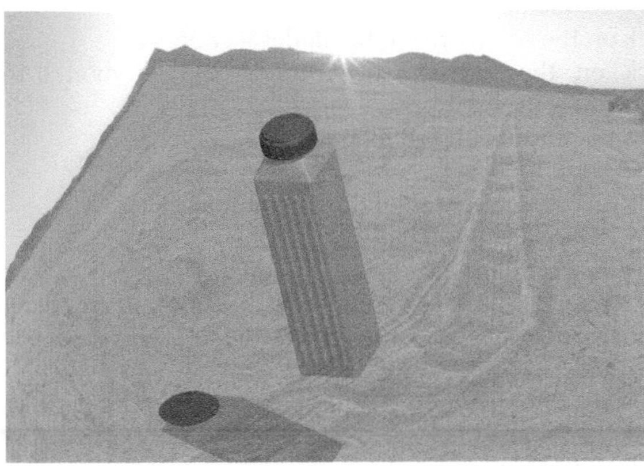

FIGURE 12.7
(See color insert.) Example of rendering of plastic bottle.

- Click on the drop-down arrow for the appearance gallery from the View tab of the ribbon interface.
- Right-click on the model section and select New.
- Enter the name of the color as 1.
- Click on Color shade in the Basic tab.
- Click anywhere on the color wheel to select the color or enter the following value in the RGB field.
  - R = 201.5
  - G = 92.1
  - B = 53.9
- Click OK to close the color editor window.
- Click Close to close the model appearance window.
- Click on the drop-down arrow for the appearance gallery from the View tab of the ribbon interface.
- Click on the color 1 from the model section.
- Click on the part Plastic_Bottle_Cap from the model tree or graphically.
- Press the middle button to assign the color.

- Similarly, click on the drop-down arrow for the appearance gallery from the View tab of the ribbon interface.
- Right-click on the model section and select New.
- Enter the name of the color as 2.
- Click on Color shade in the Basic tab.
- Click anywhere on the color wheel to select the color or enter the following value in the RGB field.
  - R = 104.8
  - G = 255.0
  - B = 247.8
- Click OK to close the color editor window.
- Click Close to close the model appearance window.
- Click on the drop-down arrow for the appearance gallery from the View tab of the ribbon interface.
- Click on the color 1 from the model section.
- Click on the part Plastic_Bottle from the model tree or graphically.
- Press the middle button to assign the color.

- Click on the Render tab of the ribbon interface.
- Click on Scene.
- Click on the Room tab.
- In the room appearance palette, double-click on any palette.
- In the room appearances, select the default ceiling appearance.
- Close the room appearance editor window.
- Similarly, double-click on other palettes and set default ceiling.
- Close the scene window.
- Click on the render window.
- The render model shall be displayed as shown in Figure 12.7.
- Save and exit the file.

Hands-On Exercise, Example 12.5: Spring Support Illustration (Software Used: Creo Parametric; Figure 12.8)

SPRING SUPPORT

Steps:

- Open the assembly file Spring_Support from Chapter 7.
- Click on tree filters as shown in Figure 12.2.

- Check the box features to display in the model tree.
- Click OK to close the box of model tree items.

- Click on the drop-down arrow to manage views and select View manager.
- A view manager window appears; click on the Explode tab.
- Click on New. A default exploded view named Exp0001 will be added. Rename it to Complete and press Enter.
- Click on Edit and select Edit position.
- Select Translate.
- Click on the Reference tab.
- Select all the components from the model tree except Spring_Base_Plate.

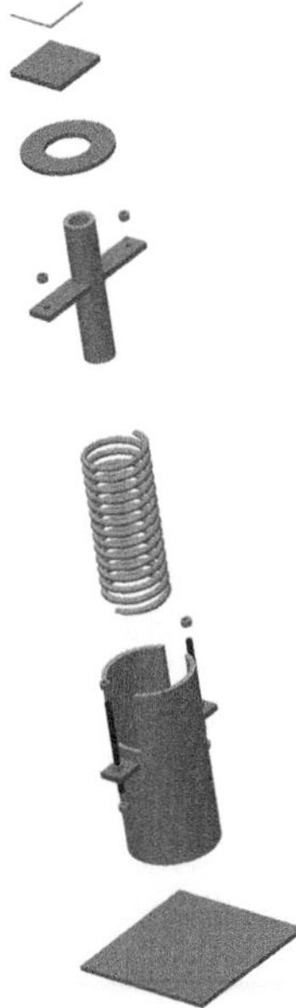

FIGURE 12.8
Exploded view of spring support.

- Click on the movement reference area.
- Select the top face of Spring_Base_Plate.
- Drag the upward arrow in the graphics area and place all the components sufficiently above.

- Click on the Reference tab again.
- Select the component Spring_Cage and move downward graphically.
- Similarly, position all the components as displayed in Figure 12.8.
- The Save display elements window will appear.
- The exploded view check box shall be automatically checked.
- Click OK to save the view.

Hands-On Exercise, Example 12.6: Spring Support Animation (Software Used: Creo Parametric; Figure 12.8)

SPRING SUPPORT

Steps:

- Open the assembly file Spring_Support from Chapter 7.
- Click on tree filters as shown in Figure 12.2.

- Check the box features to display in the model tree.
- Click OK to close the box of model tree items.

- Click on the drop-down arrow to manage views and select View manager.
- A view manager window appears; click on the Explode tab.
- Click on New. A default exploded view named Exp0001 will be added. Press Enter.
- Click on Edit and select Edit position.
- Select Translate.
- Click on the Reference tab.
- Select all the components from the model tree except Spring_Base_Plate.
- Click on the movement reference area.
- Select the top face of Spring_Base_Plate.
- Drag the upward arrow in the graphics area and place all the components sufficiently above.
- Click on OK to create the exploded view.
- In the view manager, click Edit and save to create the exploded view.
- The Save display elements window will appear.
- The exploded view check box shall be automatically checked.
- Click OK to save the view.

- Select the view exp0001. Click on Edit and select Copy.
- Copy it as exp0002. Click OK to create the copy.
- Double-click on the view exp0002 to activate it.

- Click Edit and select Edit position.
- Select Translate.
- Click on the Options tab.
- Select copy position.
- Click components to move.
- Select the component Spring_Cage part.
- Click on Copy position from.
- Select the Spring_Base_Plate.
- Click on Apply and close the copy position window.
- Click OK to create the exploded view exp0002.
- Select the exploded view exp002. Click on Edit and select Save.
- The Save display elements window will appear.
- The exploded view check box shall be automatically checked.
- Click OK to save the view.

QUESTIONS

1. Why are presentations required?
2. Why are graphical presentations required?
3. How can graphical presentations save time?
4. Describe the use of symbolic presentation and graphical presentation.
5. Classify various types of graphical presentations.
6. Describe the applications of various types of graphical presentations.
7. What is rendering?
8. Describe the uses of rendering.
9. Describe the uses of animation.
10. Differentiate the use of rendering and animation.

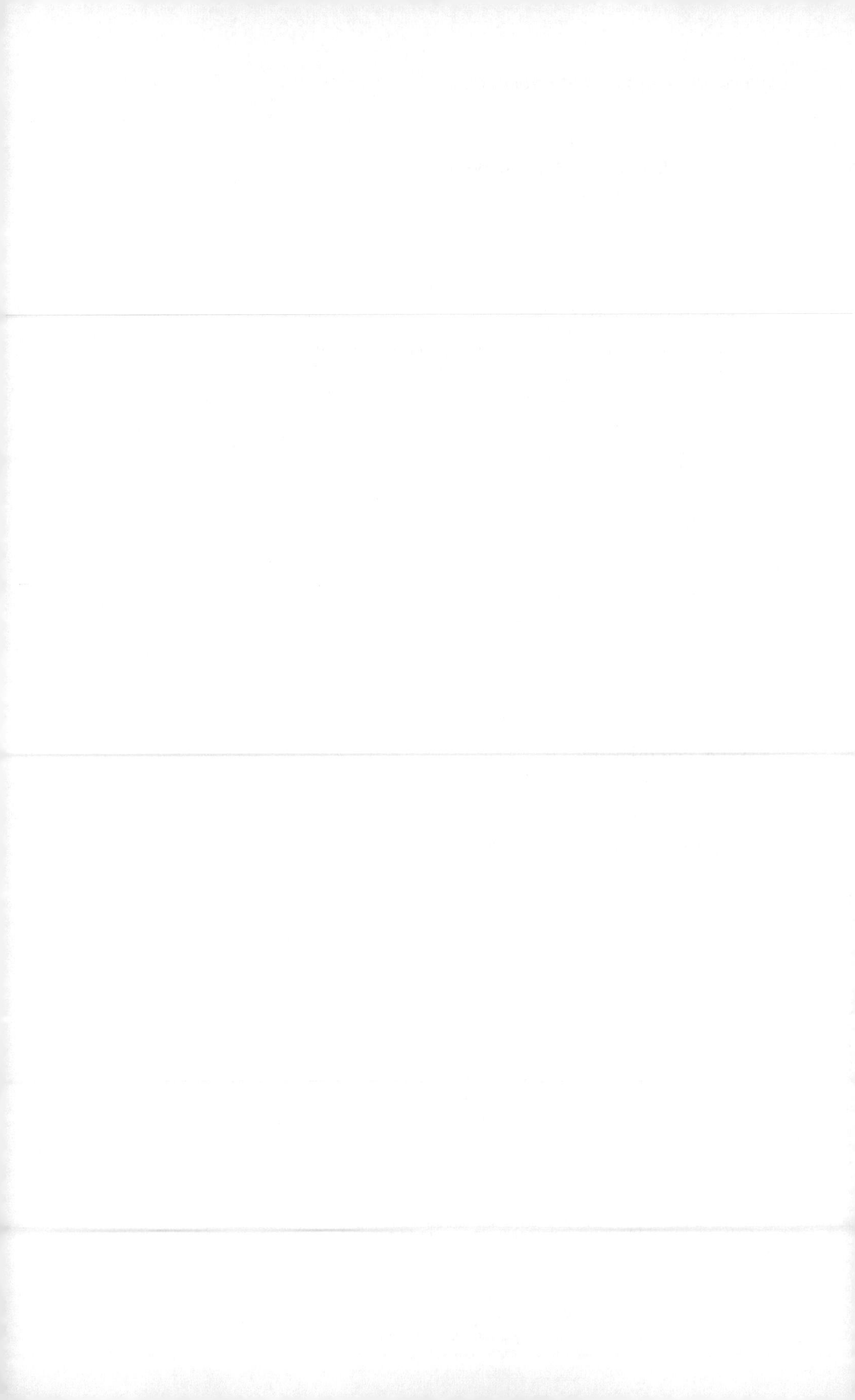

13

Automation

13.1 Introduction

Automation is the process of reduction of time and effort in drawing and designing by using customized programming and templates in the CAD software. Designers create a design based on prespecified design criteria. The design criteria are considered as input variables in the automation. Now the design is a combination of new ideas with many existing formulas and standard practices. It is obvious that the formulas are well entered into the CAD software through programming and automation. Besides the formula, standard practices are also predefined through automation into the CAD software. Therefore, automation reduces the effort and time required to do calculation with known formulas and standard practices.

Automation can be carried out in a small portion of a product design in order to help complete the entire design.

Small automation with less programming is useful to users and can be done by the users themselves when using CAD. However, when full automation of a product is needed, a team of people with expertise on automation may be required. If automation has to be made such that no further checking is needed, then a comparatively large program is required.

There is some advanced CAD software that provides many useful tools to users for carrying out small automation in modeling and drawing generation.

There are companies that manufacture standard engineering components like bearings, gears, and fasteners where every dimension can be predefined through standard design practices and formulas. Full automation on those products is made possible through advance CAD software. Users only need to provide the required values in case of full automation.

For large products in which there are more variables and combinations of formulas and variables needs a comparatively large program, full automation may not be profitable. In those cases, partial automation becomes successful, where most of the work in the design or drawing is generated through automation and the rest of the work done manually.

It is obvious that a new idea that has never been used before on a product or in a company can only come from the mind of a user or a designer.

Automation's scope is limited within the programming existing in standard design practice. Thus, in case of full automation, there will be no chance of using new ideas in the generated product. If any chances of using a new idea exist in the existing design, full automation is not possible (see Figure 13.1a).

Below are the advantages of automation:

1. It reduces the time and effort in creating design and drawing of similar products.
2. Automation reduces the chances of human error and thus increases accuracy.

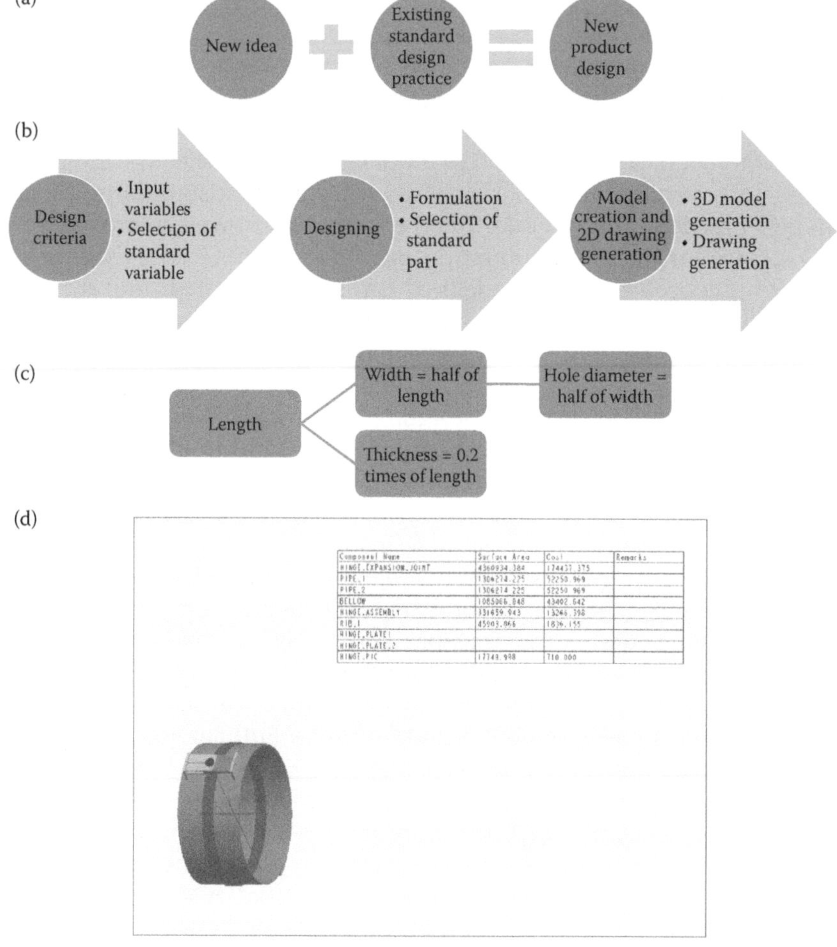

Component Name	Surface Area	Cost	Remarks
HINGE_EXPANSION_JOINT	4360934.384	174437.375	
PIPE_1	1306274.225	52250.969	
PIPE_2	1306214.225	52250.969	
BELLOW	1085066.848	43402.642	
HINGE_ASSEMBLY	331459.943	13246.398	
RIB_1	45903.866	1836.155	
HINGE_PLATE1			
HINGE_PLATE_2			
HINGE_PIC	17749.998	710.000	

FIGURE 13.1
(a) Idea of creating a new product. (b) Process of automation. (c) Sample of formula. (d) Relation in drawing.

3. Equal standard of drawings is maintained by automation, which does not depend on users.
4. It reduces the frustration of users on executing similar jobs.

The following are the disadvantages of automation:

1. Sometimes it needs a huge amount of time to create automation for a certain product that has many variables.
2. In case of wrong inputs, automation may create an unusual design or drawing. It is recommended to ensure the final design and drawing from automation output when many variables are involved.
3. Modification of automated programs requires good technical expertise and time.
4. Automation is worthwhile when a large number of repetitive product designs exist, and it may not be economical if the need is only for a few.

It is obvious that automation in CAD software is limited to the use of existing standard design practice.

13.2 Identifying the Variables

As shown in Figure 13.1a, when a need for a particular design or drawing of a particular product arises, the first point is to identify the variables. In most cases, a component has a few variables required by a customer. Those obviously form the main input variable sheet. There are some other variables selected depending on the input variables from the standard code of practice. Those variables are selected through automation, in case full automation is feasible; however, they may be selected by a user when full automation is not viable.

13.3 Parameter Creation

Parameters are designated variables within the CAD software that are defined by users and used in the creation of relation and formula, and thus automation. Every dimension is defined by ids, which are unique and created by the CAD software itself. These ids can or cannot be edited by users. Also these ids are named based on their own sequence of creation. Most of the CAD software allows users to define programs from automation by those ids. When a large number of variables are involved, programmings through these dimensional ids are not very convenient. Also there are certain variables that drive the geometry of a model; however, they are not part of the geometry. Therefore, those variables cannot be expressed and stored without parameters. When creating parameters for programming, it must be kept in mind that there are two types of parameters. In a particular automation, there can be a large number of parameters. There are some parameters used mostly in the design, and most of the geometries are dependent on those parameters. We can call those parameters as major parameters. It is advised that major parameters be created first before creation of any geometry. Other parameters aside from major parameters may be termed as auxiliary parameters. Sometimes it is also not possible to determine all the auxiliary parameters at the beginning.

13.3.1 Major Parameters

Major parameters are those that are very important with respect to product specification and automation. Every user needs to be very careful about the major parameters. For example, when creating automation for a tank, it is obvious that tank diameter and length will be the major parameters with respect to function and automation.

13.3.2 Auxiliary Parameters

Auxiliary parameters are those that are important with respect to automation criteria; however, they are not essential to specify in product specification. For example, in tank automation, the thickness of the ribs of the saddle is an obvious criterion for automation; however, it is not so essential to be specified in the product specification.

13.4 Relation and Formula Creation

Once input variables and parameters are defined, relations between them shall be defined. It can be seen that when writing relations, more constants are required. Those constants can also be written in relations in the form of parameters. This allows a user to change the constants without editing the actual relations. A note from understanding each section of a relation will be good practice, which allows users to better understand a particular section of the relation.

13.5 Use of Family Tables

Family tables are often required to be used in automation.

Example 13.1

Automation of a simple block of having one hole placed at its center. The relationship of the dimensions of this model is such that

- The length of the plate is five times the thickness and two times the width.
- The diameter of the hole is half of the width.

Hence, the parameters are as follows:

a. Length
b. Width
c. Thickness
d. Hole diameter

As shown in Figure 13.1c, if one of the above-mentioned parameters is known, then all the other parameters can be derived. Hence, any of the above-mentioned parameters can be the driving parameters of the model. In this case, we assumed that the length parameter is the driving parameter; therefore, it becomes the variable in the subject case. All other parameters are to be linked with length.

The parameters to be created are length, width, thickness, and Hole_Diameter; these are major parameters that are really needed for the design. There can also be some dimensions that are not so important from the requirement point of view; however, they are required in product design. In the subject case, the fillet radius is a dimension that can be termed as an auxiliary parameter. Users or designers need to carefully identify and select these kinds of auxiliary parameters since they often generate failure if they are not given proper attention. In the subject case, the fillet radius can be fixed at 3 mm. It can be kept without any relationship so that it is preserved in all cases of variations in the model.

Once the parameters are created, the dimensions are to be linked with the parameters. It can be done by a relationship-making window in most of the advanced software packages. Some software packages allow one to enter a parameter directly when creating a sketch or when modeling. Users are suggested to identify and create all the major parameters first before starting to create a model.

Now, in this case, the thickness is 0.2 times of the length, so it may come to a fraction or a dimension of which the standard plate may not be available in the market. In that case, a suitable program has to be developed based on available standard plate thicknesses.

The logic shall be modified as follows:

Thickness = if(Length*0.2<6,6,if(length*0.2<8,8,10)).

13.6 Template Component Generation

A programmed part can be used as a template part for further creation of any product. In this way, the time and effort can be saved for the creation of common features and relationships. An example is provided in Section 14.1.2.

In template component generation, the a component with common customizable feature or properties or relations is used as template so that repeated work can be reduced.

13.7 Relation in Drawing

In 3-D parametric software, all the relationships and formulas are created in the model itself. So when any change is required, one has to change the parametric model with the help of parameters, and the associated drawing automatically gets updated. The above-mentioned procedure is well maintained in the dimensions of the model; however, there are cases in which some additional changes are required involving drawing annotations, which are the only part of the drawing that cannot be done in a parametric model, for example, details in a name plate, certain additional views that may be required by a product for better understanding, etc. It is obvious that if the model dimensions change to a considerable amount, the annotations associated with it may get jumbled or placed over another, which cannot be controlled in automation. Therefore, 100% automation in drawing may not be possible, or an extremely large effort may be required, which is not economical with respect to time and cost.

It is well advised to automate the 3-D model, bill of materials, name plate, etc., for which the standard functionality is provided by the CAD system. Adjustment of notes, dimension, text size, etc., is left to be done manually, which can easily be done once the automated drawing is created. Therefore, a little effort is required for the touch-up of the drawing. A common engineering drawing for any engineering product mostly contains the following:

1. 3-D complete view of the product
2. Sectional and auxiliary views

3. Bill of materials
4. Annotation tables such as machining symbol parameters
5. Name plate and sheets
6. Dimensions
7. Symbols

The automation of the above-mentioned items is possible and is detailed as follows:

1. *3-D complete view of the product*: A 3-D view is generated from the parametric model. Assuming that the model is prepared, one has to determine the number of views required to completely specify the model for the required purpose. Sometimes, plan view, elevation, and side view are good enough to specify comparatively simple components. Also the variation of the size of the component plays a big role in the drawing automation. If the drawing model size is changed too much, a different scale is required to adjust the views in the drawing sheet. If the model requires a dimension that is too detailed, the sheet size or number may need to be changed. If the sheet size changes, it is obvious that name plates and sheet templates change. Also if the number of sheets increases, the additional views are to be placed. Therefore, if automation is being prepared, a plan or automation sequence has to be prepared, where similar components shall be categorized for the size shape and details. When a component varies largely, the component's dimension can be classified into more than one group, and different drawing templates can be assigned to each group. This increases the drawing automation.
2. *Sectional and auxiliary views*: Sectional and auxiliary views shall be placed suitably such that if the variables change to their largest or smallest limit, the views shall not overlap each other or do not become too small.
3. *Bill of materials*: Bill of materials is generated from 3-D models themselves. Certain formulation can also be possible in the bill of materials; however, it is recommended to generate the bill of materials from the model itself in order to avoid any mismatch.
4. *Annotation tables such as machining symbol parameters*: The symbol tables usually do not change with the detail of the model. Therefore, a table can be used in drawing.
5. *Dimensions*: All the dimensions shall not overlap with each other, which is a true problem of automation when generating models. A touch-up is necessary in the final drawing after generating any model instances through automation.

Hands-On Exercise, Example 13.1: Relation in Drawing (Software Used: Creo Parametric; Figure 13.1d)

Steps:

- Open the assembly file Hinge_Exapnsion_Joint.asm from the Chapter 7 folder.
- Create a drawing file and rename it as surface_area. (Other names may be given, but it is advisable to provide a relevant name.)
- The template selection box will appear.
- Select Empty.
- Select the standard size sheet A4.
- Click on the Place view.
- Click on the left top portion of the sheet as the view placement area.
- Click OK on the template view instruction window.
- Click on the Table tab of the ribbon interface.
- Click on Table and select a table with a 4 columns × 2 rows arrangement.
- Place the table at the top right corner.
- Click on Repeat region.
- A table region menu manager will appear.
- Click on Add to add the repeat region.
- Click on the left bottom cell and the right bottom cell once.
- Press the middle button to close the table region menu manager.
- Double click on the left bottom cell and select the parameter asm > mbr > name.
- Similarly, double click on the bottom second cell and select the parameter asm > mbr > user_defined.
- Write PRO_MP_AREA.
- Click on Repeat region, select Attribute, and select the region.
- Select No Duplicate/Level and Recursive.
- Click on Update table and notice the changes.
- Click on Relation from the Tools tab of the ribbon interface.
- In the Look in field, select Region.
- Click on the region.
- Write the following relations:
- X = asm_mbr_pro_mp_area/1000 (X defines as the quantity of barrel per 1000 mm^2)
- COST = X*40 (40 is the unit cost per barrel)
- Click OK to close the relation window.
- Similarly, double click on the bottom third cell and select the parameter rpt > rel > user_defined.
- Write Cost.
- Rename the upper boxes of the table as shown in Figure 13.2.
- Update the table and notice the changes.
- Save and exit the file.

Component name	Surface area	Cost	Remarks

FIGURE 13.2
Renaming of sample table.

13.8 Output Checking and Program Revision

An important part of automation is output checking and program revision. Automation is a program created to reduce human effort with the use of some intelligence provided in the system. Therefore, the accuracy of automation depends on the intelligence input in the program creation. The intelligence into the systems is in the form of relations, variables, and constant, which are entered into the system by human effort. Therefore, there will always be a chance of error in writing the relations. In addition to that, the relations may be written with the objective of a certain limit of variables. However, due to a certain requirement that necessitates the limits of variables to be extended, all relations may need to be changed to accommodate the extended limits. Therefore, after creating the automation, it is always advisable to check the program output with the extent of all variable limits. Many errors may appear during the first trial of the output. Therefore, programs need to be revised accordingly. The output checking and program revision need to be carried out again and again until the program output meets the quality requirement.

13.9 Project Work

Hands-On Exercise, Example 13.2: Ladder (Software Used: Creo Parametric; Figure 13.3)

LADDER

Steps:

- Create an assembly file and rename it as Ladder.asm. (Other names may be given, but it is advisable to provide a relevant name.)
- Select Design on the right-hand side of the box. This will activate the standard assembly design mode and environment.

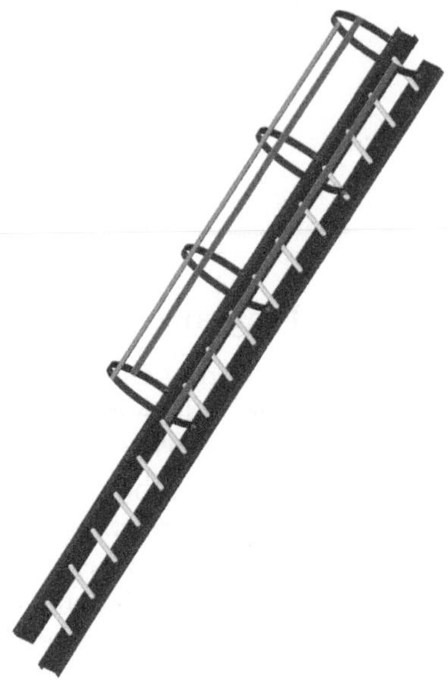

FIGURE 13.3
3-D view of ladder.

- Uncheck the box of the default template and click OK to create a new assembly design file.
- The template selection box will appear.
- Select mmns_asm_Solid. This will allow the use of the solid template file of the following units:
    - mm—millimeter for length unit
    - n—newton for force unit
    - s—second for time unit
- Once the assembly file is created, go to File > Prepare > Model properties to check the units. In the upper sections, the units of the model will be displayed.
    - Note: The units are usually displayed in the summary section of any CAD software. The setting of the units is usually in the tools or settings.
- Click on tree filters as shown in Figure 13.4.

- Check the box features to display in the model tree.
- Click on Apply. Notice that the features are displayed in the model tree.

- Right click on the plane ASM_RIGHT and select Rename. Change the name to Ladder_Start.
- Click OK datum plane from the Model tab of the ribbon interface.
- Select Ladder_Start plane as the reference plane.
- Set the offset value 5000 mm towards down.

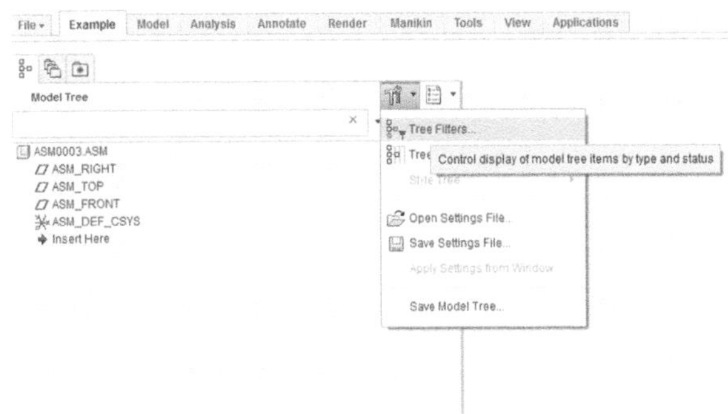

FIGURE 13.4
Model tree filter.

- Click OK to create the datum plane.
- Rename the plane as Ladder_End.

- Click on the Part create on assembly mode icon from the model tab of the ribbon interface.
- Change the name as Channel_Section_Left.
- Select Solid in the sub type.
- Click OK.
- The creation option window will appear.
- Select the Locate default datums option.
- This will allow creating three default datums in the selected references.
- Select Three Planes in the locate datums method area.
- Click OK and select ASM_RIGHT, ASM_TOP, and ASM_FRONT plane, respectively.
- Notice that the part Pipe_support_section_1 is added and activated.
- The ribbon interface will change to part interface.

- Click on Extrude from the Model tab in the ribbon interface.
- Click on the Placement tab and select Define to define the sketch.
- Select the plane Ladder_Start as the sketch plane.
- The Plane ASM_TOP shall be automatically taken as the reference and orientation as right.
- Click on Sketch.
- The sketcher model will be activated.
- Project the sketch edges as displayed in Figure 13.5.

- Click OK to create and exit the sketch.
- Select the Up to entity extrusion option.
- Select the plane ladder_End.
- Click on OK to create the extrude feature.
- Right click on the Main assembly and select Activate.

- Right click on the Main assembly Ladder and select Activate.

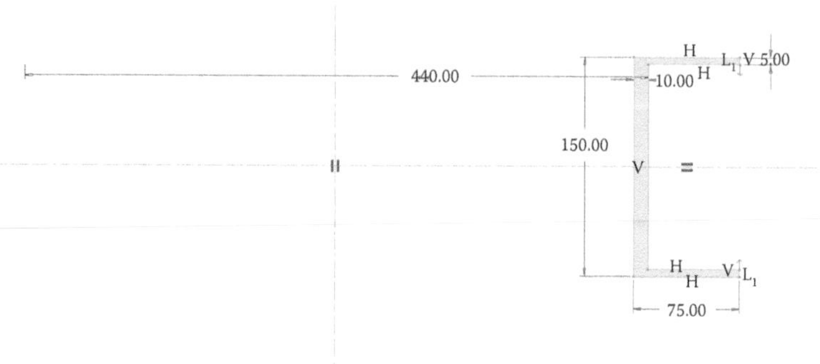

FIGURE 13.5
Constructional sketch of ladder C section.

- Click on the Part create on assembly mode icon from the Model tab of the ribbon interface.
- Change the name as Channel_section_Right.
- Select Mirror in the sub type.
- Click OK.
- The Mirror Part creation option window will appear.
- Select the Mirror geometry only option.
- Check the box for geometry and placement dependency.
- Select any surface of the part Channel_Section_Left for part reference.
- Select the plane ASM_TOP for the planar reference.

- Right click on the Main assembly Ladder and select Activate.

- Click on Extrude from the Model tab in the ribbon interface.
- Click on the Placement tab and select Define to define the sketch.
- Select the top plane as the sketch plane.
- The right plane shall be automatically taken as the reference and orientation as right.
- Click on Sketch.
- The sketcher model will be activated.
- Draw a circle of diameter 40 mm as shown in Figure 13.6.
- Click OK to create and exit the sketch.
- Select Through all extrusion for side 1 and side 2.
- Material cut shall be automatically activated.
- Click on OK to create the extrusion.

- Right click on the lastly created feature extrusion 1 and select Pattern.
- Select Direction and select the plane Ladder_Start as the directional reference.
- Make sure the direction is towards the material side of the channel section.

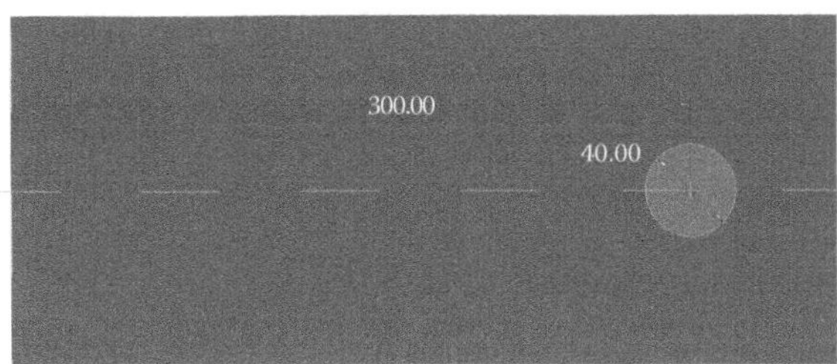

FIGURE 13.6
Constructional sketch of hole in ladder.

- Set the value 300 as the increment value and 2 as the pattern component quantity.
- Click OK to create the feature.

- Create a part file and rename it as Rung. (Other names may be given, but it is advisable to provide a relevant name.)
- Select Solid on the right-hand side of the box. This will activate the solid mode and environment.
- Uncheck the box of the default template and click OK to create a new solid file.
- The template selection box will appear.
- Select mmns_part_Solid. This will allow the use of the solid template file of the following units:
  - mm—millimeter for length unit
  - n—newton for force unit
  - s—second for time unit
- Once the solid part is created, go to File > Prepare > Model properties to check the units. In the upper sections, the units of the model will be displayed.
  - Note: The units are usually displayed in the summary section of any CAD software. The setting of the units is usually in the tools or settings.

- Click on Extrude from the Model tab in the ribbon interface.
- Click on the Placement tab and select Define to define the sketch.
- Select the right plane as the sketch plane.
- The top plane shall be automatically taken as the reference and orientation as left.
- Click on Sketch.
- The sketcher model will be activated.
- Draw a circle of diameter 33.4 mm.
- Click OK to create and exit the sketch.
- Select the Mid plane extrusion.

- Enter the value 450 mm for the extrusion depth.
- Click on OK to create the extrusion.

- Click again on Extrude from the Model tab in the ribbon interface.
- Click on the Placement tab and select Define to define the sketch.
- Select the right plane surface of the lastly created extrusion.
- The top plane shall be automatically taken as the reference and orientation as left.
- Click on Sketch.
- The sketcher model will be activated.
- Draw the geometry as shown in Figure 13.7.
- Click OK to create and exit the sketch.
- Select Up to entity extrusion.
- Select the opposite plane surface of the lastly created extrusion.
- Click on OK to create the extrusion.

- Select the lastly created feature extrusion 2 in the model tree.
- Right click and select Pattern.
- Select Axis in the type of pattern selection field.
- Select the central axis.
- Enter the quantity 12 for the patterned feature.
- Select the option to make the pattern equally spaced within 360°.
- Click OK to create the patterned feature.

- Click on Shell from the Model tab in the ribbon interface.
- Set the value 3 as the shell value.
- Select both the plane surfaces for removal.
- Click OK to create the shell feature.
- Save and exit the file.

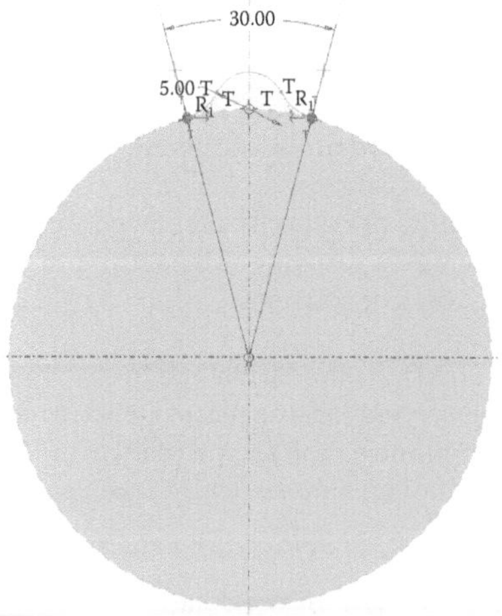

FIGURE 13.7
Constructional sketch of rung of ladder.

- Click on Assembly from the Model tab of the ribbon interface.
- Select the file Rung.
- Click on the Placement tab.
- Select the right plane of the Rung part as the component reference.
- Select the ASM_TOP plane as the assembly reference.
- Select Coincide in the component reference relation field
- Right click on the graphical area and select New constraint.
- Select the Inside cylindrical surface of the Rung part as the component reference.
- Select the Inside cylindrical surface of the first patterned feature extrude 2 as the assembly reference.
- Select Coincide in the component reference relation field.
- The Allow assumption box shall be automatically checked.
- Click OK to assemble the part.
- Right click on the part and select Pattern.
- Pattern reference type shall be automatically selected.
- Click OK to create the pattern.

- Click on the Tools tab in the ribbon interface.
- Click on Relations.
- The relation window will appear.
- Add the following relations:
  - p = LADDER_LENGTH/300
  - p8 = p
  - if(p8 = =p)
  - p8 = p-1
  - endif
- Click on Verify on the right upper side of the relation window.
- A notification box with a message "Relations have been successfully verified" should appear.
- Click OK to close the relation window.
- Note that the above-mentioned p8 in the relation is the parameter for patterned quantity; if the parameter name is not p8, the users shall select the pattern feature graphically and select the quantity parameter.
- Click on Regenerate; the rungs shall be patterned as required.

- Click on the Part create on assembly mode icon from the Model tab of the ribbon interface.
- Change the name as Flat.
- Select Sheet metal in the sub type.
- Click OK.
- The creation option window will appear.
- Select the Locate default datums option.
- This will allow creating three default datums in the selected references.
- Select Three Planes in the Locate datums method area.
- Click OK and select ASM_RIGHT, ASM_TOP, and ASM_FRONT plane, respectively.

- Notice that the part Pipe_support_section_1 is added and activated.
- The ribbon interface will change to part interface.

- Click OK datum plane from the Model tab of the ribbon interface.
- Select Ladder_Start plane as the reference plane.
- Set the offset value 2000 mm towards down.
- Click OK to create the datum plane.

- Click on Extrude from the Model tab in the ribbon interface.
- Click on the Placement tab and select Define to define the sketch.
- Select the lastly created plane as the sketch plane.
- The Plane ASM_TOP shall be automatically taken as the reference and orientation as right.
- Click on Sketch.
- The sketcher model will be activated.
- Project the sketch edges as displayed in Figure 13.8.
- Click OK to create and exit the sketch.
- Select the Mid plane extrusion.
- Set the value 50 for the extrusion depth.
- Set the value 5 as the thickness.
- Flip the material orientation if required so that the feature shall not interfere with channel_left or right section.
- Click on the Option tab.
- Check the box for Add bends on sharp edges under the sheet metal option.
- Select the radius thickness for the inside radius.
- Click on OK to create the extrude feature.
- Right click on the main assembly and select Activate.

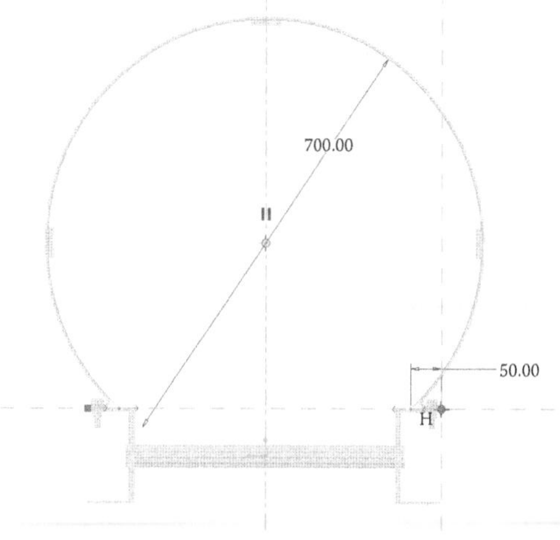

FIGURE 13.8
Constructional sketch of cage of ladder.

- Right click on the part flat and select Pattern.
- Select Direction and select the plane Ladder_Start as the directional reference.
- Make sure that the direction is towards the material side of the channel section.
- Set the value 300 as the increment value and 2 as the pattern component quantity.
- Click OK to create the feature.

- Click OK datum axis from the model tab of the ribbon interface.
- Select the Inside cylindrical surface of the first flat component.
- Click OK to create the datum axis.

- Click on the Part create on assembly mode icon from the Model tab of the ribbon interface.
- Change the name as Flat1.
- Select Sheet metal in the sub type.
- Click OK.
- The creation option window will appear.
- Select the Locate default datums option.
- This will allow creating three default datums in the selected references.
- Select Three Planes in the Locate datums method area.
- Click OK and select ASM_RIGHT, ASM_TOP, and ASM_FRONT plane, respectively.
- Notice that the part Pipe_support_section_1 is added and activated.
- The ribbon interface will change to part interface.

- Click on Extrude from the Model tab in the ribbon interface.
- Click on the Placement tab and select Define to define the sketch.
- Select Front face of flat as the sketch plane.
- The Plane ASM_TOP shall be automatically taken as the reference and orientation as right.
- Click on Sketch.
- The sketcher model will be activated.
- Project the sketch edges as displayed in Figure 13.9.
- Click OK to create and exit the sketch.
- Select the Mid plane extrusion.
- Set the value 150 for the extrusion depth.
- Set the value 5 as the thickness.
- Flip the material orientation if required so that the feature shall not interfere with the flat part, and the flat creation is towards the plane Ladder_End.
- Click on the Option tab.
- Check the box for Add bends on sharp edges under the sheet metal option.
- Select the radius thickness for the inside radius.
- Click on OK to create the extrude feature.
- Right click on the main assembly and select Activate.

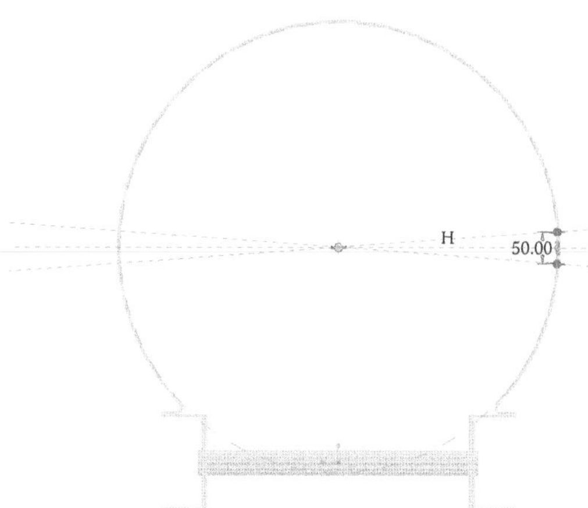

FIGURE 13.9
Constructional sketch of cage parts of ladder.

- Right click on the part flat1 and select Pattern.
- Set the axis, and select the lastly created axis in the center of the inside cylindrical surface of the flat part.
- Set the pattern quantity as 3 and the incremental angular value as 90.
- Flip the orientation so that the patterned feature is created on the flat part side.
- Click OK to create the patterned feature.

- Click on the Tools tab in the ribbon interface.
- Click on Relations.
- The relation window will appear.
- Add the following relations:
  - pc = floor(((LADDER_LENGTH-2000))/1000)
  - p41 = pc+1
  - d39 = ((LADDER_LENGTH-2000)-100)/pc
  - d1:18 = (d39*PC)+50
- Click on Verify on the right upper side of the relation window.
- A notification box with a message "Relations have been successfully verified" should appear.
- Click OK to close the relation window.
- Note that the above-mentioned p41 in the relation is the parameter for patterned quantity for the pattern feature of the flat part; if the parameter name is not p41, the users shall select the pattern feature graphically and select the quantity parameter.
- Similarly, the above-mentioned d39 in the relation is the parameter for the patterned increment for the pattern feature of the flat part; if the parameter name is not d39, the users shall select the pattern feature graphically and select the increment parameter.

- Similarly, the above-mentioned d1:18 in the relation is the parameter for the extrusion depth of the flat1 part for the pattern feature of the flat part; if the parameter name is not d1:18, the users shall select the extrusion depth of the flat1 feature graphically.

- Click on Regenerate; the flat shall be patterned as required.

- Create a part file and rename it as NUT_M10. (Other names may be given, but it is advisable to provide a relevant name.)
- Select Solid on the right-hand side of the box. This will activate the solid mode and environment.
- Uncheck the box of the default template and click OK to create a new solid file.
- The template selection box will appear.
- Select mmns_part_Solid. This will allow the use of the solid template file of the following units:
  - mm—millimeter for length unit
  - n—newton for force unit
  - s—second for time unit
- Once the solid part is created, go to File > Prepare > Model properties to check the units. In the upper sections, the units of the model will be displayed.
  - Note: The units are usually displayed in the summary section of any CAD software. The setting of the units is usually in the tools or settings.
- Click on Extrude from the Model tab in the ribbon interface.
- Click on the Placement tab and select Define to define the sketch.
- Select the front plane as the sketch plane.
- The right plane shall be automatically taken as the reference and orientation as right.
- Click on Sketch.
- The sketcher model will be activated.
- Draw the sketch as shown in Figure 13.10.
- Click OK to create and exit the sketch.
- Select the Mid plane extrusion.
- Enter the value 10 mm for the extrusion depth.
- Click on OK to create the extrusion.

- Click on Revolve from the Model tab in the ribbon interface.
- Click on the Placement tab and select Define to define the sketch.
- Activate the material removal feature option.
- Select the front plane as the sketch plane.
- The right plane shall be automatically taken as the reference and orientation as right.
- Click on Sketch.
- The sketcher model will be activated.
- Draw the sketch as displayed in Figure 13.11.
- Click OK to create and exit the sketch.
- Ensure that the revolve angle shall be 360°.
- Click on OK to create the revolve feature.

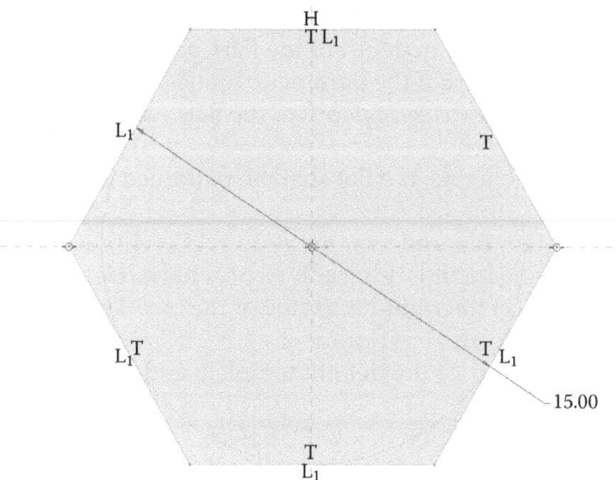

FIGURE 13.10
Constructional sketch of hexagonal extrusion feature of Nut of size M10.

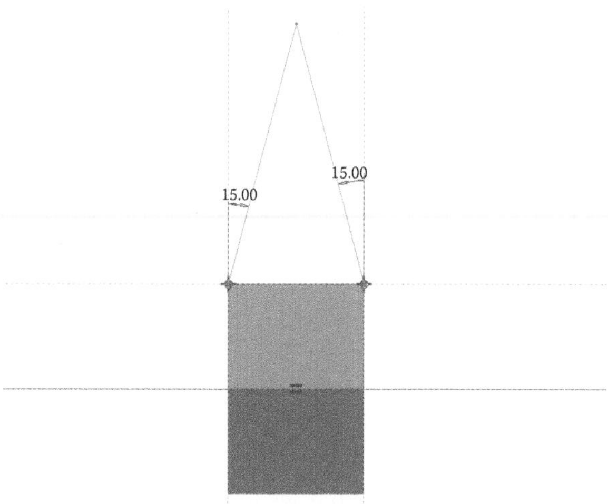

FIGURE 13.11
Constructional sketch of revolve cut feature of Nut of size M10.

- Click on Hole from the Model tab of the ribbon interface.
- Click on Activate the simple hole.
- Set the depth to Through all.
- Enter the hole diameter as 10 mm.
- Select the top flat surface of the nut.
- Click on To select offset reference and select both the front and right planes as offset references.
- Set aligned in the offset value setting box instead of distance.
- Click OK to create the hole.
- Save and exit the file.

- Create a part file and rename it as Bolt_M10. (Other names may be given, but it is advisable to provide a relevant name.)
- Select Solid on the right-hand side of the box. This will activate the solid mode and environment.
- Uncheck the box of the default template and click OK to create a new solid file.
- The template selection box will appear.
- Select mmns_part_Solid. This will allow the use of the solid template file of the following units:
  - mm—millimeter for length unit
  - n—newton for force unit
  - s—second for time unit
- Once the solid part is created, go to File > Prepare > Model properties to check the units. In the upper sections, the units of the model will be displayed.
  - Note: The units are usually displayed in the summary section of any CAD software. The setting of the units is usually in the tools or settings.
- Click on Extrude from the Model tab in the ribbon interface.
- Click on the Placement tab and select Define to define the sketch.
- Select the front plane as the sketch plane.
- The right plane shall be automatically taken as the reference and orientation as right.
- Click on Sketch.
- The sketcher model will be activated.
- Draw the sketch as shown in Figure 13.12.
- Click OK to create and exit the sketch.
- Select the Mid plane extrusion.
- Enter the value 10 mm for the extrusion depth.
- Click on OK to create the extrusion.

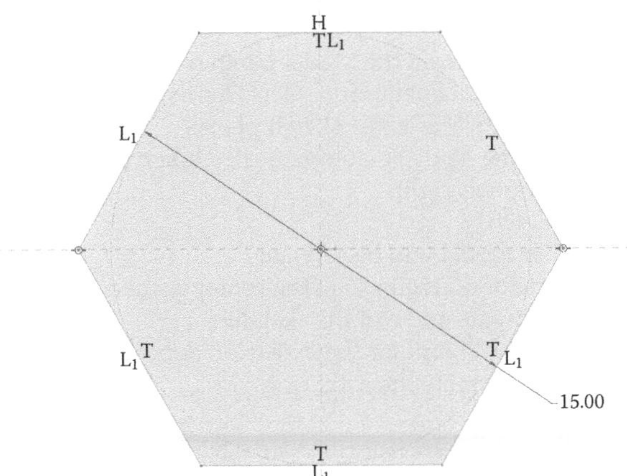

FIGURE 13.12
Constructional sketch of hexagonal extrusion feature of bolt of size M10.

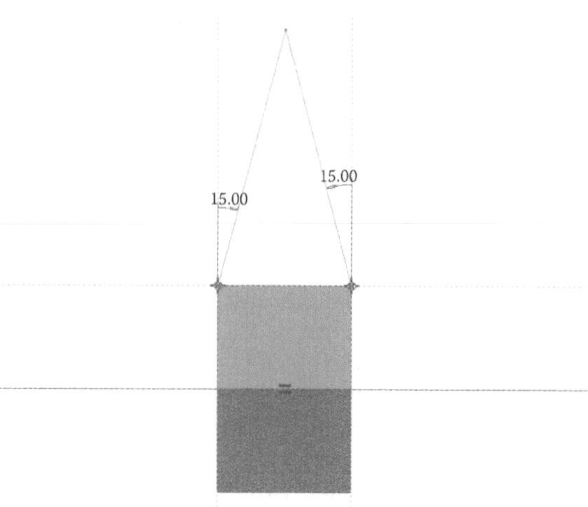

FIGURE 13.13
Constructional sketch of revolve cut feature of bolt of size M10.

- Click on Revolve from the Model tab in the ribbon interface.
- Click on the Placement tab and select Define to define the sketch.
- Activate the material removal feature option.
- Select the front plane as the sketch plane.
- The right plane shall be automatically taken as the reference and orientation as right.
- Click on Sketch.
- The sketcher model will be activated.
- Draw the sketch as displayed in Figure 13.13.
- Click OK to create and exit the sketch.
- Ensure that the revolve angle shall be 360°.
- Click on OK to create the revolve feature.

- Click on Extrude from the Model tab in the ribbon interface.
- Click on the Placement tab and select Define to define the sketch.
- Select the front plane as the sketch plane.
- The right plane shall be automatically taken as the reference and orientation as right.
- Click on Sketch.
- The sketcher model will be activated.
- Draw a circle of diameter 10 at the center of the extrusion.
- Click OK to create and exit the sketch.
- Enter the value 35 mm for the extrusion depth.
- Click on OK to create the extrusion.

- Click on Round feature from the Model tab of the ribbon interface.
- Set the value of fillet to 2 mm.
- Select all four internal edges of the newly created feature as shown in Figure 13.14.
- Click OK to create the feature.

FIGURE 13.14
Constructional view of fillet of bolt of size M10.

- Click on Chamfer from the Model tab of the ribbon interface.
- Set the value 0.5 as the chamfer value with DXD configuration.
- Select the edge as displayed in Figure 13.15.
- Click OK to create the chamfer feature.

- Click on Engineering from the Model tab of the ribbon interface.
- Select Cosmetic thread.
- Select the outer cylindrical surface of the lastly created extrusion as the thread surface.
- Select the bottom bolt surface as the start surface.
- Set the depth value to 30 mm.
- Set the value 9 as the diameter.
- Click OK to create the cosmetic thread feature.

- Save and exit the file.
- Click on the hole feature from the Model tab of the ribbon interface.
- Set the diameter value as Through all.

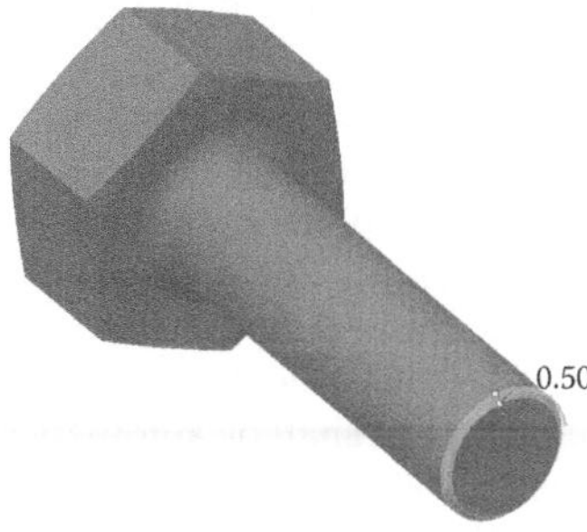

FIGURE 13.15
Constructional view of edge fillet feature bolt of size M10.

FIGURE 13.16
Constructional view of hole in ladder.

- Set the references as shown in Figure 13.16.
- Click OK to create the hole feature.
- Select Pattern to pattern the hole feature.
- Set the direction and select the plane Ladder_Start as the reference.
- Flip the orientation so that the patterned feature is to be created towards the plane Ladder_End.
- Click OK to create the pattern feature.
- Similarly, create another hole on the other side of flat1 and pattern the features.

- Click on Assembly from the Model tab of the ribbon interface.
- Select the file Bolt_M10.
- Click on the Placement tab.
- Select the cylindrical surface of the Bolt_m10 part as the component reference.
- Select the inside cylindrical hole surface as the assembly reference.
- Select Coincide in the component reference relation field.
- Right click on the graphical area and select New constraint.
- Select the bottom surface of the Bolt_M10 part as the component reference.
- Select the top surface of the flat part, where the holes were created, of the first patterned feature extrude 2 as the assembly reference.
- Select Coincide in the component reference relation field.
- The Allow assumption box shall be automatically checked.
- Click OK to assemble the part.
- Right click on the part and select Pattern.
- The pattern reference type shall be automatically selected.
- Click OK to create the pattern.
- Similarly, assemble the part Nut_M10 suitably.
- Select both parts Bolt_M10 and Nut_M10; right click and select Group.

- Right click on the group and select Pattern.
- The reference pattern shall be automatically selected.
- Click OK to create the patterned feature.

- Click on the Tools tab in the ribbon interface.
- Click on Relations.
- The relation window will appear.
- Add the following relations:
  - p = LADDER_LENGTH/300
  - p8 = p
  - if(p8 = =p)
  - p8 = p-1
  - endif
- Click on Verify on the right upper side of the relation window.
- A notification box with a message "Relations have been successfully verified" should appear.
- Click OK to close the relation window.
- Note that the above-mentioned p8 in the relation is the parameter for patterned quantity; if the parameter name is not p8, the users shall select the pattern feature graphically and select the quantity parameter.
- Click on Regenerate; the group shall be patterned as required.

Hands-On Exercise, Example 13.3: Pipe Support (Software Used: Creo Parametric; Figure 13.17)

PIPE_SUPPORT

Steps:

- Create an assembly file and rename it as Pipe_Support.asm. (Other names may be given, but it is advisable to provide a relevant name.)
- Select Design on the right-hand side of the box. This will activate the standard assembly design mode and environment.

FIGURE 13.17
3-D view of pipe support.

- Uncheck the box of the default template and click OK to create a new assembly design file.
- The template selection box will appear.
- Select mmns_asm_Solid. This will allow the use of the solid template file of the following units:
  - mm—millimeter for length unit
  - n—newton for force unit
  - s—second for time unit
- Once the assembly file is created, go to File > Prepare > Model properties to check the units. In the upper sections, the units of the model will be displayed.
  - Note: The units are usually displayed in the summary section of any CAD software. The setting of the units is usually in the tools or settings.
- Click on tree filters as shown in Figure 13.4.

- Check the box features to display in the model tree.
- Click on Apply. Notice that the features are displayed in the model tree.

- Click on the Part create on assembly mode icon from the Model tab of the ribbon interface.
- Change the name as pipe_Support_Section_1.
- Select Solid in the sub type.
- Click OK.
- The creation option window will appear.
- Select the Locate default datums option.
- This will allow creating three default datums in the selected references.
- Select Three Planes in the Locate datums method area.
- Click OK and select ASM_RIGHT, ASM_TOP, and ASM_FRONT plane, respectively.
- Notice that the part Pipe_support_section_1 is added and activated.
- The ribbon interface will change to part interface.

- Click on Extrude from the Model tab in the ribbon interface.
- Click on the Placement tab and select Define to define the sketch.
- Select the plane DTM1 as the sketch plane.
- The plane DTM2 shall be automatically taken as the reference and orientation as right.
- Click on Sketch.
- The sketcher model will be activated.
- Project the sketch edges as displayed in Figure 13.18.
- Click OK to create and exit the sketch.
- Select the Mid plane extrusion.
- Enter the value 100 mm for the extrusion depth.
- Click on OK to create the extrude feature.
- Right click on the main assembly and select Activate.

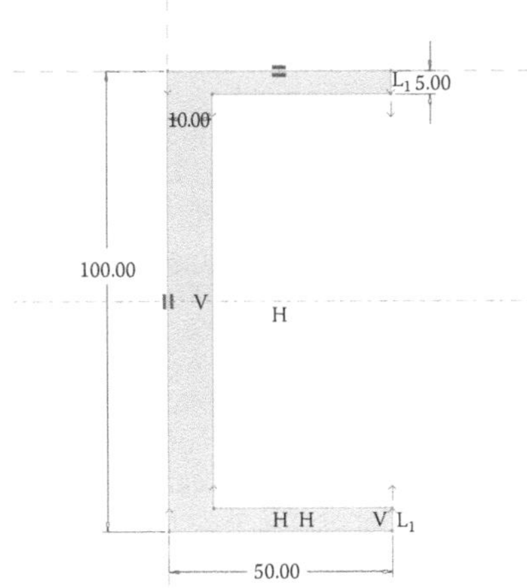

FIGURE 13.18
Constructional view of pipe support.

- Create a part file and rename it as U_Bolt. (Other names may be given, but it is advisable to provide a relevant name.)
- Select Solid on the right-hand side of the box. This will activate the solid mode and environment.
- Uncheck the box of the default template and click OK to create a new solid file.
- The template selection box will appear.
- Select mmns_part_Solid. This will allow the use of the solid template file of the following units:
    - mm—millimeter for length unit
    - n—newton for force unit
    - s—second for time unit
- Once the solid part is created, go to File > Prepare > Model properties to check the units. In the upper sections, the units of the model will be displayed.
    - Note: The units are usually displayed in the summary section of any CAD software. The setting of the units is usually in the tools or settings.
- U_Bolt can be created by the sweep method.
- Principally, it requires one cross-sectional sweep profile and sweep trajectory.
- Click on Sweep from the Model tab of the ribbon interface.
- The sweep trajectory will be created first followed by sweep operation.
- The trajectory can also be created inside the sweep operation; however, the sketch will not be available if the sweep feature is deleted.
- Select Sketch from the Model tab of the ribbon interface.
- A sketch window for the selection of the sketch plane will appear.

- Select the front plane as the sketch plane.
- The right plane shall automatically be selected as the reference and the orientation is to the right.
- Click on Sketch.
- The sketcher window will open up, and the reference selection window appears for the selection of references.
- The two references, i.e., top plane as the horizontal reference and right plane as the vertical reference, shall be automatically taken; if not, please select the same options.
- Create a sketch as displayed in Figure 13.19.
- Click OK to create and exit the sketch.
- Click on Sweep from the Model tab of the ribbon interface.
- Graphically select the sketch as trajectory.
- The system assumes the profile sketch point and shows this by displaying the directional arrow.
- Graphically, click on the arrows of the feature creation to draw the sweep profile in other control points.
- Click on Sketch to sketch the profile.
- The sketcher mode is activated with the sketcher environment.
- Draw a circle in such a way that the profile is at the outer edge of the circle, as shown in Figure 13.20.
- Change the diameter to 16 mm.
- Click OK to create and exit the sketch.
- Click OK to finish the sweep.

- Click on the Tools tab in the ribbon interface.
- Click on Relations.
- The relation window will appear.
- Click on Local parameters to display the local parameters dialog box.

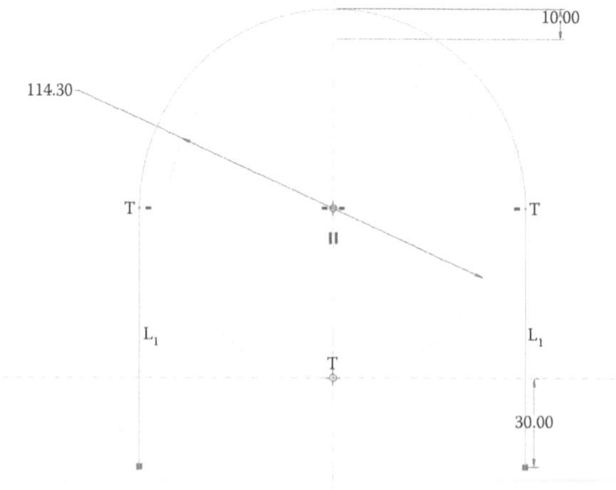

FIGURE 13.19
Constructional sketch of sweep path of U_bolt.

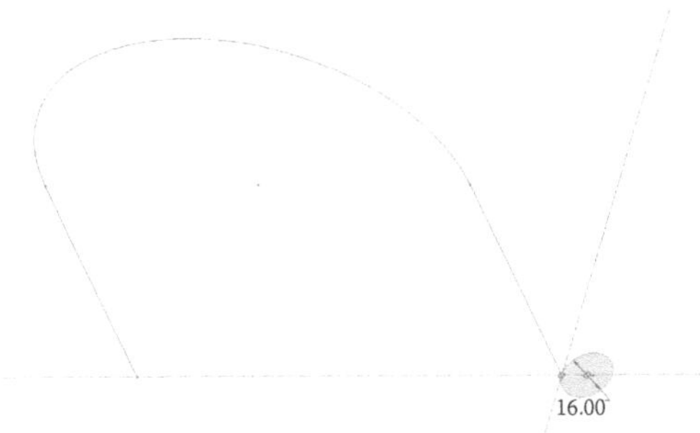

FIGURE 13.20
Constructional sketch of sweep profile of U_bolt.

- Add a parameter and rename it as Dia.
- Keep Real number as the type.
- Enter the Dia parameter value as 114.3.
- Add a parameter and rename it as Rod_Dia.
- Keep Real number as the type.
- Enter the Rod_Dia parameter value as 16.
- Add a parameter and rename it as Ext.
- Keep Real number as the type.
- Enter the Ext parameter value as 25.
- Add a parameter and rename it as Size.
- Keep the type as String.
- Enter the Size parameter value as 50.
- Click on the relation creation window.
- Add the following relations:
    - if(size = =65)
    - dia = 73
    - ext = 25
    - rod_dia = 10
    - endif
    - if(size = =80)
    - dia = 88.9
    - ext = 30
    - rod_dia = 12
    - endif
    - if(size = =100)
    - dia = 114.3
    - ext = 30
    - rod_dia = 16
    - endif
    - if(size = =150)
    - dia = 168.3
    - ext = 30
    - rod_dia = 16

    - endif
    - if(size = =200)
    - dia = 219
    - ext = 30
    - rod_dia = 16
    - endif
    - d4 = DIA
    - d5 = EXT
    - d9 = ROD_DIA
- Click on Verify on the right upper side of the relation window.
- A notification box with a message "Relations have been successfully verified" should appear.
- Click OK to close the box.
- Save and exit the file.

- Create a part file and rename it as NUT. (Other names may be given, but it is advisable to provide a relevant name.)
- Select Solid on the right-hand side of the box. This will activate the solid mode and environment.
- Uncheck the box of the default template and click OK to create a new solid file.
- The template selection box will appear.
- Select mmns_part_Solid. This will allow the use of the solid template file of the following units:
    - mm—millimeter for length unit
    - n—newton for force unit
    - s—second for time unit
- Once the solid part is created, go to File > Prepare > Model properties to check the units. In the upper sections, the units of the model will be displayed.
    - Note: The units are usually displayed in the summary section of any CAD software. The setting of the units is usually in the tools or settings.
- Click on Extrude from the Model tab in the ribbon interface.
- Click on the Placement tab and select Define to define the sketch.
- Select the front plane as the sketch plane.
- The right plane shall be automatically taken as the reference and orientation as right.
- Click on Sketch.
- The sketcher model will be activated.
- Draw the sketch as shown in Figure 13.21.
- Click OK to create and exit the sketch.
- Select the Mid plane extrusion.
- Enter the value 16 mm for the extrusion depth.
- Click on OK to create the extrusion.

- Click on Revolve from the Model tab in the ribbon interface.
- Click on the Placement tab and select Define to define the sketch.
- Activate the material removal feature option.
- Select the front plane as the sketch plane.

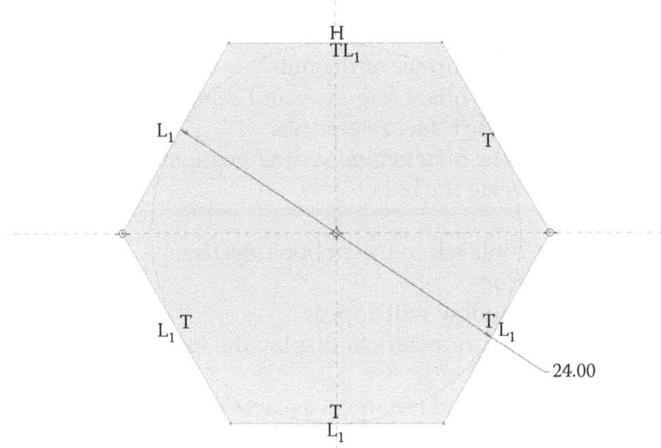

FIGURE 13.21
Constructional sketch of hexagonal extrusion feature of nut.

- The right plane shall be automatically taken as the reference and orientation as right.
- Click on Sketch.
- The sketcher model will be activated.
- Draw the sketch as displayed in Figure 13.22.
- Click OK to create and exit the sketch.
- Ensure that the revolve angle shall be 360°.
- Click on OK to create the revolve feature.

- Click on Hole from the Model tab of the ribbon interface.
- Click on Activate the simple hole.

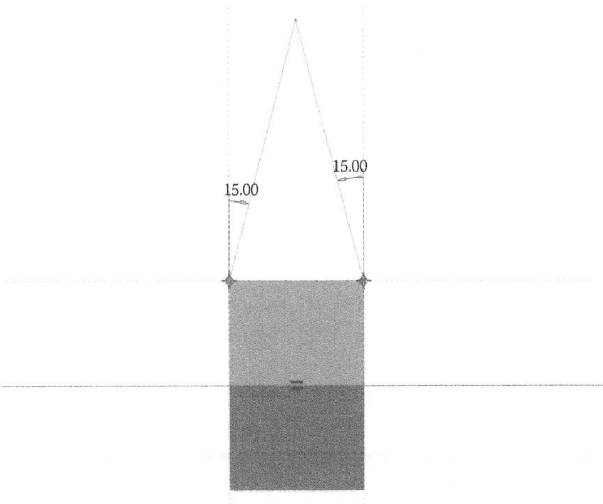

FIGURE 13.22
Constructional sketch of revolve cut feature of nut.

- Set depth to Through all.
- Enter the hole diameter as 16 mm.
- Select the top flat surface of the nut.
- Click on To select offset reference and select both the front and right planes as the offset references.
- Set aligned in the offset value setting box instead of distance.
- Click OK to create the hole.

- Click on the Tools tab in the ribbon interface.
- Click on Relations.
- The relation window will appear.
- Click on Local parameters to display the local parameters dialog box.
- Add a parameter and rename it as Hole Size.
- Keep Real number as the type.
- Enter the Hole size parameter value as 16.
- Add the following relations:
    - d4 = 1.5*hole_size
    - d3 = hole_size
    - d48 = hole_size
- Click on Verify on the right upper side of the relation window.
- A notification box with a message "Relations have been successfully verified" should appear.
- Click OK to close the box.

- Click on Assembly from the Model tab of the ribbon interface.
- Select the file U-bolt.
- Click on the Placement tab.
- Select the top plane of the U-bolt part as the component reference.
- Select the top surface of the existing channel as the assembly reference.
- Select Coincide in the component reference relation field.
- Right click on the graphical area and select New constraint.
- Select the front plane of the U-bolt part as the component reference.
- Select the top surface of the outer flange surface channel as the assembly reference.
- Enter the value 25 mm as the offset distance so that the nozzle is protruded inside the flange.
- Right click on the graphical area and select New constraint.
- Select the right plane of the U-bolt part as the component reference.
- Select the plane ASM_RIGHT as the assembly reference.
- Select Coincide in the component reference relation field.

- Click on Assembly from the Model tab of the ribbon interface.
- Select the file Nut.
- Click on the Placement tab.
- Select the top plane of the Nut part as the component reference.

- Select the top surface of the existing channel as the assembly reference.
- Select Coincide in the component reference relation field.
- Right click on the graphical area and select New constraint.
- Select the inside cylindrical surface of the Nut part as the component reference.
- Select the cylindrical surface of the U_bolt part as the assembly reference.
- Select Coincide in the component reference relation field.

- Add another three U_bolt by following the above method.
- Click on the Tools tab in the ribbon interface.
- Click on Relations.
- The relation window will appear.
- Click on Local parameters to display the local parameters dialog box.
- Add a parameter and rename it as pipe Size.
- Keep Integer as the type.
- Enter the Hole size parameter value as 100.
- Add the following relations:
  - a = 300
  - if(PIPE_SIZE = =150)
  - a = 240
  - endif
  - if(PIPE_SIZE = =100)
  - a = 190
  - endif
  - if(PIPE_SIZE = =80)
  - a = 160
  - endif
  - if(PIPE_SIZE = =65)
  - a = 125
  - endif
  - if(PIPE_SIZE = =50)
  - a = 110
  - endif
  - d5:0 = A
- Click on Verify on the right upper side of the relation window.
- A notification box with a message "Relations have been successfully verified" should appear.
- Select Part in the Look in the dialog box.
- Select the part U_bolt.
- Add the following relation:
  - SIZE = PIPE_SIZE:1
- Click on Verify on the right upper side of the relation window.
- A notification box with a message "Relations have been successfully verified" should appear.
- Select the Nut part and add the following relations:
  - IF(PIPE_SIZE:1 = =50)
  - HOLE_SIZE = 10

- ENDIF
- IF(PIPE_SIZE:1 = =65)
- HOLE_SIZE = 10
- ENDIF
- IF(PIPE_SIZE:1 = =80)
- HOLE_SIZE = 12
- ENDIF
- IF(PIPE_SIZE:1 = =100)
- HOLE_SIZE = 16
- ENDIF
- IF(PIPE_SIZE:1 = =150)
- HOLE_SIZE = 16
- ENDIF
- IF(PIPE_SIZE:1 = =200)
- HOLE_SIZE = 16
- ENDIF
- d4 = 1.5*hole_size
- d3 = hole_size
- d48 = hole_size

- Click on Verify on the right upper side of the relation window.
- A notification box with a message "Relations have been successfully verified" should appear.
- Close the relation dialog box.
- Click on Family table from the Tools tab of the ribbon interface.
- Add the parameter Pipe_size as the variables of the family table.
- Enter the family table as shown in Figure 13.23.
- Click on Verify to verify the family table instances.
- Save and exit the file.

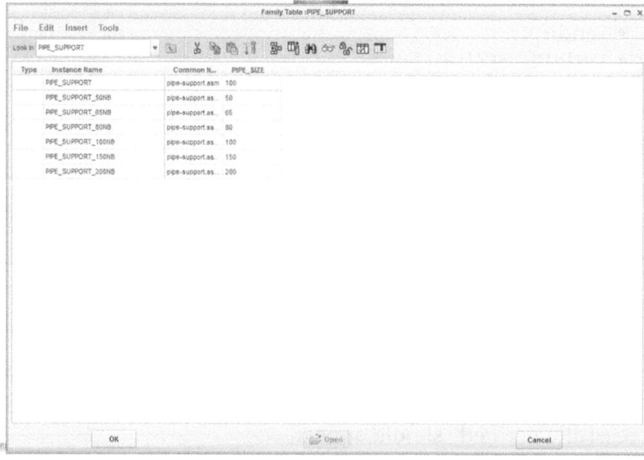

FIGURE 13.23
View of family table.

- Click Open to open the assembly file and select pipe_support .asm.
- In the instances selector box, select different instances and see that the supports are exactly as required.
- Save and exit the file.

Hands-On Exercise, Example 13.4: Shell Length (Software Used: Creo Parametric; Figure 13.24)

Shell length: Tank shells are often made by welding rolling sheets side by side. The number of circumferential weldings depends on the width of available sheets. These sheets are normally available in 1500 or 2000 mm standard width. Therefore, any given shell must be welded from a multiple of the number of standard sheets and cut sheets. This method shows how the shells can be automated to assemble the required quantity of sheets with standard or cut size.

Steps:

- Create an assembly file and rename it as Shell_Plate. (Other names may be given, but it is advisable to provide a relevant name.)
- Select Design on the right-hand side of the box. This will activate the standard assembly design mode and environment.
- Uncheck the box of the default template and click OK to create a new assembly design file.
- The template selection box will appear.
- Select mmns_asm_Solid. This will allow the use of the solid template file of the following units:
  - mm—millimeter for length unit
  - n—newton for force unit
  - s—second for time unit

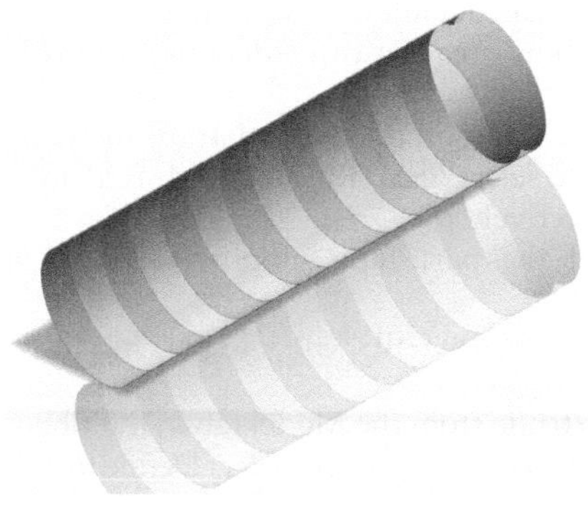

FIGURE 13.24
3-D view of shell length automation.

- Once the assembly file is created, go to File > Prepare > Model properties to check the units. In the upper sections, the units of the model will be displayed.
    - Note: The units are usually displayed in the summary section of any CAD software. The setting of the units is usually in the tools or settings.
- Click on tree filters as shown in Figure 13.4.
- Check the box features to display in the model tree.
- Click on Apply. Notice that the features are displayed in the model tree.

- Click on Relations from the Tools tab of the ribbon interface.
- Click on Parameters and add the following parameters:
    - Parameter name: Plate-Width, Type-Real Number, value—1500
    - Parameter name: Shell_Length, Type-Real Number, value—6700
- In the relations field, write the following relations:
    - qty_plate1 = floor(shell_length/PLATE_WIDTH)
    - width_plate_2 = shell_length-(plate_width*qty_plate1)
- Click on Verify on the right upper side of the relation window.
- A notification box with a message "Relations have been successfully verified" should appear.
- Click OK to close the box.
- Notice that this will add two parameters: qty_plate1—this is the number of standard plates with available width of 1500.
- Width_plate_2 is the cut sheet width.
- Click OK to close the relation window.

- Click on the Part create on assembly mode icon from the Model tab of the ribbon interface.
- Change the name as Plate_1.
- Select Sheet metal in the sub type.
- Click OK.
- The creation option window will appear.
- Select the Locate default datums option.
- This will allow creating three default datums in the selected references.
- Select Three Planes in the locate datums method area.
- Click OK and select ASM_RIGHT, ASM_TOP, and ASM_FRONT plane, respectively.
- Notice that the part Pipe_support_section_1 is added and activated.
- The ribbon interface will change to part interface.

- Click on Extrude from the Model tab in the ribbon interface.
- Click on the Placement tab and select Define to define the sketch.
- Select the plane ASM_FRONT as the sketch plane.
- The Plane ASM_Right shall be automatically taken as the reference and orientation as right.

- Click on Sketch.
- The sketcher model will be activated.
- Draw a circle of 2000 mm placing the center at the intersection of the horizontal and vertical references.
- Click OK to create and exit the sketch.
- Enter the value 1500 mm for the extrusion depth.
- Enter the value 10 as the thickness.
- Flip the material orientation inside.
- Click on ok to create the extrude feature.
- Right click on the main assembly and select activate.

- Click on relations from the tools tab of ribbon interface.
- In the relations field write the following relations:
  - d0:0 = PLATE_WIDTH
- Click on Verify on the right upper side of the relation window.
- A notification box with a message "Relations have been successfully verified" should appear.
- Click OK to close the box.
  - Note: d0:0 is the extrusion depth of the shell plate. If the symbols differ, users need to select the dimension graphically.
- Right click on the part plate_1 and select Pattern.
- Select Direction and select the plane ASM_FRONT as the directional reference.
- Set any value for incremental dimension.
- Click OK to create the feature.

- Click on Relations from the Tools tab of the ribbon interface.
- In the relations field, write the following relations:
  - p8 = QTY_PLATE1
  - d6 = PLATE_WIDTH
- Click on Verify on the right upper side of the relation window.
- A notification box with a message "Relations have been successfully verified" should appear.
- Click OK to close the box.
  - Note: p8 is the pattern quantity parameter, and d6 is the incremental dimension of the pattern. If the symbols differ, users need to select the dimension graphically.

- Create a part file and rename it as Plate2. (Other names may be given, but it is advisable to provide a relevant name.)
- Select Sheet metal on the right-hand side of the box. This will activate the sheet metal mode and environment.
- Uncheck the box of the default template and click OK to create a new sheet metal file.
- The template selection box will appear.
- Select mmns_part_Solid. This will allow the use of the solid template file of the following units:
  - mm—millimeter for length unit
  - n—newton for force unit
  - s—second for time unit

- Once the Solid part is created, go to File > Prepare > Model properties to check the units. In the upper sections, the units of the model will be displayed.
    - Note: The units are usually displayed in the summary section of any CAD software. The setting of the units is usually in the tools or settings.
- Click on Extrude from the Model tab in the ribbon interface.
- Click on the Placement tab and select Define to define the sketch.
- Select the front plane as the sketch plane.
- The right plane shall be automatically taken as the reference and orientation as right.
- Click on Sketch.
- The sketcher model will be activated.
- Draw a circle of diameter 2000.
- Click OK to create and exit the sketch.
- Enter the value 10 as the thickness.
- Enter the value 200 mm for the extrusion depth.
- Click on OK to create the extrusion.
- Save and exit the file.

- Click on Assembly from the Model tab of the ribbon interface.
- Select the file Plate2.
- Click on the Placement tab.
- Select the cylindrical surface of the plate 2 part as the component reference.
- Select the cylindrical surface of plate 1 of the existing channel as the assembly reference.
- Select Coincide in the component reference relation field.
- Right click on the graphical area and select New constraint.
- Select the flat end surface of the plate 2 part as the component reference.
- Select the plane ASM_FRONT as the assembly reference.
- Select Coincide in the component reference relation field.
- Click OK to assemble the part plate 2.

- Click on the Tools tab in the ribbon interface.
- Click on Relations.
- The relation window will appear.
- Add the following relations:
    - d10 = PLATE_WIDTH*QTY_PLATE1
    - d3:2 = SHELL_LENGTH-(PLATE_WIDTH*QTY_PLATE1)
- Click on Verify on the right upper side of the relation window.
- A notification box with a message "Relations have been successfully verified" should appear.
- Select Part in the Look in the dialog box.
- Click OK to close the dialog box.
    - Note: d10 is the offset dimension of the plate 2 flat surface from the plane ASM_FRONT. If the d10 symbol is different, users are suggested to select the offset dimension graphically.

- Similarly, d3:2 is the plate 2 extrusion depth. If required, users are suggested to select the offset dimension graphically.
- Click OK to close the relation window.
- Click on Generate to generate the shell successfully.
- Change the parameter shell length to 7500 and see the difference.

Hands-On Exercise, Example 13.5: Spring Support (Software Used: Creo Parametric; Figure 13.25)

SPRING_SUPPORT

Steps:

- Create a part file and rename it as Spring_Base_Plate. (Other names may be given, but it is advisable to provide a relevant name.)
- Select Solid on the right-hand side of the box. This will activate the solid mode and environment.
- Uncheck the box of the default template and click OK to create a new solid file.
- The template selection box will appear.
- Select mmns_part_Solid. This will allow the use of the solid template file of the following units:
    - mm—millimeter for length unit
    - n—newton for force unit
    - s—second for time unit
- Once the solid part is created, go to File > Prepare > Model properties to check the units. In the upper sections, the units of the model will be displayed.
    - Note: The units are usually displayed in the summary section of any CAD software. The setting of the units is usually in the tools or settings.
- Click on Extrude from the Model tab in the ribbon interface.

FIGURE 13.25
(See color insert.) 3-D view of spring support.

- Click on the Placement tab and select Define to define the sketch.
- Select the top plane as the sketch plane.
- The right plane shall be automatically taken as the reference and orientation as right.
- Click on Sketch.
- The sketcher model will be activated.
- Draw a rectangle of 250 × 250 symmetrically about the horizontal and vertical references.
- A centerline may need to be drawn along the reference to draw the above-mentioned rectangle.
- Click OK to create and exit the sketch.
- Flip the extrusion towards down.
- Enter the value 8 mm for the extrusion depth.
- Click on OK to create the extrusion.
- Save and exit the file.

- Create a part file and rename it as Spring_Cage_Part. (Other names may be given, but it is advisable to provide a relevant name.)
- Select Solid on the right-hand side of the box. This will activate the solid mode and environment.
- Uncheck the box of the default template and click OK to create a new solid file.
- The template selection box will appear.
- Select mmns_part_Solid. This will allow the use of the solid template file of the following units:
    - mm—millimeter for length unit
    - n—newton for force unit
    - s—second for time unit
- Once the solid part is created, go to File > Prepare > Model properties to check the units. In the upper sections, the units of the model will be displayed.
    - Note: The units are usually displayed in the summary section of any CAD software. The setting of the units is usually in the tools or settings.
- Click on Extrude from the Model tab in the ribbon interface.
- Click on the Placement tab and select Define to define the sketch.
- Select the top plane as the sketch plane.
- The right plane shall be automatically taken as the reference and orientation as right.
- Click on Sketch.
- The sketcher model will be activated.
- Draw a circle of diameter 150 placing the center at the intersection of the horizontal and vertical references.
- Click OK to create and exit the sketch.
- Flip the extrusion towards up.
- Enter the value 300 mm for the extrusion depth.
- Select the thicken part option.
- Enter 8 as the thickness value.

- Flip the material side inward.
- Click on OK to create the extrusion.

- Click on Extrude again from the Model tab in the ribbon interface.
- Click on the Placement tab and select Define to define the sketch.
- Select the top flat surface as the sketch plane.
- The right plane shall be automatically taken as the reference and orientation as right.
- Click on Sketch.
- The sketcher model will be activated.
- Draw the sketch as displayed in Figure 13.26.
- Click OK to create and exit the sketch.
- Flip the extrusion towards up.
- Enter the value 100 mm for the extrusion depth.
- Flip the material towards bottom.
- Activate the material removal option.
- Click on OK to create the extrusion.
- Save and exit the file.

- Create a part file and rename it as Spring_Cage_top. (Other names may be given, but it is advisable to provide a relevant name.)
- Select Solid on the right-hand side of the box. This will activate the solid mode and environment.
- Uncheck the box of the default template and click OK to create a new solid file.
- The template selection box will appear.
- Select mmns_part_Solid. This will allow the use of the solid template file of the following units:
    - mm—millimeter for length unit
    - n—newton for force unit
    - s—second for time unit

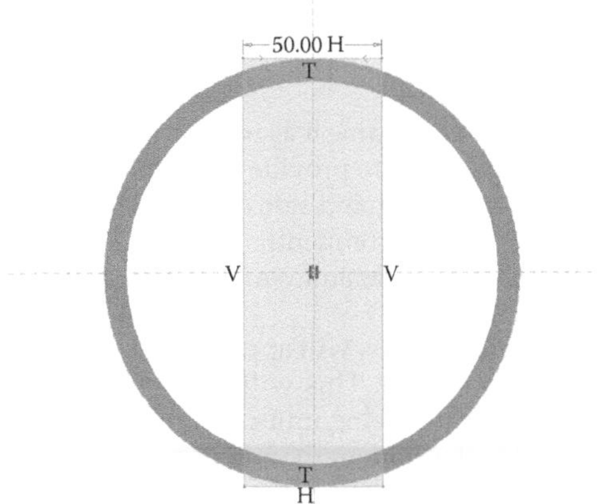

FIGURE 13.26
Constructional sketch of cage of spring support.

- Once the solid part is created, go to File > Prepare > Model properties to check the units. In the upper sections, the units of the model will be displayed.
  - Note: The units are usually displayed in the summary section of any CAD software. The setting of the units is usually in the tools or settings.
- Click on Extrude from the Model tab in the ribbon interface.
- Click on the Placement tab and select Define to define the sketch.
- Select the top plane as the sketch plane.
- The right plane shall be automatically taken as the reference and orientation as right.
- Click on Sketch.
- The sketcher model will be activated.
- Draw a circle of diameter 150 placing the center at the intersection of the horizontal and vertical references.
- Click OK to create and exit the sketch.
- Flip the extrusion towards up.
- Enter the value 8 mm for the extrusion depth.
- Flip the material side upward.
- Click on OK to create the extrusion.

- Click on Extrude again from the Model tab in the ribbon interface.
- Click on the Placement tab and select Define to define the sketch.
- Select the top flat surface as the sketch plane.
- The right plane shall be automatically taken as the reference and orientation as right.
- Click on Sketch.
- The sketcher model will be activated.
- Draw a circle of diameter 75, placing the center at the intersection of the horizontal and vertical references.
- Click OK to create and exit the sketch.
- Flip the material towards bottom.
- Activate the material removal option.
- Click on OK to create the extrusion.
- Save and exit the file.

- Create a part file and rename it as Spring. (Other names may be given, but it is advisable to provide a relevant name.)
- Select Solid on the right-hand side of the box. This will activate the solid mode and environment.
- Uncheck the box of the default template and click OK to create a new solid file.
- The template selection box will appear.
- Select mmns_part_Solid. This will allow the use of the solid template file of the following units:
  - mm—millimeter for length unit
  - n—newton for force unit
  - s—second for time unit

- Once the solid part is created, go to File > Prepare > Model properties to check the units. In the upper sections, the units of the model will be displayed.
    - Note: The units are usually displayed in the summary section of any CAD software. The setting of the units is usually in the tools or settings.
- Click on the helical sweep option from the drop-down menu of Sweep from the Model tab of the ribbon interface.
- Click on the References tab and select Define to define the helix sweep profile.
- Select the front plane as the sketch plane.
- The right plane shall be automatically taken as the reference and orientation as right.
- Click on Sketch.
- The sketcher model will be activated.
- Draw a geometry centerline on the vertical reference line in the middle.
- Draw a vertical line at a distance of 50 mm from the vertical centerline.
- Set the length of the line as 265 mm.
- Click OK to create the sketch and exit the sketcher window.
- Click Sketch to create the profile sketch above the Reference tab.
- Create a circle of 10 mm diameter placing the center at the bottom end of the vertical line.
- Click OK to create and exit the sketch.
- Enter the pitch value of 20 mm.
- Click on the Left-handed rule icon to create the helical spring clockwise downward.
- Click OK to create the helical spring.

- Click on Extrude from the Model tab in the ribbon interface.
- Click on the Placement tab and select Define to define the sketch.
- Select the front plane as the sketch plane.
- The right plane shall be automatically taken as the reference and orientation as right.
- Click on Sketch.
- The sketcher model will be activated.
- Draw the sketch as shown in Figure 13.27.
- Click OK to create and exit the sketch.
- Click on the Options tab.
- Select Through all for side 1 and side 2.
- Activate the material removal option.
- Click on OK to create the extrusion.
- Save and exit the file.

- Click on the Tools tab in the ribbon interface.
- Click on Relations.
- The relation window will appear.

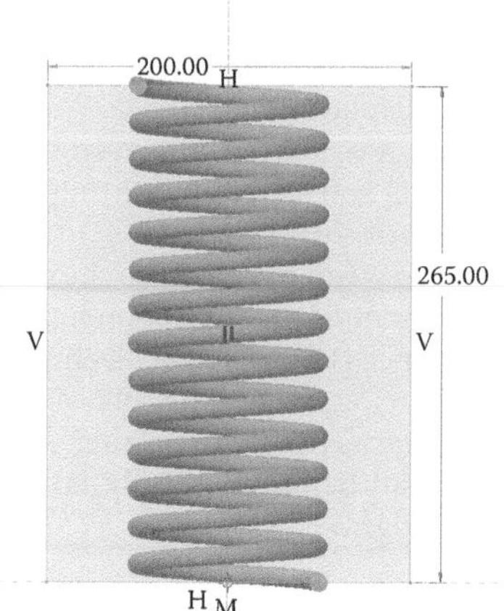

FIGURE 13.27
Constructional view of cut feature of spring.

- Click on Local parameters to display the local parameters dialog box.
- Add a parameter and rename it as Dia.
- Keep Real number as the type.
- Enter the Dia parameter value as 100.
- Add a parameter and rename it as Rod_Dia.
- Keep Real number as the type.
- Enter the Rod_Dia parameter value as 10.
- Add a parameter and rename it as Free_Length.
- Keep Real number as the type.
- Enter the Free_Length parameter value as 270.
- Add a parameter and rename it as Free_Pitch.
- Keep Real number as the type.
- Enter the Size parameter value as 20.
- Add a parameter and rename it as Load.
- Keep Real number as the type.
- Enter the Size parameter value as 500.

- Click on the relation creation window.
- *Add the following relations:*
  - d8 = rod_dia (Note: d8 is the dimension of the circle of spring profile; users may need to select it graphically.)
  - d3 = DIA (Note: d3 is the mean diameter of the spring; users may need to select it graphically.)
  - d4 = LENGTH (Note: d4 is the length of the spring; users may need to select it graphically.)
  - d5 = PITCH (Note: d5 is the pitch of the spring; users may need to select it graphically.)
  - d11 = LENGTH (Note: d11 is the length of the material removal extrusion; users may need to select it graphically.)

- d10 = DIA*2 (Note: d10 is the diameter of the material removal extrusion; users may need to select it graphically.)
- N = FREE_LENGTH/FREE_PITCH
- X = (FREE_LENGTH-(LOAD/K))/N
- IF(X<10)
- X = 10
- ENDIF
- PITCH = X
- LENGTH = X*N
- if(length<(free_length-ALLOW_MOVEMENT:1))
- X = (FREE_LENGTH-ALLOW_MOVEMENT:1)/N
- endif
- PITCH = X
- LENGTH = X*N

- Click on Verify on the right upper side of the relation window.
- A notification box with a message "Relations have been successfully verified" should appear.
- Click OK to close the box.
- Close the relation window.
- Click on Regenerate to regenerate the model.
- Save and exit the file.

- Create a part file and rename it as Lug1. (Other names may be given, but it is advisable to provide a relevant name.)
- Select Solid on the right-hand side of the box. This will activate the solid mode and environment.
- Uncheck the box of the default template and click OK to create a new solid file.
- The template selection box will appear.
- Select mmns_part_Solid. This will allow the use of the solid template file of the following units:
  - mm—millimeter for length unit
  - n—newton for force unit
  - s—second for time unit
- Once the solid part is created, go to File > Prepare > Model properties to check the units. In the upper sections, the units of the model will be displayed.
  - Note: The units are usually displayed in the summary section of any CAD software. The setting of the units is usually in the tools or settings.
- Click on Extrude from the Model tab in the ribbon interface.
- Click on the Placement tab and select Define to define the sketch.
- Select the top plane as the sketch plane.
- The right plane shall be automatically taken as the reference and orientation as right.
- Click on Sketch.
- The sketcher model will be activated.
- Draw the sketch as displayed in Figure 13.28.
- Click OK to create and exit the sketch.
- Select the Mid plane extrusion.

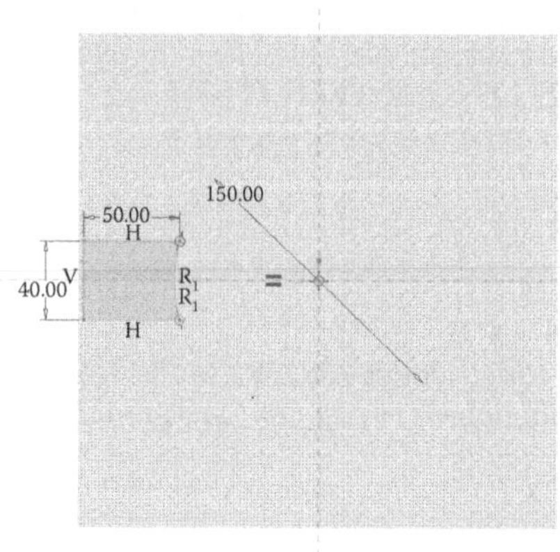

FIGURE 13.28
Constructional sketch of lugs of spring support.

- Enter the value 10 mm for the extrusion depth.
- Flip the material side upward.
- Click on OK to create the extrusion.

- Click on Extrude again from the Model tab in the ribbon interface.
- Click on the Placement tab and select Define to define the sketch.
- Select the top flat surface as the sketch plane.
- The right plane shall be automatically taken as the reference and orientation as right.
- Click on Sketch.
- The sketcher model will be activated.
- Draw the sketch as displayed in Figure 13.29.

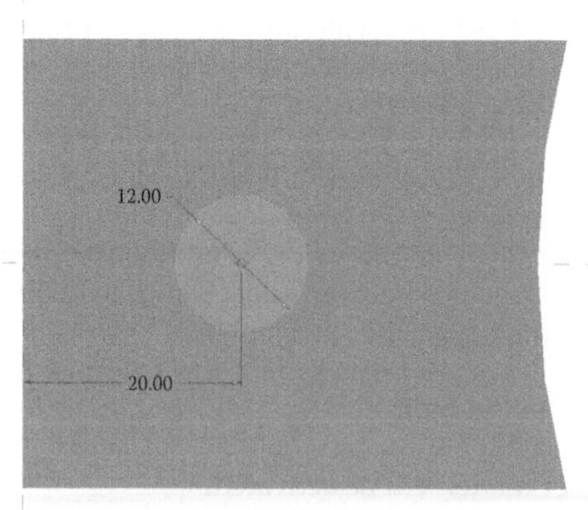

FIGURE 13.29
Constructional sketch of cut feature of spring support.

- Click OK to create and exit the sketch.
- Flip the material towards the bottom.
- Select the option Through all for the extrusion depth.
- Activate the material removal option.
- Click on OK to create the extrusion.
- Save and exit the file.

- Create a part file and rename it as Shaft. (Other names may be given, but it is advisable to provide a relevant name.)
- Select Solid on the right-hand side of the box. This will activate the solid mode and environment.
- Uncheck the box of the default template and click OK to create a new solid file.
- The template selection box will appear.
- Select mmns_part_Solid. This will allow the use of the solid template file of the following units:
  - mm—millimeter for length unit
  - n—newton for force unit
  - s—second for time unit
- Once the solid part is created, go to File > Prepare > Model properties to check the units. In the upper sections, the units of the model will be displayed.
  - Note: The units are usually displayed in the summary section of any CAD software. The setting of the units is usually in the tools or settings.
- Click on Extrude from the Model tab in the ribbon interface.
- Click on the Placement tab and select Define to define the sketch.
- Select the top plane as the sketch plane.
- The right plane shall be automatically taken as the reference and orientation as right.
- Click on Sketch.
- The sketcher model will be activated.
- Draw a circle of diameter 50.
- Click OK to create and exit the sketch.
- Select the Mid plane extrusion.
- Enter the value 240 mm for the extrusion depth.
- Flip the material side upward.
- Select Thicken sketch and set 8 as the thickness.
- Flip the material inward.
- Click on OK to create the extrusion.
- Save and exit the file.

- Create a part file and rename it as Lug2. (Other names may be given, but it is advisable to provide a relevant name.)
- Select Solid on the right-hand side of the box. This will activate the solid mode and environment.
- Uncheck the box of the default template and click OK to create a new solid file.
- The template selection box will appear.

- Select mmns_part_Solid. This will allow the use of the solid template file of the following units:
  - mm—millimeter for length unit
  - n—newton for force unit
  - s—second for time unit
- Once the solid part is created, go to File > Prepare > Model properties to check the units. In the upper sections, the units of the model will be displayed.
  - Note: The units are usually displayed in the summary section of any CAD software. The setting of the units is usually in the tools or settings.
- Click on Extrude from the Model tab in the ribbon interface.
- Click on the Placement tab and select Define to define the sketch.
- Select the top plane as the sketch plane.
- The right plane shall be automatically taken as the reference and orientation as right.
- Click on Sketch.
- The sketcher model will be activated.
- Draw the sketch as displayed in Figure 13.30.
- Click OK to create and exit the sketch.
- Select the Mid plane extrusion.
- Enter the value 10 mm for the extrusion depth.
- Flip the material side upward.
- Click on OK to create the extrusion.

- Select Hole from the Model tab of the ribbon interface.
- Select Simple hole and enter 12 as the diameter.
- Select the Through all option for the hole depth.
- Click on Placement.
- Select the top surface.
- Click on Offset reference.

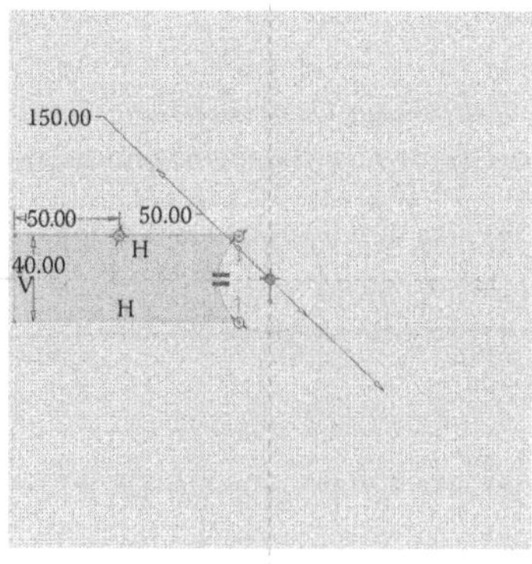

FIGURE 13.30
Constructional sketch of lug2 of spring support.

- Select the front plane and the straight surface (perpendicular to the front plane) by pressing the Control key.
- Click OK to create the hole.
- Save and exit the file.

- Create a part file and rename it as Plate1. (Other names may be given, but it is advisable to provide a relevant name.)
- Select Solid on the right-hand side of the box. This will activate the solid mode and environment.
- Uncheck the box of the default template and click OK to create a new solid file.
- The template selection box will appear.
- Select mmns_part_Solid. This will allow the use of the solid template file of the following units:
  - mm—millimeter for length unit
  - n—newton for force unit
  - s—second for time unit
- Once the solid part is created, go to File > Prepare > Model properties to check the units. In the upper sections, the units of the model will be displayed.
  - Note: The units are usually displayed in the summary section of any CAD software. The setting of the units is usually in the tools or settings.
- Click on Extrude from the Model tab in the ribbon interface.
- Click on the Placement tab and select Define to define the sketch.
- Select the top plane as the sketch plane.
- The right plane shall be automatically taken as the reference and orientation as right.
- Click on Sketch.
- The sketcher model will be activated.
- Draw a rectangle of 100 × 100 symmetrically about the horizontal and vertical references.
- A centerline may need to be drawn along the reference to draw the above-mentioned rectangle.
- Click OK to create and exit the sketch.
- Select the Mid plane extrusion.
- Enter the value 10 mm for the extrusion depth.
- Click on OK to create the extrusion.
- Save and exit the file.

- Create a part file and rename it as Wear_Plate. (Other names may be given, but it is advisable to provide a relevant name.)
- Select Solid on the right-hand side of the box. This will activate the solid mode and environment.
- Uncheck the box of the default template and click OK to create a new solid file.
- The template selection box will appear.
- Select mmns_part_Solid. This will allow the use of the solid template file of the following units:

    - mm—millimeter for length unit
    - n—newton for force unit
    - s—second for time unit
- Once the solid part is created, go to File > Prepare > Model properties to check the units. In the upper sections, the units of the model will be displayed.
    - Note: The units are usually displayed in the summary section of any CAD software. The setting of the units is usually in the tools or settings.
- Click on Extrude from the Model tab in the ribbon interface.
- Click on the Placement tab and select Define to define the sketch.
- Select the top plane as the sketch plane.
- The right plane shall be automatically taken as the reference and orientation as right.
- Click on Sketch.
- The sketcher model will be activated.
- Draw a rectangle of 80 × 80 symmetrically about the horizontal and vertical references.
- A centerline may need to be drawn along the reference to draw the above-mentioned rectangle.
- Click OK to create and exit the sketch.
- Select the Mid plane extrusion.
- Enter the value 5 mm for the extrusion depth.
- Click on OK to create the extrusion.
- Save and exit the file.

- Create a part file and rename it as Stud. (Other names may be given, but it is advisable to provide a relevant name.)
- Select Solid on the right-hand side of the box. This will activate the solid mode and environment.
- Uncheck the box of the default template and click OK to create a new solid file.
- The template selection box will appear.
- Select mmns_part_Solid. This will allow the use of the solid template file of the following units:
    - mm—millimeter for length unit
    - n—newton for force unit
    - s—second for time unit
- Once the solid part is created, go to File > Prepare > Model properties to check the units. In the upper sections, the units of the model will be displayed.
    - Note: The units are usually displayed in the summary section of any CAD software. The setting of the units is usually in the tools or settings.
- Click on Extrude from the Model tab in the ribbon interface.
- Click on the Placement tab and select Define to define the sketch.
- Select the top plane as the sketch plane.
- The right plane shall be automatically taken as the reference and orientation as right.
- Click on Sketch.

- The sketcher model will be activated.
- Draw a circle of diameter 10 placing the center at the intersection of the horizontal and vertical references.
- Click OK to create and exit the sketch.
- Flip the material direction upward.
- Enter the value 160 mm for the extrusion depth.
- Click on OK to create the extrusion.

- Click on Chamfer from the Model tab of the ribbon interface.
- Select both circular edges.
- Enter 0.5 as the chamfer dimension into DXD configuration.
- Click OK to create the feature.

- Save and exit the file.

- Create a part file and rename it as NUT_M10. (Other names may be given, but it is advisable to provide a relevant name.)
- Select Solid on the right-hand side of the box. This will activate the solid mode and environment.
- Uncheck the box of the default template and click OK to create a new solid file.
- The template selection box will appear.
- Select mmns_part_Solid. This will allow the use of the solid template file of the following units:
    - mm—millimeter for length unit
    - n—newton for force unit
    - s—second for time unit
- Once the solid part is created, go to File > Prepare > Model properties to check the units. In the upper sections, the units of the model will be displayed.
    - Note: The units are usually displayed in the summary section of any CAD software. The setting of the units is usually in the tools or settings.
- Click on Extrude from the Model tab in the ribbon interface.
- Click on the Placement tab and select Define to define the sketch.
- Select the front plane as the sketch plane.
- The right plane shall be automatically taken as the reference and orientation as right.
- Click on Sketch.
- The sketcher model will be activated.
- Draw the sketch as shown in Figure 13.31.
- Click OK to create and exit the sketch.
- Select the Mid plane extrusion.
- Enter the value 10 mm for the extrusion depth.
- Click on OK to create the extrusion.

- Click on Revolve from the Model tab in the ribbon interface.
- Click on the Placement tab and select Define to define the sketch.
- Activate the material removal feature option.
- Select the front plane as the sketch plane.

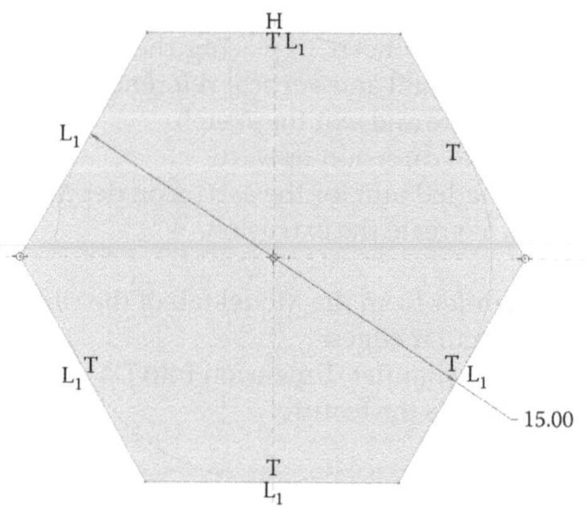

FIGURE 13.31
Constructional sketch of hexagonal extrusion feature of nut for spring support.

- The right plane shall be automatically taken as the reference and orientation as right.
- Click on Sketch.
- The sketcher model will be activated.
- Draw the sketch as displayed in Figure 13.32.
- Click OK to create and exit the sketch.
- Ensure that the revolve angle shall be 360°.
- Click on OK to create the revolve feature.

- Click on Hole from the Model tab of the ribbon interface.

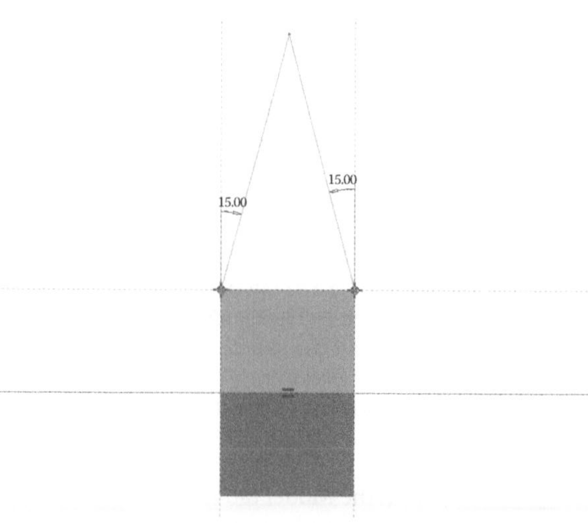

FIGURE 13.32
Constructional sketch of revolve cut feature of nut for spring support.

- Click on Activate the simple hole.
- Set the depth to Through all.
- Enter the hole diameter as 10 mm.
- Select the top flat surface of Nut.
- Click on To select offset reference and select both the front and right planes as the offset references.
- Set aligned in the offset value setting box instead of distance.
- Click OK to create the hole.
- Save and exit the file.

- Create an assembly file and rename it as Pipe_Support.asm. (Other names may be given, but it is advisable to provide a relevant name.)
- Select Design on the right-hand side of the box. This will activate the standard assembly design mode and environment.
- Uncheck the box of the default template and click OK to create a new assembly design file.
- The template selection box will appear.
- Select mmns_asm_Solid. This will allow the use of the solid template file of the following units:
    - mm—millimeter for length unit
    - n—newton for force unit
    - s—second for time unit
- Once the assembly file is created, go to File > Prepare > Model properties to check the units. In the upper sections, the units of the model will be displayed.
    - Note: The units are usually displayed in the summary section of any CAD software. The setting of the units is usually in the tools or settings.
- Click on tree filters as shown in Figure 13.4.

- Check the box features to display in the model tree.
- Click on Apply. Notice that the features are displayed in the model tree.

- Click on Assembly from the Model tab of the ribbon interface.
- Select the file Shaft.
- Click on the Placement tab.
- Select Default in the constraint type selection option box.
- Click OK to assemble the component.

- Click on Assembly from the Model tab of the ribbon interface.
- Select the file Lug2.
- Click on the Placement tab.
- Select the cylindrical surface of the Lug2 part as the component reference.
- Select the outer cylindrical surface of the shaft as the assembly reference.
- Select Coincide in the component reference relation field.
- Right click on the graphical area and select New constraint.

- Select the bottom face of the Lug2 part as the component reference.
- Select the bottom surface of the shaft as the assembly reference.
- Enter the value 100 mm as the offset distance.
- Right click on the graphical area and select New constraint.
- Select the front plane of the Lug2 part as the component reference.
- Select the plane ASM_FRONT as the assembly reference.
- Select Coincide in the component reference relation field.
- Click OK to assemble the component.

- Click on Assembly from the Model tab of the ribbon interface.
- Select the file Lug2 again.
- Click on the Placement tab.
- Select the cylindrical surface of the Lug2 part as the component reference.
- Select the outer cylindrical surface of the shaft as the assembly reference.
- Select Coincide in the component reference relation field.
- Right click on the graphical area and select New constraint.
- Select the bottom face of the Lug2 part as the component reference.
- Select the bottom surface of the preassembled Lug2 as the assembly reference.
- Select Coincide in the component reference relation field.
- Right click on the graphical area and select New constraint.
- Select the front plane of the Lug2 part as the component reference.
- Select the plane ASM_FRONT as the assembly reference.
- Select Coincide in the component reference relation field.
- Flip the orientation so that this component is placed opposite to the preassembled Lug2.
- Click OK to assemble the component.

- Click on Assembly from the Model tab of the ribbon interface.
- Select the file Plate1.
- Click on the Placement tab.
- Select the bottom surface of the Plate1 part as the component reference.
- Select the top surface of the shaft as the assembly reference.
- Select Coincide in the component reference relation field.
- Right click on the graphical area and select New constraint.
- Select the right plane of the Plate1 part as the component reference.
- Select the right plane of the shaft as the assembly reference.
- Select Coincide in the component reference relation field.
- Right click on the graphical area and select New constraint.
- Select the front plane of the Plate1 part as the component reference.
- Select the F plane of the shaft as the assembly reference.
- Select Coincide in the component reference relation field.
- Click OK to assemble the component.

- Click on Assembly from the Model tab of the ribbon interface.

- Select the file wear_plate.
- Click on the Placement tab.
- Select the bottom surface of the wear_plate part as the component reference.
- Select the top surface of the Plate1 shaft as the assembly reference.
- Select Coincide in the component reference relation field.
- Right click on the graphical area and select New constraint.
- Select the right plane of the wear_plate part as the component reference.
- Select the right plane of Plate1 as the assembly reference.
- Select Coincide in the component reference relation field.
- Right click on the graphical area and select New constraint.
- Select the front plane of the wear_plate part as the component reference.
- Select the F plane of Plate1 as the assembly reference.
- Select Coincide in the component reference relation field.
- Click OK to assemble the component.

- Save and exit the file.

- Create an assembly file and rename it as Spring_Support.asm. (Other names may be given, but it is advisable to provide a relevant name.)
- Select Design on the right-hand side of the box. This will activate the standard assembly design mode and environment.
- Uncheck the box of the default template and click OK to create a new assembly design file.
- The template selection box will appear.
- Select mmns_asm_Solid. This will allow the use of the solid template file of the following units:
    - mm—millimeter for length unit
    - n—newton for force unit
    - s—second for time unit
- Once the assembly file is created, go to File > Prepare > Model properties to check the units. In the upper sections, the units of the model will be displayed.
    - Note: The units are usually displayed in the summary section of any CAD software. The setting of the units is usually in the tools or settings.
- Click on tree filters as shown in Figure 13.4.

- Check the box features to display in the model tree.
- Click on Apply. Notice that the features are displayed in the model tree.

- Click on Assembly from the Model tab of the ribbon interface.
- Select the file Spring_Base_Plate.
- Click on the Placement tab.
- Select Default in the constraint type selection option box.
- Click OK to assemble the component.

- Click on Assembly from the Model tab of the ribbon interface.
- Select the file Spring_Cage.
- Click on the Placement tab.
- Select Default in the constraint type selection option box.
- Click OK to assemble the component.

- Select Default in the constraint type selection option box.
- Click OK to assemble the component.
- Click on Assembly from the Model tab of the ribbon interface.
- Select the file Sring_Cage_Top.
- Click on the Placement tab.
- Select the bottom surface of the Spring_Cage_Top part as the component reference.
- Select the top surface of the Spring_Cage shaft as the assembly reference.
- Select Coincide in the component reference relation field.
- Right click on the graphical area and select New constraint.
- Select the outer cylindrical surface of the Spring_Cage_Top part as the component reference.
- Select the outer cylindrical surface of Spring_Cage as the assembly reference.
- Select Coincide in the component reference relation field.
- Click OK to assemble the component.

- Click on Assembly from the Model tab of the ribbon interface.
- Select the file Spring.
- Click on the Placement tab.
- Select the Right plane of the spring part as the component reference.
- Select the plane ASM_RIGHT as the assembly reference.
- Select Coincide in the component reference relation field.
- Right click on the graphical area and select New constraint.
- Select the front plane of the spring part as the component reference.
- Select the plane ASM_FRONT as the assembly reference.
- Select Coincide in the component reference relation field.
- Right click on the graphical area and select New constraint.
- Select the bottom surface of the spring part as the component reference.
- Select the plane top surface of Spring_Base_Plate as the assembly reference.
- Select Coincide in the component reference relation field.

- Click OK to assemble the component.

- Click on Assembly from the Model tab of the ribbon interface.
- Select the file Lug1.
- Click on the Placement tab.
- Select the cylindrical surface of the Lug1 part as the component reference.

- Select the outer cylindrical surface of Spring_Cage as the assembly reference.
- Select Coincide in the component reference relation field.
- Right click on the graphical area and select New constraint.
- Select the front plane of the Lug1 part as the component reference.
- Select the plane right of Spring_Cage as the assembly reference.
- Select Coincide in the component reference relation field.
- Right click on the graphical area and select New constraint.
- Select the top surface of the Lug1 part as the component reference.
- Select the cut surface of Spring_Cage as the assembly reference as shown in Figure 13.33.

- Select Coincide in the component reference relation field.
- Click OK to assemble the component.

- Click on Assembly from the Model tab of the ribbon interface.
- Select the file Shaft_Assembly.
- Click on the Placement tab.
- Select the hole surface of Lug2 of the spring part as the component reference.
- Select the hole surface of Lug1 as the assembly reference.
- Select Coincide in the component reference relation field.
- Right click on the graphical area and select New constraint.
- Select the hole surface of opposite Lug2 of the spring part as the component reference.
- Select the hole surface of opposite Lug1 as the assembly reference.
- Select Coincide in the component reference relation field.
- Right click on the graphical area and select New constraint.

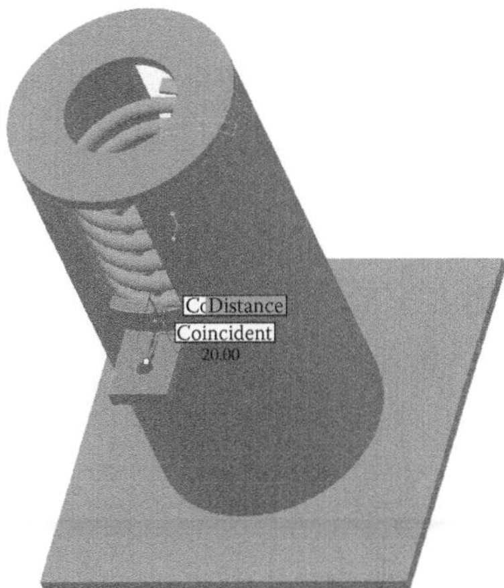

FIGURE 13.33
Constructional view of spring support assembly.

- Select the bottom surface of any Lug2 part as the component reference.
- Select the plane top surface of Spring as the assembly reference.
- Select Coincide in the component reference relation field.

- Click OK to assemble the component.

- Click on Assembly from the model tab of the ribbon interface.
- Select the file Stud.
- Click on the Placement tab.
- Select the cylindrical surface of Stud of the spring part as the component reference.
- Select the hole surface of Lug1 as the assembly reference.
- Select Coincide in the component reference relation field.
- Right click on the graphical area and select New constraint.
- Select the bottom surface of Stud of the spring part as the component reference.
- Select the bottom surface of Lug1 as the assembly reference.
- Select Distance in the component reference relation field.
- Enter 30 as the distance.

- Click OK to assemble the component.

- Click on Assembly from the Model tab of the ribbon interface.
- Select the file Stud.
- Click on the Placement tab.
- Select the cylindrical surface of Stud of the spring part as the component reference.
- Select the hole surface of another Lug1 as the assembly reference.
- Select Coincide in the component reference relation field.
- Right click on the graphical area and select New constraint.
- Select the bottom surface of Stud of the spring part as the component reference.
- Select the bottom surface of Lug1 as the assembly reference.
- Select Distance in the component reference relation field.
- Enter 30 as the distance.

- Click OK to assemble the component.

- Click on Assembly from the Model tab of the ribbon interface.
- Select the file Nut_M10.
- Click on the Placement tab.
- Select the internal cylindrical surface of Nut_M10 of the spring part as the component reference.
- Select the outer cylindrical surface of any of the studs as the assembly reference.
- Select Coincide in the component reference relation field.
- Right click on the graphical area and select New constraint.
- Select the top surface of Nut_M10 of the spring part as the component reference.

- Select the bottom surface of the Lug1, which is lying with the stud as the assembly reference.
- Select Coincide in the component reference relation field.

- Click OK to assemble the component.

- Click OK to assemble the component.

- Click OK to assemble the component.

- Click on Assembly from the Model tab of the ribbon interface.
- Select the file Nut_M10.
- Click on the Placement tab.
- Select the internal cylindrical surface of Nut_M10 of the spring part as the component reference.
- Select the outer cylindrical surface of the same stud where the previous nut is assembled as the assembly reference.
- Select Coincide in the component reference relation field.
- Right click on the graphical area and select New constraint.
- Select the bottom surface of Nut_M10 of the spring part as the component reference.
- Select the top surface of Lug1, which is lying with the stud, as the assembly reference.
- Select Coincide in the component reference relation field.

- Click OK to assemble the component.

- Click on Assembly from the Model tab of the ribbon interface.
- Select the file Nut_M10.
- Click on the Placement tab.
- Select the internal cylindrical surface of Nut_M10 of the spring part as the component reference.
- Select the outer cylindrical surface of the other stud as the assembly reference.
- Select Coincide in the component reference relation field.
- Right click on the graphical area and select New constraint.
- Select the top surface of Nut_M10 of the spring part as the component reference.
- Select the bottom surface of Lug1, which is lying with the stud, as the assembly reference.
- Select Coincide in the component reference relation field.

- Click OK to assemble the component.

- Click on Assembly from the Model tab of the ribbon interface.
- Select the file Nut_M10.
- Click on the Placement tab.
- Select the internal cylindrical surface of Nut_M10 of the spring part as the component reference.

- Select the outer cylindrical surface of the same stud where the previous nut is assembled as the assembly reference.
- Select Coincide in the component reference relation field.
- Right click on the graphical area and select New constraint.
- Select the bottom surface of Nut_M10 of the spring part as the component reference.
- Select the top surface of Lug1, which is lying with the stud, as the assembly reference.
- Select Coincide in the component reference relation field.

- Click OK to assemble the component.

- Click on Assembly from the Model tab of the ribbon interface.
- Select the file Nut_M10.
- Click on the Placement tab.
- Select the internal cylindrical surface of Nut_M10 of the spring part as the component reference.
- Select the outer cylindrical surface of the other stud as the assembly reference.
- Select Coincide in the component reference relation field.
- Right click on the graphical area and select New constraint.
- Select the top surface of Nut_M10 of the spring part as the component reference.
- Select the top surface of Lug1, which is lying with the stud, as the assembly reference.
- Select Distance in the component reference relation field.
- Enter 60 as the distance.
- Click OK to assemble the component.

- Click on Assembly from the Model tab of the ribbon interface.
- Select the file Nut_M10.
- Click on the Placement tab.
- Select the internal cylindrical surface of Nut_M10 of the spring part as the component reference.
- Select the outer cylindrical surface of another stud where the previous nut is assembled as the assembly reference.
- Select Coincide in the component reference relation field.
- Right click on the graphical area and select New constraint.
- Select the top surface of Nut_M10 of the spring part as the component reference.
- Select the top surface of Lug1, which is lying with the stud, as the assembly reference.
- Select Distance in the component reference relation field.
- Enter 60 as the distance.

- Click OK to assemble the component.

- Click OK to assemble the component.

- Click on Assembly from the Model tab of the ribbon interface.

- Select the file Nut_M10.
- Click on the Placement tab.
- Select the internal cylindrical surface of Nut_M10 of the spring part as the component reference.
- Select the outer cylindrical surface of any stud as the assembly reference.
- Select Coincide in the component reference relation field.
- Right click on the graphical area and select New constraint.
- Select the bottom surface of Nut_M10 of the spring part as the component reference.
- Select the top surface of the mid Nut_M10 on that stud, which is lying with the stud, as the assembly reference.
- Select Distance in the component reference relation field.
- Enter 40 as the distance.

- Click OK to assemble the component.

- Click on Assembly from the Model tab of the ribbon interface.
- Select the file Nut_M10.
- Click on the Placement tab.
- Select the internal cylindrical surface of Nut_M10 of the spring part as the component reference.
- Select the outer cylindrical surface of the other stud as the assembly reference.
- Select Coincide in the component reference relation field.
- Right click on the graphical area and select New constraint.
- Select the bottom surface of Nut_M10 of the spring part as the component reference.
- Select the top surface of the mid Nut_M10 on that stud, which is lying with the stud, as the assembly reference.
- Select Distance in the component reference relation field.
- Enter 40 as the distance.

- Select Summary tools from the Analysis tab of the ribbon interface. Select the option for distance measurement.
    - Note: The distance tool can also be selected from the drop-down menu of Summary tools.
- Select the top surface of Lug1 and the top surface of Spring_Base_Plate.
- The measured distance is 180 mm.
- Click on the Save icon to create the measurement as a feature.
- The name of the feature shall be MEASURE_DISTANCE_1.

- Click on the Tools tab in the ribbon interface.
- Click on Relations.
- The relation window will appear.
- Click on Local parameters to display the local parameters dialog box.
- Add a parameter and rename it as Allow_Movement.
- Keep Real number as the type.

- Enter 30 as the Allow_Movement parameter value.
- Add a parameter and rename it as Load.
- Keep Real number as the type.
- Enter 500 as the Load parameter value.

- Click on the relation creation window.
- *Add the following relations:*
    - d17 = free_length:CID_45-allow_movement-distance:FID_MEASURE_DISTANCE_1 (Note: d17 is the mating distance between the mid NUT_M10 and the top surface of the Lug; users may need to select it graphically.)
    - d18 = ALLOW_MOVEMENT+10 (Note: d18 is the mating distance between the top NUT_M10 and the top surface of the mid NUT_M10; users may need to select it graphically.)
- Click on Verify on the right upper side of the relation window.
- A notification box with a message "Relations have been successfully verified" should appear.
- Click OK to close the box.
- Select the Part in the look in field and select the part Spring.
- Add the relation "load = load:1" in before "N = FREE_LENGTH/FREE_PITCH".
- Click on Verify on the right upper side of the relation window.
- A notification box with a message "Relations have been successfully verified" should appear.
- Click OK to close the box.
- Close the relation window.
- Click on Regenerate to regenerate the model.
- Change the load to any value and click Regenerate. Multiple regenerations may be required to regenerate the model properly.
- When the load becomes excessive, the spring will be compressed and be blocked at the mid NUT_M10, which can be considered as a cold setting nut.
- Save and exit the file.

Hands-On Exercise, Example 13.6: Shell and Tube Heat Exchanger (Software Used: Creo Parametric; Figure 13.34)

SHELL AND TUBE HEAT EXCHANGER

Steps:

- Create a part file and rename it as Head. (Other names may be given, but it is advisable to provide a relevant name.)
- Select Solid on the right-hand side of the box. This will activate the solid mode and environment.
- Uncheck the box of the default template and click OK to create a new solid file.
- The template selection box will appear.

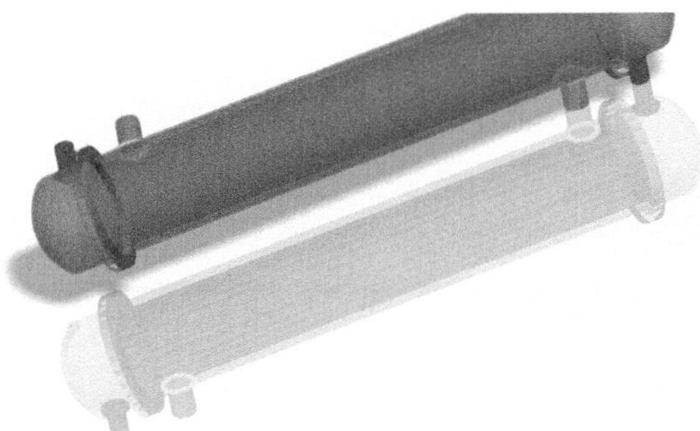

FIGURE 13.34
3-D view of shell and tube heat exchanger.

- Select mmns_part_Solid. This will allow the use of the solid template file of the following units:
  - mm—millimeter for length unit
  - n—newton for force unit
  - s—second for time unit
- Once the solid part is created, go to File > Prepare > Model properties to check the units. In the upper sections, the units of the model will be displayed.
  - Note: The units are usually displayed in the summary section of any CAD software. The setting of the units is usually in the tools or settings.
- Click on Revolve from the Model tab in the ribbon interface.
- Activate the thicken material option.
- Click on the Placement tab and select Define to define the sketch.
- Select the front plane as the sketch plane.
- The right plane shall be automatically taken as the reference and orientation as right.
- Click on Sketch.
- The sketcher model will be activated.
- Draw the sketch as displayed in Figure 13.35.
- Click OK to create and exit the sketch.
- Enter 8 as the thickness value.
- Flip the material orientation towards inside.
- Ensure that the revolve angle shall be 360°.
- Click on OK to create the revolve feature.
- Save and exit the file.

- Create a part file and rename it as Flange. (Other names may be given, but it is advisable to provide a relevant name.)
- Select Solid on the right-hand side of the box. This will activate the solid mode and environment.
- Uncheck the box of the default template and click OK to create a new solid file.
- The template selection box will appear.

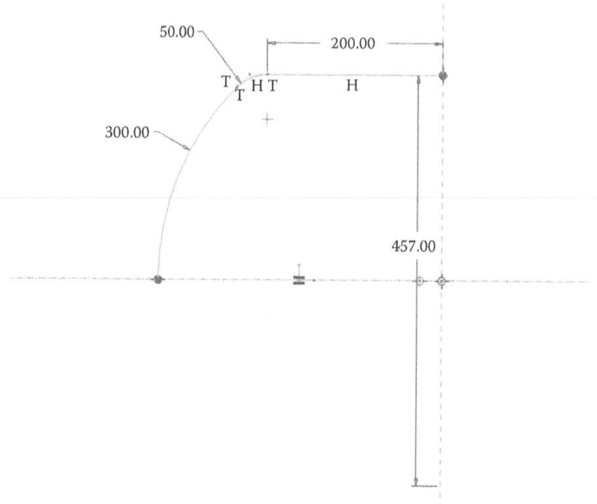

FIGURE 13.35
Constructional sketch of shell part in shell and tube heat exchanger.

- Select mmns_part_Solid. This will allow the use of the solid template file of the following units:
  - mm—millimeter for length unit
  - n—newton for force unit
  - s—second for time unit
- Once the solid part is created, go to File > Prepare > Model properties to check the units. In the upper sections, the units of the model will be displayed.
  - Note: The units are usually displayed in the summary section of any CAD software. The setting of the units is usually in the tools or settings.
- Click on Extrude from the Model tab in the ribbon interface.
- Click on the Placement tab and select Define to define the sketch.
- Select the front plane as the sketch plane.
- The right plane shall be automatically taken as the reference and orientation as right.
- Click on Sketch.
- The sketcher model will be activated.
- Draw a circle of diameter 457 placing the center at the intersection of the horizontal and vertical references.
- Draw another circle of diameter 515 placing the center at the intersection of the horizontal and vertical references.
- Click OK to create and exit the sketch.
- Flip the extrusion towards up.
- Enter the value 40 mm for the extrusion depth.
- Click on OK to create the extrusion.

- Click on Hole from the Model tab of the ribbon interface.
- Select the flat surface of the feature extrude 1.

- Select the standard screw hole M16X 1.5.
- Select the Through all option as the hole depth and remove the threaded surface.
- Select Medium fit; the diameter will be set to 18.
- Click on the Reference tab; select Diameter as the reference type.
- Select the axis of the feature extrude 1, and select the plane top.
- Set the diameter of hole PCD 515 and angle 0°.
- Click OK to create the hole feature.
- Right click on the hole and select Pattern.
- Select Axis as the type of pattern.
- Enter 24 as the pattern instances quantity and the total angle 360°.
- Click OK to create the features.
- Save and exit the file.

- Create a part file and rename it as Tube_Sheet. (Other names may be given, but it is advisable to provide a relevant name.)
- Select Solid on the right-hand side of the box. This will activate the solid mode and environment.
- Uncheck the box of the default template and click OK to create a new solid file.
- The template selection box will appear.
- Select mmns_part_Solid. This will allow the use of the solid template file of the following units:
    - mm—millimeter for length unit
    - n—newton for force unit
    - s—second for time unit
- Once the solid part is created, go to File > Prepare > Model properties to check the units. In the upper sections, the units of the model will be displayed.
    - Note: The units are usually displayed in the summary section of any CAD software. The setting of the units is usually in the tools or settings.
- Click on Extrude from the Model tab in the ribbon interface.
- Click on the Placement tab and select Define to define the sketch.
- Select the front plane as the sketch plane.
- The right plane shall be automatically taken as the reference and orientation as right.
- Click on Sketch.
- The sketcher model will be activated.
- Draw a circle of diameter 515 placing the center at the intersection of the horizontal and vertical references.
- Click OK to create and exit the sketch.
- Flip the extrusion towards up.
- Enter the value 40 mm for the extrusion depth.
- Click on OK to create the extrusion.

- Click on Hole from the Model tab of the ribbon interface.
- Select the flat surface of the feature extrude 1.
- Select the standard screw hole M16X 1.5.

- Select the Through all option as the hole depth and remove the threaded surface.
- Select Medium fit; the diameter will be set to 18.
- Click on the Reference tab; select Diameter as the reference type.
- Select the axis of the feature extrude 1, and select the plane top.
- Set the diameter of the hole PCD 515 and angle 0°.
- Click OK to create the hole feature.
- Right click on the hole and select Pattern.
- Select Axis as the type of pattern.
- Enter 24 as the pattern instances quantity and the total angle 360°.
- Click OK to create the features.

- Click on Sketch from the Model tab in the ribbon interface.
- Click on the Placement tab and select Define to define the sketch.
- Select the front plane as the sketch plane.
- The right plane shall be automatically taken as the reference and orientation as right.
- Click on Sketch.
- Draw a circle of diameter 457 placing the center at the intersection of the horizontal and vertical references.

- Click on Hole from the Model tab of the ribbon interface.
- Select the flat surface of the feature extrude 1.
- Set the hole diameter to 32 mm.
- Select the Through all option as the hole depth and remove the threaded surface.
- Select the axis of feature extrude 1 to create the hole at the axis.
- Right click on the hole and select Pattern.
- Select Axis as the type of pattern.
- Select Fill as the type and select the sketch.
- Set the space of the members in a hexagon pattern option.
- Enter 48.60 as the pitch value and 30 as the boundary dimension.
- Click OK to create the pattern features.
- Save and exit the file.

- Create a part file and rename it as Tube. (Other names may be given, but it is advisable to provide a relevant name.)
- Select Solid on the right-hand side of the box. This will activate the solid mode and environment.
- Uncheck the box of the default template and click OK to create a new solid file.
- The template selection box will appear.
- Select mmns_part_Solid. This will allow the use of the solid template file of the following units:
  - mm—millimeter for length unit
  - n—newton for force unit
  - s—second for time unit
- Once the solid part is created, go to File > Prepare > Model properties to check the units. In the upper sections, the units of the model will be displayed.

  - Note: The units are usually displayed in the summary section of any CAD software. The setting of the units is usually in the tools or settings.
- Click on Extrude from the Model tab in the ribbon interface.
- Click on the Placement tab and select Define to define the sketch.
- Select the front plane as the sketch plane.
- The right plane shall be automatically taken as the reference and orientation as right.
- Click on Sketch.
- The sketcher model will be activated.
- Draw a circle of diameter 31.75 placing the center at the intersection of the horizontal and vertical references.
- Click OK to create and exit the sketch.
- Select the Mid plane extrusion.
- Enter the value 2500 mm for the extrusion depth.
- Select the thicken option.
- Enter 2 as the thickness.
- Flip the material orientation inward.
- Click on OK to create the extrusion.

- Create a part file and rename it as Shell. (Other names may be given, but it is advisable to provide a relevant name.)
- Select Solid on the right-hand side of the box. This will activate the solid mode and environment.
- Uncheck the box of the default template and click OK to create a new solid file.
- The template selection box will appear.
- Select mmns_part_Solid. This will allow the use of the solid template file of the following units:
  - mm—millimeter for length unit
  - n—newton for force unit
  - s—second for time unit
- Once the Solid part is created, go to File > Prepare > Model properties to check the units. In the upper sections, the units of the model will be displayed.
  - Note: The units are usually displayed in the summary section of any CAD software. The setting of the units is usually in the tools or settings.
- Click on Extrude from the Model tab in the ribbon interface.
- Click on the Placement tab and select Define to define the sketch.
- Select the front plane as the sketch plane.
- The right plane shall be automatically taken as the reference and orientation as right.
- Click on Sketch.
- The sketcher model will be activated.
- Draw a circle of diameter 457 placing the center at the intersection of the horizontal and vertical references.
- Click OK to create and exit the sketch.
- Select the Mid plane extrusion.
- Enter the value 2464 mm for the extrusion depth.

- Select the thicken option.
- Enter 8 as the thickness value.
- Flip the material orientation inward.
- Click on OK to create the extrusion.

- Create an assembly file and rename it as Shell_and_Tube_Heat_Exchanger.asm. (Other names may be given, but it is advisable to provide a relevant name.)
- Select Design on the right-hand side of the box. This will activate the standard assembly design mode and environment.
- Uncheck the box of the default template and click OK to create a new assembly design file.
- The template selection box will appear.
- Select mmns_asm_Solid. This will allow the use of the solid template file of the following units:
    - mm—millimeter for length unit
    - n—newton for force unit
    - s—second for time unit
- Once the assembly file is created, go to File > Prepare > Model properties to check the units. In the upper sections, the units of the model will be displayed.
    - Note: The units are usually displayed in the summary section of any CAD software. The setting of the units is usually in the tools or settings.
- Click on tree filters as shown in Figure 13.4.

- Check the box features to display in the model tree.
- Click on Apply. Notice that the features are displayed in the model tree.

- Click on Assembly from the model tab of the ribbon interface.
- Select the file Head.
- Click on the Placement tab.
- Select Default in the constraint type selection option box.
- Click OK to assemble the component.

- Click on Assembly from the Model tab of the ribbon interface.
- Select the file Flange.
- Click on the Placement tab.
- Select the cylindrical surface of the flange part as the component reference.
- Select the outer cylindrical surface of the Head as the assembly reference.
- Select Coincide in the component reference relation field.
- Right click on the graphical area and select New constraint.
- Select the flat face of the flange part as the component reference.
- Select the flat face of the head as the assembly reference.
- Enter the value 10 mm as the offset distance as shown in Figure 13.36.
- Right click on the graphical area and select New constraint.

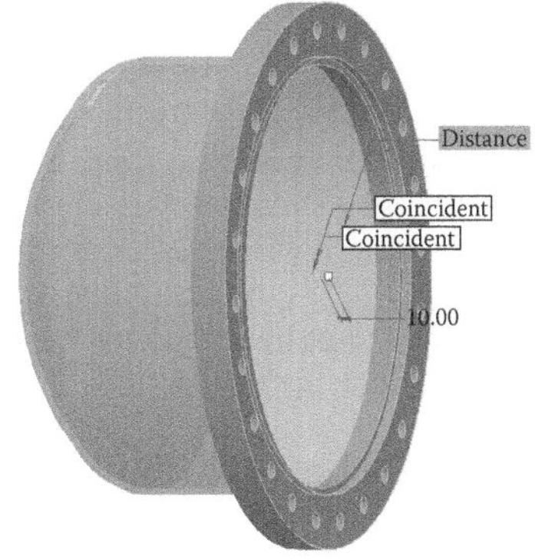

FIGURE 13.36
Assembly view of flange in shell and tube heat exchanger.

- Select the top plane of the flange part as the component reference.
- Select the plane ASM_TOP as the assembly reference.
- Select Coincide in the component reference relation field.
- Click OK to assemble the component.

- Click on Assembly from the Model tab of the ribbon interface.
- Select the file tube_sheet.
- Click on the Placement tab.
- Select the cylindrical surface of the tube sheet part as the component reference.
- Select the outer cylindrical surface of the flange as the assembly reference.
- Select Coincide in the component reference relation field.
- Right click on the graphical area and select New constraint.
- Select the flat face of the tube sheet part as the component reference.
- Select the flat surface of the flange as the assembly reference.
- Select Coincide in the component reference relation field.
- Right click on the graphical area and select New constraint.
- Select the top plane of the tube sheet part as the component reference.
- Select the plane ASM_TOP as the assembly reference.
- Select Coincide in the component reference relation field.
- Click OK to assemble the component.

- Click on Assembly from the Model tab of the ribbon interface.
- Select the file Shell.
- Click on the Placement tab.
- Select the cylindrical surface of the shell part as the component reference.
- Select the outer cylindrical surface of the tube_sheet as the assembly reference.
- Select Coincide in the component reference relation field.

- Right click on the graphical area and select New constraint.
- Select the flat face of the shell part as the component reference.
- Select the flat surface of tube_sheet as the assembly reference.
- Select Coincide in the component reference relation field.
- Right click on the graphical area and select New constraint.
- Select the top plane of the shell part as the component reference.
- Select the plane ASM_TOP as the assembly reference.
- Select Coincide in the component reference relation field.
- Click OK to assemble the component.

- Click on Assembly from the Model tab of the ribbon interface.
- Select the file tube_sheet.
- Click on the Placement tab.
- Select the cylindrical surface of the tube_sheet part as the component reference.
- Select the outer cylindrical surface of the shell as the assembly reference.
- Select Coincide in the component reference relation field.
- Right click on the graphical area and select New constraint.
- Select the flat face of the tube_sheet part as the component reference.
- Select the flat surface of the shell as the assembly reference.
- Select Coincide in the component reference relation field.
- Right click on the graphical area and select New constraint.
- Select the top plane of the tube_sheet part as the component reference.
- Select the plane ASM_TOP as the assembly reference.
- Select Coincide in the component reference relation field.
- Click OK to assemble the component.

- Click on Assembly from the Model tab of the ribbon interface.
- Select the file Flange.
- Click on the Placement tab.
- Select the cylindrical surface of the flange part as the component reference.
- Select the outer cylindrical surface of the tube_sheet as the assembly reference.
- Select Coincide in the component reference relation field.
- Right click on the graphical area and select New constraint.
- Select the flat face of the flange part as the component reference.
- Select the flat surface of tubesheet as the assembly reference.
- Select Coincide in the component reference relation field.
- Right click on the graphical area and select New constraint.
- Select the top plane of the flange part as the component reference.
- Select the plane ASM_TOP as the assembly reference.
- Select Coincide in the component reference relation field.
- Click OK to assemble the component.

- Click on Assembly from the Model tab of the ribbon interface.
- Select the file Head.

- Click on the Placement tab.
- Select the cylindrical surface of the head part as the component reference.
- Select the outer cylindrical surface of the flange as the assembly reference.
- Select Coincide in the component reference relation field.
- Right click on the graphical area and select New constraint.
- Select the flat face of the head part as the component reference.
- Select the flat face of the flange as the assembly reference.
- Enter the value 10 mm as the offset distance as shown in Figure 13.37.
- Right click on the graphical area and select New constraint.
- Select the top plane of the head part as the component reference.
- Select the plane ASM_TOP as the assembly reference.
- Select Coincide in the component reference relation field.
- Click OK to assemble the component.

- Click on Assembly from the Model tab of the ribbon interface.
- Select the file Tube.
- Click on the Placement tab.
- Select the cylindrical surface of the tube part as the component reference.
- Select the middle inner cylindrical hole surface of the tube sheet as the assembly reference.
- Select Coincide in the component reference relation field.
- Right click on the graphical area and select New constraint.
- Select the flat face of the flange part as the component reference.
- Select the flat surface of the tubesheet as the assembly reference.
- Enter 3 as the offset dimension as shown in Figure 13.38.
- Click OK to assemble the component.

- Right click on the assembled tube and select Pattern.
- The reference type shall be automatically selected.
- Click OK to create the tube pattern.

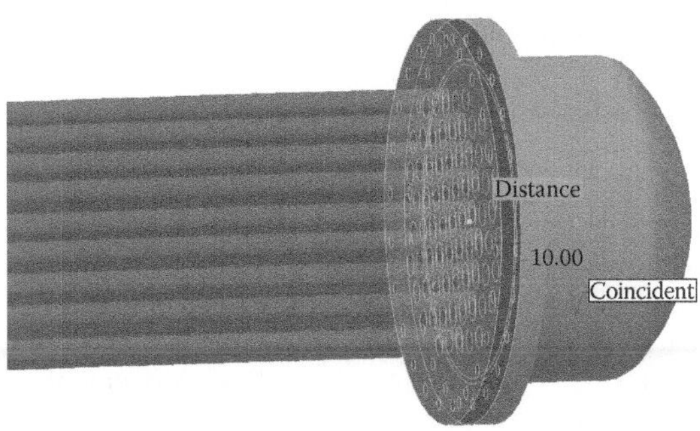

FIGURE 13.37
Constructional view of shell and tube heat exchanger.

FIGURE 13.38
(See color insert.) Assembly view of tubes in shell and tube heat exchanger.

- Click on Assembly from the Model tab of the ribbon interface.
- Select the nozzle_assembly.
- Select 80 nb.
- Click on the Placement tab.
- Select the axis of the nozzle_assembly part as the component reference.
- Select the axis as the head assembly reference.
- Select Coincide in the component reference relation field.
- Right click on the graphical area and select New constraint.
- Select the plane ASM_FRONT of the nozzle_assembly as the component reference.
- Select the ASM_RIGHT as the assembly reference.
- Enter the distance 120 in the component reference relation field.
- Right click on the graphical area and select New constraint.
- Select the plane ASM_RIGHT of the nozzle_assembly as the component reference.
- Select the front plane of the head part as the assembly reference.
- Select Coincide in the component reference relation field.
- Click OK to assemble the component.

- Similarly, attach three other nozzles in the shell and the head.

- Click on the Tools tab in the ribbon interface.
- Click on Relations.
- The relations window will appear.
- Click on Local parameters to display the local parameters dialog box.
- Add a parameter and rename it as Allow_Movement.
- Keep Real number as the type.
- Enter 30 as the Allow_Movement parameter value.
- Add a parameter and rename it as Load.

- Keep Real number as the type.
- Enter 500 as the Load parameter value.

- Click on the relation creation window.
- Add a parameter Tube_Length.
- Select real_number as the type.
- Enter 2500 mm as the tube length parameter value.

- Add the following relations:
  - d0:6 = tube_length
  - d3:8 = tube_length-d0:4-6
- Click on Verify on the right upper side of the relation window.
- A notification box with a message "Relations have been successfully verified" should appear.
- Click OK to close the box.
- Close the relation window.
- Change the tube length to 2200 and click on Regenerate; notice the change.
- Save and exit the file.

QUESTIONS

1. What is automation in CAD?
2. Describe the advantages and disadvantages of automation.
3. What are the important steps of automation?
4. How do you define the extent of automation requirement?
5. What is variable?
6. How do you identify variables?
7. Define parameters and state its uses.
8. Classify the parameters and describe their importance.
9. How do you use family table in automation?
10. How are relations created in automation?

14

Customization

14.1 Introduction

Customization is the ability to change the user interface of the software according to users' needs. There are procedures on how to use every module of any CAD system. The facilities inside the CAD such as the set of commands, icons, etc., are provided based on the users' needs. The needs are evaluated based on a common need of all users. A plant designer can use CAD software for mechanical design of a pressure vessel, which is made by fabrication. The same CAD software for mechanical design can also be used by users who model casting components. It is obvious that all the tools required for modeling casting parts are not required for modeling a pressure vessel. Also all the tools required for modeling a pressure vessel are not usable in modeling casting products. The CAD systems provide all the tools; however, accessibility of the tools is decided based on the common uses of the tools. There are some common tools that are required by all types of users who use any CAD system for mechanical design of any kind of products. So the common tools are effectively placed and made easy to access in the CAD system. Now a user who works only for casting product design may receive increased usage efficiency if all the needed tools for modeling a casting product can be made accessible to him or her in the least amount of time. Here the need for customization arises to reduce time and effort.

The advantages of customization are as follows:

1. Customization can reduce the time by increasing the ease of accessibility of all the tools needed for particular application.
2. It lessens the effort because of increased ease of accessibility.
3. It lessens the chance of human errors.

Customization can be done in the following ways:

1. Customization by manipulating the user interface of the software
2. Customization by using an existing file as a template file

14.1.1 Customization by Manipulating the User Interface of the Software

Most of the CAD software allows customization of its commands, graphical area, and other user interface. The following are the areas where customization can be possible:

1. Model tree
2. Graphical area
3. Command areas

14.1.1.1 Model Tree

The functions of the model tree are discussed in Chapter 3. The model tree can be customized to display extra properties that may be needed for a particular application apart from our regular needs. The properties include parameters, feature internal ids, derived properties, etc.

Model tree customization is useful in the following cases:

1. When the model tree becomes large because of excessive features in the model tree
2. When different parameters are needed to be set or changed, which can be done through the model tree to reduce time

Hands-On Exercise, Example 14.1: Model Tree Customization (Software Used: Creo Parametric; Figure 14.1)

MODEL TREE

Steps:

- Open the assembly file Hinge_Expansion_Joint.asm from the Chapter 3 folder.
- Click on tree filers as displayed in Figure 14.2.
- Click on the check box displaying the features.
- Click on Apply. Notice that the features are displayed in the model tree.
- Similarly, check the box for the placement folder and uncheck the box of the datum plane display.
- Click OK to apply and exit.

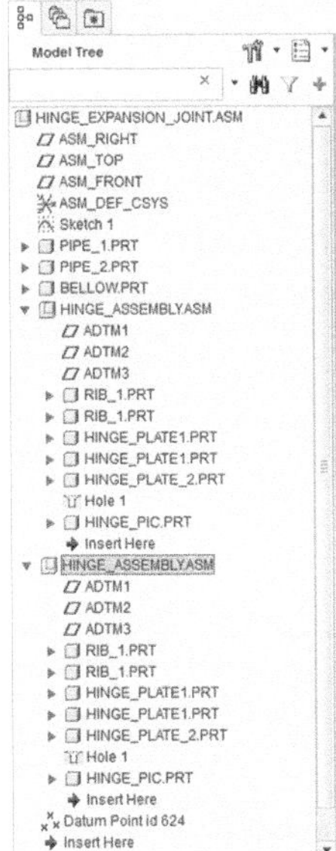

FIGURE 14.1
Model tree customization.

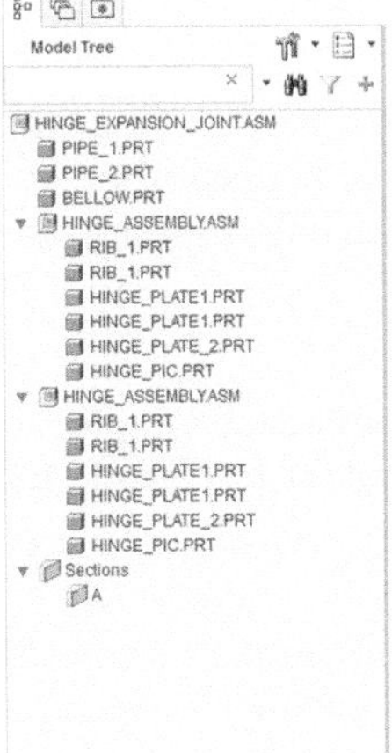

FIGURE 14.2
Model tree of an Assembly File.

- Notice that all the datum planes are no longer displayed in the model tree, and the features and placement folders are displayed.
- Click on the tree filters again as displayed in Figure 14.3.
- Make sure that the feature display box is checked.
- Uncheck all the datum features.

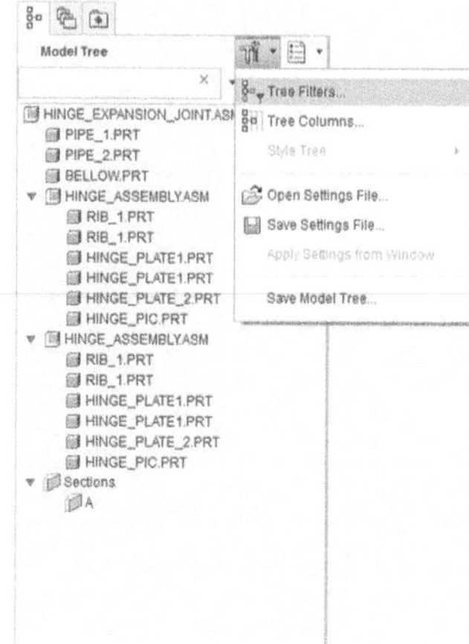

FIGURE 14.3
Tree filter settings.

- Click OK to apply and close.
- Click on Save tree settings file to save the tree structure type.
- The file extension for tree setting files is .cfg.
- Save and close the file.

14.1.1.2 Graphical Area

Apart from the size and shape change of graphical areas, many things can be done to increase the ease of accessibility. With the addition of notes and color displays, graphical area customization is sometimes useful according to users' needs.

Hands-On Exercise, Example 14.2: Graphical Customization (Software Used: Creo Parametric; Figure 14.4)

Steps:

- Open the assembly file Tank.asm from the Chapter 7 folder.
- The graphical area as shown in Figure 14.4 will be customized in this example.
- The customization shown can also be done without opening any file; however, it is done for better understanding.
- Click on File options.
- The Creo parametric options window will appear as shown in Figure 14.5.
- Select System colors on the left side of the window.
- Double click on the graphics on the right section to open individual settings of colors for respective entities or sections.
- Click on the small arrow beside Background > Gradient > More Gradient.

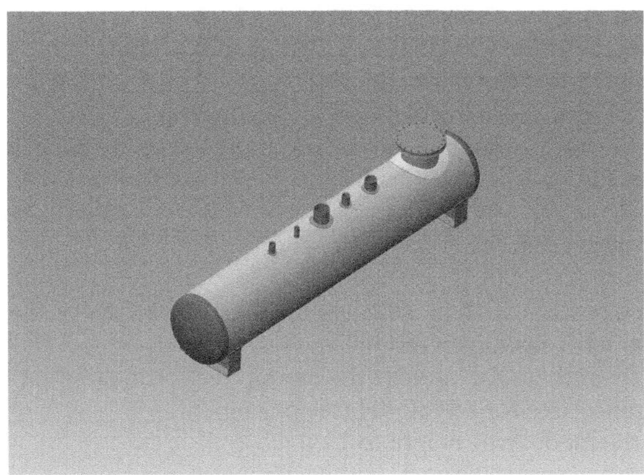

FIGURE 14.4
(See color insert.) Example of graphical customization.

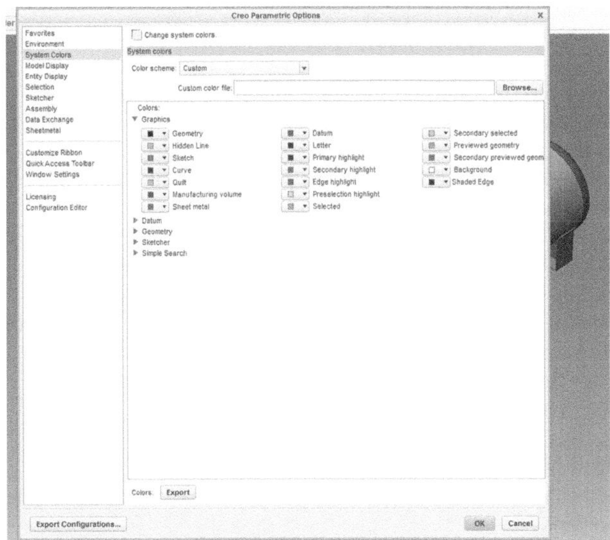

FIGURE 14.5
Parametric option window of Creo.

- Make sure that the check box for the dynamic update model window is checked.
- Click on Top to open the settings for the top section color of the background.
- Make sure that the check box RGB is checked.
- Enter the value R = 252.0, G = 83.0, and B = 43.0.
- Click OK to close the window.
- Click on Bottom to open the settings for the bottom section color of the background.

- Make sure the check box RGB is checked.
- Enter the value R = 115.0, G = 211.0, and B = 241.0.
- Click OK to close the window.
- Click OK to close the blended color window.
- On the bottom section, click on Export to export the color settings file.
- Set the name as CAD1.syscol. (*.syscol is the system color configuration file).

- Click on the window settings on the left section of the Creo parametric options window.
- On the right section under the model tree settings, select Below graphics area for the model tree placement.
- Set the model tree height to 15.
- Click on Import/Export on the right bottom corner.
- Select Export window settings.
- Set the name as creo_parametric_customization.ui (*.ui is the user customization file extension in Creo).
- Click on Save to the customization file.
- Click OK to close the Creo parametric options window.
- Click on File options again.
- The Creo parametric options window will appear.
- Select system colors on the left section.
- In the color scheme, select Default.
- Select Window settings on the left section of Creo parametric options.
- Select Restore default on the right bottom corner side.
- Click OK to close the Creo parametric options window.
- Notice that the Creo default graphical area is restored.
- Now if the customized graphical settings need to be restored, click on File options again.
- The Creo parametric options window will appear.
- Select System colors on the left section.
- In the color scheme, select Custom.
- Browse the file CAD1.syscol.
- Select Window settings on the left section of Creo parametric options.
- Click on Import/Export on the right bottom corner.
- Select Import window settings.
- Set the name as creo_parametric_customization.ui.
- Click OK to close the Creo parametric options window.
- Save and exit the file TANK.ASM.

14.1.1.3 Command Areas

Command areas are customized according to individual user needs for faster work. If frequently needed commands for daily work use can be identified, frequent command customization in the command areas may be very useful and saves time and effort. Here is an example of customization of command areas.

Hands-On Exercise, Example 14.3: Command Areas (Software Used: Creo Parametric; Figure 14.6)

Steps:

- Open the assembly file Tank.asm from the Chapter 7 folder.
- The Command area as shown in Figure 14.6 will be customized in this example.
- Command area customization can be very useful and helpful for saving time when properly used.
- Click on File options.
- The Creo parametric options window will appear as shown in Figure 14.7.
- Click on Customize ribbon on the left section of the Creo parametric options window.
- Notice that on the right section of the Creo parametric options window, there are two sections.
- The left section is where one chooses the commands, and the right section is where one customizes the ribbon.
- On the right section, select Main tabs to customize the ribbon section.
- Click on the new tab button.
- Right click on the newly created tab and select Rename.
- Rename it as Example.
- Select the newly created tab Example.
- Click on the new group button to create New group under the Example tab.
- Select the newly created group and right click on it.
- Select Rename and rename it as Create.
- On the left section (Choose commands from), select All commands.
- Make sure that the newly created group Create is selected on the right section.

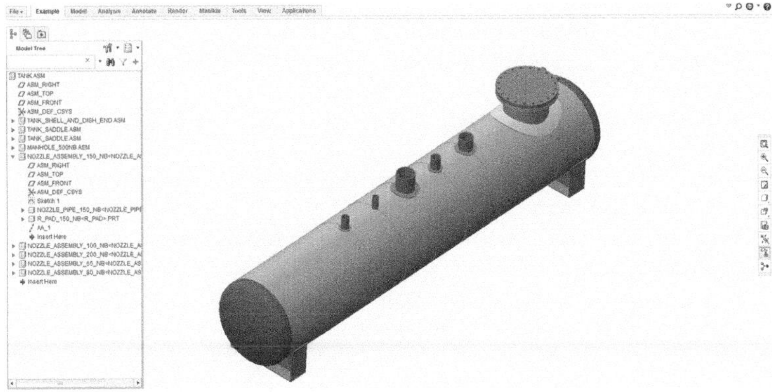

FIGURE 14.6
Example of command areas customization.

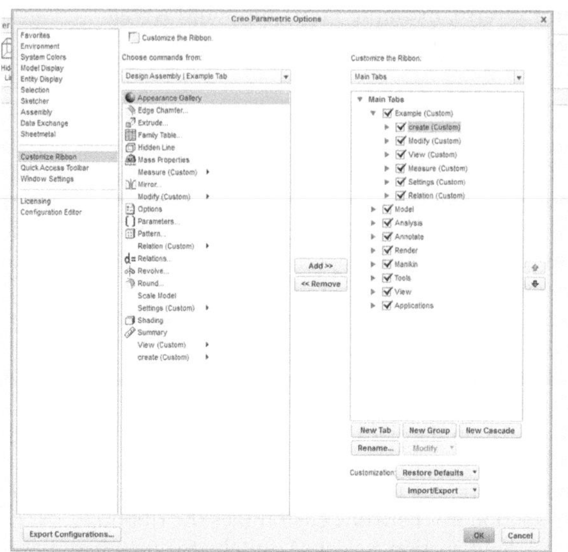

FIGURE 14.7
Parametric option window of Creo.

- Select the command Extrude from the left section of the window Choose commands from.
- Click on the right arrow to insert the command into the Create group.
- Similarly, insert the revolve command into the Create group.

- Select the newly created tab Example.
- Click on the new group button to create a new group under the Example tab.
- Select the newly created group and right click on it.
- Select Rename and rename it as Modify.
- On the left section (Choose commands from), select All commands.
- Make sure that the newly created group Modify is selected on the right section.
- Select the command Mirror from the left section of the window Choose commands from.
- Click on the right arrow to insert the command into the Create group.
- Similarly, insert Pattern, Scale Model, Edge Chamfer, and Round commands into the Create group.

- Select the newly created tab Example.
- Click on the new group button to create a new group under the Example tab.
- Select the newly created group and right click on it.

- Select Rename and rename it as View.
- On the left section (Choose commands from), select All commands.
- Make sure that the newly created group View is selected on the right section.
- Select the command Appearance Gallery from the left section of the window Choose commands from.
- Click on the right arrow to insert the command into the Create group.
- Similarly, insert Shading and Hidden Line commands into the Create group.

- Select the newly created tab Example.
- Click on the new group button to create a new group under the Example tab.
- Select the newly created group and right click on it.
- Select Rename and rename it as Measure.
- On the left section (Choose commands from), select All commands.
- Make sure that the newly created group Measure is selected on the right section.
- Select the command Mass Properties from the left section of the window Choose commands from.
- Click on the right arrow to insert the command into the Create group.
- Similarly, insert the Mass properties command into the Create group.

- Select the newly created tab Example.
- Click on the new group button to create a new group under the Example tab.
- Select the newly created group and right click on it.
- Select Rename and rename it as Relation.
- On the left section (Choose commands from), select All commands.
- Make sure that the newly created group Relation is selected on the right section.
- Select the command Relations from the left section of the window Choose commands from.
- Click on the right arrow to insert the command into the Create group.
- Similarly, insert Parameters and Family Table command into the Create group.

- Select the newly created tab Example.
- Click on the new group button to create a new group under the Example tab.
- Select the newly created group and right click on it.
- Select Rename and rename it as Settings.
- On the left section (Choose commands from), select All commands.

- Make sure that the newly created group Settings is selected on the right section.
- Select the command Options from the left section of the window Choose commands from.
- Click on the right arrow to insert the command into the Create group.

14.1.2 Customization by Using an Existing File as a Template File

Obviously, a lot of features can be common in many of the CAD files. For example, basic extrude may be the same for a number of components. There are also many datum features or parameters that can be common in some files. Using an existing file as a template file is most effective to reduce time and human error for the same set of features. Here is an example of customization by using an existing file as a template file.

Hands-On Exercise, Example 14.4: Using an Existing File as a Template File (Software Used: Creo Parametric; Refer to Figure 14.8)

Steps:

- Create a part file and rename it as Plate. (Other names may be given, but it is advisable to provide a relevant name.)
- Select Solid on the right-hand side of the box. This will activate the solid mode and environment.
- Uncheck the box of the default template and click OK to create a new solid file.

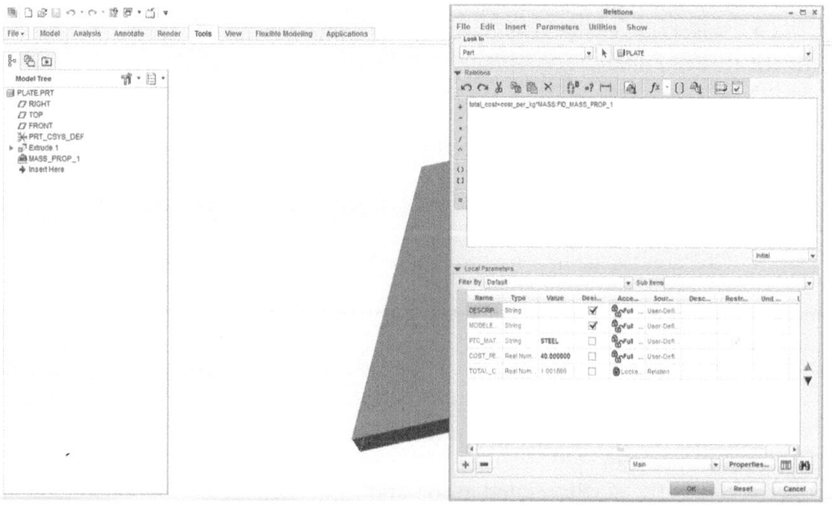

FIGURE 14.8
Relation window.

- The template selection box will appear.
- Select mmns_part_Solid. This will allow the use of the solid template file of the following units:
  - mm—millimeter for length unit
  - n—newton for force unit
  - s—second for time unit
- Once the solid part is created, go to File > Prepare > Model properties to check the units. In the upper sections, the units of the model will be displayed.
  - Note: The units are usually displayed in the summary section of any CAD software. The setting of the units is usually in the tools or settings.
- Click on Extrude from the Model tab in the ribbon interface.
- Click on the Placement tab and select Define to define the sketch.
- Select the top plane as the sketch plane.
- The right plane shall be automatically taken as the reference and orientation as right.
- Click on Sketch.
- The sketcher model will be activated.
- Draw a rectangle of side 400 × 400 mm placing the center at the intersection of the horizontal and vertical references.
- Click OK to create and exit the sketch.
- Select Mid plane extrusion.
- Enter the value 20 mm for the extrusion depth.
- Click on OK to create the extrusion.

- Click on File > Prepare > Model properties.
- Click on Change to change the material properties on the top of the model properties window.
- Double click on Steel on the left side to select and assign the material steel in the part.
- Click OK to close the materials dialog box.
- Click on Close to close the model properties window.
- Click on Mass properties from the Analysis tab of the ribbon interface.
- Click on Preview to compute the mass properties.
- Click on Drop box in quick/saved/feature box and select Feature.
- Click OK to create the mass properties analysis as features.
- Notice that the analysis feature Mass_Prop_1 is added in the model tree.
- Click on Relation from the Tools tab of the ribbon interface.
- Click on the local parameter window on the lower portion to open the local parameter dialog box.
- Add a parameter and rename it as Cost_per_kg. Set the attribute type to Real number. Set the value to 40.
- In the relation window, write the following relation:
  - total_cost = cost_per_kg*MASS:FID_MASS_PROP_1
- Click on OK on the right top corner to verify the relation.

- A small Verify relations window should appear with a message showing that "Relations have been successfully verified."
- Click OK to close the Relation verification window.
- Click OK to close the relation and local parameter window.
- Click save and exit the file plate.

- Create a part file and rename it as plate1. (Other names may be given, but it is advisable to provide a relevant name.)
- Select Solid on the right-hand side of the box. This will activate the solid mode and environment.
- Uncheck the box of the default template and click OK to create a new solid file.
- The template selection box will appear.
- Browse the file and select the last created saved file plate.prt.
- Click OK to create the file.
- Notice that all the features created in the plate file are present in the newly created file.

14.2 Customized Template Creation

Every company has their own standard practice of work. Therefore, template file requirement may also have differences. Most of the CAD software provides a standard set of templates that may not be fully useful and may need some amount of customization as per a company's standard practice of work.

Hands-On Exercise, Example 14.5: Creation of Template File (Software Used: Creo Parametric; Refer to Figure 14.9)

Steps:

- Create a part file and rename it as cad_item. (Other names may be given, but it is advisable to provide a relevant name.)
- Select Solid on the right-hand side of the box. This will activate the solid mode and environment.
- Uncheck the box of the default template and click OK to create a new solid file.
- The template selection box will appear.
- Select mmns_part_Solid. This will allow the use of the solid template file of the following units:
  - mm—millimeter for length unit
  - n—newton for force unit
  - s—second for time unit
- Once the solid part is created, go to File > Prepare > Model properties to check the units. In the upper sections, the units of the model will be displayed.

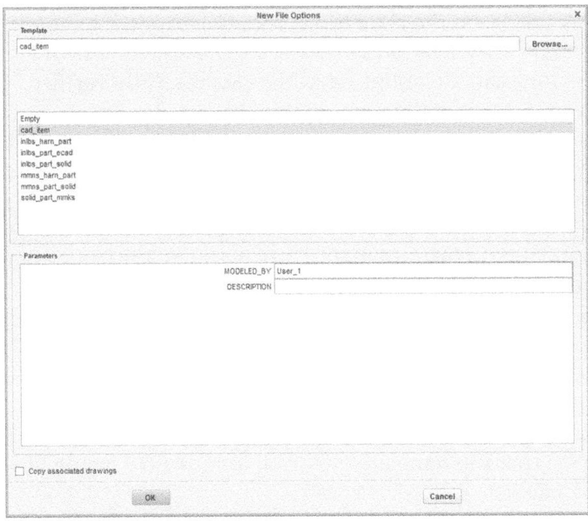

FIGURE 14.9
New file creation option.

  - Note: The units are usually displayed in the summary section of any CAD software. The setting of the units is usually in the tools or settings.
- Click on File > Prepare > Model properties.
- Click on Change to change the material properties on the top of the model properties window.
- Double click on Steel on the left side to select and assign the material steel in the part.
- Click OK to close the materials dialog box.
- Click on Close to close the model properties window.
- Click on Mass properties from the Analysis tab of the ribbon interface.
- Click on Preview to compute the mass properties.
- Click on Drop box in quick/saved/feature box and select Feature.
- Click OK to create the mass properties analysis as features.
- Notice that the analysis feature Mass_Prop_1 is added in the model tree.
- Click on Relation from the Tools tab of the ribbon interface.
- Click on the local parameter window on the lower portion to open the local parameter dialog box.
- Add a parameter and rename it as Total_weight. Set the attribute type to Real number.
- Add another parameter and rename it as ERP_CODE. Set the attribute type to String.
- Add another parameter and rename it as SPECIFICATION. Set the attribute type to String.
- In the relation window, write the following relation:
  - Total_Weight = MASS:FID_MASS_PROP_1

- Click on OK on the right top corner to verify the relation.
- A small verify relations window should appear with a message showing that "Relations have been successfully verified."
- Click OK to close the relation verification window.
- Click OK to close the relation and local parameter window.
- Click Save and save the file to the Creo default template load point location.

- Create a part file and rename it as plate1. (Other names may be given, but it is advisable to provide a relevant name.)
- Select Solid on the right-hand side of the box. This will activate the solid mode and environment.
- Uncheck the box of the default template and click OK to create a new solid file.
- The template selection box will appear.
- Notice that the template part cad_item is now available in the template list.
- Click OK to create the file.
- Notice that all the features created in the plate file are present in the newly created file.

Here is an example of customization of a drawing file.

Hands-On Exercise, Example 14.6: Creation of a Template Drawing File (Software Used: Creo Parametric; Refer to Figure 14.10)

Steps:

- Create a drawing file and rename it as "cad." (Other names may be given, but it is advisable to provide a relevant name.)

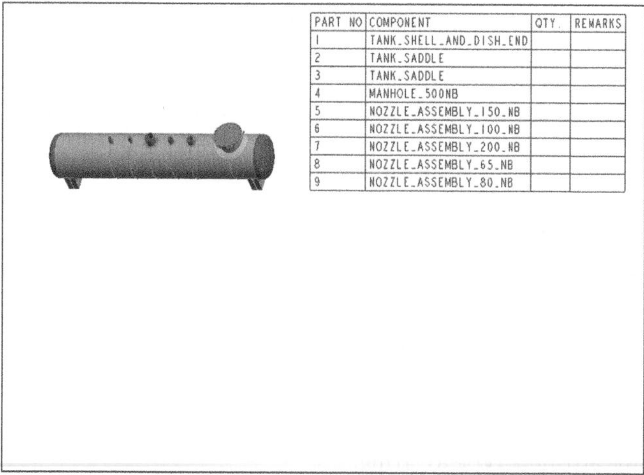

PART NO	COMPONENT	QTY.	REMARKS
1	TANK_SHELL_AND_DISH_END		
2	TANK_SADDLE		
3	TANK_SADDLE		
4	MANHOLE_500NB		
5	NOZZLE_ASSEMBLY_150_NB		
6	NOZZLE_ASSEMBLY_100_NB		
7	NOZZLE_ASSEMBLY_200_NB		
8	NOZZLE_ASSEMBLY_65_NB		
9	NOZZLE_ASSEMBLY_80_NB		

FIGURE 14.10
Sample drawing file.

- The template selection box will appear.
- Select Empty.
- Select the standard size sheet A4.
- Click on the Tools tab of the ribbon interface.
- Click on Template.
- Click on the Layout tab of the ribbon interface.
- Click on the template view button.
- Let the orientation as Front.
- Click on the Place view.
- Click on the left top portion of the sheet as the view placement area.
- Click OK on the template view instruction window.
- Click on the Table tab of the ribbon interface.
- Click on Table and select a table with a 4 columns × 2 rows arrangement.
- Place the Table at the top right corner.
- Click on Repeat region.
- A table region menu manager will appear.
- Click on Add to add the repeat region.
- Click on the left bottom cell and the right bottom cell once.
- Press the middle button to close the table region menu manager.
- Double click on the left bottom cell and select the parameter rpt > index.
- Similarly, double click on the bottom second cell and select the parameter asm > mbr > name.
- Similarly, double click on the bottom third cell and select the parameter rpt > qty.
- Rename the upper boxes of the table as shown in Figure 14.11.
- Save the file to the Creo load point > Creo standard > Template directory.
- Save and exit the file.
- Create a drawing file and rename it as "cad." (Other names may be given, but it is advisable to provide a relevant name.)
- Uncheck the box of the default template and click Ok to create a new solid file.
- The template selection box will appear.
- Browse for the template and select the file cad.drw.
- Browse for the model file and select the file TANK.ASM from the folder Chapter 7.
- Click OK to create the drawing model TANK.ASM with the template file cad.drw.
- Notice that the front view of TANK.ASM will appear with the Bill of materials at the top right corner.

PART NO	COMPONENT	QTY.	REMARKS

FIGURE 14.11
Sample of a table.

14.3 Graphical Customization

Graphical customization is mostly required in cases where similar types of models are commonly prepared for which a specific graphical code is instead more useful.

Hands-On Exercise, Example 14.7: Graphical Customization (Software Used: Creo Parametric; Refer to Figure 14.12)

Steps:

- Open the assembly file HINGE_EXPANSION_JOINT from the Chapter 7 folder.
- The graphical area as shown in Figure 14.12 will be customized in this example.
- The customization shown above can also be done without opening any file; however, it is done for better understanding.
- Click on File options.
- The Creo parametric options window will appear as shown in Figure 14.13.
- Select System colors on the left side of the window.
- Click on the Datum section.
- Select the axes and the centerline color blue.
- Click on the Graphics section.
- Click on Curve and click on More colors.
- Make sure that the check box RGB is checked.
- Enter the value R = 240.0, G = 120.3, and B = 102.2.
- Click OK to close the window.

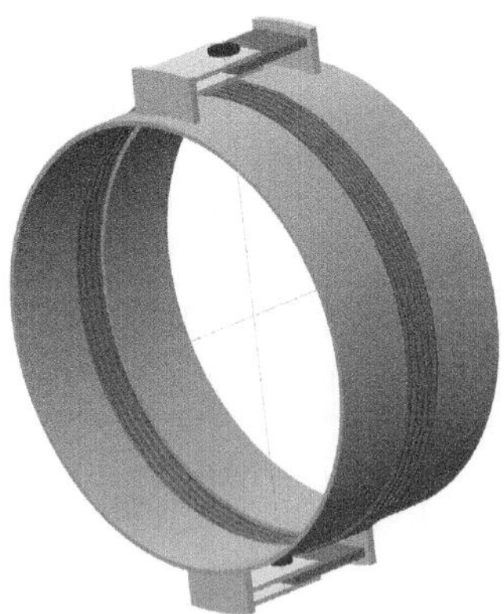

FIGURE 14.12
Example of graphical customization.

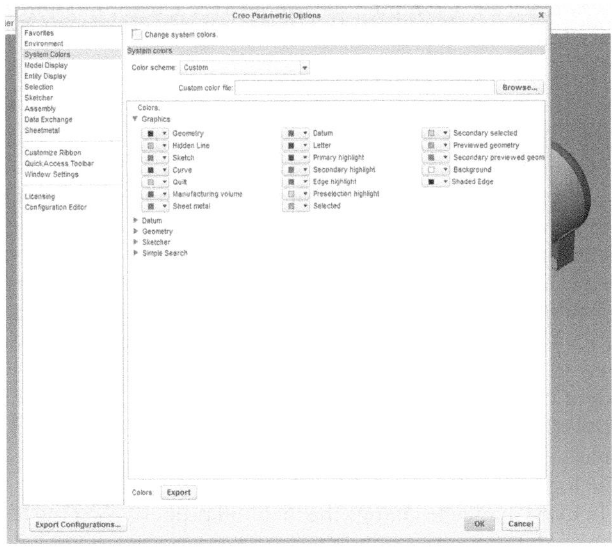

FIGURE 14.13
Parametric option window of Creo.

- Similarly, select all the colors you wish to change in the display as a company color standard.
- On the bottom section, click on Export to export the color settings file.
- Set the name as CAD1.syscol (*.syscol is the system color configuration file).

- Click on Save to the customization file.
- Click OK to close the Creo parametric options window.
- Click on File options again.
- The Creo parametric options window will appear.
- Save and exit the file HINGE_EXPANSION_JOINT.ASM.

14.4 Time Reduction through Customization

If customization is carried out in a proper way with a proper objective, a lot of time can be saved in the work along with increased accuracy.

Below is an example showing how the model tree customization reduces to check individual model size.

Hands-On Exercise, Example 14.8: Model Tree Customization 2 (Software Used: Creo Parametric; Refer to Figure 14.14)

Steps:

- Open the assembly file Tank.asm from the Chapter 7 folder.
- Click on tree columns as displayed in Figure 14.15.
- This will display the additional information in the model tree along with the component and features name.
- Select Info in the type selection window.
- Select Model size and click the right arrow to display it in the model tree. Enter the value 10 in the model tree width field.
- Click OK to display the model size in the model tree as displayed in Figure 14.16.
- Right click on the part tank shell under the TANK_SHELL_AND+DISH_END assembly.
- Select Open to open the part file TANK_SHELL.PRT.
- Click on File > Prepare > Model properties.
- Click on Change to change the material properties on the top of the model properties window.
- Double click on Steel on the left side to select and assign the material steel in the TANK_SHELL part.
- Click Close to close the model properties window.
- Save and exit the file TANK_SHELL part.
- Click on tree columns as displayed in Figure 14.16.

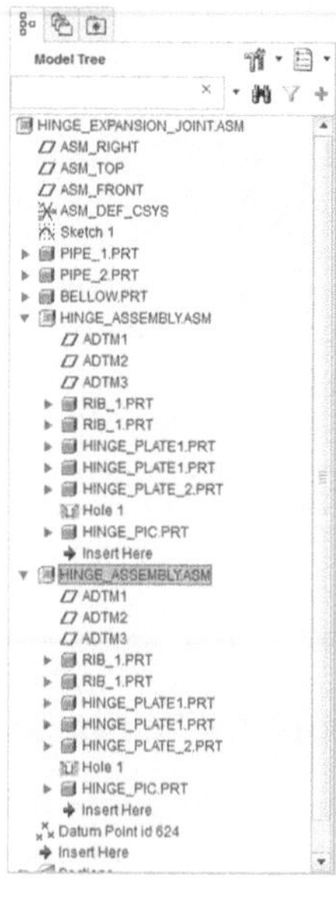

FIGURE 14.14
Model tree of an Assembly File.

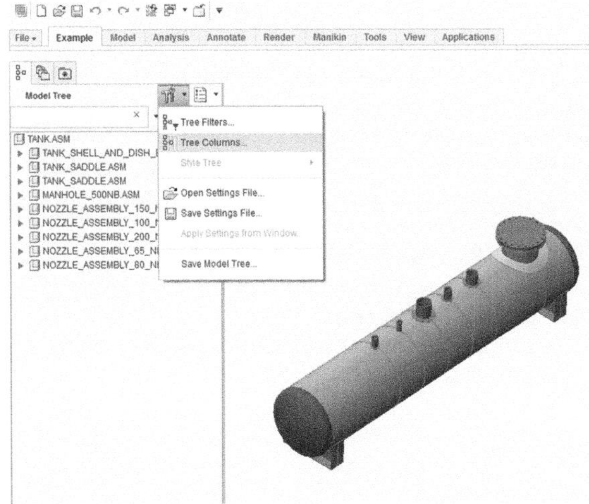

FIGURE 14.15
Tree filter settings.

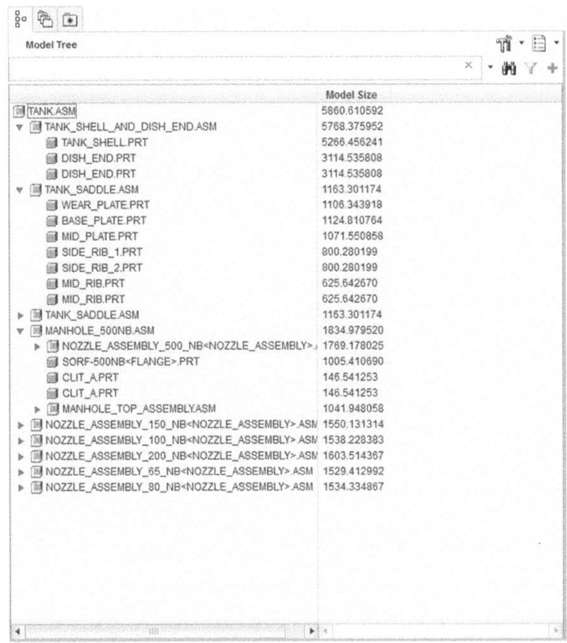

FIGURE 14.16
Example of model tree customization.

- Model size shall be on the right section of the model tree window in the displayed section.
- Double click on the Model size to remove this field from the display.
- Select Model parameters in the type field.
- In the Name section, write PTC_MATERIAL_NAME.
- Click on the right arrow to enter the parameter in the displayed section.
- Click OK to close the model tree window.
- Notice that the assigned material in the part TANK_SHELL is displayed in the model tree.
- Therefore, materials of any part can be checked from the model tree through this method.

QUESTIONS

1. What is customization?
2. Define various types of customization.
3. Why are customizations needed?
4. Describe the process of customization.
5. Describe graphical customization.
6. How can time be saved by customization?

15

Project Work

15.1 Simple Automation

Automation is already discussed in Chapter 13. However, there are many essential components required in mechanical systems (Figure 15.1); automation of those components saves time and effort with increased accuracy. Gear is a very essential component that is used for power transmission. An automation of spur gear is shown in the following example.

Hands-On Exercise, Example 15.1: Creating Gear (Software Used: Creo Parametric; Figure 15.2)

Steps:

- Create a part file and rename it as Gear. (Other names may be given, but it is advisable to provide a relevant name.)
- Select Solid on the right-hand side of the box. This will activate the solid mode and environment.
- Uncheck the box of the default template and click OK to create a new solid file.
- The template selection box will appear.
- Select mmns_part_Solid. This will allow the use of the solid template file of the following units:
  - mm—millimeter for length unit
  - n—newton for force unit
  - s—second for time unit
- Once the solid part is created, go to File > Prepare > Model properties to check the units. In the upper sections, the units of the model will be displayed.
  - Note: The units usually are displayed in the summary section of any CAD software. The setting of units usually is in the tools or settings.
- The gear to be created is shown in Figure 15.2.

- Click on Extrude from the Model tab of the ribbon interface.
- Click on the Placement tab and select Define to define the sketch.

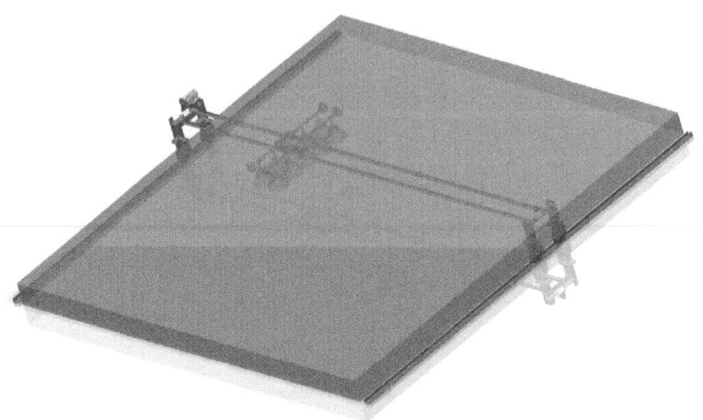

FIGURE 15.1
(See color insert.) Industrial spray system of equipment.

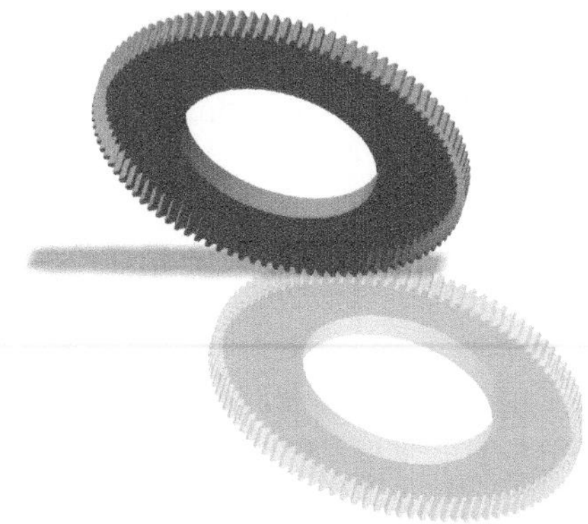

FIGURE 15.2
Three-dimensional model of Gear.

- Select the front plane as the sketch plane.
- The right plane shall be automatically taken as reference and orientation as right.
- Click on Sketch.
- The sketcher model will be activated.
- Draw a circle of any diameter at the center of the intersection of horizontal and vertical references.
- Click OK to create and exit the sketch.
- Select Mid plane extrusion.
- Enter the value 20 mm as the extrusion depth.
- Click on OK to create the extrusion.

- Click on Parameters from the tools tab of the ribbon interface to display the local parameters dialog box.
- Add a parameter and rename it as Major_Dia.

- Keep Real number as the type.
- Enter the Major_Dia parameter value as 1200.
- Add a parameter and rename it as Minor_Dia.
- Keep Real number as the type.
- Enter the Minor_Dia parameter value as 1140.
- Add a parameter and rename it as Base_Dia.
- Keep Real number as the type.
- Enter the Base_Dia parameter value as 1140.
- Add a parameter and rename it as Pitch_Dia.
- Keep Real number as the type.
- Enter the Pitch_Dia parameter value as 1170.
- Add a parameter and rename it as No_of_Teeth.
- Keep Real number as the type.
- Enter the No_of_Teeth parameter value as 100.
- Add a parameter and rename it as Shaft_Dia.
- Keep Real number as the type.
- Enter the Shaft_Dia parameter value as 500.
- Add a parameter and rename it as Face_width.
- Keep Real number as the type.
- Enter the Face_width parameter value as 75.
- Add a parameter and rename it as gamma.
- Keep Real number as the type.
- Enter the gamma parameter value as 100.
- Add a parameter and rename it as start_angle.
- Keep Real number as the type.
- Enter the start_angle parameter value as 30.
- Add a parameter and rename it as angle_f.
- Keep Real number as the type.
- Enter the angle_f parameter value as 1.64.
- Add a parameter and rename it as in_angle.
- Keep Real number as the type.
- Enter the angle_f parameter value as 0.2.
- Add a parameter and rename it as In.
- Keep Real number as the type.
- Enter the In parameter value as 131.
- Add a parameter and rename it as angle_rad.
- Keep Real number as the type.
- Enter the angle_rad parameter value as 0.06.
- Add a parameter and rename it as angle_d.
- Keep Real number as the type.
- Enter the angle_d parameter value as 3.6.
- Add a parameter and rename it as x.
- Keep Real number as the type.
- Enter the x parameter value as 36.
- Add a parameter and rename it as y.
- Keep Real number as the type.
- Enter the y parameter value as 16.75.
- Click OK to close the parameter dialog box.
- Click on the tools tab in the ribbon interface.
- Click on Relations.

- The relation window will appear.
- In the main relation window, enter the following relations:
  - d1 = major_dia
  - angle_rad = (2*pi)/no_of_teeth
  - angle_d = (180*angle_rad)/pi
  - x = 2*(pitch_dia/2)*cos(90-(angle_d/2))
  - y = x-tooth_thickness
  - angle_f = (90-ACOS(Y/pitch_dia))*2
  - in = (sqrt((pitch_dia/2)^2-(base_dia/2)^2))
  - in_angle=(in*(360/(2*pi*(base_dia/2)))-atan(in/(base_dia/2)))
- Click on Verify on the right upper side of the relation window.
- A notification box with a message "Relations have been successfully verified" should appear.
- Click OK to close the box.
- Close the relation window.

- Click on Curve > Curve from equation.
- Select Cylindrical co-ordinate system.
- Click on the reference tab and select the default co-ordinate system PRT_SYS_DEF.
- Click on Equation; the equation window shall appear.
- Write down the following equation:
  - solve
  - gamma = t*(sqrt((major_dia/2)^2-(base_dia/2)^2))
  - for gamma
  - r = sqrt(gamma^2+(base_dia/2)^2)
  - theta = start_angle+(gamma*(360/(2*pi*(base_dia/2)))-atan (gamma/(base_dia/2)))
  - z = 0
- Click on Verify on the right upper side of the relation window.
- A notification box with a message "Relations have been successfully verified" should appear.
- Click OK to close the relation window.
- Click OK to close the equation window.
- Click OK to create the curve feature.

- Click again on Curve > Curve from equation.
- Select Cylindrical co-ordinate system.
- Click on the reference tab and select the default co-ordinate system PRT_SYS_DEF.
- Click on Equation; the equation window shall appear.
- Write down the following equation:
  - solve
  - gamma = t*(sqrt((major_dia/2)^2-(base_dia/2)^2))
  - for gamma
  - r = sqrt(gamma^2+(base_dia/2)^2)
  - theta = start_angle-(angle_f-2*in_angle)-(gamma*(360/(2*pi* (base_dia/2)))-atan(gamma/(base_dia/2)))
  - z = 0
- Click on Verify on the right upper side of the relation window.

- A notification box with a message "Relations have been successfully verified" should appear.
- Click OK to close the relation window.
- Click OK to close the equation window.
- Click OK to create the curve feature.

- The gear as shown in Figure 15.2 is to be created.

- Click on Extrude from the Model tab in the ribbon interface.
- Click on the Placement tab and select Define to define the sketch.
- Select the front flat surface of the extrusion as the sketch plane.
- The right plane shall be automatically taken as reference and orientation as right.
- Click on Sketch.
- The sketcher model will be activated.
- Select both curves and project them to the sketch.
- Draw two circular segments on a major diameter and a minor diameter as shown in Figure 15.3.
- Select the material removal option.
- Select Through all option for the extrusion depth.
- Flip the orientation towards the material side.
- Click on OK to create the extrusion.

- Click on Round feature from the Model tab of the ribbon interface.
- Enter the following expression in the value field:
    - 0.25*(pitch_dia/no_of_teeth)
- Select two internal edges of the newly created feature as shown in Figure 15.4.
- Click OK to create the feature.

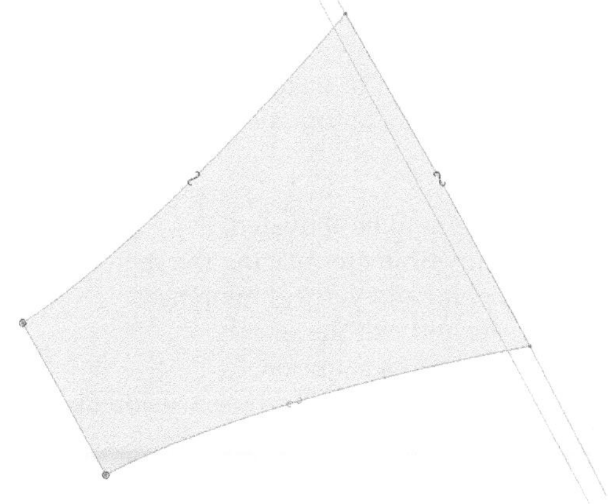

FIGURE 15.3
Sketch of the gear teeth profile.

FIGURE 15.4
Edge of the gear teeth.

- Select both the features in the model tree by pressing the control key.
- Right-click and select Group to create a group feature.

- Select the newly created group in the model tree.
- Right-click and select Pattern.
- Select Axis in the type of pattern selection field.
- Select the Central axis.
- Enter the quantity 8 for patterned feature.
- Select the option to make the pattern equally spaced within 360°.
- Click OK to create the patterned feature.

- Click on Extrude from the Model tab in the ribbon interface.
- Click on the Placement tab and select Define to define the sketch.
- Select the front plane as the sketch plane.
- The right plane shall be automatically taken as reference and orientation as right.
- Click on Sketch.
- The sketcher model will be activated.
- Draw a circle of any diameter placing the center at the intersection of the horizontal and vertical references.
- Click OK to create and exit the sketch.
- Select the material removal option.
- Select the Through all option for the extrusion depth.
- Flip the orientation towards the material side.
- Click on OK to create the extrusion.

- Click on the Tools tab in the ribbon interface.
- Click on Relations.

- The relation window will appear.
- In the main relation window, enter the following relations:
    - p13 = no_of_teeth (Note: p13 is the circular pattern quantity; the symbol may get changed. Users are advised to select graphically if needed.)
    - d16 = shaft_dia (Note: d16 is the diameter of the last extrusion; the symbol may get changed. Users are advised to select graphically if needed.)
- Click on Verify on the right upper side of the relation window.
- A notification box with a message "Relations have been successfully verified" should appear.
- Click OK to close the box.
- Close the relation window.

- Click on File > Save as > Save as a copy.
- Enter the file name as gear_dia_100.
- Click OK to save the file.

- Save and exit the file.
- Click on File > Open; open the file name gear_dia_100.
- Click on Parameters from the Tools tab of the ribbon interface.
- Change the parameter values as follows:

Base_Dia	90
Face_width	10
Major_Dia	100
Minor_Dia	90
No_of_Teeth	40
Pitch_Dia	95
Shaft_Dia	30
Tooth_Thickness	3

- Click OK to close the parameter box.
- Click on Regenerate and notice the changes.
- Save and exit the file.

15.2 Real-Time Product Development through Optimization

15.2.1 Requirement

A spray system is required for continuous spray over equipment of dimension 5000 × 4800 × 250 mm. The equipment is placed on a deck level where a rail can be fitted. The spray shall be approximately above 200 mm from the equipment top face for effective cleaning. Each nozzle is of the cone type,

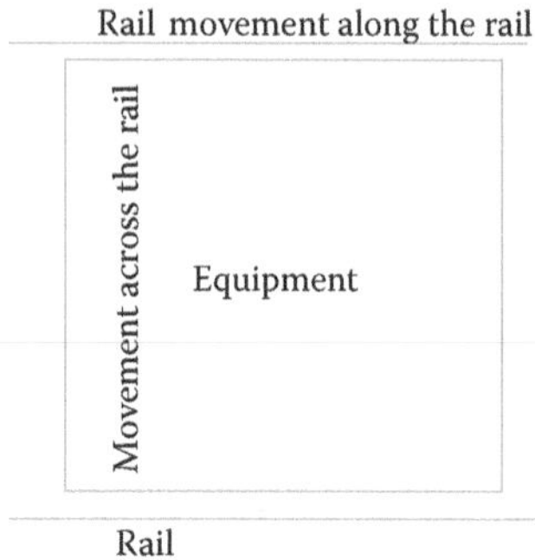

FIGURE 15.5
Sketch of the equipment.

with a 14 lpm water flow, and the nozzle spray angle is 60°. The available pump for supply of spray water is of 10 m^3/h capacity. Stepping on the equipment can damage it and hence is not advised.

Keeping in mind the above-described situation, the equipment sketch is drawn first as shown in Figure 15.5.

15.2.2 Solution

Two rails can be placed along the longer side of the equipment so that any moving unit can be guided through the rail.

Similarly, another rail system can be made on the moving unit so that another moving unit can be placed on those rails.

Therefore, there will be movements on two directions. One is along the rail and the other is across the rail. We refer to these movements as "along movement" and "across movement" from here onward for better understanding as shown in the sketch in Figure 15.5.

Since the complete system is primarily dependent on spray nozzle arrangement, we shall proceed to design the spray nozzle system first.

15.2.3 Design of Spray Nozzle Arrangement

Each nozzle required 14 lpm spray water. The spray water supply pump capacity is limited to 10 m^3/h. Hence the quantities of maximum nozzles are as follows:

$$\text{pump flow} = 10\ m^3/h = 10 \times 10^3\ L/h\ (\text{Since } 1\ m^3 = 1000\ L) = 166.667\ \text{lpm}.$$

Hence, the number of nozzles is 166.667/14 = 11 nozzles.

Let us consider 10 nozzles, since an even-numbered arrangement is simpler and easier compared to an odd-numbered one.

Each nozzle has 60° spray cone as shown in Figure 15.6.

The diameter of the base of the cone is 230.94 mm. An overlap of approximately 10 mm is suggested for ensuring spray without gap. Hence, the pitch value (230 – 10 = 220) can be taken for this spray nozzle system design.

If all the nozzles are placed in a single row, there shall be approximately 10 × 220 = 2200 mm length required for connecting the pipe line. However, if we consider five nozzles with a two-row arrangement, it will require half of the length and will result in a more compact design for this spray system. Therefore, we consider the 2 × 5 nozzle arrangement.

Now, there will be five nozzles in each nozzle arrangement. Therefore, the connected pipe diameter for each nozzle arrangement shall be calculated as follows.

Total flow in five nozzles = 5 × 14 = 70 lpm = 70×10^{-3} m^3/min = $(70 \times 10^{-3})/60$ m^3/s.

Considering a velocity of 2.3 m/s for carbon steel pipes, the diameter would be = $\sqrt{\left(\left(\left(\frac{70 \times 10^{-3}}{60}\right)/2.3\right) \times 4\right)/\pi}$ = 0.02541 m. = 25.41 mm. Hence, 25 nb pipe size is selected and is of standard schedule.

The CAD models are now prepared independently first followed by assembly modeling. Users are advised to create all the models as shown in the following hands-on exercise.

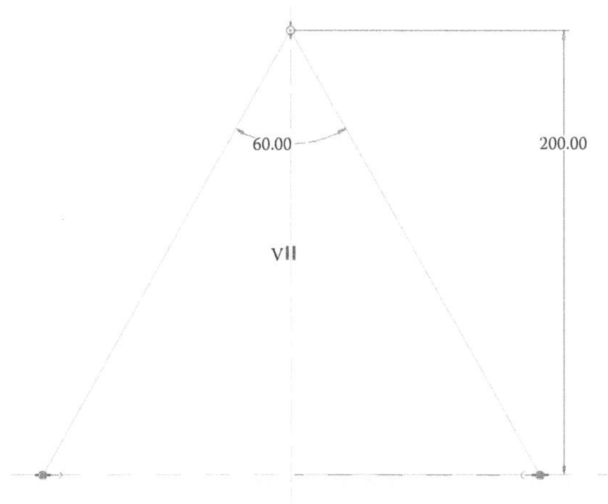

FIGURE 15.6
Sketch of the nozzle spray.

Hands-On Exercise, Example 15.2: 15 NB Adapter (Software Used: Creo Parametric; Refer to Figure 15.7)

Steps:

- Create a part file and rename it as Adapter. (Other names may be given, but it is advisable to provide a relevant name.)
- Select Solid on the right-hand side of the box. This would activate the solid mode and environment.
- Uncheck the box of default template and Click OK to create a new solid file.
- The template selection box will appear.
- Select mmns_part_Solid. This will allow the use of the solid template file of the following units:
  - mm—millimeter for length unit
  - n—newton for force unit
  - s—second for time unit
- Once the solid part is created, go to File > Prepare > Model properties to check the units. In the upper sections, the units of the model will be displayed.
  - Note: The units usually are displayed in the summary section of any CAD software. The setting of the unit usually is in the tools or settings.
- The adapter as shown in Figure 15.7 is to be created.
- Click on Revolve from the Model tab in the ribbon interface.
- Click on the placement tab and select Define to define the sketch.
- Select the front plane as the sketch plane.
- The right plane shall be automatically taken as reference and orientation as right.
- Click on Sketch.
- The sketcher model will be activated.
- Draw the sketch as displayed in Figure 15.8.
- Click OK to create and exit the sketch.
- Ensure that the revolve angle shall be 360°.
- Click on OK to create the revolve feature.

FIGURE 15.7
Three-dimensional view of a 15-NB adapter.

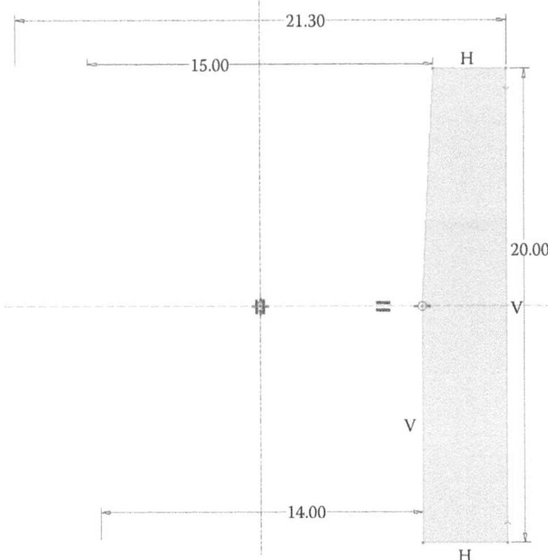

FIGURE 15.8
Constructional sketch of an adapter.

- Click on the helical sweep option from the drop-down menu of Sweep from the Model tab of the ribbon interface.
- Click on the References tab and select Define to define the helix sweep profile.
- Select the front plane as the sketch plane.
- The right plane shall be automatically taken as reference and orientation as right.
- Click on Sketch.
- The sketcher model will be activated.
- Draw a geometry centerline on the vertical reference line in the middle.
- Project the internal vertical edge.
- Click OK to create the sketch and exit the sketcher window.
- Click Sketch to create the profile sketch above the reference tab.
- Draw the sketch as shown in Figure 15.9.
- Click OK to create and exit the sketch.
- Enter the pitch value of 1 mm.
- Click on the left-handed rule icon to create the helical spring clockwise downward.
- Click OK to create the thread.

- Click on the round feature from the Model tab of the ribbon interface.
- Set the value of fillet to 0.1 mm.
- Select the edge of the minimum diameter in the internal thread.
- Click OK to create the feature.

- Save and exit the file.

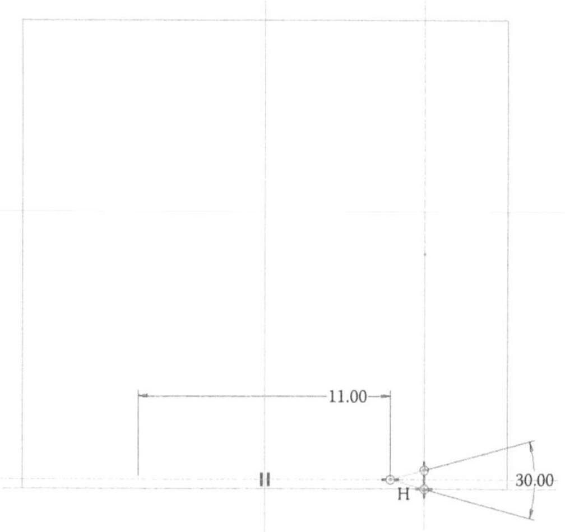

FIGURE 15.9
Constructional sketch of the thread of an adapter.

Hands-On Exercise, Example 15.3: SORF & BLRF, 25 NB Flanges (Software Used: Creo Parametric; Refer to Figure 15.10)

Steps:

- Open the file flange.prt from Chapter 7.
- Click Generic to open the generic model.
- Click on Family tables from the Tools tab of the ribbon interface.
- Add two instances and rename them as Blind_25nb and sorf_25NB, respectively.
- Enter the values indicated in Table 15.1.

- Click to verify instances.
- Click on Verify in the family tree window. Verification of all the instances must be successful in order to proceed.
- Save and exit the file.

FIGURE 15.10
Three-dimensional model of a flange.

TABLE 15.1
Family Table Data for Flanges

Instance Name	D1	D3	D2	D12	P19	D21	F648 [REVOLVE_2]	F697 [REVOLVE_3]	D27	D25	D24
Blind_25NB	110	50.8	12.7	79.4	4	14	N	N	34.5	16	49
SORF_25NB	110	50.8	12.7	79.4	4	14	Y	Y	34.5	16	49

Hands-On Exercise, Example 15.4: Hex_Bolt_M8 (Software Used: Creo Parametric; Figure 15.11)

Steps:

- Create a part file and rename it as Hex_Bolt_M8. (Other names may be given, but it is advisable to provide a relevant name.)
- Select Solid on the right-hand side of the box. This will activate the solid mode and environment.
- Uncheck the box of the default template and click OK to create a new solid file.
- The template selection box will appear.
- Select mmns_part_Solid. This will allow the use of the solid template file of the following units:
  - mm—millimeter for length unit
  - n—newton for force unit
  - s—second for time unit.
- Once the solid part is created, go to File > Prepare > Model properties to check the units. In the upper sections, the units of the model will be displayed.
  - Note: The units are usually displayed in the summary section of any CAD software. The setting of units is usually in the tools or settings.

FIGURE 15.11
Three-dimensional model of Hex Bolt of size M8.

- Click on Extrude from the Model tab in the ribbon interface.
- Click on the placement tab and select Define to define the sketch.
- Select the top plane as the sketch plane.
- The right plane shall be automatically taken as reference and orientation as right.
- Click on Sketch.
- The sketcher model will be activated.
- Draw a circle of diameter 8 placing the center at the intersection of the horizontal and vertical references.
- Click OK to create and exit the sketch.
- Flip the material orientation towards down.
- Enter the value 30 mm for the extrusion depth.
- Click on OK to create the extrusion.

- Click on Extrude from the Model tab in the ribbon interface.
- Click on the placement tab and select Define to define the sketch.
- Select the front plane as the sketch plane.
- The right plane shall be automatically taken as reference and orientation as right.
- Click on Sketch.
- The sketcher model will be activated.
- Draw the sketch as shown in Figure 15.12.
- Click OK to create and exit the sketch.
- Select Mid plane extrusion.
- Enter the value 8 mm for the extrusion depth.
- Click on OK to create the extrusion.
- Click on Revolve from the Model tab in the ribbon interface.
- Click on placement tab and select Define to define the sketch.
- Activate the material removal feature option.
- Select the front plane as the sketch plane.

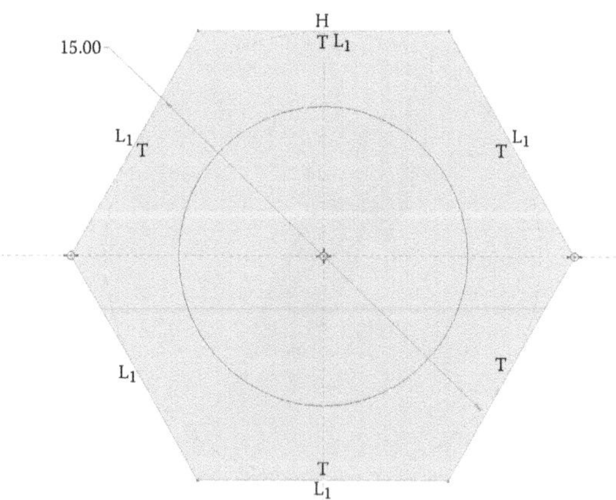

FIGURE 15.12
Constructional sketch of head of Hex Bolt of size M8.

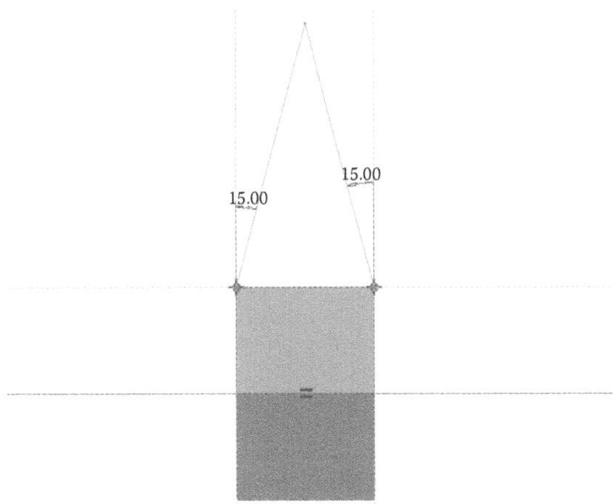

FIGURE 15.13
Constructional sketch of the material removal feature of Hex Bolt of size M8.

- The right plane shall be automatically taken as reference and orientation as right.
- Click on Sketch.
- The sketcher model will be activated.
- Draw the sketch as displayed in Figure 15.13.
- Click OK to create and exit the sketch.
- Ensure that the revolve angle shall be 360°.
- Click on OK to create the revolve feature.
- Save and exit the file.

Hands-On Exercise, Example 15.5: Nut_M_8 (Software Used: Creo Parametric; Figure 15.14)

Steps:

- Create a part file and rename it as NUT_M_8. (Other names may be given, but it is advisable to provide a relevant name.)
- Select Solid on the right-hand side of the box. This will activate the solid mode and environment.
- Uncheck the box of the default template and click OK to create a new solid file.
- The template selection box will appear.
- Select mmns_part_Solid. This will allow the use of the solid template file of the following units:
    - mm—millimeter for length unit
    - n—newton for force unit
    - s—second for time unit

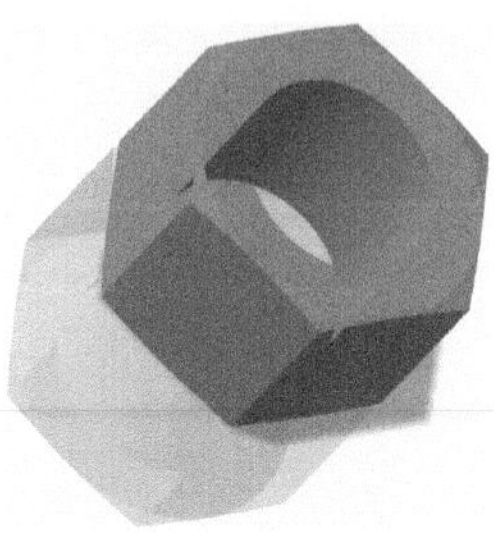

FIGURE 15.14
Three-dimensional model of Nut of size M8.

- Once the solid part is created, go to File > Prepare > Model properties to check the units. In the upper sections, the units of the model will be displayed.
  - Note: The units are usually displayed in the summary section of any CAD software. The setting of the units is usually in the tools or settings.
- Click on Extrude from the Model tab in the ribbon interface.
- Click on the placement tab and select Define to define the sketch.
- Select the front plane as the sketch plane.
- The right plane shall be automatically taken as the reference and orientation as right.
- Click on Sketch.
- The sketcher model will be activated.
- Draw the sketch as shown in Figure 15.15.
- Click OK to create and exit the sketch.

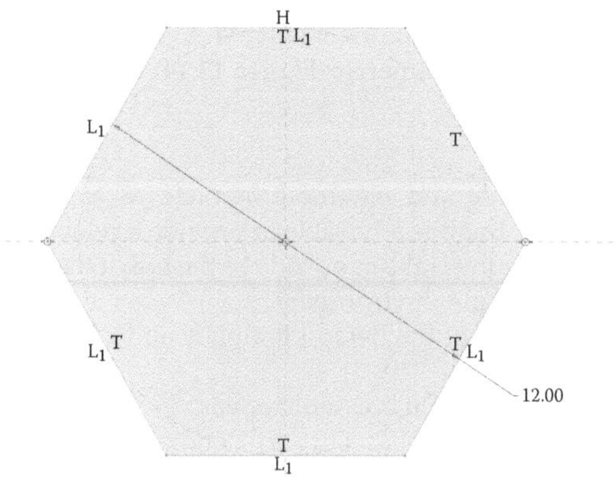

FIGURE 15.15
Constructional sketch of Nut of size M8.

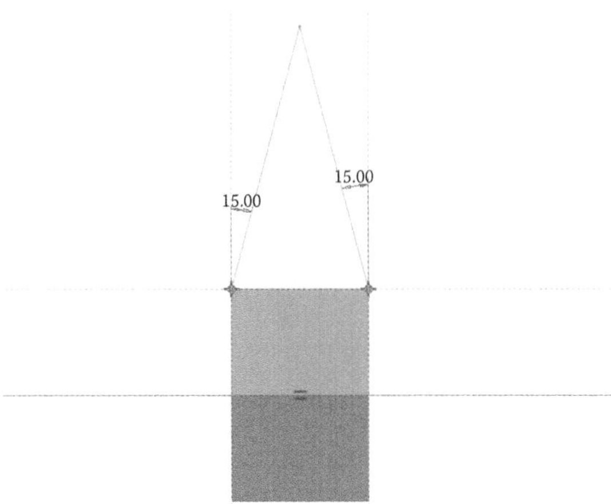

FIGURE 15.16
Constructional sketch of the material removal feature of Nut of size M8.

- Select Mid plane extrusion.
- Enter the value 8 mm for the extrusion depth.
- Click on OK to create the extrusion.

- Click on Revolve from the Model tab in the ribbon interface.
- Click on the Placement tab and select Define to define the sketch.
- Activate the material removal feature option.
- Select the front plane as the sketch plane.
- The right plane shall be automatically taken as the reference and orientation as right.
- Click on Sketch.
- The sketcher model will be activated.
- Draw the sketch as displayed in Figure 15.16.
- Click OK to create and exit the sketch.
- Ensure that the revolve angle shall be 360°.
- Click on OK to create the revolve feature.
- Save and exit the file.

Hands-On Exercise, Example 15.6: Idle_Shaft (Software Used: Creo Parametric; Figure 15.17)

Steps:

- Create a part file and rename it as Idle_Shaft. (Other names may be given, but it is advisable to provide a relevant name.)
- Select Solid on the right-hand side of the box. This will activate the solid mode and environment.

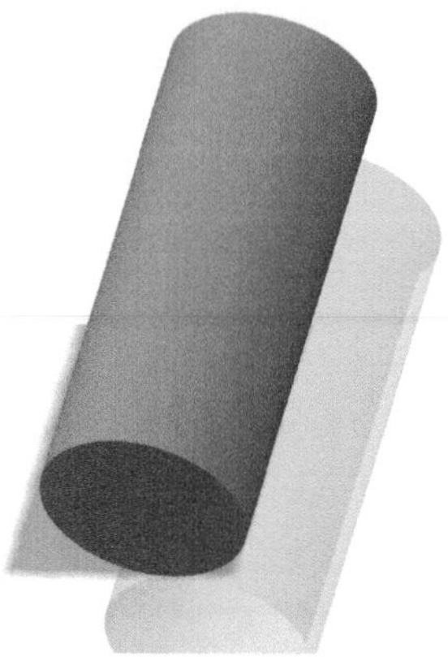

FIGURE 15.17
Three-dimensional view of Idle Shaft.

- Uncheck the box of the default template and click OK to create a new solid file.
- The template selection box will appear.
- Select mmns_part_Solid. This will allow the use of the solid template file of the following units:
    - mm—millimeter for length unit
    - n—newton for force unit
    - s—second for time unit
- Once the solid part is created, go to File > Prepare > Model properties to check the units. In the upper sections, the units of the model will be displayed.
    - Note: The units are usually displayed in the summary section of any CAD software. The setting of the units is usually in the tools or settings.
- Click on Extrude from the Model tab in the ribbon interface.
- Click on the placement tab and select Define to define the sketch.
- Select the front plane as the sketch plane.
- The right plane shall be automatically taken as the reference and orientation as right.
- Click on Sketch.
- The sketcher model will be activated.
- Draw a circle of diameter 40 placing the center point at the intersection of the horizontal and vertical references.
- Click OK to create and exit the sketch.
- Select Mid plane extrusion.
- Enter the value 100 mm for the extrusion depth.
- Click on OK to create the extrusion.
- Save and exit the file.

Hands-On Exercise, Example 15.7: Journal_1 (Software Used: Creo Parametric; Figure 15.18)

Steps:

- Create a part file and rename it as Journal_1. (Other names may be given, but it is advisable to provide a relevant name.)
- Select Solid on the right-hand side of the box. This will activate the solid mode and environment.
- Uncheck the box of the default template and click OK to create a new solid file.
- The template selection box will appear.
- Select mmns_part_Solid. This will allow the use of the solid template file of the following units:
    - mm—millimeter for length unit
    - n—newton for force unit
    - s—second for time unit
- Once the solid part is created, go to File > Prepare > Model properties to check the units. In the upper sections, the units of the model will be displayed.
    - Note: The units are usually displayed in the summary section of any CAD software. The setting of the units is usually in the tools or settings.
- Click on Extrude from the Model tab in the ribbon interface.
- Click on the Placement tab and select Define to define the sketch.
- Select the front plane as the sketch plane.
- The right plane shall be automatically taken as the reference and orientation as right.
- Click on Sketch.
- The sketcher model will be activated.
- Draw the sketch as shown in Figure 15.19.
- Click OK to create and exit the sketch.
- Select Mid plane extrusion.
- Enter the value 16 mm for the extrusion depth.
- Click on OK to create the extrusion.
- Save and exit the file.

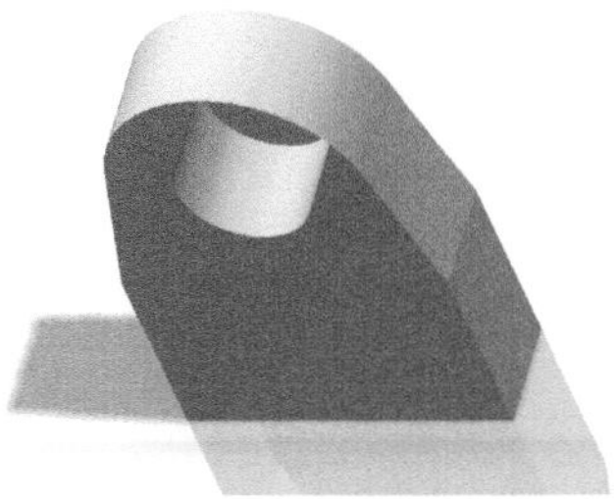

FIGURE 15.18
Three-dimensional view of a journal.

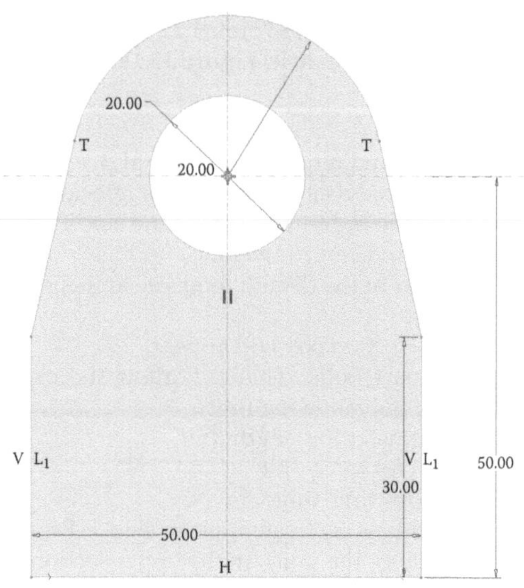

FIGURE 15.19
Constructional sketch of a journal.

Hands-On Exercise, Example 15.8: Hex_Bolt_M10 (Software Used: Creo Parametric; Figure 15.20)

Steps:

- Create a part file and rename it as Hex_Bolt_M10. (Other names may be given, but it is advisable to provide a relevant name.)
- Select Solid on the right-hand side of the box. This will activate the solid mode and environment.
- Uncheck the box of the default template and click OK to create a new solid file.
- The template selection box will appear.

FIGURE 15.20
Three-dimensional model of Hex Bolt of size M10.

- Select mmns_part_Solid. This will allow the use of the solid template file of the following units:
  - mm—millimeter for length unit
  - n—newton for force unit
  - s—second for time unit
- Once the solid part is created, go to File > Prepare > Model properties to check the units. In the upper sections, the units of the model will be displayed.
  - Note: The units are usually displayed in the summary section of any CAD software. The setting of the units is usually in the tools or settings.
- Click on Extrude from the Model tab in the ribbon interface.
- Click on the Placement tab and select Define to define the sketch.
- Select the top plane as the sketch plane.
- The right plane shall be automatically taken as the reference and orientation as right.
- Click on Sketch.
- The sketcher model will be activated.
- Draw a circle of diameter 10 placing the center at the intersection of the horizontal and vertical references.
- Click OK to create and exit the sketch.
- Flip the material orientation towards down.
- Enter the value 30 mm for the extrusion depth.
- Click on OK to create the extrusion.

- Click on Extrude from the Model tab in the ribbon interface.
- Click on the Placement tab and select Define to define the sketch.
- Select the front plane as the sketch plane.
- The right plane shall be automatically taken as the reference and orientation as right.
- Click on Sketch.
- The sketcher model will be activated.
- Draw the sketch as shown in Figure 15.21.
- Click OK to create and exit the sketch.
- Select Mid plane extrusion.
- Enter the value 8 mm for the extrusion depth.
- Click on OK to create the extrusion.

- Click on Revolve from the Model tab in the ribbon interface.
- Click on the Placement tab and select Define to define the sketch.
- Activate the material removal feature option.
- Select the front plane as the sketch plane.
- The right plane shall be automatically taken as the reference and orientation as right.
- Click on Sketch.
- The sketcher model will be activated.
- Draw the sketch as displayed in Figure 15.13.

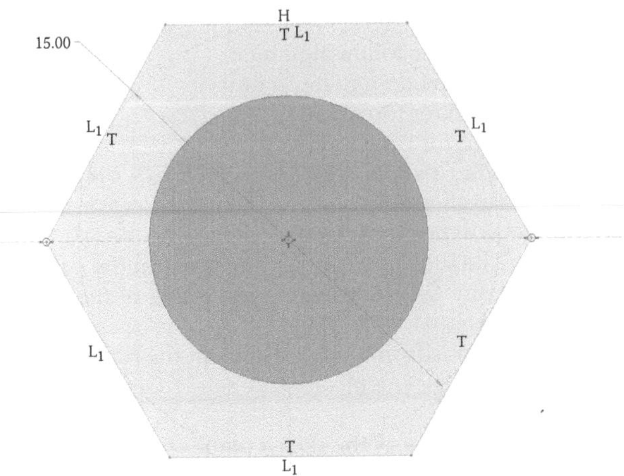

FIGURE 15.21
Constructional sketch of the head of Hex Bolt of size M10.

- Click OK to create and exit the sketch.
- Ensure that the revolve angle shall be 360°.
- Click on OK to create the revolve feature.
- Save and exit the file.

Hands-On Exercise, Example 15.9: Nut_M_10 (Software Used: Creo Parametric)

Steps:

- Create a part file and rename it as NUT_M_10. (Other names may be given, but it is advisable to provide a relevant name.)
- Select Solid on the right-hand side of the box. This will activate the solid mode and environment.
- Uncheck the box of the default template and click OK to create a new solid file.
- The template selection box will appear.
- Select mmns_part_Solid. This will allow the use of the solid template file of the following units:
    - mm—millimeter for length unit
    - n—newton for force unit
    - s—second for time unit
- Once the solid part is created, go to File > Prepare > Model properties to check the units. In the upper sections, the units of the model will be displayed.
    - Note: The units are usually displayed in the summary section of any CAD software. The setting of the units is usually in the tools or settings.
- Click on Extrude from the Model tab in the ribbon interface.

- Click on the Placement tab and select Define to define the sketch.
- Select the front plane as the sketch plane.
- The right plane shall be automatically taken as the reference and orientation as right.
- Click on Sketch.
- The sketcher model will be activated.
- Draw the sketch as shown in Figure 15.22.
- Click OK to create and exit the sketch.
- Select Mid plane extrusion.
- Enter the value 12 mm for the extrusion depth.
- Click on OK to create the extrusion.

- Click on Revolve from the Model tab in the ribbon interface.
- Click on the Placement tab and select Define to define the sketch.
- Activate the material removal feature option.
- Select the front plane as the sketch plane.
- The right plane shall be automatically taken as the reference and orientation as right.
- Click on Sketch.
- The sketcher model will be activated.
- Draw the sketch as displayed in Figure 15.13.

- Click OK to create and exit the sketch.
- Ensure that the revolve angle shall be 360°.
- Click on OK to create the revolve feature.

- Click on Hole from the Model tab of the ribbon interface.
- Click on Activate the simple hole.
- Set depth to Through all.

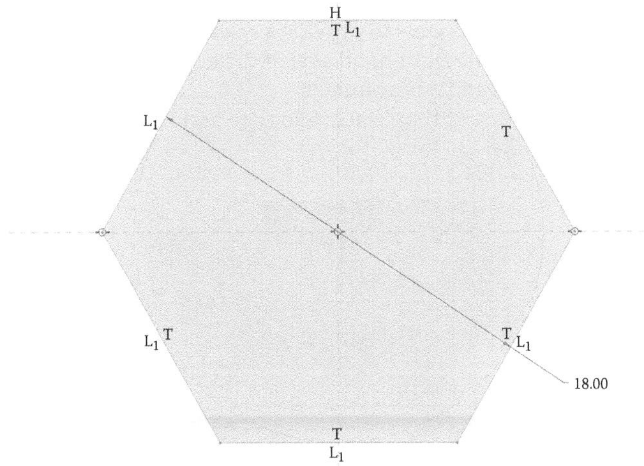

FIGURE 15.22
Constructional sketch of Nut of size M10.

- Enter the hole diameter as 10 mm.
- Select the top flat surface of the nut.
- Click on To select offset reference and select both the front and right planes as offset references.
- Set aligned in the offset value setting box instead of distance.
- Click OK to create the hole.

- Click on the tools tab in the ribbon interface.
- Click on Relations.
- Relations window will appear.
- Click on local parameters to display the local parameters dialog box.
- Add a parameter and rename it as Hole Size.
- Keep the type as real number.
- Enter the hole size parameter value as 10.
- Add the following relations:
  - d4 = 1.5*hole_size
  - d3 = hole_size
  - d48 = hole_size
- Click on Verify on the right upper side of the relation window.
- A notification box with a message "Relations have been successfully verified" should appear.
- Click OK to close the box.
- Save and exit the file.

Hands-On Exercise, Example 15.10: Motor (Software Used: Creo Parametric; Figure 15.23)

Steps:

- Create a part file and rename it as Motor. (Other names may be given, but it is advisable to provide a relevant name.)
- Select Solid on the right-hand side of the box. This will activate the solid mode and environment.
- Uncheck the box of the default template and click OK to create a new solid file.

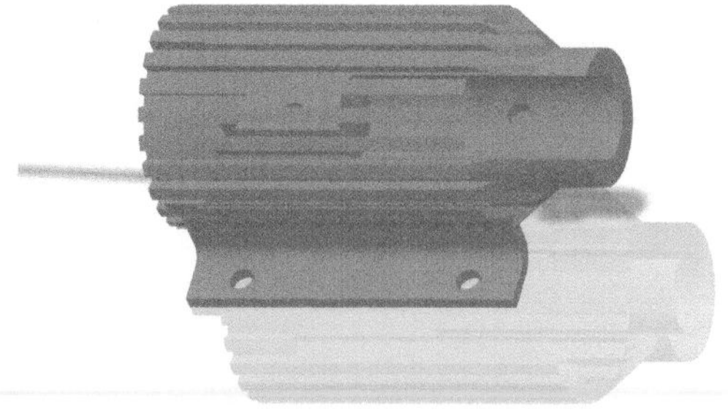

FIGURE 15.23
Three-dimensional model of the Motor.

- The template selection box will appear.
- Select mmns_part_Solid. This will allow the use of the solid template file of the following units:
  - mm—millimeter for length unit
  - n—newton for force unit
  - s—second for time unit
- Once the solid part is created, go to File > Prepare > Model properties to check the units. In the upper sections, the units of the model will be displayed.
  - Note: The units are usually displayed in the summary section of any CAD software. The setting of the units is usually in the tools or settings.
- Click on Revolve from the Model tab in the ribbon interface.
- Click on the Placement tab and select Define to define the sketch.
- Select the front plane as the sketch plane.
- The right plane shall be automatically taken as the reference and orientation as right.
- Click on Sketch.
- The sketcher model will be activated.
- Draw the sketch as displayed in Figure 15.24.
- Click OK to create and exit the sketch.
- Ensure that the revolve angle shall be 360°.
- Click on OK to create the revolve feature.

- Click on Extrude from the Model tab in the ribbon interface.
- Click on the Placement tab and select Define to define the sketch.
- Select the large circular face as the sketch plane.
- The references shall be automatically taken.
- Click on Sketch.

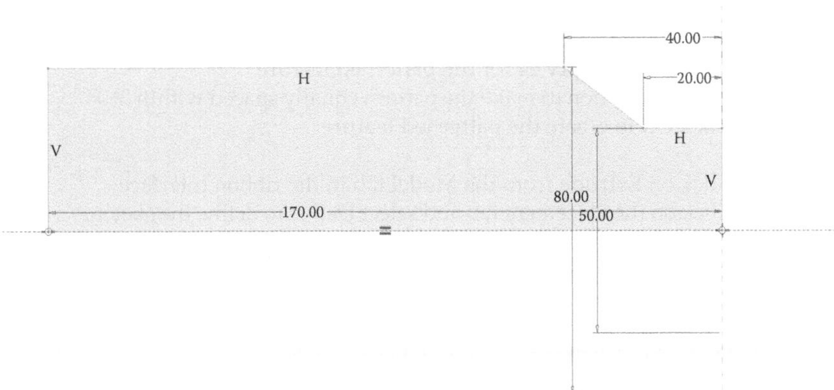

FIGURE 15.24
Constructional sketch of the Motor.

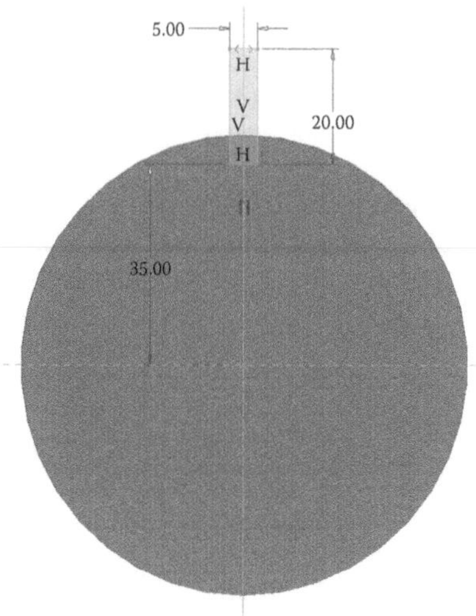

FIGURE 15.25
Constructional sketch of extrusion for creating ribs on the Motor.

- The sketcher model will be activated.
- Draw the sketches as shown in Figure 15.25.

- Click OK to create and exit the sketch.
- Select Through all option for the extrusion depth.
- Activate the material removal option.
- Flip the orientation towards the material side.
- Click on OK to create the extrusion.

- Select the last extrusion.
- Right-click and select Pattern.
- Select Axis in the type of pattern selection field.
- Select the central axis.
- Enter the quantity 24 for the patterned feature.
- Select the option to make the pattern equally spaced within 360°.
- Click OK to create the patterned feature.

- Click on Extrude from the Model tab in the ribbon interface.
- Click on the Placement tab and select Define to define the sketch.
- Select the large circular face as the sketch plane.
- The references shall be automatically taken.
- Click on Sketch.
- The sketcher model will be activated.
- Draw the sketches as shown in Figure 15.26.
- Click OK to create and exit the sketch.
- Select the Upto surface option.
- Select the reference as shown in Figure 15.27.
- Click on OK to create the extrusion.

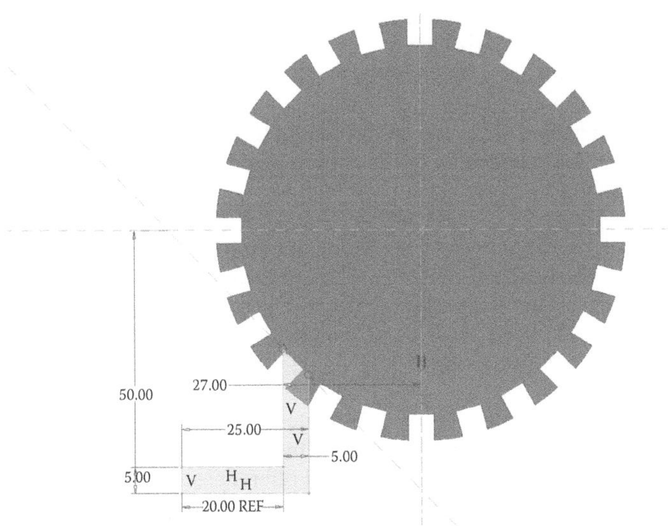

FIGURE 15.26
Constructional sketch of base plate on the Motor.

FIGURE 15.27
Constructional view of a feature of the Motor.

- Click on the round feature from the Model tab of the ribbon interface.
- Set the value of fillet to 5 mm.
- Select the internal edge of the newly created feature.
- Click OK to create the feature.

- Click on the round feature from the Model tab of ribbon interface.
- Set the value of the fillet to 10 mm.
- Select the external edge of the newly created feature.
- Click OK to create the feature.

- Select both the round feature and the last extrusion feature by pressing the control key.
- Right-click and select Group to create the group.
- Select the group.
- Select Mirror from the model tree of the ribbon interface.
- Select the plane front as the mirror plane.
- Click OK to create the mirror feature.

- Click on Extrude from the Model tab in the ribbon interface.
- Click on the Placement tab and select Define to define the sketch.
- Select the small circular face as the sketch plane.
- The right plane shall be automatically taken as the reference and orientation as right.
- Click on Sketch.
- The sketcher model will be activated.
- Draw a circle of diameter 30 placing the center at the intersection of the horizontal and vertical references.
- Click OK to create and exit the sketch.
- Activate the material removal option.
- Flip the orientation towards the material side.
- Enter the value 100 mm for the extrusion depth.
- Click on OK to create the extrusion.

- Click on Extrude from the Model tab in the ribbon interface.
- Click on the Placement tab and select Define to define the sketch.
- Select the top face of the bottom extrude 2 as the sketch plane.
- The right plane shall be automatically taken as the reference and orientation as right.
- Click on Sketch.

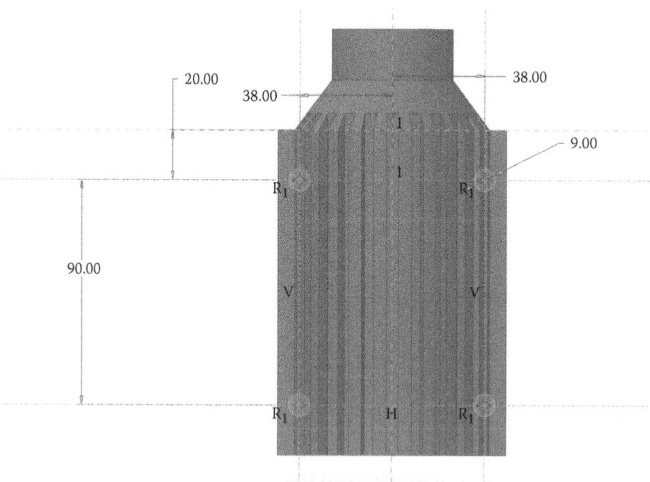

FIGURE 15.28
Constructional sketch for creation of hole in base plate of the Motor.

- The sketcher model will be activated.
- Draw the sketch as shown in Figure 15.28.
- Click OK to create and exit the sketch.
- Activate the material removal option.
- Flip the orientation towards the material side.
- Select Through all as the extrusion depth.
- Click on OK to create the extrusion.
- Save and exit the file.

Hands-On Exercise, Example 15.11: Motor Base Plate (Software Used: Creo Parametric; Refer to Figure 15.29)

Steps:

- Create a part file and rename it as Motor Base Plate. (Other names may be given, but it is advisable to provide a relevant name.)
- Select Solid on the right-hand side of the box. This will activate the solid mode and environment.
- Uncheck the box of the default template and click OK to create a new solid file.
- The template selection box will appear.
- Select mmns_part_Solid. This will allow the use of the solid template file of the following units:
    - mm—millimeter for length unit
    - n—newton for force unit
    - s—second for time unit
- Once the solid part is created, go to File > Prepare > Model properties to check the units. In the upper sections, the units of the model will be displayed.
    - Note: The units are usually displayed in the summary section of any CAD software. The setting of the units is usually in the tools or settings.
- Click on Extrude from the Model tab in the ribbon interface.

FIGURE 15.29
Three-dimensional view of the Motor Base Plate.

- Click on the Placement tab and select Define to define the sketch.
- Select the top plane as the sketch plane.
- The right plane shall be automatically taken as the reference and orientation as right.
- Click on Sketch.
- The sketcher model will be activated.
- Draw the sketch as shown in Figure 15.30.
- Click OK to create and exit the sketch.
- Flip the material orientation towards down.
- Enter the value 10 mm for the extrusion depth.
- Click on OK to create the extrusion.

- Click on Extrude from the Model tab in the ribbon interface.
- Click on the Placement tab and select Define to define the sketch.
- Select the top face as the sketch plane.
- The right plane shall be automatically taken as the reference and orientation as right.
- Click on Sketch.
- The sketcher model will be activated.
- Draw the sketch as shown in Figure 15.31.
- Click OK to create and exit the sketch.
- Activate the material removal option.
- Flip the orientation towards the material down.
- Click on OK to create the extrusion.

- Save and exit the file.

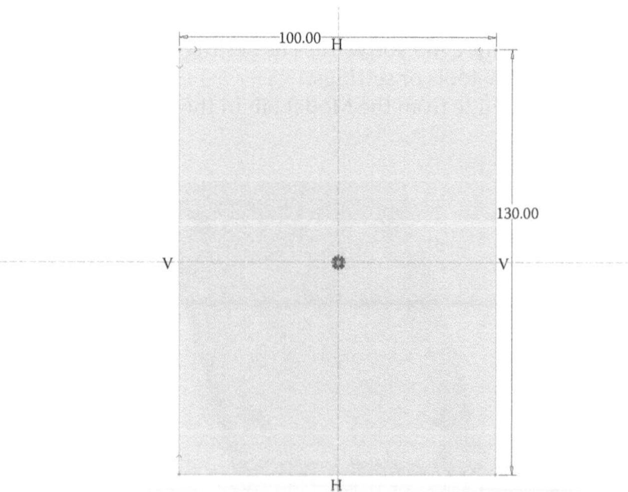

FIGURE 15.30
Constructional sketch of the Motor Base Plate.

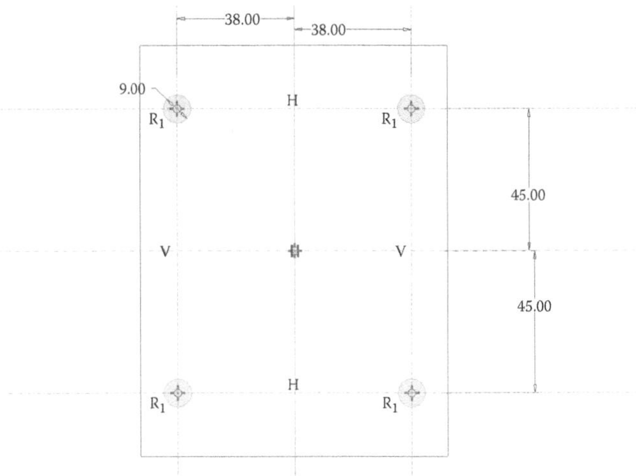

FIGURE 15.31
Constructional sketch of creating holes in Motor Base Plate.

Hands-On Exercise, Example 15.12: Motor_Pulley (Software Used: Creo Parametric)

Steps:

- Click on File > Open and open the file name gear_dia_100.
- Click on Save > Save as.
- Type the name Pulley_Shaft.
- Click OK to save as copy.
- Save and exit the file.

Hands-On Exercise, Example 15.13: Nut_M_12 (Software Used: Creo Parametric)

Steps:

- Create a part file and rename it as NUT. (Other names may be given, but it is advisable to provide a relevant name.)
- Select Solid on the right-hand side of the box. This will activate the solid mode and environment.
- Uncheck the box of the default template and click OK to create a new solid file.
- The template selection box will appear.
- Select mmns_part_Solid. This will allow the use of the solid template file of the following units:
  - mm—millimeter for length unit
  - n—newton for force unit
  - s—second for time unit

- Once the solid part is created, go to File > Prepare > Model properties to check the units. In the upper sections, the units of the model will be displayed.
  - Note: The units are usually displayed in the summary section of any CAD software. The setting of the units is usually in the tools or settings.
- Click on Extrude from the Model tab in the ribbon interface.
- Click on the Placement tab and select Define to define the sketch.
- Select the front plane as the sketch plane.
- The right plane shall be automatically taken as the reference and orientation as right.
- Click on Sketch.
- The sketcher model will be activated.
- Draw the sketch as shown in Figure 15.32.

- Click OK to create and exit the sketch.
- Select Mid plane extrusion.
- Enter the value 12 mm for the extrusion depth.
- Click on OK to create the extrusion.

- Click on Revolve from the Model tab in the ribbon interface.
- Click on the Placement tab and select Define to define the sketch.
- Activate the material removal feature option.
- Select the front plane as the sketch plane.
- The right plane shall be automatically taken as the reference and orientation as right.
- Click on Sketch.
- The sketcher model will be activated.
- Draw the sketch as displayed in Figure 15.13.

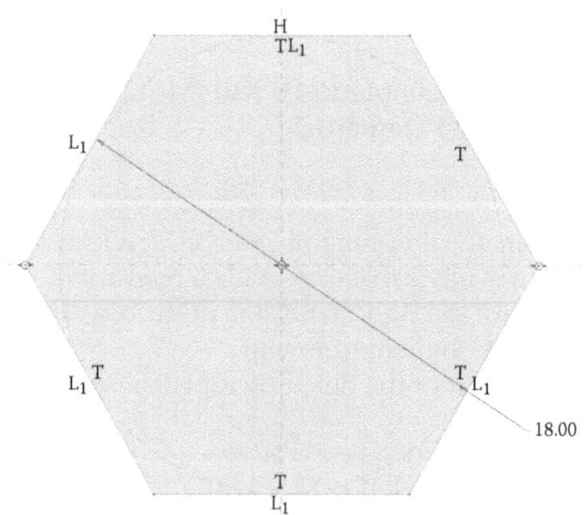

FIGURE 15.32
Constructional sketch of Nut of size M12.

- Click OK to create and exit the sketch.
- Ensure that the revolve angle shall be 360°.
- Click on OK to create the revolve feature.

- Click on Hole from the Model tab of the ribbon interface.
- Click on Activate the simple hole.
- Set depth to Through all.
- Enter the hole diameter as 12 mm.
- Select the top flat surface of the nut.
- Click on To select offset reference and select both front and right planes as the offset references.
- Set aligned in the offset value setting box instead of distance.
- Click OK to create the hole.

- Click on the tools tab in the ribbon interface.
- Click on Relations.
- The relation window will appear.
- Click on Local parameters to display the local parameters dialog box.
- Add a parameter and rename it as Hole Size.
- Keep the type as real number.
- Enter the hole size parameter value as 12.
- Add the following relations:
  - d4 = 1.5*hole_size
  - d3 = hole_size
  - d48 = hole_size
- Click on Verify on the right upper side of the relation window.
- A notification box with a message "Relations have been successfully verified" should appear.
- Click OK to close the box.
- Save and exit the file.

Hands-On Exercise, Example 15.14: Rail_ProfileC (Software Used: Creo Parametric; Figure 15. 33)

Steps:

- Create a part file and rename it as Rail_ProfileC. (Other names may be given, but it is advisable to provide a relevant name.)
- Select Solid on the right-hand side of the box. This will activate the solid mode and environment.
- Uncheck the box of the default template and click OK to create a new solid file.
- The template selection box will appear.
- Select mmns_part_Solid. This will allow the use of the solid template file of the following units:
  - mm—millimeter for length unit
  - n—newton for force unit
  - s—second for time unit

FIGURE 15.33
Three-dimensional view of the Rail Profile.

- Once the solid part is created, go to File > Prepare > Model properties to check the units. In the upper sections, the units of the model will be displayed.
  - Note: The units usually are displayed in the summary section of any CAD software. The setting of the units is usually in the tools or settings.
- The Rail_ProfileC as shown in Figure 15.33 is to be created.
- Click on Extrude from the Model tab of the ribbon interface.
- Select the thicken material option.
- Enter 5 as the thickness.
- Click on the Placement tab and select Define to define the sketch.
- Select the front plane as the sketch plane.
- The right plane shall be automatically taken as the reference and orientation as right.
- Click on Sketch.
- The sketcher model will be activated.
- Draw the sketch as shown in Figure 15.34.
- Click OK to create and exit the sketch.
- Select the Mid plane extrusion option.
- Enter 5000 as the extrusion depth.
- Click OK to create the extrusion.

- Click on the round feature from the Model tab of the ribbon interface.
- Set the value of fillet to 5 mm.
- Select all six internal edges of the newly created feature.
- Click OK to create the feature.

- Click on the round feature again from the Model tab of the ribbon interface.
- Set the value of fillet to 10 mm.

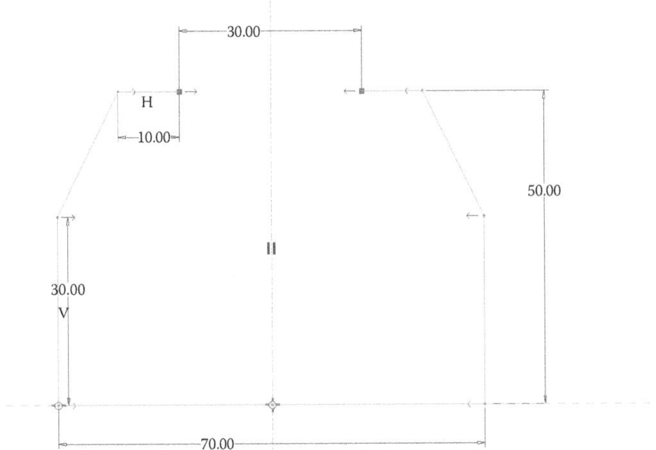

FIGURE 15.34
Constructional sketch of the Rail Profile.

- Select all six external edges of the newly created feature.
- Click OK to create the feature.

- Save and exit the file.

Hands-On Exercise, Example 15.15: Stud_M_12 (Software Used: Creo Parametric)

Steps:

- Create a part file and rename it as Stud. (Other names may be given, but it is advisable to provide a relevant name.)
- Select Solid on the right-hand side of the box. This will activate the solid mode and environment.
- Uncheck the box of the default template and click OK to create a new solid file.
- The template selection box will appear.
- Select mmns_part_Solid. This will allow the use of the solid template file of the following units:
  - mm—millimeter for length unit
  - n—newton for force unit
  - s—second for time unit
- Once the solid part is created, go to File > Prepare > Model properties to check the units. In the upper sections, the units of the model will be displayed.
  - Note: The units are usually displayed in the summary section of any CAD software. The setting of the units is usually in the tools or settings.

- Click on Extrude from the Model tab in the ribbon interface.
- Click on the Placement tab and select Define to define the sketch.
- Select the front plane as the sketch plane.
- The right plane shall be automatically taken as the reference and orientation as right.
- Click on Sketch.
- The sketcher model will be activated.
- Draw a circle of diameter 12 placing the center at the intersection of the horizontal and vertical references.
- Click OK to create and exit the sketch.
- Select Mid plane extrusion.
- Enter the value 70 mm for the extrusion depth.
- Click on OK to create the extrusion.

- Click on Chamfer from the Model tab of the ribbon interface.
- Select the circular edge from both sides.
- Enter 0.5 as the chamfer value with D × D configuration.
- Click OK to create the chamfer.
- Save and exit the file.

Hands-On Exercise, Example 15.16: Wheel (Software Used: Creo Parametric; Refer to Figure 15.35)

Steps:

- Create a part file and rename it as Wheel. (Other names may be given, but it is advisable to provide a relevant name.)
- Select Solid on the right-hand side of the box. This would activate the solid mode and environment.
- Uncheck the box of the default template and click OK to create a new solid file.

FIGURE 15.35
Three-dimensional view of the Wheel.

- The template selection box will appear.
- Select mmns_part_Solid. This will allow the use of the solid template file of the following units:
  - mm—millimeter for length unit
  - n—newton for force unit
  - s—second for time unit
- Once the solid part is created, go to File > Prepare > Model properties to check the units. In the upper sections, the units of the model will be displayed.
  - Note: The units are usually displayed in the summary section of any CAD software. The setting of the units is usually in the tools or settings.
- The Nozzle Support Wheel as shown in Figure 15.35 is to be created.
- Click on Revolve from the Model tab in the ribbon interface.
- Click on the Placement tab and select Define to define the sketch.
- Select the front plane as the sketch plane.
- The right plane shall be automatically taken as the reference and orientation as right.
- Click on Sketch.
- The sketcher model will be activated.
- Draw the sketch as displayed in Figure 15.36.
- Click OK to create and exit the sketch.
- Ensure that the revolve angle shall be 360°.
- Click on OK to create the revolve feature.

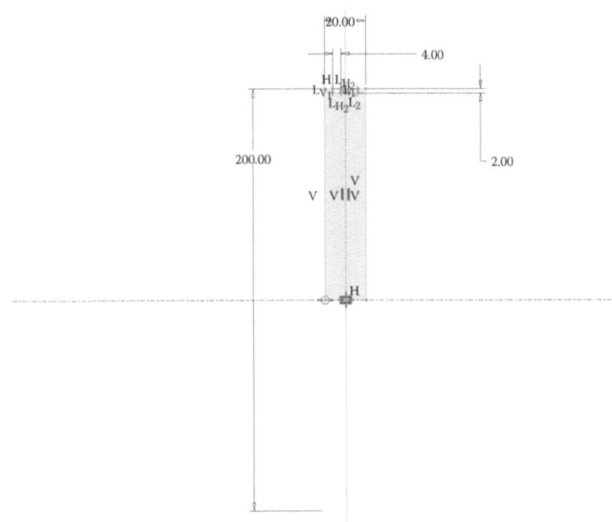

FIGURE 15.36
Constructional sketch of the Wheel.

- Click on Extrude from the Model tab in the ribbon interface.
- Click on the Placement tab and select Define to define the sketch.
- Select the front face as the sketch plane.
- The right plane shall be automatically taken as the reference and orientation as right.
- Click on Sketch.
- The sketcher model will be activated.
- Draw a circle of diameter 40 placing the center point at the intersection of the horizontal and vertical references.
- Click OK to create and exit the sketch.
- Select the Through all option.
- Activate the material removal option.
- Flip the orientation towards the material side.
- Click on OK to create the extrusion.

- Click on Extrude from the Model tab in the ribbon interface.
- Click on the Placement tab and select Define to define the sketch.
- Select the front face as the sketch plane.
- The right plane shall be automatically taken as the reference and orientation as right.
- Click on Sketch.
- The sketcher model will be activated.
- Draw the sketch as shown in Figure 15.37.
- Click OK to create and exit the sketch.
- Select the Through all option.
- Activate the material removal option.
- Flip the orientation towards the material side.
- Click on OK to create the extrusion.

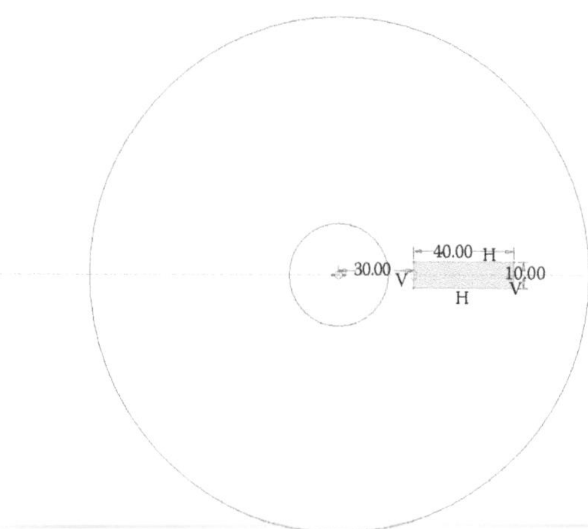

FIGURE 15.37
Constructional sketch of a feature of the Wheel.

- Click on the round feature from the Model tab of the ribbon interface.
- Set the value of fillet to 5 mm.
- Select all four internal edges of the newly created feature.
- Click OK to create the feature.

- Select the feature extrude 2 and the last round by pressing the control key.
- Right-click and select Group to create a group feature.

- Select the newly created group in the model tree.
- Right-click and select Pattern.
- Select Axis in the type of pattern selection field.
- Select the central axis.
- Enter the quantity 8 for the patterned feature.
- Select the option to make the pattern equally spaced within 360°.
- Click OK to create the patterned feature.

- Click on Extrude from the Model tab in the ribbon interface.
- Click on the Placement tab and select Define to define the sketch.
- Activate the thicken option.
- Enter 10 as the thickness.
- Select the front face of the wheel as the sketch plane.
- The right plane shall be automatically taken as the reference and orientation as right.
- Click on Sketch.
- The sketcher model will be activated.
- Draw a circle of diameter 170 placing the center at the intersection of the horizontal and vertical references.
- Click OK to create and exit the sketch.
- Activate the material removal option.
- Flip the orientation towards the material side.
- Enter the value 4 mm for the extrusion depth.
- Click on OK to create the extrusion.

- Select the extrusion, and click on the mirror from the Model tab of the ribbon interface.
- Select the front plane as the mirror plane.
- Click OK to mirror the extrusion cut feature.
- Save and exit the file.

Hands-On Exercise, Example 15.17: Wheel Base Plate (Software Used: Creo Parametric)

Steps:

- Create a part file and rename it as Wheel Base Plate. (Other names may be given, but it is advisable to provide a relevant name.)
- Select Solid on the right-hand side of the box. This will activate the solid mode and environment.

- Uncheck the box of the default template and click OK to create a new solid file.
- The template selection box will appear.
- Select mmns_part_Solid. This will allow the use of the solid template file of the following units:
    - mm—millimeter for length unit
    - n—newton for force unit
    - s—second for time unit
- Once the solid part is created, go to File > Prepare > Model properties to check the units. In the upper sections, the units of the model will be displayed.
    - Note: The units are usually displayed in the summary section of any CAD software. The setting of the units is usually in the tools or settings.
- Click on Extrude from the Model tab in the ribbon interface.
- Click on the Placement tab and select Define to define the sketch.
- Select the top plane as the sketch plane.
- The right plane shall be automatically taken as the reference and orientation as right.
- Click on Sketch.
- The sketcher model will be activated.
- Draw the sketch as shown in Figure 15.38.
- Click OK to create and exit the sketch.
- Flip the material orientation towards down.
- Enter the value 10 mm for the extrusion depth.
- Click on OK to create the extrusion.

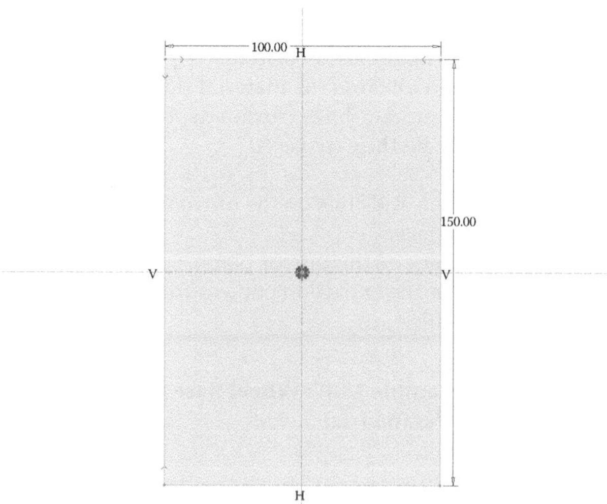

FIGURE 15.38
Constructional sketch of a feature of the Wheel Base Plate.

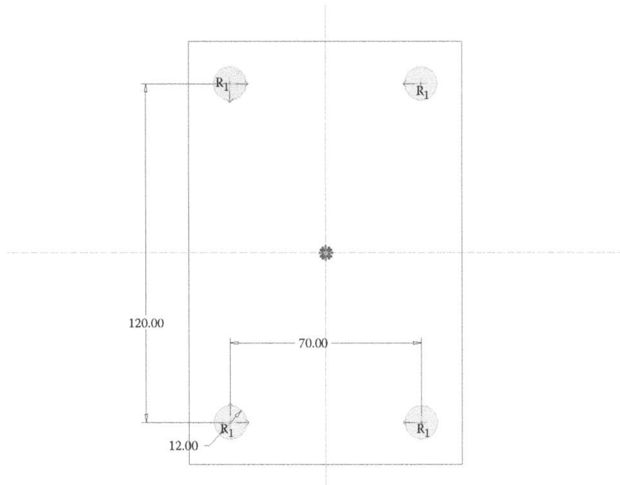

FIGURE 15.39
Constructional sketch of creating holes in the Wheel Base Plate.

- Click on Extrude from the Model tab in the ribbon interface.
- Click on the Placement tab and select Define to define the sketch.
- Select the top face as the sketch plane.
- The right plane shall be automatically taken as the reference and orientation as right.
- Click on Sketch.
- The sketcher model will be activated.
- Draw the sketch as shown in Figure 15.39.
- Click OK to create and exit the sketch.
- Activate the material removal option.
- Flip the orientation towards the material down.
- Click on OK to create the extrusion.

- Save and exit the file.

Hands-On Exercise, Example 15.18: Wheel_Bracket (Software Used: Creo Parametric; Refer to Figure 15.40)

Steps:

- Create a part file and rename it as Wheel_Bracket. (Other names may be given, but it is advisable to provide a relevant name.)
- Select Solid on the right-hand side of the box. This will activate the solid mode and environment.
- Uncheck the box of the default template and click OK to create a new solid file.
- The template selection box will appear.

FIGURE 15.40
Three-dimensional view of the Wheel bracket.

- Select mmns_part_Solid. This will allow the use of the solid template file of the following units:
  - mm—millimeter for length unit
  - n—newton for force unit
  - s—second for time unit
- Once the solid part is created, go to File > Prepare > Model properties to check the units. In the upper sections, the units of the model will be displayed.
  - Note: The units are usually displayed in the summary section of any CAD software. The setting of the units is usually in the tools or settings.
- Click on Extrude from the Model tab in the ribbon interface.
- Click on the Placement tab and select Define to define the sketch.
- Select the front plane as the sketch plane.
- The right plane shall be automatically taken as the reference and orientation as right.
- Click on Sketch.
- The sketcher model will be activated.
- Draw the sketch as shown in Figure 15.41.
- Click OK to create and exit the sketch.
- Enter the value 5 mm for the extrusion depth.
- Click on OK to create the extrusion.

- Click on Extrude from the Model tab in the ribbon interface.
- Click on the Placement tab and select Define to define the sketch.
- Select the front plane as the sketch plane.
- The right plane shall be automatically taken as the reference and orientation as right.
- Click on Sketch.
- The sketcher model will be activated.
- Take the reference of the circular arc.

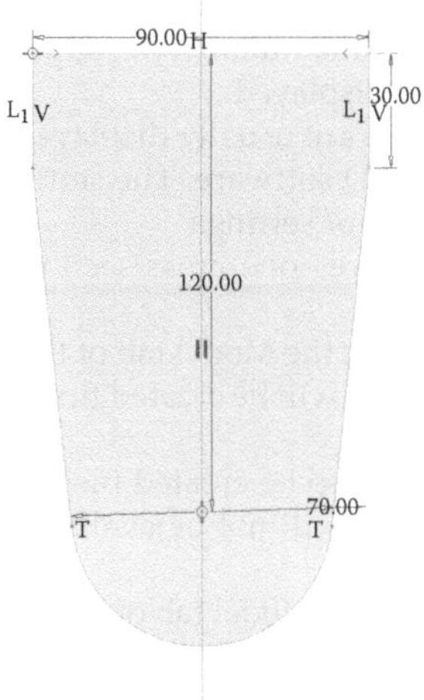

FIGURE 15.41
Constructional sketch of a feature of the Wheel bracket.

- Draw the circle of diameter 40 placing the center point at the arc center.
- Click OK to create and exit the sketch.
- Activate the material removal option.
- Select Through all option.
- Flip the orientation towards the material side.
- Click on OK to create the extrusion.

- Save and exit the file.

Hands-On Exercise, Example 15.19: U-Bolt_25 NB (Software Used: Creo Parametric)

Steps:

- Create a part file and rename it as U-Bolt_25 NB. (Other names may be given, but it is advisable to provide a relevant name.)
- Select Solid on the right-hand side of the box. This will activate the solid mode and environment.
- Uncheck the box of the default template and click OK to create a new solid file.
- The template selection box will appear.
- Select mmns_part_Solid. This will allow the use of the solid template file of the following units:
    - mm—millimeter for length unit
    - n—newton for force unit
    - s—second for time unit

- Once the solid part is created, go to File > Prepare > Model properties to check the units. In the upper sections, the units of the model will be displayed.
  - Note: The units are usually displayed in the summary section of any CAD software. The setting of the units is usually in the tools or settings.
- Principally it requires one cross-sectional sweep profile and sweep trajectory.
- Click on Sweep from the Model tab of the ribbon interface.
- The sweep trajectory will be created first followed by the sweep operation.
- The trajectory can also be created inside the sweep operation; however, the sketch will not be available if the sweep feature is deleted.
- Select Sketch from the Model tab of the ribbon interface.
- A sketch window for selection of the sketch plane will appear.
- Select the front plane as the sketch plane.
- The right plane shall automatically be selected as the reference and the orientation is right.
- Click on Sketch.
- The sketcher window will open up, and a reference selection window appears for selection of references.
- The two references, i.e., the top plane as horizontal reference and the right plane as vertical reference, shall automatically be taken; if not, please select the same options for references.
- Create a sketch as displayed in Figure 15.42.
- Click OK to create and exit the sketch.
- Click on Sweep from the Model tab of the ribbon interface.

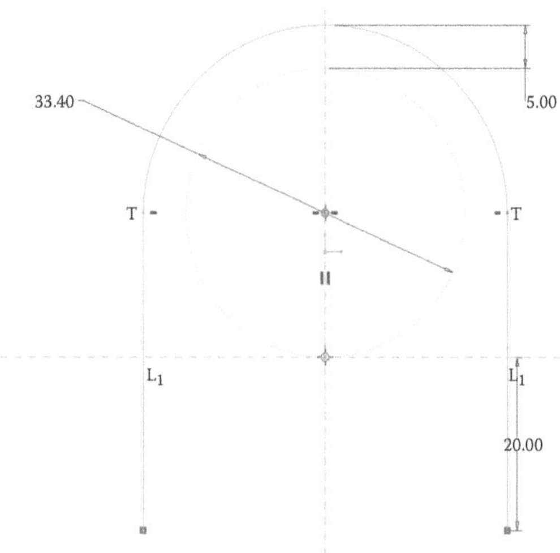

FIGURE 15.42
Constructional sketch of U-Bolt_25 NB.

- Graphically select the sketch as the trajectory.
- The system assumes the profile sketch point and shows it by displaying the directional arrow.
- Graphically click on the arrows of the feature creation to draw the sweep profile in other control points.
- Click on Sketch to sketch the profile.
- The sketcher mode is activated with the sketcher environment.
- Draw a circle in such a way that the profile should be at the outer edge of the circle as shown in Figure 15.43.
- Change the diameter to 6 mm.
- Click OK to create and exit the sketch.
- Click OK to finish the sweep.

- Click on Sketch from the Model tab in the ribbon interface.
- Click on the Placement tab and select Define to define the sketch.
- Select the front plane as the sketch plane.
- The right plane shall be automatically taken as the reference and orientation as right.
- Click on Sketch.
- The sketcher model will be activated.
- In the reference entity box, click on the internal circular edges of diameter 33.4.
- Place a geometry point in the center of the reference circle.
- Click OK to create and exit the sketch.

- Click on the datum axis from the Model tab of the ribbon interface.
- Select the point and the front plane by pressing the control key.
- Click OK to create the axis.

- Save and exit the file.

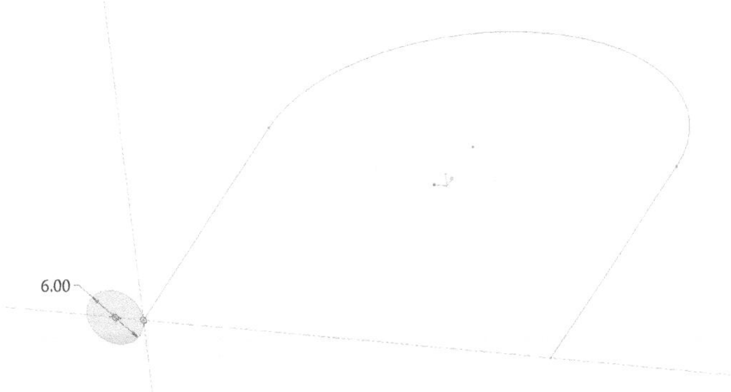

FIGURE 15.43
Constructional sketch of sweep profile in U-Bolt_25 NB.

Hands-On Exercise, Example 15.20: Pulley_Shaft (Software Used: Creo Parametric)

Steps:

- Create a part file and rename it as Pulley_Shaft. (Other names may be given, but it is advisable to provide a relevant name.)
- Select Solid on the right-hand side of the box. This would activate the solid mode and environment.
- Uncheck the box of the default template and click OK to create a new solid file.
- The template selection box will appear.
- Select mmns_part_Solid. This will allow the use of the solid template file of the following units:
    - mm—millimeter for length unit
    - n—newton for force unit
    - s—second for time unit
- Once the solid part is created, go to File > Prepare > Model properties to check the units. In the upper sections, the units of the model will be displayed.
    - Note: The units are usually displayed in the summary section of any CAD software. The setting of the units is usually in the tools or settings.
- Click on Extrude from the Model tab in the ribbon interface.
- Click on the Placement tab and select Define to define the sketch.
- Select the front plane as the sketch plane.
- The right plane shall be automatically taken as the reference and orientation as right.
- Click on Sketch.
- The sketcher model will be activated.
- Draw a circle of diameter 20 placing the center point at the intersection of the horizontal and vertical references.
- Click OK to create and exit the sketch.
- Select blind extrusion on both sides.
- Enter the value 50 mm for the extrusion depth for both sides.
- Click on OK to create the extrusion.
- Save and exit the file.

Hands-On Exercise, Example 15.21: Rail_Profile_C1 (Software Used: Creo Parametric)

Steps:

- Create a part file and rename it as Rail_Profile_C1. (Other names may be given, but it is advisable to provide a relevant name.)
- Select Solid on the right-hand side of the box. This will activate the solid mode and environment.

- Uncheck the box of the default template and click OK to create a new solid file.
- The template selection box will appear.
- Select mmns_part_Solid. This will allow the use of the solid template file of the following units:
    - mm—millimeter for length unit
    - n—newton for force unit
    - s—second for time unit
- Once the solid part is created, go to File > Prepare > Model properties to check the units. In the upper sections, the units of the model will be displayed.
    - Note: The units are usually displayed in the summary section of any CAD software. The setting of the units is usually in the tools or settings.
- Click on Extrude from the Model tab of the ribbon interface.
- Select the Thicken material option.
- Enter 3 as the thickness.
- Click on the Placement tab and select Define to define the sketch.
- Select the front plane as the sketch plane.
- The right plane shall be automatically taken as the reference and orientation as right.
- Click on Sketch.
- The sketcher model will be activated.
- Draw the sketch as shown in Figure 15.44.
- Click OK to create and exit the sketch.
- Select the Mid plane extrusion option.
- Enter 4935 as the extrusion depth.
- Click OK to create the extrusion.

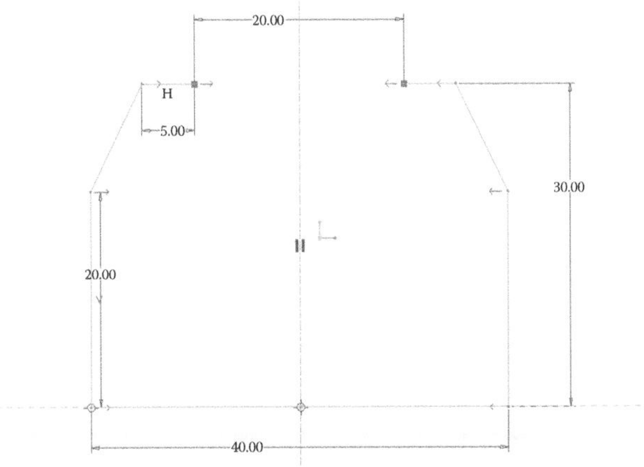

FIGURE 15.44
Constructional sketch of the Rail Profile.

- Click on the round feature from the Model tab of the ribbon interface.
- Set the value of fillet to 3 mm.
- Select all six internal edges of the newly created feature.
- Click OK to create the feature.

- Click on the round feature again from the Model tab of the ribbon interface.
- Set the value of fillet to 6 mm.
- Select all six external edges of the newly created feature.
- Click OK to create the feature.

- Save and exit the file.

Hands-On Exercise, Example 15.22: Nozzle (Software Used: Creo Parametric; Refer to Figure 15.45)

Steps:

- Create a part file and rename it as Nozzle. (Other names may be given, but it is advisable to provide a relevant name.)
- Select Solid on the right-hand side of the box. This will activate the solid mode and environment.
- Uncheck the box of the default template and click OK to create a new solid file.
- The template selection box will appear.

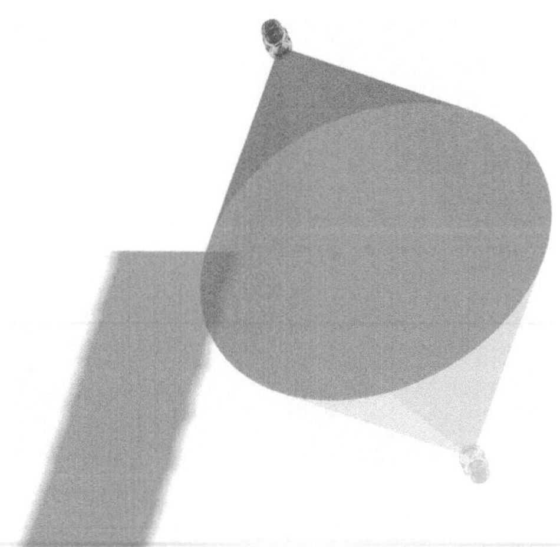

FIGURE 15.45
Three-dimensional view of the Spray Nozzle.

- Select mmns_part_Solid. This will allow the use of the solid template file of the following units:
  - mm—millimeter for length unit
  - n—newton for force unit
  - s—second for time unit
- Once the solid part is created, go to File > Prepare > Model properties to check the units. In the upper sections, the units of the model will be displayed.
  - Note: The units are usually displayed in the summary section of any CAD software. The setting of the units is usually in the tools or settings.
- The nozzle as shown in Figure 15.45 is to be created.
- Click on Extrude from the Model tab in the ribbon interface.
- Click on the Placement tab and select Define to define the sketch.
- Select the top plane as the sketch plane.
- The right plane shall be automatically taken as the reference and orientation as right.
- Click on Sketch.
- The sketcher model will be activated.
- Draw the sketch as shown in Figure 15.46.
- Click OK to create and exit the sketch.
- Flip the material orientation towards down.
- Enter the value 6 mm for the extrusion depth.
- Click on OK to create the extrusion.

- Click on Revolve from the Model tab in the ribbon interface.
- Click on the Placement tab and select Define to define the sketch.
- Select the front plane as the sketch plane.
- The right plane shall be automatically taken as the reference and orientation as right.
- Click on Sketch.
- The sketcher model will be activated.
- Draw the sketch as displayed in Figure 15.47.

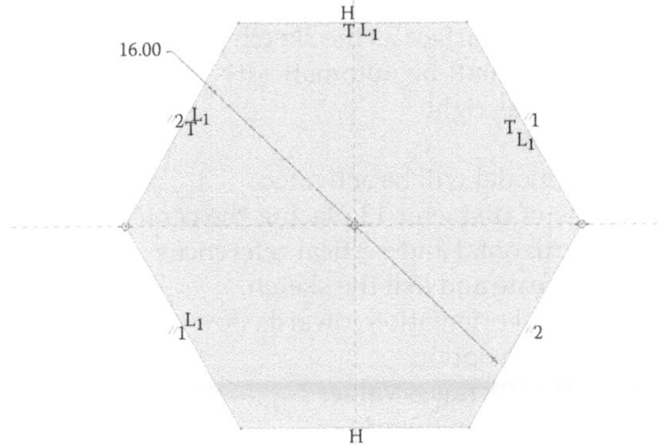

FIGURE 15.46
Constructional sketch of a hexagonal extrusion of the Spray Nozzle.

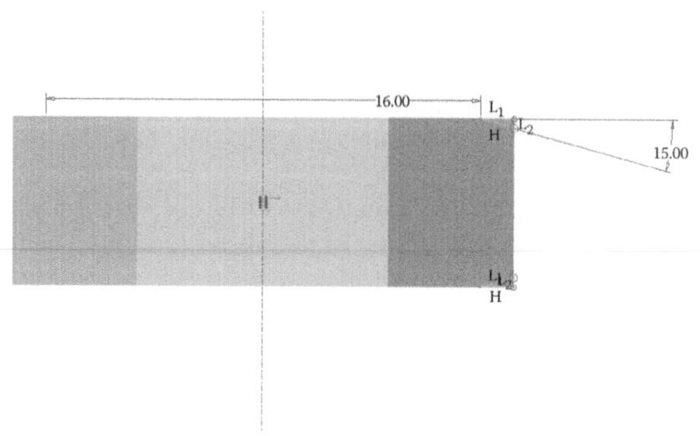

FIGURE 15.47
Constructional sketch of a revolve material removal feature of the Spray Nozzle.

- Click OK to create and exit the sketch.
- Ensure that the revolve angle shall be 360°.
- Activate the material removal option.
- Click on OK to create the revolve feature.

- Click on Revolve again from the Model tab in the ribbon interface.
- Click on the Placement tab and select Define to define the sketch.
- Select the front plane as the sketch plane.
- The right plane shall be automatically taken as the reference and orientation as right.
- Click on Sketch.
- The sketcher model will be activated.
- Draw the sketch as displayed in Figure 15.48.
- Click OK to create and exit the sketch.
- Ensure that the revolve angle shall be 360°.
- Click on OK to create the revolve feature.

- Click on Extrude from the Model tab in the ribbon interface.
- Click on the Placement tab and select Define to define the sketch.
- Select the bottom surface as the sketch plane.
- The right plane shall be automatically taken as the reference and orientation as right.
- Click on Sketch.
- The sketcher model will be activated.
- Draw a circle of diameter 13 placing the center at the intersection of the horizontal and vertical references.
- Click OK to create and exit the sketch.
- Flip the material orientation towards down.
- Select the thicken option.
- Enter 0.5 as the thickness value.
- Flip the orientation outward.
- Enter the value 1 mm for the extrusion depth.
- Activate the material removal option.
- Click on OK to create the extrusion.

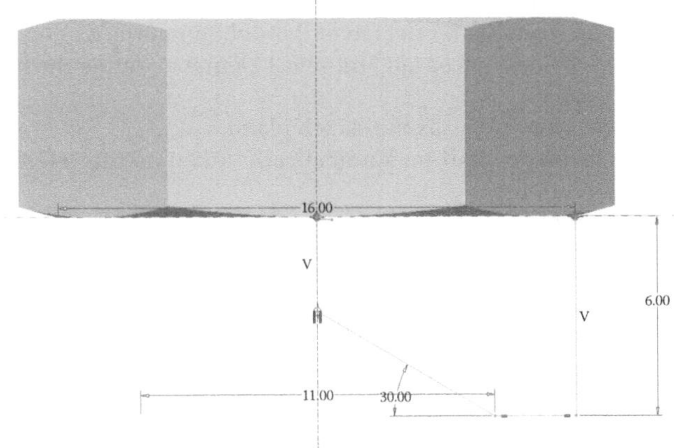

FIGURE 15.48
Constructional sketch of revolve feature of the Spray Nozzle.

- Click on Revolve again from the Model tab in the ribbon interface.
- Click on the Placement tab and select Define to define the sketch.
- Select the front plane as the sketch plane.
- The right plane shall be automatically taken as the reference and orientation as right.
- Click on Sketch.
- The sketcher model will be activated.
- Draw the sketch as displayed in Figure 15.49.
- Click OK to create and exit the sketch.
- Ensure that the revolve angle shall be 360°.
- Click on ok to create the revolve feature.

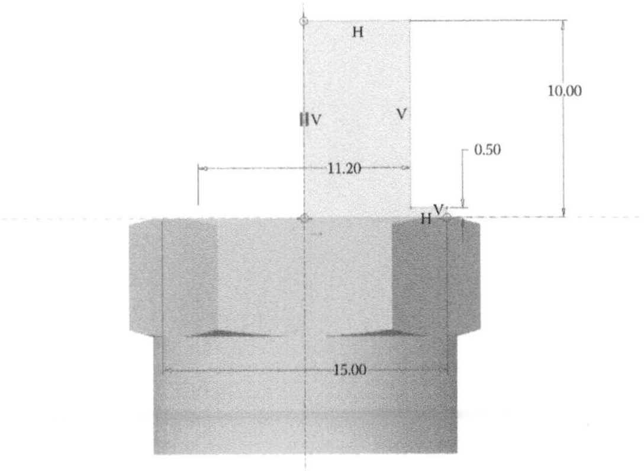

FIGURE 15.49
Constructional sketch of evolve feature for upper section of the Spray Nozzle.

- Click on the helical sweep option from the drop-down menu of the Sweep from the Model tab of the ribbon interface.
- Click on the references tab and select Define to define the helix sweep profile.
- Select the front plane as the sketch plane.
- The right plane shall be automatically taken as the reference and orientation as right.
- Click on Sketch.
- The sketcher model will be activated.
- Draw a geometry centerline on the vertical reference line in the middle.
- Project the internal vertical edge.
- Click OK to create the sketch and exit the sketcher window.
- Click Sketch to create the profile sketch above the reference tab.
- Draw the sketch as shown in Figure 15.50.
- Click OK to create and exit the sketch.
- Enter the pitch value of 1 mm.
- Click on the left-handed rule icon to create the helical spring clockwise downward.
- Click OK to create the thread.

- Click on the round feature from the Model tab of the ribbon interface.
- Set the value of fillet to 0.1 mm.
- Select the edge of the minimum diameter in the internal thread.
- Click OK to create the feature.

- Click on Revolve from the Model tab in the ribbon interface.
- Click on the Placement tab and select Define to define the sketch.
- Select the front plane as the sketch plane.

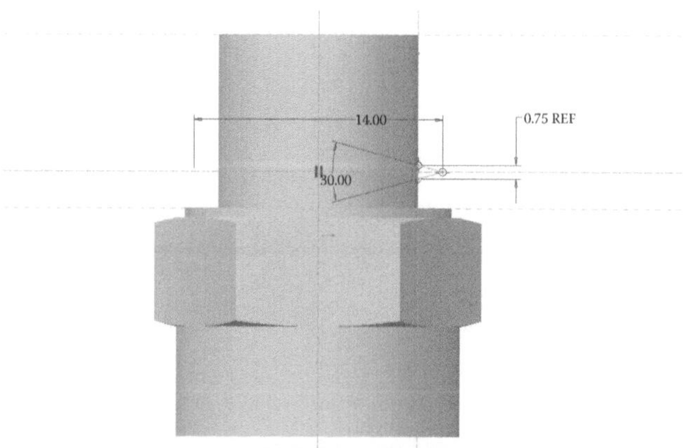

FIGURE 15.50
Constructional sketch for thread profile of the Spray Nozzle.

- The right plane shall be automatically taken as the reference and orientation as right.
- Click on Sketch.
- The sketcher model will be activated.
- Draw the sketch as displayed in Figure 15.51.
- Click OK to create and exit the sketch.
- Ensure that the revolve angle shall be 360°.
- Activate the material removal option.
- Click on OK to create the revolve feature.

- Click on the round feature from the Model tab of the ribbon interface.
- Set the value of fillet to 0.1 mm.
- Select the thread edge of the maximum diameter.
- Click OK to create the feature.

- Click on Revolve again from the Model tab in the ribbon interface.
- Click on the Placement tab and select Define to define the sketch.
- Select the front plane as the sketch plane.
- The right plane shall be automatically taken as the reference and orientation as right.
- Click on Sketch.
- The sketcher model will be activated.
- Draw the sketch as displayed in Figure 15.52.
- Click OK to create and exit the sketch.
- Ensure that the revolve angle shall be 360°.
- Click on OK to create the revolve feature.

- Save and exit the file.

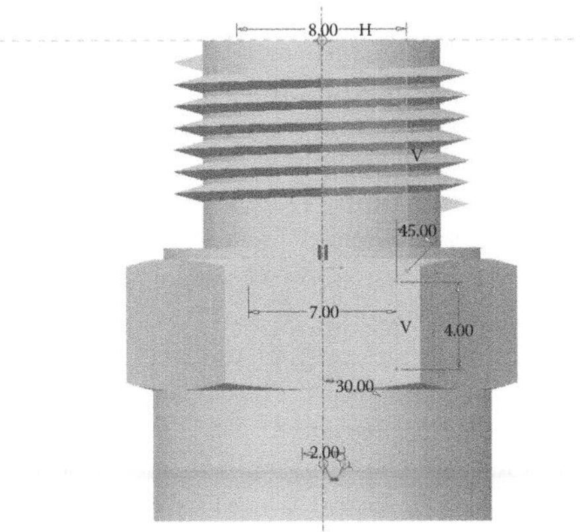

FIGURE 15.51
Constructional sketch of a central revolved material removal feature.

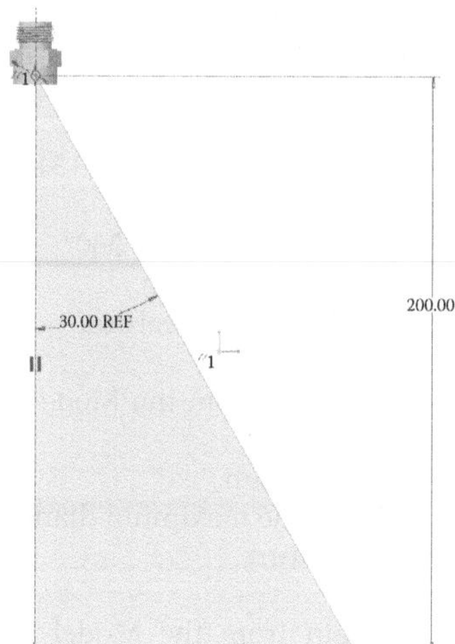

FIGURE 15.52
Constructional sketch of the Spray Nozzle.

Hands-On Exercise, Example 15.23: Motor_Shaft (Software Used: Creo Parametric)

Steps:

- Create a part file and rename it as Motor_Shaft. (Other names may be given, but it is advisable to provide a relevant name.)
- Select Solid on the right-hand side of the box. This will activate the solid mode and environment.
- Uncheck the box of the default template and click OK to create a new solid file.
- The template selection box will appear.
- Select mmns_part_Solid. This will allow the use of the solid template file of the following units:
    - mm—millimeter for length unit
    - n—newton for force unit
    - s—second for time unit
- Once the solid part is created, go to File > Prepare > Model properties to check the units. In the upper sections, the units of the model will be displayed.
    - Note: The units are usually displayed in the summary section of any CAD software. The setting of the units is usually in the tools or settings.
- Click on Extrude from the Model tab in the ribbon interface.
- Click on the Placement tab and select Define to define the sketch.
- Select the front plane as the sketch plane.
- The right plane shall be automatically taken as the reference and orientation as right.

- Click on Sketch.
- The sketcher model will be activated.
- Draw a circle of diameter 30 placing the center point at the intersection of the horizontal and vertical references.
- Click OK to create and exit the sketch.
- Select Mid plane extrusion.
- Enter the value 60 mm for the extrusion depth.
- Click on OK to create the extrusion.
- Save and exit the file.

Hands-On Exercise, Example 15.24: Creating Washer (Software Used: Creo Parametric)

Steps:

- Create a part file and rename it as Washer. (Other names may be given, but it is advisable to provide a relevant name.)
- Select Solid on the right-hand side of the box. This will activate the solid mode and environment.
- Uncheck the box of the default template and click OK to create a new solid file.
- The template selection box will appear.
- Select mmns_part_Solid. This will allow the use of the solid template file of the following units:
  - mm—millimeter for length unit
  - n—newton for force unit
  - s—second for time unit
- Once the solid part is created, go to File > Prepare > Model properties to check the units. In the upper sections, the units of the model will be displayed.
  - Note: The units are usually displayed in the summary section of any CAD software. The setting of the units is usually in the tools or settings.
- Click on Extrude from the Model tab in the ribbon interface.
- Click on the Placement tab and select Define to define the sketch.
- Select the front plane as the sketch plane.
- The right plane shall be automatically taken as the reference and orientation as right.
- Click on Sketch.
- The sketcher model will be activated.
- Draw two concentric circles placing the center at the intersection of the horizontal and vertical references.
- Change the diameter of the circle to 30 and 20, respectively.
- Click OK to create and exit the sketch.
- Select Mid plane extrusion.
- Enter the value 2 mm for the extrusion depth.
- Click on OK to create the extrusion.
- Save and exit the file.

Hands-On Exercise, Example 15.25: Nozzle_Support_Shaft (Software Used: Creo Parametric)

Steps:

- Create a part file and rename it as Nozzle_Support_Shaft. (Other names may be given, but it is advisable to provide a relevant name.)
- Select Solid on the right-hand side of the box. This will activate the solid mode and environment.
- Uncheck the box of the default template and click OK to create a new solid file.
- The template selection box will appear.
- Select mmns_part_Solid. This will allow the use of the solid template file of the following units:
  - mm—millimeter for length unit
  - n—newton for force unit
  - s—second for time unit
- Once the solid part is created, go to File > Prepare > Model properties to check the units. In the upper sections, the units of the model will be displayed.
  - Note: The units are usually displayed in the summary section of any CAD software. The setting of the units is usually in the tools or settings.
- Click on Revolve from the Model tab in the ribbon interface.
- Click on the Placement tab and select Define to define the sketch.
- Select the front plane as the sketch plane.

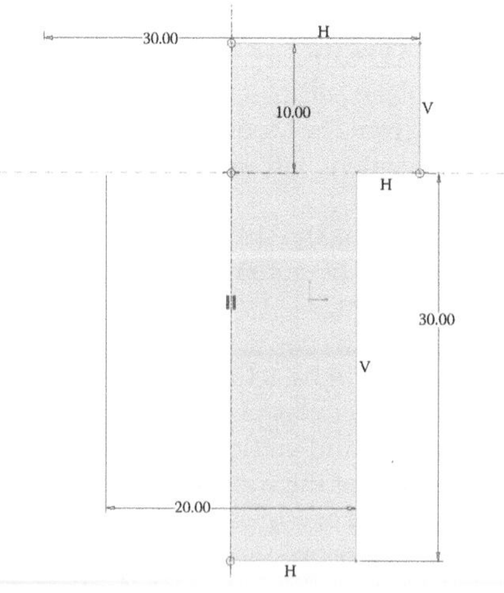

FIGURE 15.53
Constructional sketch of the Nozzle Support Shaft.

- The right plane shall be automatically taken as the reference and orientation as right.
- Click on Sketch.
- The sketcher model will be activated.
- Draw the sketch as displayed in Figure 15.53.
- Click OK to create and exit the sketch.
- Ensure that the revolve angle shall be 360°.
- Click on OK to create the revolve feature.
- Save and exit the file.

Hands-On Exercise, Example 15.26: Nozzle_Support_Wheel (Software Used: Creo Parametric; Refer to Figure 15.54)

Steps:

- Create a part file and rename it as Nozzle_Support_Wheel. (Other names may be given, but it is advisable to provide a relevant name.)
- Select Solid on the right-hand side of the box. This will activate the solid mode and environment.
- Uncheck the box of the default template and click OK to create a new solid file.
- The template selection box will appear.
- Select mmns_part_Solid. This will allow the use of the solid template file of the following units:
    - mm—millimeter for length unit
    - n—newton for force unit
    - s—second for time unit
- Once the solid part is created, go to File > Prepare > Model properties to check the units. In the upper sections, the units of the model will be displayed.

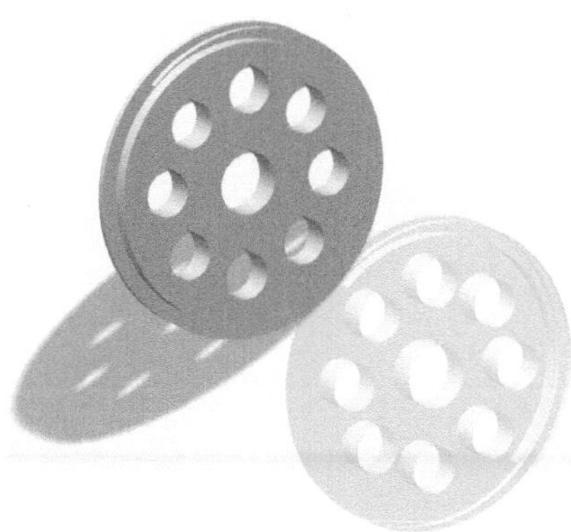

FIGURE 15.54
Three-dimensional view of the Nozzle Support Wheel.

- Note: The units are usually displayed in the summary section of any CAD software. The setting of the units is usually in the tools or settings.
- Click on Revolve from the Model tab in the ribbon interface.
- Click on the Placement tab and select Define to define the sketch.
- Select the front plane as the sketch plane.
- The right plane shall be automatically taken as the reference and orientation as right.
- Click on Sketch.
- The sketcher model will be activated.
- Draw the sketch as displayed in Figure 15.55.
- Click OK to create and exit the sketch.
- Ensure that the revolve angle shall be 360°.
- Click on OK to create the revolve feature.

- Click on Extrude from the Model tab in the ribbon interface.
- Click on the Placement tab and select Define to define the sketch.
- Select the front face as the sketch plane.
- The right plane shall be automatically taken as the reference and orientation as right.
- Click on Sketch.
- The sketcher model will be activated.
- Draw a circle of diameter 20 placing the center point at the intersection of the horizontal and vertical references.
- Click OK to create and exit the sketch.
- Select the Through all option.

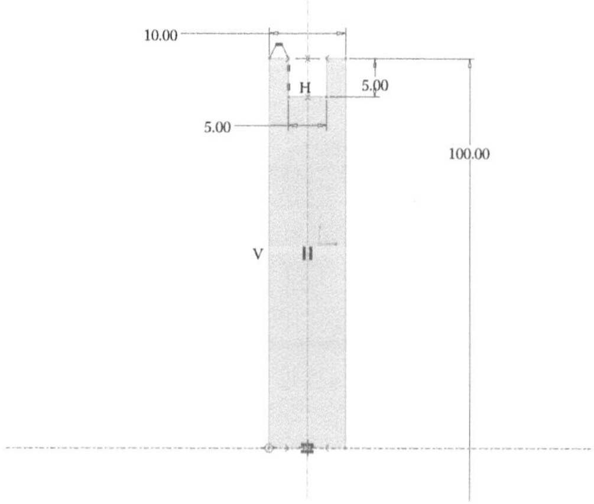

FIGURE 15.55
Constructional sketch of the Nozzle Support Wheel.

- Activate the material removal option.
- Flip the orientation towards the material side.
- Click on OK to create the extrusion.

- Click on Extrude from the Model tab in the ribbon interface.
- Click on the Placement tab and select Define to define the sketch.
- Select the front face as the sketch plane.
- The right plane shall be automatically taken as the reference and orientation as right.
- Click on Sketch.
- The sketcher model will be activated.
- Draw sketch as shown in Figure 15.56.
- Click OK to create and exit the sketch.
- Select the Through all option.
- Activate the material removal option.
- Flip the orientation towards the material side.
- Click on OK to create the extrusion.

- Select the last extrusion feature.
- Right-click and select the pattern.
- Select Axis in the type of pattern selection field.
- Select the central axis.
- Enter the quantity 8 for the patterned feature.
- Select the option to make the pattern equally spaced within 360°.
- Click OK to create the patterned feature.
- Save and exit the file.

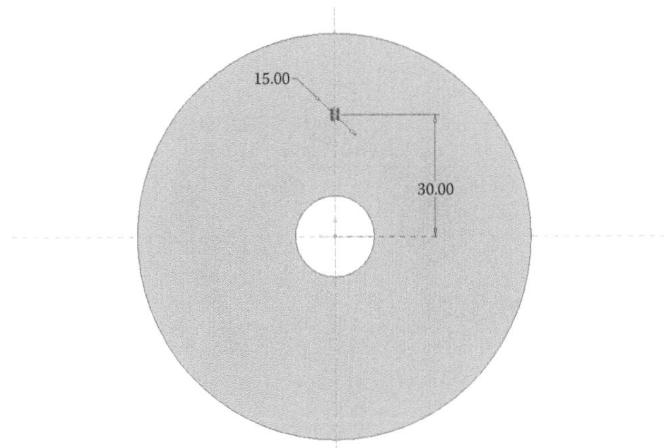

FIGURE 15.56
Constructional sketch of material removal feature of the Nozzle Support Wheel.

Hands-On Exercise, Example 15.27: Bearing_Block (Software Used: Creo Parametric; Figure 15.57)

Steps:

- Create an assembly file and rename it as Bearing_Block.asm. (Other names may be given, but it is advisable to provide a relevant name.)
- Select Design on the right-hand side of the box. This will activate the standard assembly design mode and environment.
- Uncheck the box of the default template and click OK to create a new assembly design file.
- The template selection box will appear.
- Select mmns_asm_Solid. This will allow the use of the solid template file of the following units:
  - mm—millimeter for length unit
  - n—newton for force unit
  - s—second for time unit
- Once the assembly file is created, go to File > Prepare > Model properties to check the units. In the upper sections, the units of the model will be displayed.
  - Note: The units are usually displayed in the summary section of any CAD software. The setting of the units is usually in the tools or settings.
- Click on tree filers as displayed in Figure 15.58.
- Click on the check box displaying the features.
- Click on Apply. Notice that the features are displayed in the model tree.

- Click on Assembly from the Model tab of the ribbon interface.
- Select the file Journal_1.

FIGURE 15.57
Three-dimensional view of the Bearing Block.

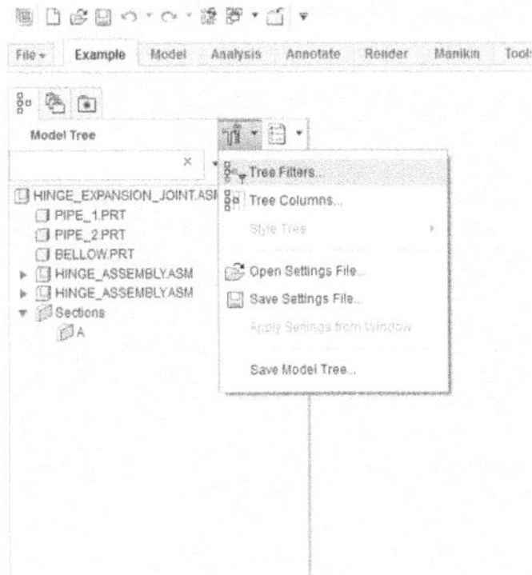

FIGURE 15.58
Model Tree Display.

- Click on the Placement tab.
- Select the front face of the Journal_1 part as the component reference.
- Select the plane ASM_FRONT as the assembly reference.
- Select the distance in the component reference relation field.
- Enter 20 as the distance.
- Right-click on the graphical area and select New constraint.
- Select the plane Right of Journal_1 part as the component reference.
- Select the plane ASM_RIGHT as the assembly reference.
- Select Coincide in the component reference relation field.
- Right-click on the graphical area and select New constraint.
- Select the bottom surface of the Journal_1 part as the component reference.
- Select the plane ASM_TOP as the assembly reference.
- Select Coincide in the component reference relation field.
- Click OK to assemble the component.
- Similarly, assemble another Journal_1 on the other side of the plane ASM_RIGHT.

- Click on Assembly from the Model tab of the ribbon interface.
- Select the file Pulley_Shaft.
- Click on the Placement tab.
- Select the cylindrical surface of Pulley_Shaft as the component reference.
- Select the cylindrical surface of Journal_1 as the assembly reference.
- Select Coincide in the component reference relation field.
- Right-click on the graphical area and select New constraint.
- Select the plane Front of Journal_1 part as the component reference.

- Select the plane ASM_FRONT as the assembly reference.
- Select Coincide in the component reference relation field.
- Click OK to assemble the component.

- Click on Assembly from the Model tab of the ribbon interface.
- Select the file Gear_100_Dia.asm.
- Select Pin in the connection field.
- Click on the Placement tab.
- Select the cylindrical surface of Gear_100_Dia as the component reference.
- Select the cylindrical surface of Pulley_Shaft as the assembly reference.
- Select Coincide in the component reference relation field.
- Right-click on the graphical area and select New constraint.
- Select the plane Front of Gear_100_Dia as the component reference.
- Select the plane Front of Pulley_Shaft as the assembly reference.
- Select Coincide in the component reference relation field.
- Click OK to assemble the component.

- Click on Part create on assembly mode icon from the Model tab of the ribbon interface.
- Change the name as Bearing_Block_Base_Plate.
- Select Solid in the sub type.
- Click OK.
- The creation option window will appear.
- Select the locate default datums option.
- This will allow creating three default datums in the selected references.
- Select Three Planes in the locate datums method area.
- Click OK and select ASM_RIGHT, ASM_TOP, and ASM_FRONT plane, respectively.
- Notice that the part Nozzle_Beam is added and activated.
- The ribbon interface will change to part interface.

- Click on Extrude from the Model tab in the ribbon interface.
- Click on the Placement tab and select Define to define the sketch.
- Select the ASM_TOP plane as the sketch plane.
- The ASM_RIGHT plane shall be automatically taken as the reference and orientation as right.
- Click on Sketch.
- The sketcher model will be activated.
- Draw the sketch as shown in Figure 15.59.
- Click OK to create and exit the sketch.
- Flip the material orientation towards down.
- Enter the value 10 mm for the extrusion depth.
- Click on OK to create the extrusion.
- Right-click on the main assembly and select Activate.

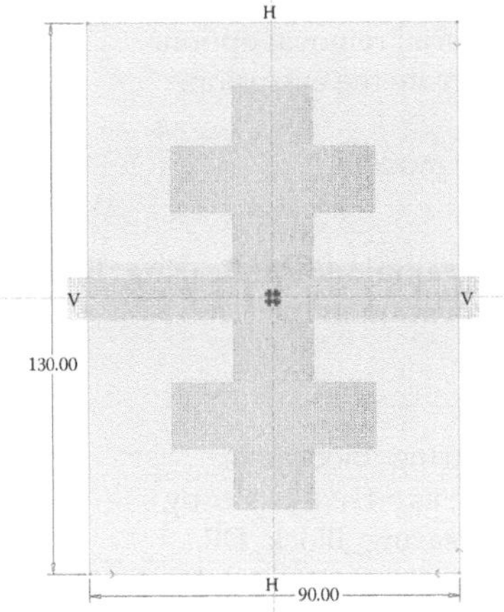

FIGURE 15.59
Constructional sketch of the Bearing Block.

- Click on Extrude from the Model tab in the ribbon interface.
- Click on the Placement tab and select Define to define the sketch.
- Select the ASM_TOP plane as the sketch plane.
- The ASM_RIGHT plane shall be automatically taken as the reference and orientation as right.
- Click on Sketch.
- The sketcher model will be activated.
- Draw the sketch as displayed in Figure 15.60.
- Click OK to create and exit the sketch.

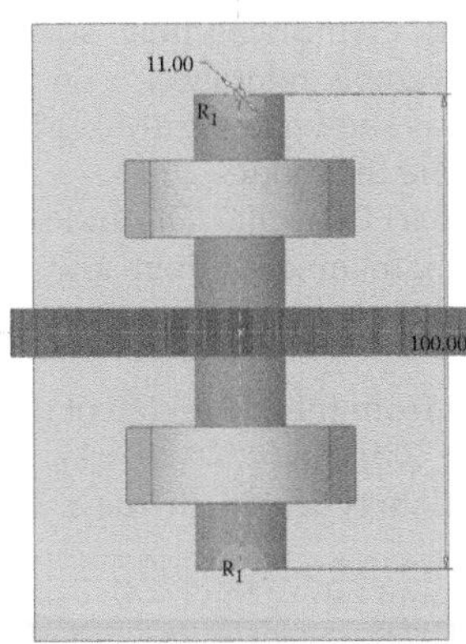

FIGURE 15.60
Constructional sketch of creating holes in the Bearing Block.

- Select Through all option.
- Activate the material removal option.
- Click on OK to create the extrusion.

- Save and exit the model.

Hands-On Exercise, Example 15.28: Bearing_Block_DR (Software Used: Creo Parametric)

Steps:

- Open the file Bearing_Block.asm.
- Click on File save as > Save as a copy.
- Type the name Bearing_Block_DR.
- Click on OK.
- The Save as copy window opens up.
- Select all the components.
- Enter the suffix _DR.
- Click on Generate new names.
- Click on Save a copy.
- Close the file Bearing_Block.asm.
- Open the file Bearing_Block_DR.
- Double click on the part Pulley_Shaft_Dr in the model tree.
- Right-click on the feature extrude and select Edit.
- Double click on the dimension 50 on the left side.
- Enter 90 as the dimension.
- Click on Regenerate.
- Right-click on the Pulley shaft and select Edit definition.
- Delete all the mating conditions.
- Select Pin in the mechanism definition window.
- Select the cylindrical surface of the Pulley_Shaft as the component reference.
- Select the internal cylindrical hole surface of the second Journal_1 as the assembly reference.
- Select the surfaces as shown in Figure 15.61.
- Click OK to assemble the model.
- Right-click on the part Gear_100_Dia and select Edit definition.
- Click on the mechanism definition and change it to user_defined from the pin.

- Click on Assembly from the Model tab of the ribbon interface.
- Select the file Gear_100_Dia.
- Select the connection type pin in the mechanism definition selection field.
- Click on the Placement tab.
- Select the internal cylindrical surface of the Gear_100_Dia as the component reference.
- Select the cylindrical surface of the second Pulley_Shaft as the assembly reference.

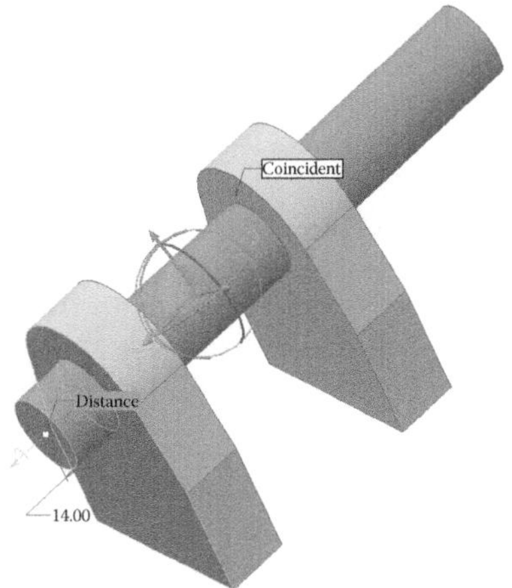

FIGURE 15.61
Constructional view for assembly of shaft in the Bearing Block.

- Select Coincide in the component reference relation field.
- Right-click on the graphical area and select New constraint.
- Select the surface as shown in Figure 15.62.
- Select Distance in the component reference relation field.
- Enter 15 as the distance.
- Click OK to assemble the component.
- Save and exit the file.

FIGURE 15.62
Constructional view for assembly of gear in the Bearing Block.

Hands-On Exercise, Example 15.29: Motor_Assembly (Software Used: Creo Parametric; Figure 15.63)

Steps:

- Create an assembly file and rename it as Motor_Assembly.asm. (Other names may be given, but it is advisable to provide a relevant name.)
- Select Design on the right-hand side of the box. This will activate the standard assembly design mode and environment.
- Uncheck the box of the default template and click OK to create a new assembly design file.
- The template selection box will appear.
- Select mmns_asm_Solid. This will allow the use of the solid template file of the following units:
  - mm—millimeter for length unit
  - n—newton for force unit
  - s—second for time unit
- Once the assembly file is created, go to File > Prepare > Model properties to check the units. In the upper sections, the units of the model will be displayed.
  - Note: The units are usually displayed in the summary section of any CAD software. The setting of the units is usually in the tools or settings.
- Click on tree filers as displayed in Figure 15.58.
- Click on the check box displaying the features.
- Click on Apply. Notice that the features are displayed in the model tree.

FIGURE 15.63
Three-dimensional view of the Motor Assembly.

- Click on Assembly from the Model tab of the ribbon interface.
- Select the file Motor_Base_Plate.
- Click Default in the Placement constraint tab.
- Click OK to assemble the component.

- Click on Assembly from the Model tab of the ribbon interface.
- Select the file Motor.
- Click on the Placement tab.
- Select the bottom face of the Motor part as the component reference.
- Select the top face Motor_Base_Plate as the assembly reference.
- Select Coincide in the component reference relation field.
- Right-click on the graphical area and select New constraint.
- Select any cylindrical hole surface of the Motor_Base_Plate part as the component reference.
- Select the similar cylindrical hole surface of Motor_Base_Plate as the assembly reference.
- Select Coincide in the component reference relation field.
- Right-click on the graphical area and select New constraint.
- Select the other cylindrical hole surface of the Motor_Base_Plate part as the component reference.
- Select the similar cylindrical hole surface of Motor_Base_Plate as the assembly reference.
- Select Coincide in the component reference relation field.
- Click OK to assemble the component.

- Click on Assembly from the Model tab of the ribbon interface.
- Select the file Motor_Shaft.
- Select pin joint in the mechanism definition field.
- Select the plane front of Motor_Shaft as the component reference.
- Select the front surface of Motor as the assembly reference.
- Select Coincide in the component reference relation field.
- Click OK to assemble the component.

- Click on Assembly from the Model tab of the ribbon interface.
- Select the file Motor_Pulley.asm.
- Click on the Placement tab.
- Select the cylindrical surface of Motor_Pulley as the component reference.
- Select the cylindrical surface of Motor_Pulley shaft as the assembly reference.
- Select Coincide in the component reference relation field.
- Right-click on the graphical area and select New constraint.
- Select the surfaces as shown in Figure 15.64.
- Select Distance in the component reference relation field.
- Enter 10 as the distance.
- Click OK to assemble the component.

FIGURE 15.64
Constructional view for assembly of gear in the Motor Assembly.

- Click OK datum plane from the model tab of the ribbon interface.
- Select the front end surface as the reference plane as shown in Figure 15.65.
- Set the offset value 50 mm towards the other end of the motor.
- Click OK to create the datum plane.

- Click on Part create on assembly mode icon from the Model tab of the ribbon interface.
- Change the name as ISMC_100_1.
- Select Solid in the sub type.

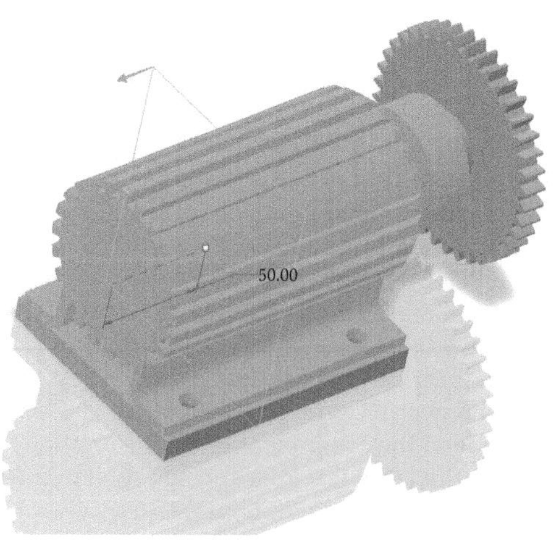

FIGURE 15.65
Constructional view of the Motor Assembly.

- Click OK.
- The creation option window will appear.
- Select Locate default datums option.
- This would allow creating three default datums in the selected references.
- Select Three Planes in the locate datums method area.
- Click OK and select ASM_RIGHT, ASM_TOP, and ASM_FRONT plane, respectively.
- Notice that the part Nozzle_Beam is added and activated.
- The ribbon interface will change to part interface.
- Click on Sketch from the Model tab in the ribbon interface.
- Click on the Placement tab and select Define to define the sketch.
- Select the plane ADTM1 as the sketch plane.
- The ASM_RIGHT plane shall be automatically taken as the reference and orientation as right.
- Click on Sketch.
- The sketcher model will be activated.
- Draw the sketch as shown in Figure 15.66.
- Click OK to create and exit the sketch.

- Click Sweep from the Model tab of the ribbon interface.
- Select the left vertical line of the assembly sketch.
- Click on Create sweep section.
- Notice that sketcher mode is activated.
- Draw the figure as shown in Figure 15.67.
- Click OK to create and exit the sketch.
- Click OK to create the sweep.
- Right-click on the main assembly and select Activate.

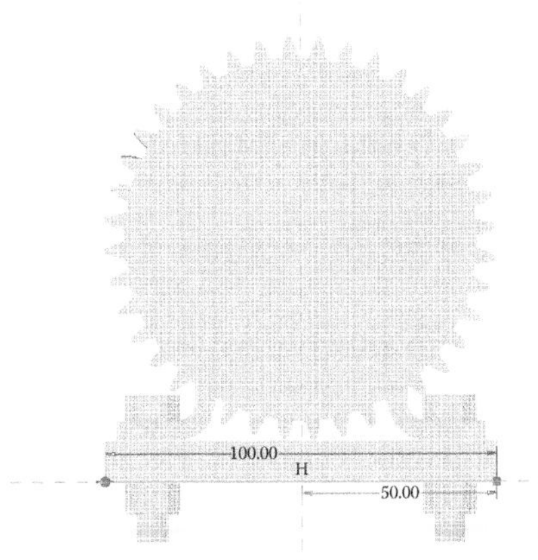

FIGURE 15.66
Constructional sketch of the Motor Assembly.

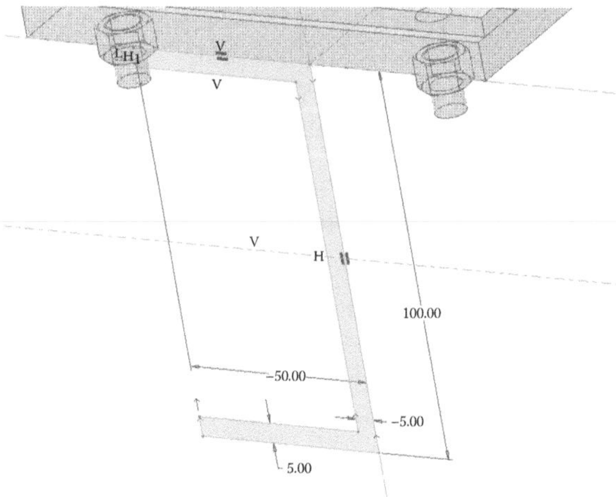

FIGURE 15.67
Constructional sketch of support for the Motor Assembly.

- Click on Assembly from the Model tab of the ribbon interface.
- Select the file Hex_Bolt_M_8.
- Click on the Placement tab.
- Select the cylindrical surface of the Hex_Bolt_M_8 part as the component reference.
- Select the inside cylindrical surface of the Motor base flange as the assembly reference.
- Select Coincide in the component reference relation field.
- Right-click on the graphical area and select New constraint.
- Select the sitting surface of the Hex_Bolt_M_8 part as the component reference.
- Select the top surface of the Motor base flange of the main assembly as the assembly reference.
- Click OK to assemble the component.

- Click on Assembly from the Model tab of the ribbon interface.
- Select the file NUT_M_12.
- Click on the Placement tab.
- Select the cylindrical surface of the NUT_M_12 part as the component reference.
- Select the inside cylindrical surface of the Hex_Bolt_M_12 as the assembly reference.
- Select Coincide in the component reference relation field.
- Right-click on the graphical area and select New constraint.
- Select the sitting surface of the NUT_M_12 part as the component reference.
- Select the bottom surface of the Motor_Base_Plate as the assembly reference.
- Click OK to assemble the component.

- Similarly, assemble Hex_Bolt_M_8 and Nut_M_8 in all other holes of Motor_Base_Plate.
- Save and exit the file.

Hands-On Exercise, Example 15.30: Creating Nozzle_Beam (Software Used: Creo Parametric; Refer to Figure 15.68)

Steps:

- Create an assembly file and rename it as Nozzle_Beam.asm. (Other names may be given, but it is advisable to provide a relevant name.)
- Select Design on the right-hand side of the box. This will activate the standard assembly design mode and environment.
- Uncheck the box of the default template and click OK to create a new assembly design file.
- The template selection box will appear.
- Select mmns_asm_Solid. This will allow the use of the solid template file of the following units:
  - mm—millimeter for length unit
  - n—newton for force unit
  - s—second for time unit
- Once the assembly file is created, go to File > Prepare > Model properties to check the units. In the upper sections, the units of the model will be displayed.
  - Note: The units are usually displayed in the summary section of any CAD software. The setting of the units is usually in the tools or settings.
- Click on tree filers as displayed in Figure 15.68.
- Click on the check box displaying the features.
- Click on Apply. Notice that the features are displayed in the model tree.

FIGURE 15.68
Three-dimensional view of the Nozzle Assembly.

- Click OK datum plane from the Model tab of the ribbon interface.
- Select the ASM_TOP plane as the reference plane.
- Set the offset value 241 mm towards up.
- Click OK to create the datum plane.

- Click on Sketch from the Model tab of the ribbon interface.
- Click on the Placement tab and select Define to define the sketch.
- Select the front plane as the sketch plane.
- The right plane shall be automatically taken as the reference and orientation as right.
- Click on Sketch.
- The sketcher model will be activated.
- Draw the sketch as shown in Figure 15.69.
- Click OK to create and exit the sketch.

- Select the sketch from the model tree.
- Right-click and select Pattern.
- Select Direction in the type of pattern selection field.
- Select the plane ASM_RIGHT.
- Enter the quantity 5 for the patterned feature.
- Enter 220 as the increments.
- Flip the orientation towards right.
- Click OK to create the patterned feature.

- Click on Part create on assembly mode icon from the Model tab of the ribbon interface.
- Change the name as Nozzle_Beam.
- Select Solid in the sub type.
- Click OK.

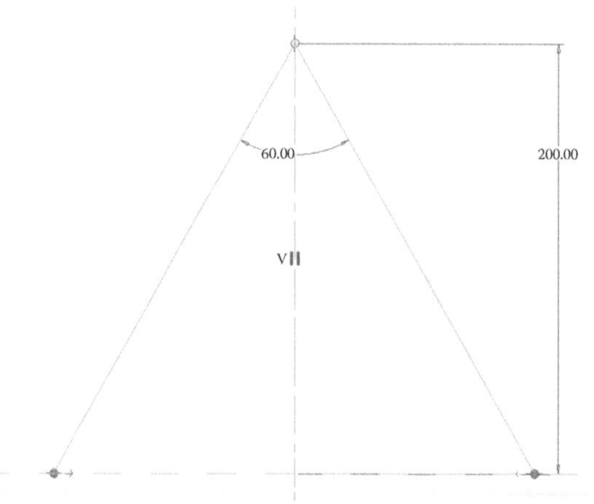

FIGURE 15.69
Constructional sketch of the Nozzle Assembly.

- The creation option window will appear.
- Select Locate default datums option.
- This will allow creating three default datums in the selected references.
- Select Three Planes in the locate datums method area.
- Click OK and select ASM_RIGHT, ASM_TOP, and ASM_FRONT plane, respectively.
- Notice that the part Nozzle_Beam is added and activated.
- The ribbon interface will change to part interface.

- Click on Sketch from the Model tab of the ribbon interface.
- Click on the Placement tab and select Define to define the sketch.
- Select the top plane as the sketch plane.
- The right plane shall be automatically taken as the reference and orientation as right.
- Click on Sketch.
- The sketcher model will be activated.
- Draw the sketch as shown in Figure 15.70.
- Click OK to create and exit the sketch.
- Click Sweep from the Model tab of the ribbon interface.
- Select the curve as shown in Figure 15.70.

- Click to activate the thicken option.
- Enter 3.4 as the thickness.
- Click on Create sweep section.
- Notice that sketcher mode is activated.
- Draw a circle of diameter 33.4 placing the center point in the intersection of the horizontal and vertical references.
- Click OK to create and exit the sketch.
- Flip the material orientation inward.
- Click OK to create the sweep.

- Similarly, create another sweep section having the same thickness and profile while selecting the other curves of the sketches.

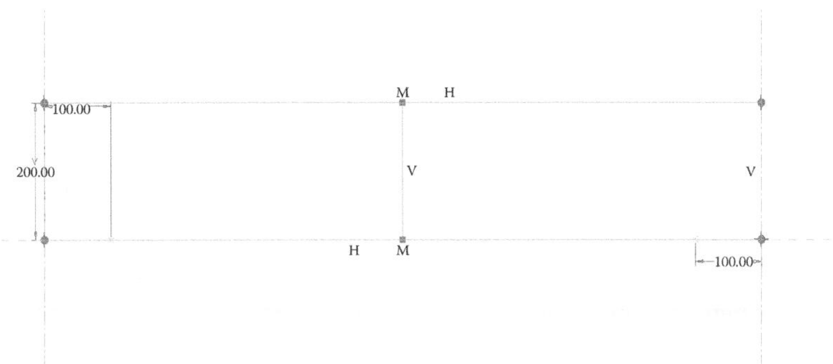

FIGURE 15.70
Constructional sketch of the Nozzle Assembly.

- Click Datum axis from the Model tab of the ribbon interface.
- Select any of the three cylindrical surfaces.
- Click OK to create the axis.
- Similarly, create other datum axes by selecting other two cylindrical surfaces.
- Right-click on the main assembly, i.e., Nozzle_Beam, and select Activate.

- Click on Assembly from the model tab of the ribbon interface.
- Select the file Adapter.
- Click on the Placement tab.
- Select the plane Right of Adapter part as the component reference.
- Select the plane ASM_RIGHT as the assembly reference.
- Select Coincide in the component reference relation field.
- Right-click on the graphical area and select New constraint.
- Select the plane Front of Adapter part as the component reference.
- Select the plane ASM_FRONT as the assembly reference.
- Select Coincide in the component reference relation field.
- Right-click on the graphical area and select New constraint.
- Select the top surface of the Adapter part as the component reference.
- Select the plane SPRAY_Distance as the assembly reference.
- Select Distance in the component reference relation field.
- Enter the offset value 12 mm.
- Right-click on the graphical area and select New constraint.
- Note: Adapter is kept approximately in position; however, the interference is ignored.
- Click OK to assemble the component.

- Right-click on the component adapter and select Pattern.
- Select Direction as the reference type.
- Select the plane ASM_Right as the first directional reference.
- Enter 5 as the patterned component and 220 as the increment on the first direction.
- Select the plane ASM_FRONT.
- Enter 2 as the patterned component and 200 as the increment on the first direction.
- Flip the orientation so that the adapters are kept on the nozzle beam.

- Click on Assembly from the model tab of the ribbon interface.
- Select the file Nozzle.
- Click on the Placement tab.
- Select the cylindrical surface of the nozzle part as the component reference.
- Select the cylindrical surface of the adapter as the assembly reference.
- Select Coincide in the component reference relation field.
- Right-click on the graphical area and select New constraint.

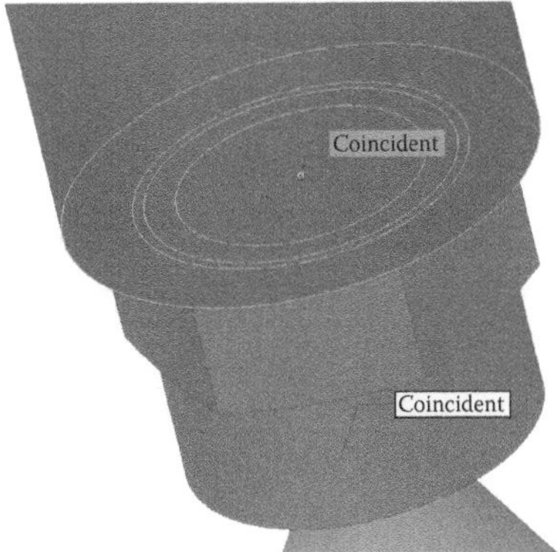

FIGURE 15.71
Constructional view of the Nozzle Assembly.

- Select the surface as shown in Figure 15.71.
- Select Coincide in the component reference relation field.
- Click OK to assemble the component.

- Right-click on the component nozzle and select Pattern.
- Select Direction as the reference type.
- Select the plane ASM_Right as the first directional reference.
- Enter 5 as the patterned component and 220 as the increment on the first direction.
- Select the plane ASM_FRONT.
- Enter 2 as the patterned component and 200 as the increment on the first direction.
- Flip the orientation so that the adapters are kept on the nozzle beam.

- Click on Assembly from the model tab of the ribbon interface.
- Select the file Flange.
- Select the instance SORF_25NB.
- Click on the Placement tab.
- Select the cylindrical surface of the flange part as the component reference.
- Select the cylindrical surface of the pipe as the assembly reference.
- Select Coincide in the component reference relation field.
- Right-click on the graphical area and select New constraint.
- Select the surface as shown in Figure 15.72.
- Enter the distance as 10 mm.
- Click OK to assemble the component.

- Similarly, assemble the four instances of SORF_25NB flange at the other three pipe ends.

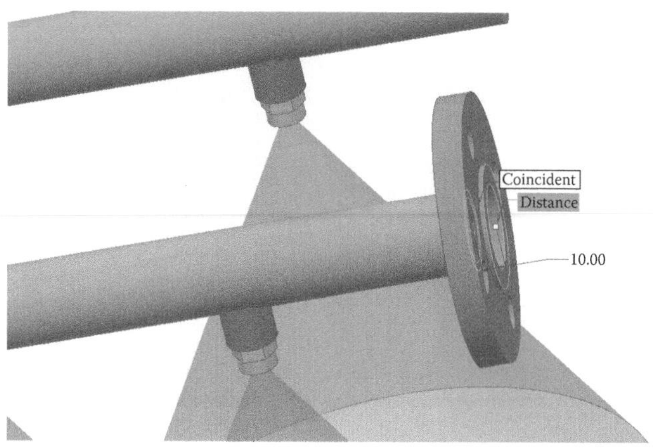

FIGURE 15.72
Constructional view of assembly of flange of the Nozzle Assembly.

- Click on Assembly from the Model tab of the ribbon interface.
- Select the file Flange.
- Select the instance BLIND_25NB.
- Click on the Placement tab.
- Select the cylindrical surface of the BLIND_25NB flange part as the component reference.
- Select the cylindrical surface of the SORF_25NB flange as the assembly reference.
- Select Coincide in the component reference relation field.
- Right-click on the graphical area and select New constraint.
- Select the raised face of the BLIND_25NB flange part as the component reference.
- Select the raised face of the SORF_25NB flange part as the component reference.
- Select the cylindrical surface of any hole of the BLIND_25NB flange part as the component reference.
- Select the cylindrical surface of any hole of the SORF_25NB flange as the assembly reference.
- Select Coincide in the component reference relation field.
- Click OK to assemble the component.

- Click on Assembly from the Model tab of the ribbon interface.
- Select the file Stud_M_12.
- Click on the Placement tab.
- Select the cylindrical surface of the Stub_M_12 part as the component reference.
- Select the cylindrical hole surface of the SORF_25NB flange as the assembly reference.
- Select Coincide in the component reference relation field.
- Right-click on the graphical area and select New constraint.

- Select the top face of the Stub_M_12 part as the component reference.
- Select the back face of the SORF_25NB flange as the assembly reference.
- Select Distance in the component reference relation field.
- Enter 20 as the distance.
- Click OK to assemble the component.

- Click on Assembly from the Model tab of the ribbon interface.
- Select the file Nut_M_12.
- Click on the Placement tab.
- Select the cylindrical surface of the Nut_M_12 part as the component reference.
- Select the cylindrical hole surface of the Stub_M_12 flange as the assembly reference.
- Select Coincide in the component reference relation field.
- Right-click on the graphical area and select New constraint.
- Select the top face of the Stub_M_12 part as the component reference.
- Select the back face of the SORF_25NB flange as the assembly reference.
- Select Coincide in the component reference relation field.
- Click OK to assemble the component.

- Similarly, assemble Nut_M_12 again on the other side of the blind flange into that same stud.
- Select the parts Stud_M_12 and both the nuts by pressing the control key.
- Right-click and select Pattern.
- The reference pattern shall be selected automatically.
- Click OK to create the pattern.
- Notice that the stud and the nuts will be assembled in all the holes of the same flange joint.

- Similarly, assemble the stud and the nuts for all the flange joints.

- Save and exit the file.

Hands-On Exercise, Example 15.31: Nozzle_Support (Software Used: Creo Parametric; Figure 15.73)

Steps:

- Create an assembly file and rename it as Nozzle_Support.asm. (Other names may be given, but it is advisable to provide a relevant name.)
- Select Design on the right-hand side of the box. This will activate the standard assembly design mode and environment.
- Uncheck the box of the default template and click OK to create a new assembly design file.

FIGURE 15.73
Three-dimensional view of the Nozzle_Support.

- The template selection box will appear.
- Select mmns_asm_Solid. This will allow the use of the solid template file of the following units:
  - mm—millimeter for length unit
  - n—newton for force unit
  - s—second for time unit
- Once the assembly file is created, go to File > Prepare > Model properties to check the units. In the upper sections, the units of the model will be displayed.
  - Note: The units are usually displayed in the summary section of any CAD software. The setting of the units is usually in the tools or settings.
- Click on tree filers as displayed in Figure 15.58.
- Click on the check box displaying the features.
- Click on Apply. Notice that the features are displayed in the model tree.

- Click OK datum plane from the Model tab of the ribbon interface.
- Select the ASM_RIGH plane as the reference plane.
- Set the offset value 93 mm towards up.
- Click OK to create the datum plane.

- Click on Part create on assembly mode icon from the Model tab of the ribbon interface.
- Change the name as Angle_2.
- Select Solid in the sub type.
- Click OK.
- The creation option window will appear.
- Select Locate default datums option.
- This will allow creating three default datums in the selected references.
- Select Three Planes in the locate datums method area.
- Click OK and select ASM_RIGHT, ASM_TOP, and ASM_FRONT plane, respectively.
- Notice that the part angle_2 is added and activated.
- The ribbon interface will change to part interface.

- Click on Extrude from the Model tab of the ribbon interface.
- Select the thicken material option.
- Enter 6 as the thickness.
- Click on the Placement tab and select Define to define the sketch.
- Select the top plane as the sketch plane.
- The right plane shall be automatically taken as the reference and orientation as right.
- Click on Sketch.
- The sketcher model will be activated.
- Draw the sketch as shown in Figure 15.74.
- Click OK to create and exit the sketch.
- Select the Mid plane extrusion option.
- Enter 400 as the extrusion depth.
- Click OK to create the extrusion.
- Right-click on the main assembly and select Activate.

- Click on Extrude from the Model tab of the ribbon interface.
- Click on the Placement tab and select Define to define the sketch.
- Select the top plane as the sketch plane.
- The right plane shall be automatically taken as the reference and orientation as right.
- Click on Sketch.
- The sketcher model will be activated.
- Draw the sketch as shown in Figure 15.75.
- Click OK to create and exit the sketch.
- Select the Though all option.
- Activate the material removal option.
- Flip the orientation towards the material side.

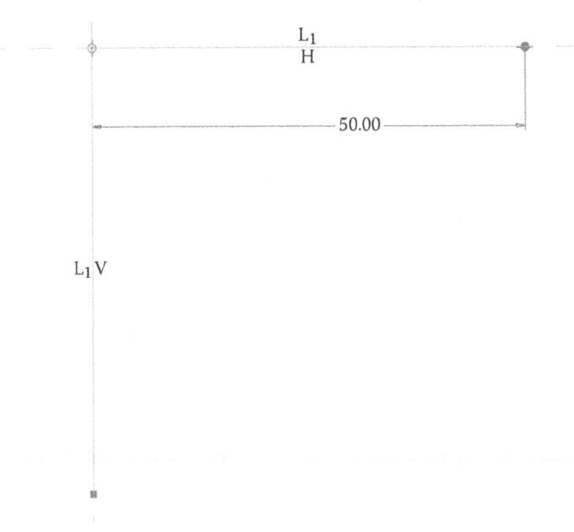

FIGURE 15.74
Constructional sketch of the Nozzle_Support.

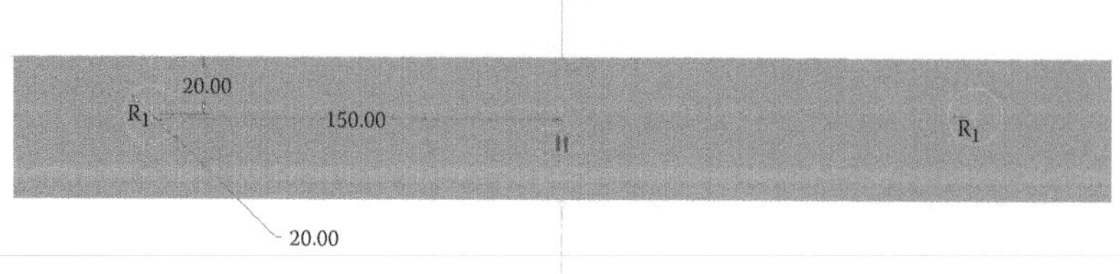

FIGURE 15.75
Constructional sketch of creating holes in Nozzle_Support.

- Click on Intersect tab.
- Uncheck the box for automatic update.
- Set the display level to Part level.
- Check the box for check geometry.
- Click OK to create the extrusion.

- Right-click on the main assembly and select Activate.
- Click on Save to save the file.
- Click on Assembly from the Model tab of the ribbon interface.
- Select the file angle_2.
- Click on the Placement tab.
- Select the plane DTM2 of Angle_2 part as the component reference.
- Select the plane DTM2 of Angle_2 as the assembly reference.
- Select Coincide in the component reference relation field.
- Right-click on the graphical area and select New constraint.
- Select the bottom surface of the Angle_2 part as the component reference.
- Select the bottom surface of Angle_2 as the assembly reference.
- Select Coincide in the component reference relation field.
- Right-click on the graphical area and select New constraint.
- Select the surface as shown in Figure 15.76.
- Enter the offset value 186 mm.
- Right-click on the graphical area and select New constraint.
- Click OK to assemble the component.

- Click on Assembly from the Model tab of the ribbon interface.
- Select the file Washer.
- Click on the Placement tab.
- Select the cylindrical surface of the washer as the component reference.
- Select the plane inside the cylindrical surface of any hole of the part angle_2 as the assembly reference.
- Select Coincide in the component reference relation field.
- Right-click on the graphical area and select New constraint.
- Select the flat surface of the washer as the component reference.
- Select the flat surface of the part angle_2 as the assembly reference.

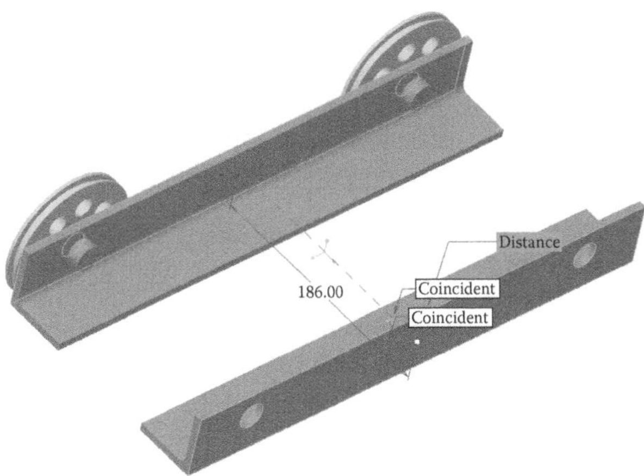

FIGURE 15.76
Constructional view of the Nozzle_Support.

- Select Coincide in the component reference relation field.
- Flip the orientations suitably so that the washer does not have any interference with the part angle_2.
- Right-click on the graphical area and select New constraint.
- Click OK to assemble the component.

- Click on Assembly from the Model tab of the ribbon interface.
- Select the file Nozzle_Support_Wheel.
- Click on the Placement tab.
- Select Pin joint in the mechanism definition.
- Select the cylindrical surface of Nozzle_Support_Wheel as the component reference.
- Select the plane inside the cylindrical surface of the washer as the assembly reference.
- Select Coincide in the component reference relation field.
- Right-click on the graphical area and select New constraint.
- Select the flat surface of Nozzle_Support_Wheel as the component reference.
- Select the flat surface of the washer as the assembly reference.
- Select Coincide in the component reference relation field.
- Flip the orientations suitably so that the washer does not have any interference.
- Right-click on the graphical area and select New constraint.
- Click OK to assemble the component.

- Click on Assembly from the model tab of the ribbon interface.
- Select the file washer.
- Click on the Placement tab.
- Select the cylindrical surface of the washer as the component reference.

- Select the plane inside the cylindrical surface of Nozzle_Support_Wheel as the assembly reference.
- Select Coincide in the component reference relation field.
- Right-click on the graphical area and select New constraint.
- Select the flat surface of the washer as the component reference.
- Select the flat surface of the part Nozzle_Support_Wheel as the assembly reference.
- Select Coincide in the component reference relation field.
- Flip the orientations suitably so that the washer does not have any interference.
- Right-click on the graphical area and select New constraint.
- Click OK to assemble the component.

- Click on Assembly from the Model tab of the ribbon interface.
- Select the file Nozzle_Support_Shaft.
- Click on the Placement tab.
- Select the cylindrical surface of Nozzle_Support_Shaft as the component reference.
- Select the plane inside the cylindrical surface of the washer as the assembly reference.
- Select Coincide in the component reference relation field.
- Right-click on the graphical area and select New constraint.
- Select the flat surface of Nozzle_Support_Shaft as the component reference.
- Select the flat surface of the washer as the assembly reference.
- Select Coincide in the component reference relation field.
- Flip the orientations suitably so that the washer does not have any interference.
- Right-click on the graphical area and select New constraint.
- Click OK to assemble the component.

- Similarly, assemble the parts Washer, Nozzle_Support_Wheel, Washer, and Nozzle_Support_Shaft, respectively, on all the other holes in the part angle_2.

- Create a part file and rename it as Angle_3. (Other names may be given, but it is advisable to provide a relevant name.)
- Select Solid on the right-hand side of the box. This will activate the solid mode and environment.
- Uncheck the box of the default template and click OK to create a new solid file.
- The template selection box will appear.
- Select mmns_part_Solid. This will allow the use of the solid template file of the following units:
  - mm—millimeter for length unit
  - n—newton for force unit
  - s—second for time unit
- Once the solid part is created, go to File > Prepare > Model properties to check the units. In the upper sections, the units of the model will be displayed.

  - Note: The units are usually displayed in the summary section of any CAD software. The setting of the units is usually in the tools or settings.
- Click on Extrude from the Model tab of the ribbon interface.
- Select Thicken material option.
- Enter 6 as the thickness.
- Click on the Placement tab and select Define to define the sketch.
- Select the top plane as the sketch plane.
- The right plane shall be automatically taken as the reference and orientation as right.
- Click on Sketch.
- The sketcher model will be activated.
- Draw the sketch as shown in Figure 15.77.
- Click OK to create and exit the sketch.
- Select the Mid plane extrusion option.
- Enter 170 as the extrusion depth.
- Click OK to create the extrusion.
- Save and exit the file.

- Click on Assembly from the Model tab of the ribbon interface.
- Select the file Angle_3.
- Click on the Placement tab.
- Select the DTM_3 of Angle_3 as the component reference.
- Select the plane ADTM1 as the assembly reference.
- Select Coincide in the component reference relation field.
- Right-click on the graphical area and select New constraint.
- Select the other references as shown in Figure 15.78.

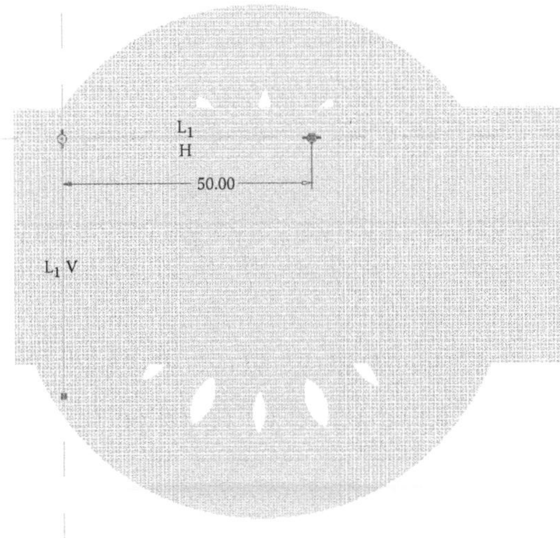

FIGURE 15.77
Constructional sketch of the Nozzle_Support.

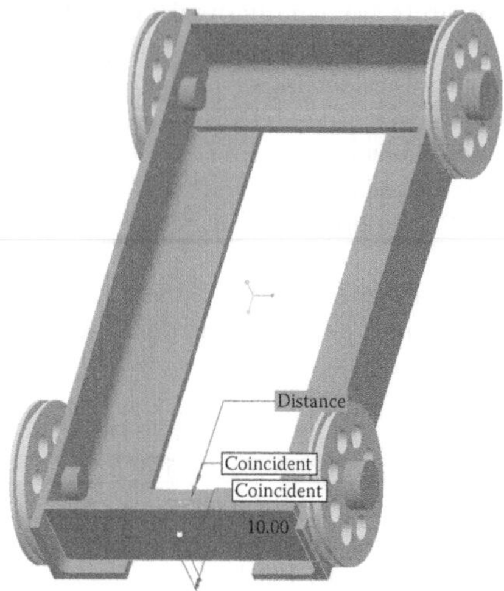

FIGURE 15.78
Constructional view for assembly of the Nozzle_Support.

- Click OK to assemble the component.
- Similarly, assemble another angle_3 in the opposite side.
- Save and exit the file.

Hands-On Exercise, Example 15.32: Nozzle_Assembly (Software Used: Creo Parametric; Figure 15.79)

Steps:

- Create an assembly file and rename it as Nozzle_Assembly.asm. (Other names may be given, but it is advisable to provide a relevant name.)

FIGURE 15.79
(See color insert.) Three-dimensional view of the Nozzle_Assembly.

- Select Design on the right-hand side of the box. This will activate the standard assembly design mode and environment.
- Uncheck the box of the default template and click OK to create a new assembly design file.
- The template selection box will appear.
- Select mmns_asm_Solid. This will allow the use of the solid template file of the following units:
    - mm—millimeter for length unit
    - n—newton for force unit
    - s—second for time unit
- Once the assembly file is created, go to File > Prepare > Model properties to check the units. In the upper sections, the units of the model will be displayed.
    - Note: The units are usually displayed in the summary section of any CAD software. The setting of the units is usually in the tools or settings.
- Click on tree filers as displayed in Figure 15.58.
- Click on the check box displaying the features.
- Click on Apply. Notice that the features are displayed in the model tree.

- Click on Assembly from the model tab of the ribbon interface.
- Select the file Nozzle_Assembly.
- Click Default in the Placement constraint tab.
- Click OK to assemble the component.

- Click on Assembly from the Model tab of the ribbon interface.
- Select the file Nozzle_Support.asm.
- Click on the Placement tab.
- Select the ASM_FRONT of Nozzle_Support as the component reference.
- Select the plane ASM_FRONT of Nozzle_Beam as the assembly reference.
- Select Coincide in the component reference relation field.
- Right-click on the graphical area and select New constraint.
- Select the plane ADTM1 of Nozzle_Support as the component reference.
- Select the plane ASM_RIGHT as the assembly reference.
- Select Distance in the component reference relation field.
- Enter the distance 440.
- Right-click on the graphical area and select New constraint.
- Select the surface as shown in Figure 15.80.
- Notice that the constant becomes the tangent.
- Click OK to assemble the component.

- Click on Assembly from the Model tab of the ribbon interface.
- Select the file U-Bolt_25NB.
- Click on the Placement tab.
- Select the axis of U-Bolt_25NB as the component reference.
- Select the axis of Nozzle_Beam as the assembly reference.

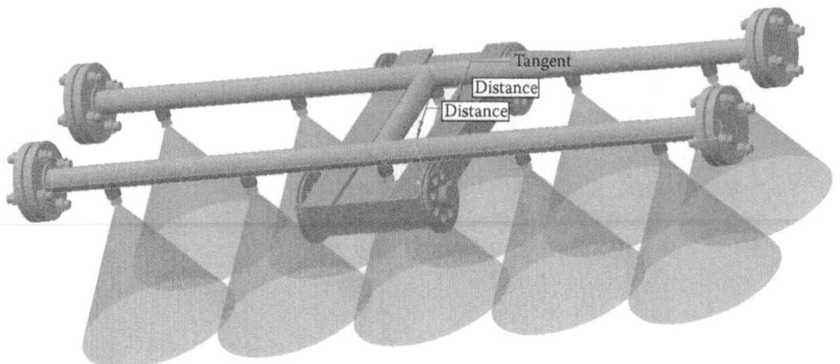

FIGURE 15.80
Constructional view for assembly of the Nozzle_Assembly.

- Select Coincide in the component reference relation field.
- Right-click on the graphical area and select New constraint.
- Select the plane FRONT of U-Bolt_25NB as the component reference.
- Select the surface of the angle as shown in Figure 15.81 as the assembly reference.
- Select Distance in the component reference relation field.
- Enter the distance 30.
- Right-click on the graphical area and select New constraint.
- Select the plane RIGHT of U-Bolt_25NB as the component reference.
- Select the ASM_FRONT of Nozzle_Support as the assembly reference.
- Select Parallel in the component reference relation field.

- Click OK to assemble the component.

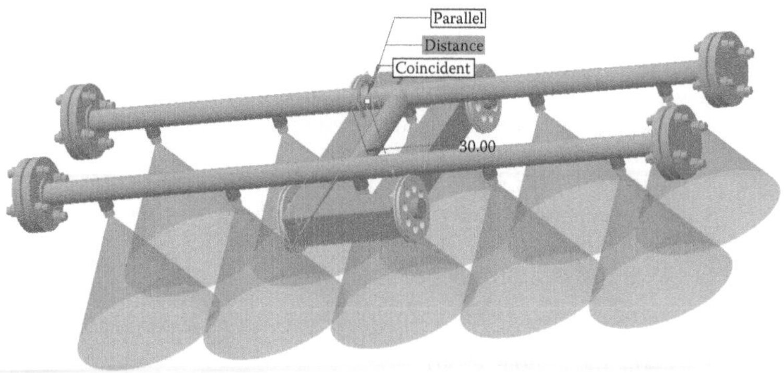

FIGURE 15.81
Constructional view for assembly of U-Bolt in the Nozzle_Assembly.

- Similarly, assemble other 4 U-Bolt as shown in Figure 15.81.
- Click on Extrude from the Model tab in the ribbon interface.
- Click on the Placement tab and select Define to define the sketch.
- Select the top surface of angle_2, where the nozzle pipes are resting as the sketch plane.
- The right plane shall be automatically taken as the reference and orientation as right.
- Click on Sketch.
- The sketcher model will be activated.
- Draw the sketch as shown in Figure 15.82.
- Click OK to create and exit the sketch.
- Activate the material removal option.
- Flip the orientation towards down.
- Select up to the entity option. Select the bottom surface of angle_2.
- Click on OK to create the extrusion.

- Click on Assembly from the Model tab of the ribbon interface.
- Select the file Nut_M_6.
- Click on the placement tab.
- Select the internal cylindrical hole surface of Nut_M_6 as the component reference.
- Select the external cylindrical surface of the U-Bolt as the assembly reference.
- Select Coincide in the component reference relation field.
- Right-click on the graphical area and select New constraint.
- Select the bottom surface of the Nut_M_6 as the component reference.
- Select the top surface of the angle_2 as the assembly reference.
- Select Coincide in the component reference relation field.

- Click OK to assemble the component.

- Similarly, attach all the Nut_m_6 on both sides of angle_2 on the u_bolt.

- Save and exit the file.

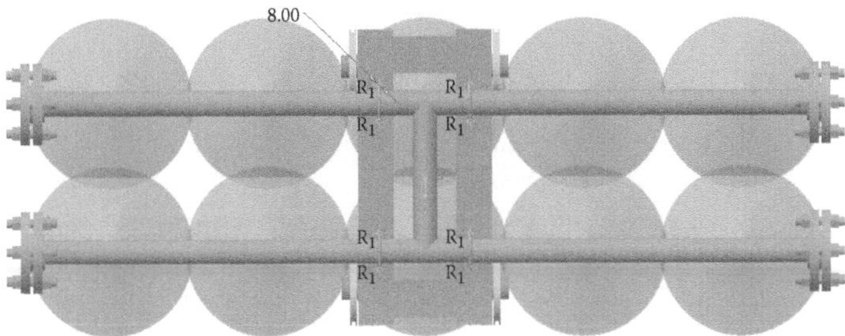

FIGURE 15.82
Constructional sketch of creating holes in U-Bolt in the Nozzle_Assembly.

Hands-On Exercise, Example 15.33: Wheel_Assembly (Software Used: Creo Parametric; Figure 15.83)

Steps:

- Create an assembly file and rename it as Wheel_Assembly.asm. (Other names may be given, but it is advisable to provide a relevant name.)
- Select Design on the right-hand side of the box. This will activate the standard assembly design mode and environment.
- Uncheck the box of the default template and click OK to create a new assembly design file.
- The template selection box will appear.
- Select mmns_asm_Solid. This will allow the use of the solid template file of the following units:
  - mm—millimeter for length unit
  - n—newton for force unit
  - s—second for time unit
- Once the assembly file is created, go to File > Prepare > Model properties to check the units. In the upper sections, the units of the model will be displayed.
  - Note: The units are usually displayed in the summary section of any CAD software. The setting of the units is usually in the tools or settings.
- Click on tree filers as displayed in Figure 15.58.
- Click on the check box displaying the features.
- Click on Apply. Notice that the features are displayed in the model tree.

- Click on Assembly from the Model tab of the ribbon interface.
- Select the file Wheel_Base_Plate.

FIGURE 15.83
Three-dimensional view of the Wheel_Assembly.

- Click Default in the Placement constraint tab.
- Click OK to assemble the component.

- Click on Assembly from the Model tab of the ribbon interface.
- Select the file Wheel_Bracket.asm.
- Click on the Placement tab.
- Select the top surface of Wheel_Bracket as the component reference.
- Select the bottom surface of Wheel_Base_Plate as the assembly reference.
- Select Coincide in the component reference relation field.
- Right-click on the graphical area and select New constraint.
- Select the plane Right of Wheel_Bracket as the component reference.
- Select the plane ASM_RIGHT of Wheel_Base_Plate as the assembly reference.
- Select Coincide in the component reference relation field.
- Right-click on the graphical area and select New constraint.
- Select the surface as shown in Figure 15.84.
- Select Distance in the component reference relation field.
- Enter 30 as the distance.
- Click OK to assemble the component.

- Similarly, assemble another wheel_bracket on the opposite side of the plane ASM_FRONT.

- Click on Assembly from the Model tab of the ribbon interface.
- Select the file Idle_Shaft.asm.
- Click on the Placement tab.
- Select the cylindrical surface of Idle_Shaft as the component reference.

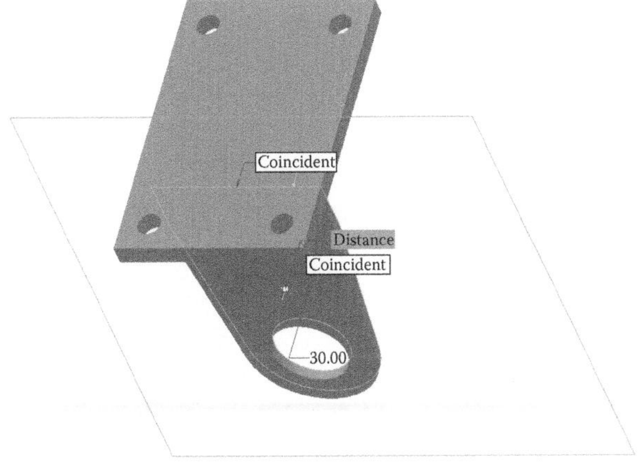

FIGURE 15.84
Constructional view of the Wheel_Assembly.

- Select the internal cylindrical hole surface of Wheel_Bracket as the assembly reference.
- Select Coincide in the component reference relation field.
- Right-click on the graphical area and select New constraint.
- Select the plane FRONT of Idle_Shaft as the component reference.
- Select the plane ASM_FRONT as the assembly reference.
- Select Coincide in the component reference relation field.
- Click OK to assemble the component.

- Click on assembly from the model tab of the ribbon interface.
- Select the file Wheel.asm.
- Select Pin in the connection field.
- Click on the Placement tab.
- Select the cylindrical surface of Wheel as the component reference.
- Select the cylindrical surface of Idle_Shaft as the assembly reference.
- Select Coincide in the component reference relation field.
- Right-click on the graphical area and select New constraint.
- Select the plane FRONT of Wheel as the component reference.
- Select the plane FRONT of Idle_Shaft as the assembly reference.
- Select Coincide in the component reference relation field.
- Click OK to assemble the component.

- Save and exit the file.

Hands-On Exercise, Example 15.34: Support (Software Used: Creo Parametric; Figure 15.85)

Steps:

- Create an assembly file and rename it as Support.asm. (Other names may be given, but it is advisable to provide a relevant name.)

FIGURE 15.85
(See color insert.) Three-dimensional view of the support.

- Select Design on the right-hand side of the box. This will activate the standard assembly design mode and environment.
- Uncheck the box of the default template and click OK to create a new assembly design file.
- The template selection box will appear.
- Select mmns_asm_Solid. This will allow the use of the solid template file of the following units:
    - mm—millimeter for length unit
    - n—newton for force unit
    - s—second for time unit
- Once the assembly file is created, go to File > Prepare > Model properties to check the units. In the upper sections, the units of the model will be displayed.
    - Note: The units are usually displayed in the summary section of any CAD software. The setting of the units is usually in the tools or settings.
- Click on tree filers as displayed in Figure 15.58.
- Click on the check box displaying the features.
- Click on Apply. Notice that the features are displayed in the model tree.

- Click on Sketch from the Model tab of the ribbon interface.
- Click on the Placement tab and select Define to define the sketch.
- Select the plane ASM_FRONT plane as the sketch plane.
- The right plane shall be automatically taken as the reference and orientation as right.
- Click on Sketch.
- The sketcher model will be activated.
- Draw the sketch as shown in Figure 15.86.
- Click OK to create and exit the sketch.

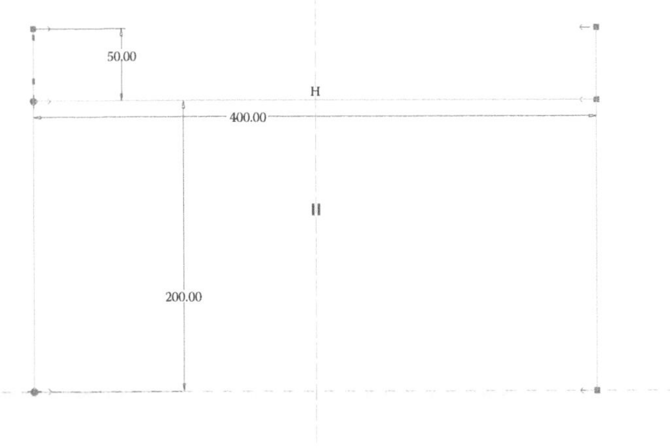

FIGURE 15.86
Constructional sketch of the support.

- Click on Part create on assembly mode icon from the Model tab of the ribbon interface.
- Change the name as C_Profile.
- Select solid In the sub type.
- Click OK.
- The creation option window will appear.
- Select Locate default datums option.
- This will allow creating three default datums in the selected references.
- Select Three Planes in the locate datums method area.
- Click OK and select ASM_RIGHT, ASM_TOP, and ASM_FRONT plane, respectively.
- Notice that the part Nozzle_Beam is added and activated.
- The ribbon interface will change to part interface.

- Click Sweep from the Model tab of the ribbon interface.
- Select the left vertical line of the assembly sketch.
- Click on Create sweep section.
- Notice that sketcher mode is activated.
- Draw the figure as shown in Figure 15.87.
- Click OK to create and exit the sketch.
- Click OK to create the sweep.

- Right-click on the main assembly and select Activate.

- Click on Part create on assembly mode icon from the Model tab of the ribbon interface.
- Change the name as C_Profile1.
- Select Mirror in the sub type.
- Click OK.

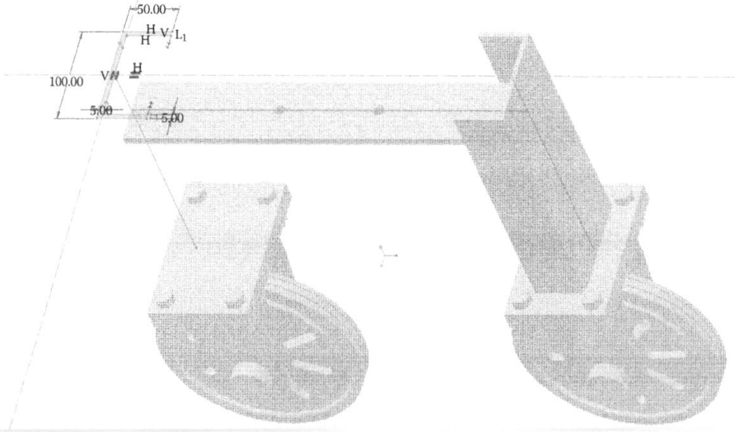

FIGURE 15.87
Constructional sketch for creation of the C_Profile in support.

- The creation option window will appear.
- Select the part C_Profile as part reference.
- Select the plane ASM_RIGHT as the planner reference for the mirror plane.
- Click OK to create the part.

- Click on Part create on assembly mode icon from the Model tab of the ribbon interface.
- Change the name as Angle.
- Select Solid in the sub type.
- Click OK.
- The creation option window will appear.
- Select Locate default datums option.
- This will allow creating three default datums in the selected references.
- Select Three Planes in Locate datums method area.
- Click OK and select ASM_RIGHT, ASM_TOP, and ASM_FRONT plane, respectively.
- Notice that the part Nozzle_Beam is added and activated.
- The ribbon interface will change to part interface.

- Click Sweep from the Model tab of the ribbon interface.
- Select Thicken option.
- Enter 8 as the thickness.
- Select the middle horizontal line of the assembly sketch.
- Click on Create sweep section.
- Notice that sketcher mode is activated.
- Draw the figure as shown in Figure 15.88.
- Click OK to create and exit the sketch.

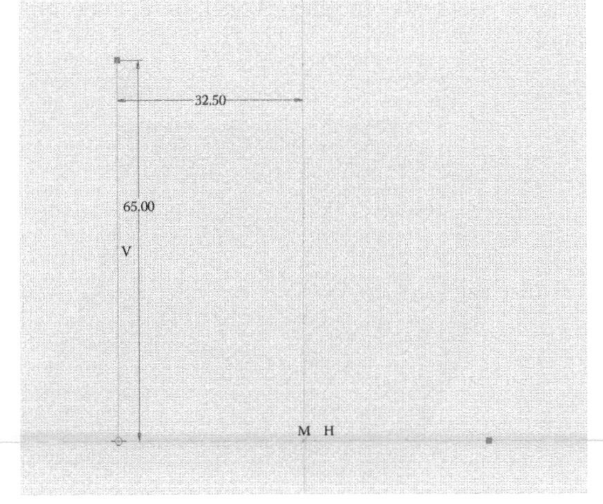

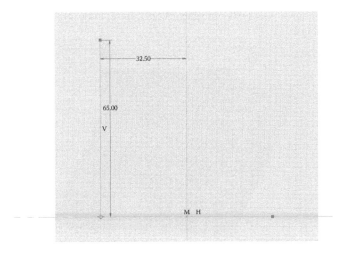

FIGURE 15.88
Constructional sketch of the support.

- Flip the material orientation inward.
- Click OK to create the sweep.

- Click on Round feature from the Model tab of the ribbon interface.
- Set the value of fillet to 8 mm.
- Select Internal edge of the newly created feature.
- Click OK to create the feature.

- Right-click on Main assembly and select Activate.

- Click on Assembly from the Model tab of the ribbon interface.
- Select the file Wheel_Base_Plate.
- Click on the Placement tab.
- Select the plane FRONT of Wheel_Base_Plate part as the component reference.
- Select the plane ASM_FRONT as the assembly reference.
- Select Coincide in the component reference relation field.
- Right-click on the graphical area and select New constraint.
- Select the plane top surface of the Wheel_Base_Plate part as the component reference.
- Select the plane bottom surface of the channel as the assembly reference.
- Select Coincide in the component reference relation field.
- Right-click on the graphical area and select New constraint.
- Select the plane and the surface as shown in Figure 15.89.
- Select Distance in the component reference relation field.
- Enter the offset value 25 mm.
- Right-click on the graphical area and select New constraint.
- Click OK to assemble the component.

- Similarly, assemble another Wheel_Base_Plate on the other channel.

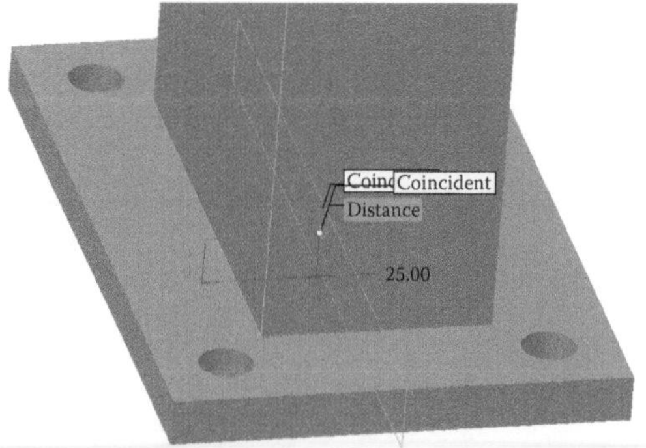

FIGURE 15.89
Constructional view for assembly of Wheel Base Plate in the assembly support.

- Click on Assembly from the Model tab of the ribbon interface.
- Select the file Wheel_Assembly.
- Click on the Placement tab.
- Select the surfaces of Wheel_Base_Plate as shown in Figure 15.90.
- Select Coincide in the component reference relation field.
- Right-click on the graphical area and select New constraint.
- Select the hole surface of Wheel_Base_Plate in the wheel assembly as the component reference.
- Select the similar hole surface of Wheel_Base_Plate in the main assembly as the assembly reference.
- Select Coincide in the component reference relation field.
- Right-click on the graphical area and select New constraint.
- Select another hole surface of Wheel_Base_Plate in the wheel assembly as the component reference.
- Select the similar hole surface of Wheel_Base_Plate in the main assembly as the assembly reference.
- Select Oriented in the component reference relation field.
- Click OK to assemble the component.
- Similarly, assemble another Wheel_Assembly on the other side.

- Click on Extrude from the Model tab in the ribbon interface.
- Click on the Placement tab and select Define to define the sketch.
- Select the top face of the angle as the sketch plane.
- The right plane shall be automatically taken as the reference and orientation as right.
- Click on Sketch.
- The sketcher model will be activated.
- Draw the sketch as shown in Figure 15.91.
- Click OK to create and exit the sketch.
- Select the Upto entity option.

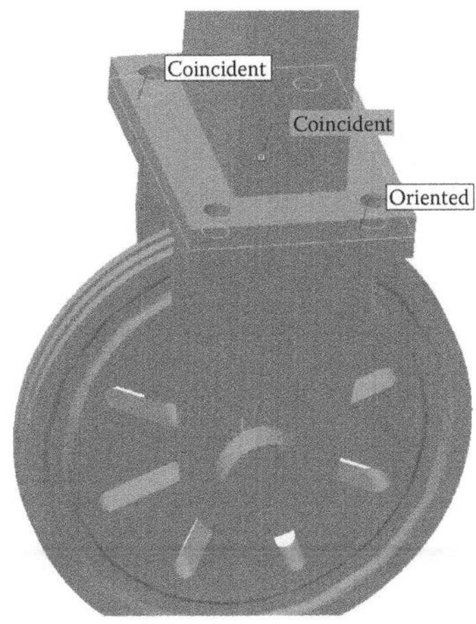

FIGURE 15.90
Constructional view of the support.

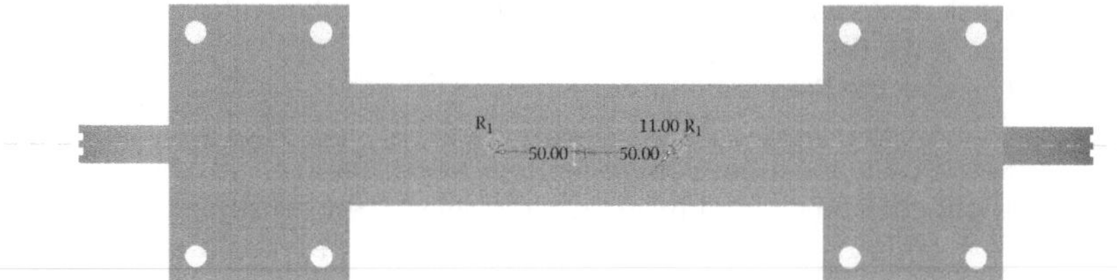

FIGURE 15.91
Constructional view for creation of holes in the support.

- Select the bottom surface of the angle.
- Click on OK to create the extrusion.

- Click on Assembly from the Model tab of the ribbon interface.
- Select the file Hex_Bolt_M_12.
- Click on the Placement tab.
- Select the cylindrical surface of the Hex_Bolt_M_12 part as the component reference.
- Select the inside cylindrical surface of the Wheel_Base_Plate as the assembly reference.
- Select Coincide in the component reference relation field.
- Right-click on the graphical area and select New constraint.
- Select the sitting surface of the Hex_Bolt_M_12 part as the component reference.
- Select the top surface of the Wheel_Base_Plate of the main assembly as the assembly reference.
- Click OK to assemble the component.

- Click on Assembly from the Model tab of the ribbon interface.
- Select the file NUT_M_12.
- Click on the Placement tab.
- Select the cylindrical surface of the NUT_M_12 part as the component reference.
- Select the inside cylindrical surface of the Hex_Bolt_M_12 as the assembly reference.
- Select Coincide in the component reference relation field.
- Right-click on the graphical area and select New constraint.
- Select the sitting surface of the NUT_M_12 part as the component reference.
- Select the bottom surface of Wheel_Base_Plate of Wheel_Assembly as the assembly reference.
- Click OK to assemble the component.

- Similarly, assemble Hex_Bolt_M_12 and Nut_M_12 in all other holes of Wheel_Base_Plates.
- Save and exit the file.

Hands-On Exercise, Example 15.35: Spray_System (Software Used: Creo Parametric; Figure 15.92)

Steps:

- Create an assembly file and rename it as Spray_System.asm. (Other names may be given, but it is advisable to provide a relevant name.)
- Select design on the right-hand side of the box. This will activate the standard assembly design mode and environment.
- Uncheck the box of the default template and click OK to create a new assembly design file.
- The template selection box will appear.
- Select mmns_asm_Solid. This will allow the use of the solid template file of the following units:
  - mm—millimeter for length unit
  - n—newton for force unit
  - s—second for time unit
- Once the assembly file is created, go to File > Prepare > Model properties to check the units. In the upper sections, the units of the model will be displayed.
  - Note: The units are usually displayed in the summary section of any CAD software. The setting of the units is usually in the tools or settings.
- Click on tree filers as displayed in Figure 15.58.
- Click on the check box displaying the features.
- Click on Apply. Notice that the features are displayed in the model tree.

- Click on Assembly from the Model tab of the ribbon interface.
- Select the file Rail_ProfileC.asm.
- Click on the Placement tab.
- Select the top of Rail_ProfileC as the component reference.
- Select the plane ASM_TOP of Nozzle_Beam as the assembly reference.
- Select Coincide in the component reference relation field.

FIGURE 15.92
Three-dimensional view of the spray system.

- Right-click on the graphical area and select New constraint.
- Select the right of Rail_ProfileC as the component reference.
- Select the plane ASM_RIGHT of Nozzle_Beam as the assembly reference.
- Select Distance in the component reference relation field.
- Enter the distance 2500.
- Right-click on the graphical area and select New constraint.
- Select the front of Rail_ProfileC as the component reference.
- Select the plane ASM_FRONT of Nozzle_Beam as the assembly reference.
- Select Coincide in the component reference relation field.

- Click OK to assemble the component.

- Similarly, assemble another Rail_ProfileC on the other side of the plane ASM_RIGHT at a distance 2500.

- Click on Part create on assembly mode icon from the model tab of the ribbon interface.
- Change the name as Equipment.
- Select Solid in the sub type.
- Click OK.
- The creation option window will appear.
- Select Locate default datums option.
- This will allow creating three default datums in the selected references.
- Select Three Planes in Locate datums method area.
- Click OK and select ASM_RIGHT, ASM_TOP, and ASM_FRONT plane, respectively.
- Notice that the part angle_2 is added and activated.
- The ribbon interface will change to part interface.

- Click on Extrude from the Model tab in the ribbon interface.
- Click on the Placement tab and select Define to define the sketch.
- Select the DTM3 plane as the sketch plane.
- The DTM1 plane shall be automatically taken as the reference and orientation as right.
- Click on Sketch.
- The sketcher model will be activated.
- Draw the sketch as shown in Figure 15.93.
- Click OK to create and exit the sketch.
- Flip the orientation along the length of the rail.
- Enter the value 5000 mm for the extrusion depth.
- Click on OK to create the extrusion.
- Right-click on the main assembly and select Activate.

- Click on Assembly from the model tab of the ribbon interface.
- Select the file Support.
- Select the planner on the Connection tab.

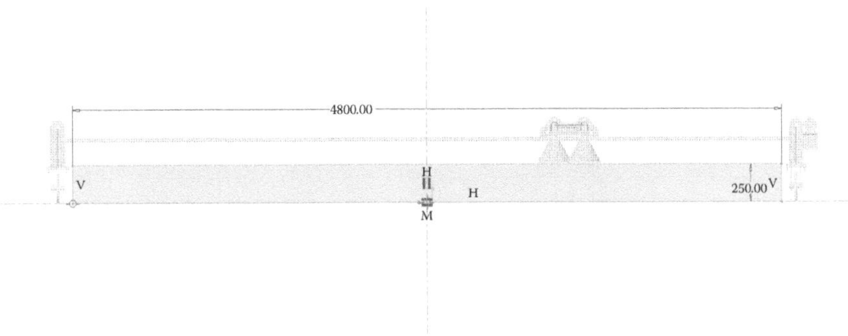

FIGURE 15.93
Constructional sketch of the spray system.

- Select the plane ASM_FRONT of Support.asm as the component reference.
- Select the plane right of the first Rail_ProfileC as the assembly reference.
- Select Coincide in the component reference relation field.
- Click OK to assemble the component.

- Click on the Application tab of the ribbon interface.
- Click on Mechanism to activate the mechanism environment.
- Click on CAM to define the cam follower geometry.
- Select the cylindrical surface of the wheel of maximum diameter as the cam1 geometry.
- (Note: Select Total surface on the perimeter by pressing the control key.)
- Click on CAM 2 tab in the cam-follower connection definition window.
- Select the inside flat surface of the rail.
- Click OK; notice that it prompts to define boundaries.
- Select the front and the rear point on the surface.
- Click OK to create the cam follower connection.
- Similarly, create another cam follower connection on the other wheel to the same rail surface.

- Click on Assembly from the Model tab of the ribbon interface.
- Select the file Rail_Profile_C1.asm.
- Click on the Placement tab.
- Select the surfaces as shown in Figure 15.94.
- Select Distance in the component reference relation field.
- Enter 10 as the distance.
- Make sure that the rail shall be below that selected surface of the angle.
- Right-click on the graphical area and select New constraint.
- Select the plane right of Rail_Profile_C1 as the component reference.

FIGURE 15.94
Constructional view for assembly of Rail_Profile_C1 in the spray system.

- Select the plane ASM_RIGHT of support as the assembly reference.
- Select Distance in the component reference relation field.
- Enter 100 as the distance.
- Right-click on the graphical area and select New constraint.

- Select the surfaces as shown in Figure 15.95.
- Select Coincide in the component reference relation field.
- Click OK to assemble the component.

- Click on Assembly from the Model tab of the ribbon interface.
- Select the file Support.
- Select the planner on the connection tab.

FIGURE 15.95
Constructional view of the spray system.

- Select the plane ASM_FRONT of Support.asm as the component reference.
- Select the plane Right of the second Rail_ProfileC as the assembly reference.
- Select Coincide in the component reference relation field.
- Click OK to assemble the component.
- Create cam connection on the new support to the second rail as described in the first support.

- Assemble another Rail_Profile_C1 on the other side of the plane ASM_RIGHT.

- Click on Assembly from the Model tab of the ribbon interface.
- Select the file Nozzle-Assembly.
- Select Planner on the Connection tab.
- Select the plane ADTM1 of Nozzle_Support.asm as the component reference.
- Select the plane DTM1 of the first Rail_Profile_C1 as the assembly reference.
- Select Coincide in the component reference relation field.
- Click OK to assemble the component.

- Click on cam connection between nozzle support wheels of Nozzle_Assembly outer surfaces to the Rail_Profile_C1 inner surface by following the cam connection defining the method as shown before.

- Click on Assembly from the Model tab of the ribbon interface.
- Select the file Bearing_Block.
- Click on the Placement tab.
- Select the bottom surface of the Bearing_Block_Base_Plate of Bearing_Block as the component reference.
- Select the top surface of the angle of the second support as the assembly reference.
- Select Coincide in the component reference relation field.
- Right-click on the graphical area and select New constraint.
- Select the plane ASM_FRONT of the Bearing_Block as the component reference.
- Select the plane DTM1 of the angle of the second support as the assembly reference.
- Select Coincide in the component reference relation field.
- Right-click on the graphical area and select New constraint.
- Select the plane ASM_RIGHT of the Bearing_Block as the component reference.
- Select the plane ASM_FRONT of the second support as the assembly reference.
- Click OK to assemble the component.

- Click on Assembly from the Model tab of the ribbon interface.
- Select the file Bearing_Block_DR.

- Assemble the component to the first support by following the method for assembling Bearing_Block.

- Click on Assembly from the Model tab of the ribbon interface.
- Select the file Motor_Assembly.
- Click on the Placement tab.
- Select the surface as shown in Figure 15.96.
- Select Coincide in the component reference relation field.
- Right-click on the graphical area and select New constraint.
- Select the surface as shown in Figure 15.97.
- Select Distance in the component reference relation field.
- Enter 50 as the distance.
- Right-click on the graphical area and select New constraint.

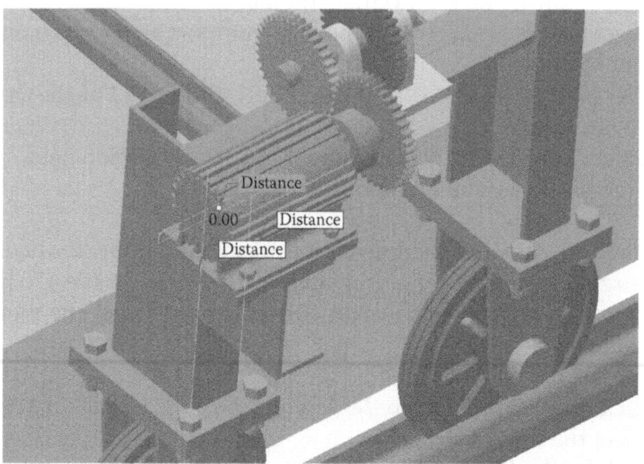

FIGURE 15.96
Constructional view of the spray system.

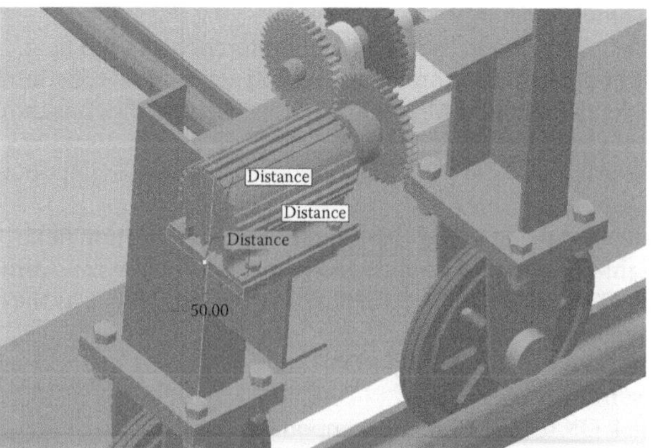

FIGURE 15.97
Constructional view of the spray system.

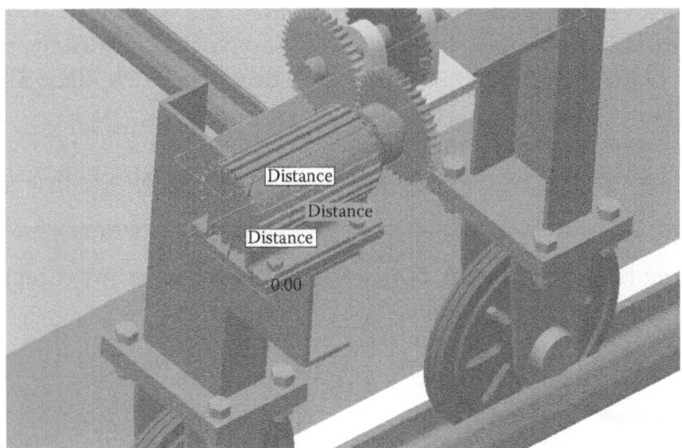

FIGURE 15.98
Constructional view of the spray system.

- Select the surfaces as shown in Figure 15.98.
- Select Coincide in the component reference relation field.

- Click OK to assemble the component.

- Click on Assembly from the Model tab of the ribbon interface.
- Select the file Hex_Bolt_M10.
- Click on the Placement tab.
- Select the cylindrical surface of Hex_Bolt_M10 as the component reference.
- Select the cylindrical surface of Bearing_Block_Base_Plate as the assembly reference.
- Select Coincide in the component reference relation field.
- Right-click on the graphical area and select New constraint.
- Select the sitting surface of Hex_Bolt_M10 as the component reference.
- Select the top surface of Bearing_Block_Base_Plate as the assembly reference.
- Click OK to assemble the component.

- Click on Assembly from the Model tab of the ribbon interface.
- Select the file Nut_M10.
- Click on the Placement tab.
- Select the inside cylindrical hole surface of Nut_M10 as the component reference.
- Select the cylindrical surface of Hex_Bolt_M10 as the assembly reference.
- Select Coincide in the component reference relation field.
- Right-click on the graphical area and select New constraint.
- Select the sitting surface of Nut_M10 as the component reference.
- Select the bottom surface angle or the corresponding support as the assembly reference.

- Click OK to assemble the component.
- Similarly, assemble Hex_Bolt_M10 and Nut_M10 in all the holes of Bearing_Block_Base_Plate and Bearing_Block_Base_Plate_DR.
- Click on drag components from the model tab of the ribbon interface.
- Drag the nozzle beam; notice that it can move in any direction.

15.3 Mechanism Design Project

Mechanism design consists of defining connection and joints followed by assigning a motor, conducting an analysis, and tracing the results.

The mechanism creation for the above-mentioned example is described here.

Hands-On Exercise, Example 15.36: Spray_System Mechanism (Software Used: Creo Parametric; Refer to Figure 15.99)

Steps:

- Open the file Spray_System.asm.
- Click on the application tab of the ribbon interface.
- Click on Mechanism to activate the mechanism environment.
- Click on Gear connection.
- Select Spur as the type.
- Click on the Gear 1 tab.
- Select the rotational axis of the motor to Motor_Pulley connection.
- Enter 95 as the diameter.
- Click on the Gear 2 tab.
- Select the rotational axis of the outer Gear_100_Dia.
- Enter 95 as the diameter.

FIGURE 15.99
Three-dimensional view of the mechanized spray system.

- Change the name to Motor on the top.
- Click OK to create the gear connection.

- Click on Gear connection again.
- Select Rack & Pinion as the type.
- Click on the Pinion tab.
- Select the rotational axis of the wheel of Wheel_Assembly connection.
- Enter 200 as the diameter.
- Click on the Rack tab.
- Select the translation axis from the connection of support to Rail_ProfileC. Users may need to list the connection names first if facing difficulties in graphical selection.
- Change the name to Bottom_Rail4 on the top.
- Click OK to create the gear connection.

- Similarly, create other Rack & Pinion connections to all the four wheels to Rail_ProfileC connections.

- Click on Gear connection again.
- Select Rack & Pinion as the type.
- Click on the Pinion tab.
- Select the rotational axis of the Nozzle_Support_Wheel of the Nozzle support connection.
- Enter 100 as the diameter.
- Click on the Rack tab.
- Select the translation axis from the connection of Nozzle_Assembly to Rail_Profile_C1. Users may need to list the connection names first if facing difficulties in graphical selection.
- Click OK to create the gear connection.

- Similarly, create other Rack & Pinion connections to all the four Nozzle_Support_Wheel to Rail_Profile_C1 connections.
- Click Drag components and select every movable part; ensure that all the wheels are rotating in the right direction. Otherwise, edit the corresponding connections and flip the relative rotational direction.

- Click again on Gear connection.
- Select Generic as the type.
- Click on the Gear 1 tab.
- Select any of the rotational axis of the Nozzle_Support_Wheel of Nozzle support connection.
- Enter 100 as the diameter.
- Click on the Gear 2 tab.
- Select the rotational axis from the connection of gear_100_dia of Bearing_Block_Assembly. Users may need to list the connection names first if facing difficulties in graphical selection.
- Enter 100 as the diameter.
- Click OK to create the gear connection.

- Click on Belt from the mechanism tab of the ribbon interface.
- Click the Reference tab.
- Select the outer cylindrical surface of the part Gear_100_Dia of Bearing_Block assembly and the part Gear_100_Dia_Dr of Bearing_Block_Dr assembly.
- Select the plane ASM_FRONT of Bearing_Block assembly as the Belt Plane.
- Click OK to create the belt connection.

- Click on the Servo Motors from the mechanism tab of the ribbon interface.
- Select the rotational axis of Motor_Pulley from the Motor_Assembly.
- Click on the Profile tab.
- Select Velocity on the specification area.
- Select Cosine in the Magnitude selection area.
- Enter 100 in the field A and 10 in the field Time.
- Click on the graph to check the velocity profile.
- Click OK to assign the motor.
- Click on Mechanism analysis from the Mechanism tab of the ribbon interface.
- Click Run to run the analysis.
- Save and exit the file.

The velocity adjustment is required to be set by the application frequency of spray, which is beyond the scope of this book. Users are advised to select various velocity profiles and run the analysis to see the results.

15.4 Animation Project

Animation is a powerful tool that can be used to demonstrate the function of any equipment.

Hands-On Exercise, Example 15.37: Spray_System Animation (Software Used: Creo Parametric)

Steps:

- Open the file Spray_System.asm.
- Assign the color suitably as described in Chapter 12.
- Click on the Mechanism tab from the Application tab of the ribbon interface.
- Click on Mechanism Analysis.
- Run the analysis.
- Click Play back.
- Save the result as analysis1.pbk.

- Close the Mechanism tab of the ribbon interface.
- Click on Animation from the Application tab of the ribbon interface.
- Click on New animation drop-down arrow from the Animation tab of the ribbon interface.
- Select Import from MDO.
- Click on Import the file icon.
- Select the file analysis1.pbk.
- Click OK to create the animation.
- Click on Play back to play the animation.
- Click Save besides playback to save the animation.

Hands-On Exercise, Example 15.38: Nozzle Spray Simulation (Software Used: Creo Parametric; Refer to Figure 15.100)

Steps:

- Create a part file and rename it as Spray_1. (Other names may be given, but it is advisable to provide a relevant name.)
- Select Solid on the right-hand side of the box. This will activate the solid mode and environment.
- Uncheck the box of the default template and click OK to create a new solid file.
- The template selection box will appear.
- Select mmns_part_Solid. This will allow the use of the solid template file of the following units:
    - mm—millimeter for length unit
    - n—newton for force unit
    - s—second for time unit

FIGURE 15.100
Three-dimensional view of the nozzle spray simulation.

- Once the Solid part is created, go to File > Prepare > Model properties to check the units. In the upper sections, the units of the model will be displayed.
  - Note: The units are usually displayed in the summary section of any CAD software. The setting of the units is usually in the tools or settings.
- The Nozzle as shown in Figure 15.100 is to be created.
- Click on Revolve from the Model tab in the ribbon interface.
- Click on the Placement tab and select Define to define the sketch.
- Select the front plane as the sketch plane.
- The right plane shall be automatically taken as the reference and orientation as right.
- Click on Sketch.
- The sketcher model will be activated.
- Draw the sketch as displayed in Figure 15.101.
- Click OK to create and exit the sketch.
- Ensure that the revolve angle shall be 360°.
- Click on OK to create the revolve feature.

- Click on Extrude from the Model tab in the ribbon interface.
- Click on the Placement tab and select Define to define the sketch.
- Select the front plane as the sketch plane.
- The right plane shall be automatically taken as the reference and orientation as right.
- Click on Sketch.
- The sketcher model will be activated.
- Draw the sketch as shown in Figure 15.102.

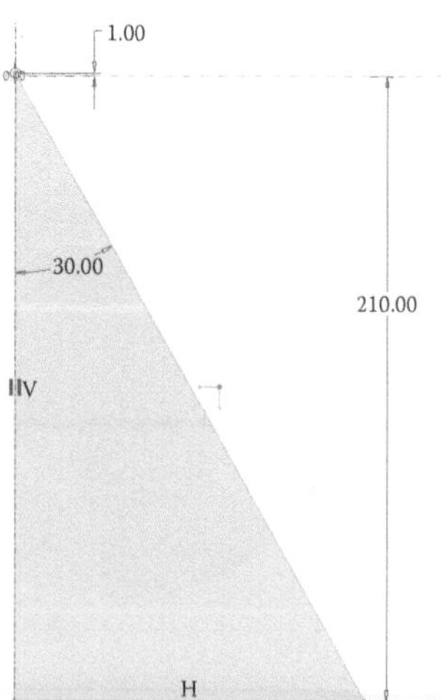

FIGURE 15.101
Constructional sketch of the nozzle spray simulation.

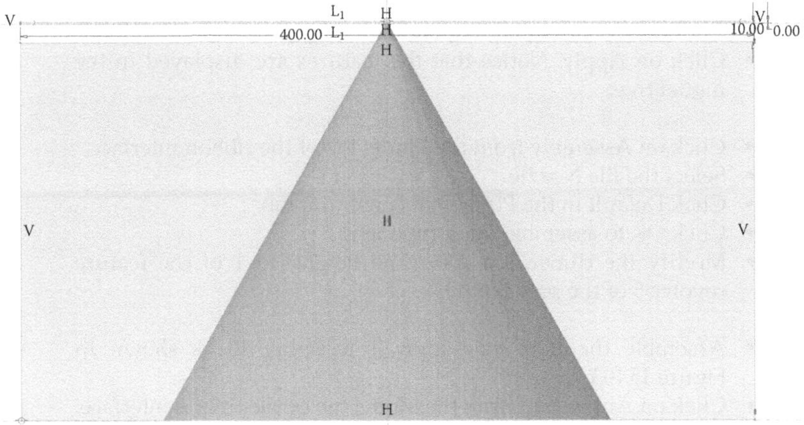

FIGURE 15.102
Constructional sketch of removal feature for nozzle spray simulation.

- Click OK to create and exit the sketch.
- Click on the Options tab.
- Select the Through all option for side 1 and side 2.
- Activate the material removal option.
- Click on OK to create the extrusion.
- Save and exit the file.

- Similarly, create other 9 nos part file with name sequence Spray 2 to Spray 10 with the dimension varying from 20 to 100 of the feature extrude 1.

- Create an assembly file and rename it as Nozzle_Spray.asm. (Other names may be given, but it is advisable to provide a relevant name.)
- Select Design on the right-hand side of the box. This will activate the standard assembly design mode and environment.
- Uncheck the box of the default template and click OK to create a new assembly design file.
- The template selection box will appear.
- Select mmns_asm_Solid. This will allow the use of the solid template file of the following units:
  - mm—millimeter for length unit
  - n—newton for force unit
  - s—second for time unit
- Once the assembly file is created, go to File > Prepare > Model properties to check the units. In the upper sections, the units of the model will be displayed.
  - Note: The units are usually displayed in the summary section of any CAD software. The setting of the units is usually in the tools or settings.

- Click on tree filers as displayed in Figure 15.58.
- Click on the check box displaying the features.
- Click on Apply. Notice that the features are displayed in the model tree.

- Click on Assembly from the Model tab of the ribbon interface.
- Select the file Nozzle.
- Click Default in the Placement constraint tab.
- Click OK to assemble the component.
- Modify the dimension 200 (cone height) to 1 of the feature revolve 5 of the part nozzle.

- Assemble the part files spray 1 to spray 10 as shown in Figure 15.103.
- Click on Animation from the Model tab of the ribbon interface.
- Click on Trans@Time from the Animation tab of the ribbon interface.
- Rename it as Transparency1.
- Set the time to 0 with the part spray_1 opaque.
- Click OK to create the step.
- Click on Trans@Time from the Animation tab of the ribbon interface.
- Rename it as Transparency1a.
- Set the time to 0 with the part spray_2 80% transparent.
- Click OK to create the step.
- Click on Trans@Time from the Animation tab of the ribbon interface.
- Rename it as Transparency1b.

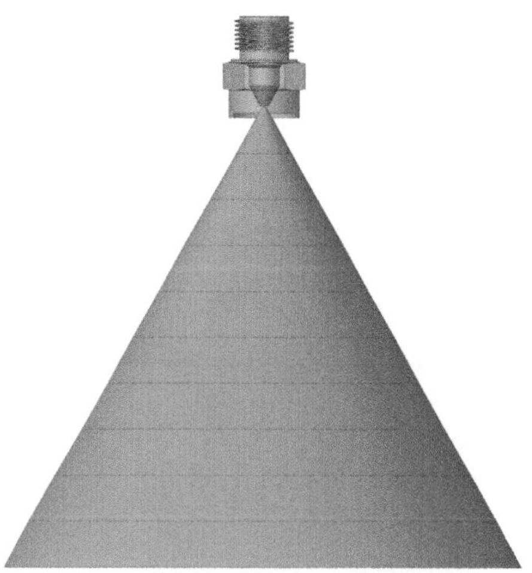

FIGURE 15.103
Constructional view of the nozzle spray simulation.

- Set the time to 0 with the part spray_3 to spray_4 50% transparent.
- Click OK to create the step.
- Similarly, create all other transparency steps with the next part being opaque, the nearest part being 80% transparent, and all the other spray parts being 50% transparent.
- Play the animation to create the animation.
- Save and exit the file.

Index

Page numbers followed by f and t indicate figures and tables, respectively.

N

T

U

V

W

Y